ENCYCLOPEDIA OF
ENVIRONMENTAL ISSUES

ENCYCLOPEDIA OF ENVIRONMENTAL ISSUES

REVISED EDITION

Volume 2

Dioxin—Lake Erie

Editor

Craig W. Allin

Cornell College

SALEM PRESS

Pasadena, California   Hackensack, New Jersey

Editor in Chief: Dawn P. Dawson

Editorial Director: Christina J. Moose   *Photo Editor:* Cynthia Breslin Beres
Development Editor: R. Kent Rasmussen   *Production Editor:* Joyce I. Buchea
Project Editor: Judy Selhorst   *Graphics and Design:* James Hutson
Acquisitions Manager: Mark Rehn   *Layout:* William Zimmerman
Research Supervisor: Jeffry Jensen

Cover photo: ©Radius Images/CORBIS

Library of Congress Cataloging-in-Publication Data

Encyclopedia of environmental issues / editor, Craig W. Allin. — Rev. ed.
    Planned in 4 v.
    Includes bibliographical references and index.
    ISBN 978-1-58765-735-1 (set : alk. paper) — ISBN 978-1-58765-736-8 (vol. 1 : alk. paper) —
ISBN 978-1-58765-737-5 (vol. 2 : alk. paper) — ISBN 978-1-58765-738-2 (vol. 3 : alk. paper) —
ISBN 978-1-58765-739-9 (vol. 4 : alk. paper)
    1. Environmental sciences—Encyclopedias. 2. Pollution—Encyclopedias.
I. Allin, Craig W. (Craig Willard)
    GE10.E523 2011
    363.7003—dc22
                                                                                    2011004176

PRINTED IN THE UNITED STATES OF AMERICA

Contents

Complete List of Contents

Volume 1

Volume 2

Contents xxxix
Complete List of Contents xliii

Volume 3

Contents.

Volume 4

Appendixes

Indexes

ENCYCLOPEDIA OF
ENVIRONMENTAL ISSUES

Dioxin

CATEGORY: Pollutants and toxins

DEFINITION: Toxic chemical that is a by-product of certain manufacturing processes

SIGNIFICANCE: Dioxin has become an environmental concern because exposure to the chemical, even in trace amounts, can cause severe health problems in humans and other organisms.

Approximately seventy-five different types of dioxins exist, but the term "dioxin" is commonly used to refer to a variety known as 2,3,7,8-tetrachlorodibenzo-p-dioxin (TCDD), a highly toxic chemical that has caused great concern among environmentalists. Dioxin can be destroyed by exposure to direct sunlight in the presence of hydrogen, but the chemical can remain under the surface of the ground for ten years or longer.

One of the earliest documented cases of dioxin exposure occurred in West Germany in 1957, when thirty-one workers at a chemical plant developed chloracne, a skin disease that is one of the characteristic effects of exposure. In 1977 investigators in the Netherlands discovered dioxin in fly ash from a municipal incinerator. By 1980 scientists had determined that dioxin is produced when practically any organic substance is burned.

At first it was thought that chloracne was the only effect of exposure to dioxin. As time went on, however, experiments with animals revealed that dioxin is highly toxic. Researchers found that guinea pigs could be killed by as little as 1 microgram of dioxin per kilogram of body weight, but hamsters could take a dose of 5,000 micrograms per kilogram. Further experimentation showed that the organs within different animals were also affected differently by the chemical. Such differences among animal species and organs had never been found with any previously tested substance, and they invalidated the usual methods used in testing on animals. Because of this, dioxin became known as the first environmental hormone. That is, it acts like a hormone in animals and plants in that it has strange effects on various organs.

Scientists also began to suspect that dioxin could be involved in causing various cancers. The most famous exposure incident occurred between January, 1965, and April, 1970, when the U.S. military used the defoliant Agent Orange during the Vietnam War to kill trees and plants that provided hiding places and food for North Vietnamese soldiers. One of the herbicides used was contaminated with dioxin during the manufacturing process. After the war, the National Academy of Sciences reviewed more than six thousand studies of dioxin exposure and came to the conclusion that four kinds of cancer could be linked to the chemical.

Other studies of dioxin exposure have produced mixed results. Thirty-seven thousand people were exposed to dioxin in Seveso, Italy, in 1976 after a factory explosion. Although no immediate human fatalities were traceable to the accident, a significant number of male births in the region were interrupted before term during the four months following exposure. Dioxin was therefore suspected of being a factor in the decline of male births worldwide. Two hundred workers were exposed to dioxin in 1949 at Nitro, West Virginia, and several died from cancer. Sweden reported six times the normal cancer rate in people exposed to dioxin, but Finland traced nineteen hundred cases of exposure and found no harmful effects. A Veterans Administration study of eighty-five thousand Vietnam veterans found lower-than-normal rates of cancer. Many researchers have explained such disparate findings by noting that dioxin does not seem to cause cancer itself; rather, it acts as an influence on other cancer-causing chemicals.

Another well-publicized dioxin incident occurred in 1983 in Times Beach, Missouri, when a local resident tried to settle the dust on the town's roads by wetting the roads with a spray of recycled oil that was unknowingly contaminated with dioxin. The federal government bought the entire town and moved everyone out to prevent any further exposure to the chemical, even though at the time it was not known that dioxin could cause any of the cancers later linked to exposure.

In a September, 1994, report, the Environmental Protection Agency (EPA) stated that perhaps no more than 14 kilograms (30 pounds) of dioxin are released into the U.S. environment annually. Even this trace amount is unacceptable, however, given that dioxin is suspected of being an endocrine disrupter, which means that trace amounts can disrupt the effects of other hormones and cause numerous disorders. An EPA lab worker was quoted as saying, "I don't know a hormone system that dioxin doesn't like to disrupt." To make matters worse, approximately seventy other chemicals in the environment are also suspected of being disrupters, including dichloro-diphenyl-trichloroethane (DDT) and other insecticides that were used for many years. The 1994 EPA report also stated that dioxin may be responsible for damaging immune systems and creating other hormone-related diseases, such as diabetes. For example, mice treated with dioxin readily die after exposure to viruses that ordinarily would have no effect on them.

Scientific investigation of dioxin has expanded the view of environmental chemicals to include environmental hormones and hormone disrupters. These seventy or more substances are like no other chemicals known in the past, raising the critical question of how they are to be regulated when so little is known about their effects. All previous ways of measuring are invalid because these substances do not have uniform effects. One animal is killed by a tiny exposure and another is seemingly unharmed, while both may have internal effects that remain undiscovered.

Robert B. Bechtel

FURTHER READING

Friis, Robert H. "Pesticides and Other Organic Chemicals." In *Essentials of Environmental Health*. Sudbury, Mass.: Jones and Bartlett, 2007.

Schecter, Arnold, and Thomas Gasiewicz, eds. *Dioxins and Health*. 2d ed. New York: Taylor & Francis, 2002.

Young, Alvin Lee. *The History, Use, Disposition, and Environmental Fate of Agent Orange*. New York: Springer, 2009.

SEE ALSO: Agent Orange; Chloracne; Italian dioxin release; Pesticides and herbicides; Times Beach, Missouri, evacuation.

Diquat

CATEGORY: Pollutants and toxins
DEFINITION: Contact herbicide
SIGNIFICANCE: As is the case for many herbicides, diquat has the potential to contaminate supplies of drinking water if it is released into wastewater and subsequently absorbed into soil.

The herbicide diquat, which has been used in the United States since the 1950's, is potentially toxic if swallowed or inhaled. It is not produced in the United States, but nearly one million pounds of the compound are imported each year. Approximately two-thirds of this amount is utilized as desiccant or defoliant, with another one-third used for aquatic weed control. Diquat is readily adsorbed into clay particles in the soil, sediment in water, and the surfaces of weeds. While it is biodegradable, if it is adsorbed into plant life and subject to photodegradation by sunlight, when bound to sediment diquat remains stable for weeks or even months. Ultimately, however, the herbicide undergoes degradation through the action of soil flora.

The extensive overgrowth of weeds affecting both crops and waterways has created significant problems in management. Regulations pertaining to use of herbicides remain under the auspices of the Environmental Protection Agency (EPA) and its various offices. Cutbacks in funding, however, have hampered research into the short- and long-term effects of using herbicides in the control of weed problems; therefore, the long-term effects of diquat are unclear.

Since diquat is a nonselective herbicide, its use is limited. It is used as a growth regulator in suppressing the flowering of sugarcane and to control aquatic weeds in the absence of endangered plant species. However, the most important use of diquat is in the desiccation of potato haulm or seed crops such as clover and alfalfa. Such practices are generally carried out to prepare crops for harvest. Application of diquat results in loss of moisture from the leaf, usually killing the plant. Desiccation may also be utilized in prevention of seed loss resulting from scattering of seed upon opening of the pod. The application of diquat prior to harvest of crops such as alfalfa greatly reduces seed loss.

The primary metabolic effect of diquat seems to be its ability to divert electrons activated during photosynthesis into the production of toxic compounds such as hydrogen peroxide. Peroxide production in turn results in membrane damage to those parts of the plant in contact with the herbicide. Human exposure to diquat is primarily occupational, with agricultural workers who use the chemical at highest risk. Most actual poisonings have been intentional, with diquat used as a means of suicide. Nevertheless, though diquat is not as toxic as some herbicides, the level of its toxicity is mainly a function of degree. Exposure to significant levels of the chemical may cause damage to the central nervous system and kidneys, while cataract formation is the most common effect of chronic exposure.

Richard Adler

Further Reading

Manahan, Stanley E. "Water Pollution." In *Fundamentals of Environmental Chemistry.* 2d ed. Boca Raton, Fla.: CRC Press, 2001.

Udeh, Patrick J. "Organic Contaminants in Drinking Water." In *A Guide to Healthy Drinking Water: All You Need to Know About the Water You Drink.* Lincoln, Nebr.: iUniverse, 2004.

See also: Agricultural chemicals; Groundwater pollution; Hazardous and toxic substance regulation; Pesticides and herbicides; Water pollution.

Disposable diapers

Category: Waste and waste management

Definition: Single-use diapers made from paper and plastic products

Significance: Every year an estimated 27 billion or more soiled diapers end up in landfills in the United States alone. Since the 1970's, environmentalists have debated whether reusable cloth diapers are more earth-friendly than disposables. The convenience of disposables has led to their gaining and holding market dominance in the United States, and their use has spread even to developing nations.

The modern disposable diaper typically consists of a plastic layer treated with a surfactant that wicks moisture into the diaper and away from the skin; an absorbent layer of wood pulp mixed with sodium polyacrylate, a polymer that assumes a gel-like form when it is exposed to liquid; and a backing layer made from a plastic such as polyethylene. Leg cuffs and the top of the diaper are made from polypropylene or another water-resistant material to minimize leakage. Petroleum- and wood-derived adhesives are applied as a hot melt to hold the layers together. Some disposable diapers feature a reusable outer cover and a disposable or flushable liner. Other disposable absorbent goods, such as adult incontinence products and feminine sanitary products, have a composition similar to that of disposable diapers.

Although the wood pulp and human waste contained in disposable diapers are biodegradable, in conventional disposables these materials are sealed

---

Components of a Disposable Diaper

- Polyethylene, or clothlike film

- Tissue

- Hot melts (mixture of resins, oils, and tackifiers, used as a glue)

- Hydrophobic nonwoven sheet (polypropylene resin with surface surfactants)

- Hydrophilic nonwoven sheet (polypropylene resin without surface surfactants)

- Elastic (polyurethane or polyester foam, synthetic rubber, or Lycra)

- Lateral tape and frontal tapes

- Cellulose

- Acquisition and distribution layer

- Sodium polyacrylate (super absorbent polymer)

- Top-sheet surface lotions

- Decorated film

- Wetness indicators

---

within the diapers' nonbiodegradable plastic backing. Furthermore, landfills do not contain the oxygen and water needed for the biodegradable components of disposable diapers to break down. It has been estimated that a disposable diaper would take 550 years to decompose in a landfill and 450 years to disintegrate in the ocean.

A 2005 industry report estimated that, on average, a baby uses a total of 3,700 to 4,200 diapers before toilet training is completed. Economic and cultural factors influence how many diapers a baby goes through before toilet training. In Japan, for example, diapers are changed more frequently than they are in the United States. However, babies in the United States take longer to toilet train, in part because of fears that rushing this phase of a child's development will cause psychological stress. The United States consumes more diapers per baby than do other countries. Some 96 percent of American babies wear disposable diapers. Since 2003, single-use disposable diapers have accounted for at least 2 percent of the discarded material in municipal waste streams in the United States. The tonnage of diapers in the waste stream has increased more than tenfold since 1970.

HISTORY

Disposable diapers began to appear on the market during the 1940's. The earliest disposables were high-cost, low-performance items intended for use primarily during travel. The diaper brand Pampers, which offered improved performance, was first marketed in 1961 but did not catch on quickly. The only convenience the disposables offered then was that they could be thrown away; caregivers still had to fold and pin them, just as they had to fold and pin cloth diapers. As the popularity of Pampers gained momentum, more brands were marketed, and the competition prompted improvements, such as prefolded and shaped diapers, greater absorbency, leak guards, resealable closures, different designs for boys and girls, and decorative plastic backings. By 1980 disposable diapers accounted for 1.4 percent of the discarded material in municipal waste streams.

While manufacturers worked on improving the convenience of the product for consumers, they also strove to address environmental concerns. They experimented with "green" disposables, which were made from plastic films that eventually broke down in water. Others had backings made from a cornstarch-polyethylene blend to promote degradation after they were discarded.

Procter & Gamble, the maker of Pampers, conducted two separate pilot programs during the late 1980's and early 1990's. One was intended to create a partially flushable diaper. This was partly in response to complaints that fecal matter should enter the sewage system, not landfills. The program was discontinued after only three years, however, because consumers did not want to bother with removing the flushable inner section and discarding the plastic cover. Some municipal officials also claimed that diapers would overtax sewage treatment systems. Finally, the diaper was not compatible with water-saving, low-flush toilets.

In another program, Procter & Gamble set up centers to recycle diapers. The centers separated organic materials from the plastics, which were processed and recycled into other products. The company found that products made from recycled diapers had low marketability, however. The project was deemed a technical success but an economic failure.

In early 2008 *Time* magazine reported that cloth diapers had experienced sales increases of 25 to 50 percent in the preceding years. In some communities—including Westmount, Quebec; Vienna, Austria; and Munich, Germany—municipal solid waste departments had begun subsidizing cloth-diaper purchases to minimize the number of diapers discarded in landfills. The increasing appeal of reusables arose not only because of environmental considerations but also because of concerns regarding infant health. A 1999 study had found that rats exposed to disposable diapers suffered eye, nose, and throat irritation. A study published the following year linked disposables with high testicular temperatures in boys and subsequent low sperm production.

Improved design also helped boost cloth diaper sales. Borrowing features from their disposable counterparts, modern cloth diapers included Velcro, button, and snap closures, elasticized legs, and liquid-resistant covers. "Hybrid" diapers were developed that combined reusability and disposability, featuring washable coverings fitted with flushable or compostable inserts.

Meanwhile, the disposable diaper industry made its own advances. Some manufacturers employed a new plastics additive technology to enable diapers to break down after sufficient exposure to light, heat, and mechanical stress. In some brands, chlorine bleaching was eliminated. By 2010 high-temperature composting methods, advanced plastics recycling technologies, and waste-to-energy schemes fueled by diapers were being explored.

DISPOSABLE VERSUS CLOTH

Proponents of disposable diapers argue that diapers in landfills are not the real issue; rather, the problem is garbage in general and the limited space for landfills. They maintain that the amount of garbage going into landfills during a baby's diapering life is minimal when compared with the total amount of garbage adults send to landfills each day. They assert that human urine and feces that travel to landfills with diapers are not really an issue, either, given that pet fecal matter also ends up in landfills, as does medical waste containing fecal matter and urine.

Those who object to disposable diapers because they contribute to the volume of solid waste sent to landfills argue that cloth diapers are the better alternative. They cite the potential for human fecal matter in landfills to contaminate groundwater with bacteria and viruses. Proponents of disposables counter that, whether parents launder their own cloth diapers or have a diaper service perform this chore, energy is burned and pollution is added to the environment. Hot water, bleach, and energy are used in washing and drying diapers. Diaper services consume additional energy by transporting soiled and cleaned diapers in fossil-fuel-burning motor vehicles, not to mention adding to traffic congestion; however, services have been shown to consume less wash water per diaper than does home laundering.

In regard to the energy and resources used to manufacture both types of diapers, neither appears to be an ideal option from the standpoint of the environment. Cloth diapers are typically made from cotton (although diapers made from other renewable resources, such as hemp and bamboo fiber, have become increasingly popular), and growing cotton consumes huge amounts of resources. Cotton is one of the most water-intensive crops grown, and vast quantities of pesticides are used on cotton unless it is grown organically. In addition, harvesting and transporting plant fiber consumes energy, and the chlorine bleach used in processing cotton produces toxic by-products. Spinning the fiber into cloth and fashioning it into diapers uses still more energy.

Conventional disposable diapers are made from wood pulp; although this is a renewable resource derived from trees, it does not replenish itself as quickly as cotton. Disposables also contain plastics, which are derived from nonrenewable petrochemical resources. Their manufacture consumes water and energy and involves chemical bleaching agents. Effluents from the pulp, paper, and plastics industries have a toxicity that rivals that of cotton-growing and -processing effluents.

The ways in which cloth diapers are washed and dried have a major influence on the environmental impacts of reusables. Home laundering tends to consume more water than do the laundering processes used by diaper services, but home launderers can minimize the environmental impacts of the use of cloth diapers by washing the diapers at lower temperatures and by line drying them instead of tumble drying them.

Lisa A. Wroble
Updated by Karen N. Kähler

FURTHER READING

Aumônier, Simon, Michael Collins, and Peter Garrett. *An Updated Lifecycle Assessment Study for Disposable and Reusable Nappies.* Bristol, England: Environment Agency, 2008.

Brower, Michael, and Warren Leon. *The Consumer's Guide to Effective Environmental Choices: Practical Advice from the Union of Concerned Scientists.* New York: Three Rivers Press, 1999.

Management Institute for Environment and Business. "Procter & Gamble Inc.: Disposable and Reusable Diapers—A Life-Cycle Analysis." In *Environmental Management: Readings and Cases,* edited by Michael V. Russo. 2d ed. Thousand Oaks, Calif.: Sage, 2008.

SEE ALSO: Composting; Landfills; Plastics; Pulp and paper mills; Refuse-derived fuel.

Dodo bird extinction

CATEGORY: Animals and endangered species

IDENTIFICATION: Extinction of a species of flightless birds that were native to the island Mauritius in the Indian Ocean

SIGNIFICANCE: The extinction of the dodo bird is an example of the potential effects of overhunting and the introduction to an area of nonnative animal species.

Dodo birds were indigenous to the Mascarene Islands in the Indian Ocean east of Madagascar. The largest island of the group is Mauritius, 765 kilo-

Nineteenth century engraving of a dodo. (©Duncan Walker/iStockphoto.com)

meters (475 miles) east of Madagascar. The other large islands are Réunion, 225 kilometers (140 miles) southwest of Mauritius, and Rodrigues, 644 kilometers (400 miles) east of Mauritius. The Mauritius species *Raphus cucullatus* is, strictly, the dodo bird, but some people consider the closely related "solitaire," *Raphus solitarius*, of Réunion Island and some *Pezophaps solitaria* of Rodrigues Island also to be dodo birds. These species were pigeons that, in isolation, had evolved to large size and odd forms.

The dodo bird of Mauritius was flightless and has been described as having a rather globular body supported on legs that were, for a bird, short and stout. It had a short neck crowned with a large head ending in a long, heavy beak that was hooked downward at the end. The nostrils were well forward on the beak. The similar birds of Réunion and Rodrigues Islands were somewhat slimmer and had longer legs.

The male dodo is reported to have weighed about 22.7 kilograms (50 pounds), much larger than an adult male North American wild turkey, which ranges from 6.8 to 9 kilograms (15 to 20 pounds). The female dodo was reportedly somewhat smaller. The bill was about 22.9 centimeters (9 inches) long. The wings were rudimentary, displaying only three or four feathers, and the tail was equally rudimentary, consisting of a few short, curly feathers. The body feathers of the Mauritius dodo were blue-gray, and the sparse wing and tail feathers were white with a splash of yellow. Overall, the dodo appears to have been a rather unattractive bird. Dodos apparently made their nests on the ground.

The dodo evolved in the absence of humans and predators and initially showed no great fear of sailors who hunted them for food, giving rise to the expression "dumb as a dodo." They were slow moving, and a human could outrun a dodo on open ground. One observer reported that Dutch sailors from one ship caught twenty-four dodos one day and twenty the next.

Some visitors to the islands reported that the meat of dodos was hard and tough, even after prolonged boiling. Others claimed that the breast and belly were good to eat, while the rest of the bird was tough. It may have been a matter of season, as one source said that the Rodrigues solitaire was extremely fat and good to eat from March to September.

Seafarers first visited the Mascarene Islands and reported the existence of dodos about 1507. The dodo of Mauritius was extinct by 1681, the solitaires of Réunion had disappeared by 1746, and those of Rodrigues were gone by 1790. The extinctions were not sudden: The Mauritius dodo survived for 174 years beyond first human contact. Macaques, which are fond of eggs, and pigs were introduced to Mauritius in the sixteenth century; these changes, along with heavy predation by humans, were probably important factors in the ultimate extinction of the birds.

Robert E. Carver

FURTHER READING
Pinto-Correia, Clara. *Return of the Crazy Bird: The Sad, Strange Tale of the Dodo.* New York: Copernicus Books, 2003.
Quammen, David. *The Song of the Dodo: Island Biogeography in an Age of Extinctions.* 1996. Reprint. New York: Charles Scribner's Sons, 2004.

SEE ALSO: Extinctions and species loss; Hunting; Passenger pigeon extinction.

Dolly the sheep

CATEGORY: Biotechnology and genetic engineering
IDENTIFICATION: First mammal successfully cloned from adult cells
DATES: Born July 5, 1996; died February 14, 2003
SIGNIFICANCE: The birth of Dolly and subsequently cloned animals raised a host of ethical issues and opened the door to possible means of improving human health and the environment. Dolly's comparatively short life span, about half that of the typical sheep of her breed, may have been related to her clone origins.

On February 5, 1997, Ian Wilmut of the Roslin Institute in Edinburgh, Scotland, announced the birth of Dolly the sheep, the first clone produced from a cell taken from an adult mammal. Scientists and the general public were shocked at this announcement, because it was believed that cloning a mammal from an adult cell was, at the time, technically impossible. Dolly was seven months old before her birth was made public. Because she was cloned from mammary cells, the research team named her after well-endowed country-western singer Dolly Parton.

How Dolly Was Created

A clone is an organism developed from a single cell isolated from another organism. The cell donor and the clone are genetically identical. Prior to the creation of Dolly, no attempts at cloning a mammal from adult cells had been unequivocally successful. In the early 1980's scientists had created clones of mammals by using donor cells from young embryos. The research team that cloned Dolly first cloned a pair of sheep, Megan and Morag, from embryonic cells grown in the laboratory. Adult cells and embryonic cells have identical genetic material, or deoxyribonucleic acid (DNA); however, adult cells produce proteins specific to the type of cell they become. For example, brain cells produce neurotransmitters and do not produce hemoglobin, even though they possess the hemoglobin gene. Scientists believed that the structure of the DNA in an adult cell was irreversibly altered during the process of maturation to gain this specificity and therefore could not be used to produce a clone.

Wilmut and his colleagues at the Roslin Institute used a novel approach to clone Dolly. The donor cells were sheep udder cells from a six-year-old pregnant ewe. The cells had been frozen for about three years, and the donor was long deceased. The researchers believed that prior attempts at cloning mammals from adult cells had failed because the cells were too active or in the wrong phase of their life. To make the cells quiescent, the research team starved them for several days. Meanwhile, the researchers removed genetic material of eggs from a different breed of sheep, a process called enucleation. They then fused the starved cells with the enucleated donor eggs and implanted them into surrogate mother sheep of a third breed. Of 277 attempts, only a single egg went full-term, resulting in the birth of Dolly. Dolly looked strikingly different from the breed of the egg donor or the surrogate mother but identical to the breed that donated the adult DNA, an observation that provides suggestive evidence that she developed from the donor DNA.

Initially many questions surrounded the validity of the experiment that produced Dolly. Some scientists believed that she could have been the result of a contaminating fetal cell, and the results of the experiment were not easily replicated by other researchers. However, in July, 1998, Japanese scientists announced the birth of two calves cloned from adult cow uterus cells, and researchers in Hawaii successfully produced more than fifty mouse clones from adult mouse ovarian cells. Also, DNA analysis of Dolly confirmed that she was indeed the first clone of an adult mammal. After reaching sexual maturity, Dolly mated naturally and on April 13, 1998, gave birth to a lamb, showing that she was a healthy young adult whose ability to reproduce was not compromised by her unusual origin. She would have two more pregnancies and bear five more lambs during her life.

Implications

In May, 1999, researchers found that the telomeres in Dolly's cells were shorter than those in other mammals of similar age, a finding borne out in other cloning research. Telomeres, which are sequences at the ends of chromosomes, become progressively shorter as an organism ages, but it was not clear whether

Dolly the sheep, the first animal to be cloned from adult cells. (AP/Wide World Photos)

Dolly's life span would be shorter than usual. By January, 2002, Dolly was showing signs of arthritis; whether the condition was the result of premature aging or the amount of time she had spent indoors on a hard floor was unclear. She subsequently developed a progressive lung disease.

The ailing Dolly was euthanized in February, 2003, at the age of six and a half. Sheep of her breed typically live to be about twelve. A postmortem examination confirmed that she had arthritis in her hind legs and that her lung ailment was sheep pulmonary adenomatosis (a lung tumor associated with a retrovirus), a fatal disease common in older sheep, particularly those living indoors. The examination revealed no other abnormalities. During her lifetime, she had tended toward stoutness, but her keepers attributed Dolly's weight problem not to her clone origins but to her living mostly indoors and being fed treats by her many visitors. Dolly's remains were donated to the National Museum of Scotland in Edinburgh, where her preserved body was placed on exhibit.

There are numerous potential applications for the cloning technology that produced Dolly. Scientists envision using cloning in tandem with genetic engineering to create animals with organs suitable for transplant into humans, or ones that produce human proteins for use in pharmaceuticals. In fact, Wilmut's research was sponsored by the Scottish pharmaceutical company PPL Therapeutics, Ltd. His subsequent research involved sheep cloned from fetal cells that had been genetically altered to carry a human gene that caused the animals' milk to contain a blood-clotting protein with the potential for use in treating human hemophilia.

Some researchers envision entire herds of genetically identical cattle. Because it is very difficult to produce prize milk- or meat-producing animals consistently with traditional breeding methods, repeated cloning of one prize breeding animal would greatly speed the process. A potentially serious problem with genetically identical herds, however, is that genetic diversity allows species to survive changes in their environment and attacks by disease. Diseases affecting only a few individuals of a genetically diverse species may become rampant in a genetically identical one. If genetic diversity is lost, it could lead to the extinction of that species.

Cloning has long been regarded as a possible means of bringing endangered animal species back from the brink of extinction, particularly those species that do not breed well in captivity. However, the expense and failure rate of the technique may be too great for it to be practicable. The year 2001 saw the birth of the world's first cloned endangered wild animals. A cloned gaur (a type of Southeast Asian ox) was successfully brought to term by a domestic cow that served as its surrogate mother, but the newborn succumbed to common dysentery two days after its birth. A cloned European mouflon, one of the world's smallest wild sheep species, fared better. Born to a domestic sheep, the mouflon was raised at a wildlife center in Sardinia. Other endangered species that have since been successfully cloned include the Javan banteng (a type of wild cattle) and the African wildcat. The first clone of an extinct animal was born in 2009. Cloned from the frozen skin of a bucardo, or Pyrenean ibex, a subspecies of wild goat that went extinct in 2000, the animal died minutes after birth.

Karen E. Kalumuck
Updated by Karen N. Kähler

FURTHER READING

Einsiedel, Edna. "Brave New Sheep: The Clone Named Dolly." In *Biotechnology: The Making of a Global Controversy*, edited by Martin W. Bauer and George Gaskell. New York: Cambridge University Press, 2002.

Franklin, Sarah. *Dolly Mixtures: The Remaking of Genealogy*. Durham, N.C.: Duke University Press, 2007.

Kolata, Gina Bari. *Clone: The Road to Dolly, and the Path Ahead*. New York: HarperCollins, 1998.

Morgan, Rose M. *The Genetics Revolution: History, Fears, and Future of a Life-Altering Science*. Westport, Conn.: Greenwood Press, 2006.

Wilmut, Ian, Keith Campbell, and Colin Tudge. *The Second Creation: Dolly and the Age of Biological Control*. Cambridge, Mass.: Harvard University Press, 2001.

Wilmut, Ian, and Roger Highfield. *After Dolly: The Uses and Misuses of Human Cloning*. New York: W. W. Norton & Company, 2006.

SEE ALSO: Biotechnology and genetic engineering; Cloning; Genetically modified organisms; Wilmut, Ian.

Dolphin-safe tuna

CATEGORIES: Agriculture and food; animals and endangered species

DEFINITION: Commercially produced food products containing tuna fish caught using methods that minimize dolphin mortality

SIGNIFICANCE: In the mid-twentieth century, massive numbers of dolphins died in tuna nets. Concerns regarding the magnitude of the dolphin mortality led to the concept of dolphin-safe tuna, and laws, standards, and industry practices were established with the intent of protecting dolphins during tuna fishing.

The eastern tropical Pacific tuna fishery covers the ocean from California to Hawaii and south to the equator and Chile. Dolphins and yellowfin tuna frequently feed in the same areas within this region. During the mid-twentieth century, the tuna fishing industry adopted the practice of locating schools of tuna by spotting associated dolphin populations. Dolphins are mammals and must remain near the surface in order to breathe. The tuna fishing industry exploited this knowledge, using dolphins to locate yellowfin tuna swimming below the surface. During the late 1950's technological developments enabled fishing boats that had been limited to catching individual tuna with poles and lines to capture entire schools of the fish in purse-seine nets. It became common industry practice to encircle dolphin schools with huge purse seines in order to capture the associated tuna schools. Dolphins and tuna together were herded with speedboats and helicopters into the nets. The dolphins, of no commercial value to the tuna fishing industry, were sometimes released alive, but often they became entangled in the nets and drowned.

By the early 1970's the number of dolphins killed annually in the eastern Pacific Ocean had risen to more than 300,000. A 2008 estimate by the U.S. National Oceanic and Atmospheric Administration (NOAA) Fisheries Service placed the number of dolphins killed by tuna fishing since the 1950's at more than 6 million.

Public awareness of the plight of the dolphins began as early as the mid-1960's, leading to the 1972 passage of the U.S. Marine Mammal Protection Act (MMPA). Intended to prevent exploitation of dolphins and related aquatic animals, the act limited the number of dolphins that could be killed annually to 20,500. Unsatisfied, conservationists and animal rights activists pushed for a tuna boycott, a move that

Cans of tuna with labels displaying a dolphin-safe claim. (Yasmine Cordoba)

succeeded largely because of the widespread popularity of dolphins. By 1977 annual dolphin deaths in purse seines had declined to about 25,450. The respite was brief, however: As U.S. fishing fleets declined during the early 1980's, foreign vessels began to enter the eastern Pacific fishery. This led to an increase in dolphin mortality; by 1986, the numbers had risen to 133,000 deaths per year.

In 1990 declining sales and concern about public opinion led the three major canners of tuna sold in the United States to stop buying tuna that had been caught through the intentional setting of nets around dolphins. At the same time, much of the U.S. tuna fishing fleet moved into the western tropical Pacific Ocean, where tuna and dolphin do not habitually swim in the same waters. In November, 1990, the United States enacted the Dolphin Protection Consumer Information Act (DPCIA), which set conditions for protecting dolphins during purse-seine tuna fishing but included few provisions for enforcement. In 1994 provisions of the 1992 U.S. International Dolphin Conservation Act went into effect. Under the act, only dolphin-safe tuna—tuna caught using methods that do not involve chasing, encircling, or killing dolphins—could be sold in the United States, the country constituting the largest market for canned tuna.

Critics of the legislation claimed that canneries were simply using the term "dolphin-safe" as a marketing tool and that use of the term was no true guarantee of dolphin-friendly fishing practices. Criticism escalated after the MMPA was amended by the 1997 International Dolphin Conservation Act (IDCA), which redefined the dolphin-safe label. Under the new, looser guidelines, dolphins could be caught in nets as long as they were not physically harmed. An individual dolphin could thus legally be chased, captured, and released many times during its life.

In 1999, by which time reported dolphin mortality had fallen to fewer than 3,000 per year, the Agreement on the International Dolphin Conservation Program (AIDCP) went into effect. This binding document, which replaced a voluntary international program established during the early 1990's, adopted most of the IDCA's provisions. Over the next decade the United States, the European Union, and eleven other major fishing countries ratified or acceded to the AIDCP.

A 2004 decision in a lawsuit filed by the Earth Island Institute, the U.S. Humane Society, and other organizations restored the original definition of "dolphin-safe" for tuna sold in the United States. Tuna caught through methods that involve chasing, encircling, and capturing dolphins without physically harming them can legally be sold in the United States but cannot be labeled dolphin-safe. By 2008, according to figures from the NOAA Fisheries Service, the number of dolphins captured unintentionally by the tuna industry had been reduced to an estimated 1,000 per year.

P. S. Ramsey
Updated by Karen N. Kähler

FURTHER READING

Bonanno, Alessandro, and Douglas Constance. *Caught in the Net: The Global Tuna Industry, Environmentalism, and the State.* Lawrence: University Press of Kansas, 1996.

Kennelly, Steven J., and Matt K. Broadhurst. "By-Catch Begone: Changes in the Philosophy of Fishing Technology." *Fish and Fisheries* 3, no. 4 (2002): 340-355.

White, Thomas I. *In Defense of Dolphins: The New Moral Frontier.* Malden, Mass.: Blackwell, 2007.

SEE ALSO: Commercial fishing; Fisheries; Gill nets and drift nets; Marine Mammal Protection Act; Ocean pollution.

Donora, Pennsylvania, temperature inversion

CATEGORIES: Disasters; atmosphere and air pollution

THE EVENT: Temperature inversion resulting in the first outdoor air-pollution disaster in U.S. history

DATES: October 26-31, 1948

SIGNIFICANCE: The temperature inversion event that took place in Donora helped to prompt an increase in research concerning air pollution and raised awareness of the importance of the association between certain meteorological conditions and air-pollution episodes.

Donora is an industrial town of 14,000 located 40 kilometers (25 miles) southeast of Pittsburgh, Pennsylvania, in the Monongahela River Valley. It is surrounded by hills that are 107 meters (350 feet) in elevation. In 1948 four industries with 6-meter (20-

foot) smokestacks were located along a 4.8-kilometer (3-mile) stretch of the Monongahela's banks: a steel mill, which used high-sulfur coke; a zinc smelting plant, which used high-sulfur ores; a wire manufacturing plant; and a sulfuric acid plant. In addition, high-sulfur coal was used to generate the area's electricity and to heat most homes and businesses. Therefore, sulfur dioxide fumes and hydrocarbon particulates were common in the air of Donora.

On the morning of Tuesday, October 26, 1948, a surface high-pressure weather cell with light winds settled over the eastern United States. A blocking ridge in the upper air and the westward position of the jet stream prevented any other weather systems from moving into the area. That night, the cool, heavy air from the mountain slopes drained into Donora's valley, and this air combined with the sinking air of the high-pressure cell to create a strong surface temperature inversion.

A temperature inversion occurs whenever the temperature of the atmosphere increases with height. Because cool air is heavier than warm air, a temperature inversion discourages surface air from rising and dissipating surface air pollutants into the upper air. The air within Donora's valley, which was filled with the pollutants being emitted by industries, shops, and homes, became trapped at the surface with no way of dispersing either vertically, because of the temperature inversion, or horizontally, because of the surrounding mountains. To complicate matters, radiation fog, which had begun to form during the afternoon, blocked sunlight from reaching the surface of the earth, preventing surface heating from breaking the inversion.

This meteorological situation remained over Donora for five days. The result was an unprecedented air-pollution episode. By Wednesday morning, visibility in Donora was so poor that local residents were becoming lost. On Friday, respiratory-related health complaints from residents began to pour in; four thousand people became ill from the polluted air between 6:00 A.M. and midnight.

The first death occurred early Saturday morning, with sixteen more dead and two thousand more ill by midnight. Three more deaths occurred on Sunday even though relief came as an approaching front pushed the high-pressure cell out of the area and broke the temperature inversion. Subsequent rain washed most of the pollutants out of the air, the wind swept the smoke away, and the disaster was over.

Active responses to the lessons learned from this event were slow to materialize, but the experience in Donora did eventually prompt an increase in air-pollution research. It also caused many people to question the acceptance of billowing smokestacks as a necessary part of economic progress and to recognize the importance of the association between certain meteorological conditions and air-pollution episodes.

Kay R. S. Williams

FURTHER READING

Magoc, Chris J. "The Donora Disaster and the Problem of Air Pollution." In *Environmental Issues in American History: A Reference Guide with Primary Documents.* Westport, Conn.: Greenwood Press, 2006.

Markowitz, Gerald, and David Rosner. *Deceit and De-*

News photograph taken in the main business district of Donora, Pennsylvania, on October 30, 1948, shows that sunlight was virtually obliterated by thick, low-hanging pollution during the inversion event. (AP/Wide World Photos)

nial: *The Deadly Politics of Industrial Pollution*. Berkeley: University of California Press, 2002.

Snyder, Lynne Page. "Revisiting Donora, Pennsylvania's 1948 Air Pollution Disaster." In *Devastation and Renewal: An Environmental History of Pittsburgh and Its Region*, edited by Joel A. Tarr. Pittsburgh: University of Pittsburgh Press, 2003.

SEE ALSO: Air pollution; Air-pollution policy; Black Wednesday; Coal-fired power plants; London smog disaster; Smog.

Dredging

CATEGORY: Water and water pollution

DEFINITION: Removal by excavation or suction of sedimentary materials from the bottoms of lakes, rivers, estuaries, and coastal areas

SIGNIFICANCE: Dredging is performed to maintain navigation systems, exploit mineral resources, or remove polluted bottom sediments. The activity of dredging disturbs bottom-dwelling organisms and has negative impacts on water quality. It also generates waste sediment, which must be used or disposed of in an environmentally responsible manner.

Natural erosion processes, sometimes enhanced or exacerbated by human activity, break down rock and soil into sediments that ultimately migrate downstream. When suspended sediments eventually sink to the bottom of a lake, river, estuary, or bay, they form deposits that can impede navigable waterways or increase the likelihood of flooding. Dredging, or removing underwater sediments with digging or suction equipment, keeps ports, harbors, and waterways open for commerce, security, flood-control, and recreational purposes.

Dredging also serves as a mining method. Where suspended sediments contain ore minerals, the dense ore particles sink more readily than nonore materials and form concentrated deposits. Diamonds, cassiterite (a major tin mineral), ilmenite (a titanium ore mineral), and gold are among the minerals mined by dredging. Other resources obtained through dredging include sand and gravel, which are used in construction. In areas where land is expensive, dredging river bottoms for these materials is an alternative to quarrying them on land. For example, dredging

along the Kansas River in the Kansas City, Missouri, metropolitan area has provided a local source of sand and gravel for many years.

Another reason for dredging is to remove polluted bottom sediments from a body of water. Contaminant removal by dredging generates its own problems, as disturbing the sediments allows toxic materials that have settled out of suspension to reenter the overlying water, where they can affect aquatic life or be transported to another location. Maintenance dredging for navigation purposes is similarly problematic when it occurs in bodies of water where sewage or industrial pollutants have contaminated bottom sediments.

Large quantities of sediment are displaced during dredging operations. In the United States alone, maintenance dredging of navigable waters generates several hundred million cubic yards of sediment per year. In Southeast Asia, large dredges used to mine tin deposits handle more than 5 million cubic meters (177 million cubic feet) of sediment annually.

Dredging can mobilize existing contaminants but is not in itself a significant source of pollutants. However, dredging disrupts the habitats of bottom-dwelling organisms and can kill them outright. In addition, the dredging process and associated dumping of waste sediments in open waters causes turbidity (water cloudiness caused by suspended particles), which can harm or kill organisms that need clear water to survive. Oysters, for example, are detrimentally affected by turbid waters. Dredging can also affect erosion patterns, river flow, and ocean currents. In the Kansas River, dredging has removed far more sediment than can be replenished naturally. This has resulted in increased riverbank erosion, a wider and deeper river channel, a lowered water surface, a steeper bed gradient, and loss of vegetation and farmland along the river.

Clean dredged sediments can be employed for beneficial purposes, among them replenishing beaches and creating wetlands habitat. They can also be dried and used in construction materials or as fill dirt. Heavily polluted waste sediment must be sent to a secure disposal facility equipped to handle the contaminants in question.

In the United States, concerns about environmental degradation and watershed management issues related to dredging led to the adoption of a national dredging policy in 1995. The National Dredging Team (NDT), a multidisciplinary group made up of representatives from several governmental agencies and cochaired by the Environmental Protection Agency and the Army Corps of Engineers, serves as a forum for implementation of the policy. The NDT works to ensure that timely dredging operations keep navigable waterways open while protecting, conserving, and restoring the coastal environment. A priority of the NDT is the beneficial use of waste sediments.

Robert L. Cullers
Updated by Karen N. Kähler

FURTHER READING

Bray, R. N., ed. *Environmental Aspects of Dredging.* London: Taylor & Francis, 2008.

National Dredging Team. *Dredged Material Management: Action Agenda for the Next Decade.* Washington, D.C.: U.S. Environmental Protection Agency, 2003.

Palermo, Michael R., et al. *Technical Guidelines for Environmental Dredging of Contaminated Sediments.* Vicksburg, Miss.: U.S. Army Corps of Engineers, Environmental Laboratory, 2008.

SEE ALSO: Clean Water Act and amendments; Coastal Zone Management Act; Ocean dumping; Ocean pollution; Sedimentation; Water pollution.

Drift nets. *See* Gill nets and drift nets

Drinking water

CATEGORY: Water and water pollution
DEFINITION: Water supplies that are safe for human consumption
SIGNIFICANCE: Many areas of the world lack sufficient safe drinking water to support their populations. International, national, and local efforts have been undertaken to conserve existing water supplies, to clean up polluted water supplies with appropriate treatment, and to find new ways of distributing safe drinking water to those who need it.

In many parts of the world human beings lack access to safe drinking water. More than one million people die each year because of a lack of drinking water or because their water supplies are contaminated owing to unsanitary conditions. Even in developed countries where water supplies are sufficient, drinking water may contain dangerous contaminants, including such toxic chemicals as lead and arsenic, harmful microorganisms, and even radioactive compounds such as radon.

QUALITY, ACCESSIBILITY, AND REGULATION

Most problems with water quality involve contamination with disease-causing (pathogenic) microorganisms. Safe, clean drinking water is free of all pathogenic microbes. A number of dangerous microorganisms can be transmitted by water, including viruses (for example, hepatitis A), bacteria (*Salmonella* and *Vibrio cholera*), protozoa (*Giardia*, *Cryptosporidium*, and *Entamoeba*), and parasitic worms, or helminths (*Ascaris lumbricoides*). Pollution and microbial contamination make the waters of most rivers, lakes, and ponds (surface water) unfit for human consumption without prior purifying treatment. The most common method of treating surface water to produce safe drinking water is disinfection by chlorination. Chloramines and ultraviolet light are also used for treatment.

A sufficient quantity of good-quality drinking water exists to satisfy the needs of all human beings on the planet, but the water is not distributed in such a way that it reaches populations in all parts of the world. Population growth also has impacts on drinking-water resources, even in regions with large supplies of water. In many developing countries people have no choice but to use water that is polluted with various wastes—including human sewage, animal excrement, and a variety of pathogenic microorganisms—because no water treatment systems are in place. The United Nations estimates that one billion people around the world do not have access to unpolluted drinking water.

In many locations, drinking water is obtained from groundwater sources. Groundwater—water found beneath the ground surface in soil and rock spaces—can generally be used for drinking with minimal treatment because it has already been purified by passing through soil. Wells must be dug to reach this water. More than one and one-half billion people in the world use groundwater as a source of drinking water. In the United States, about 60 percent of drinking water comes from rivers and lakes and about 40 percent from groundwater sources. In many parts of the world, groundwater usage is increasingly outpacing the rate of groundwater replacement.

Several countries have introduced regulations intended to ensure the good quality of drinking water.

In the United States, the Environmental Protection Agency (EPA) sets standards for drinking water under the 1974 Safe Drinking Water Act. These standards cover some ninety-four possible contaminants, including biological and chemical substances. For example, the EPA sets the acceptable number of microorganisms per 1 milliliter (0.034 ounce) of water at fewer than 10, and water that is considered drinkable can contain no coliform bacteria.

DESALINATION AND CONSERVATION

The need to obtain more drinking water in some regions has led to increased use of desalination technologies as well as increasing emphasis on the conservation of existing water supplies. The desalination of ocean water produces billions of gallons of drinking water per day around the world, but this process can have some negative environmental impacts, such as thermal pollution and damage to shoreline ecosystems.

Water conservation is the most cost-effective way to reduce demand for drinking water. Both local and national governments in many nations have introduced regulations and programs aimed at encouraging water conservation on the level of individual households; among the areas targeted for reduced usage of drinking-water supplies have been lawn maintenance and toilet flushing. The practice of rainwater harvesting is encouraged in some areas, and some governments have instituted strict compulsory water metering to raise awareness of the need to conserve water.

Another option for increasing supplies of drinking water is the restoration of municipal and industrial wastewater to drinkable quality. Using technologies such as membrane bioreactor treatment (which involves treating wastewater with certain types of microorganisms and then putting it through microfiltration, followed by disinfection with ultraviolet light), communities can turn their own wastewater into drinking water.

Sergei A. Markov

FURTHER READING

Hammer, Mark J. *Water and Wastewater Technology.* Upper Saddle River, N.J.: Pearson, 2004.

Shannon, Mark A., et al. "Science and Technology for Water Purification in the Coming Decades." *Nature* 452 (March, 2008): 301-310.

Sigee, David. *Freshwater Microbiology.* New York: John Wiley & Sons, 2005.

Wright, Richard T. *Environmental Science: Toward a Sustainable Future.* 10th ed. Upper Saddle River, N.J.: Pearson, 2008.

SEE ALSO: Chlorination; Clean Water Act and amendments; Desalination; Safe Drinking Water Act; Water conservation; Water quality; Water-saving toilets.

Droughts

CATEGORY: Weather and climate

DEFINITION: Significant deviations from normal precipitation patterns lasting long enough to cause water shortages that disrupt normal human and biological activities

SIGNIFICANCE: The economic, social, and environmental effects of droughts can be devastating. In arid regions, prolonged or frequent droughts can lead to desertification.

Droughts are extreme events that are part of climate variability. In climatic terms, drought is said to occur when rainfall over a given period is substantially below the mean. The effects may be seen in diminishing water supplies for domestic and industrial users and in shortfalls in meeting water needs of rainfed agricultural systems. Four categories of definitions of the term "drought" focus on specific aspects of the phenomenon: climatological definitions compare precipitation level to long-term averages and refer to duration of a dry period, hydrologic definitions consider the effects of dry periods on water supplies such as reservoirs and rivers, agricultural definitions emphasize deficits in meeting crop water needs, and socioeconomic definitions stress the effects of water shortages on human society and economy.

DROUGHT MANAGEMENT

In contrast to common natural hazards such as floods and hurricanes, which are distinct violent events, droughts often have no easily recognized beginnings or ends. Another difference between droughts and other natural disasters is that the effects of droughts are cumulative. The aim of drought management is to reduce societal vulnerability to the effects of drought by promoting proactive planning and the adoption of appropriate techniques for risk management.

Three categories of drought risk management are generally recognized. The first is water-demand man-

A Time Line of Historic Droughts

1270-1350 SOUTHWEST: A prolonged drought destroys Anasazi Indian culture.

1585-1587 NEW ENGLAND: A severe drought destroys the Roanoke colonies of English settlers in Virginia.

1887-1896 GREAT PLAINS: Droughts drive out many early settlers.

1899 INDIA: The lack of monsoons results in many deaths.

1910-1915 SAHEL, AFRICA: First in a series of recurring droughts.

1933-1936 GREAT PLAINS: Extensive droughts in the southern Great Plains destroy many farms and create the Dust Bowl during the worst U.S. drought in more than 300 years.

1968-1974 SAHEL, AFRICA: Intense period of drought; 22 million affected in four countries, 200,000-500,000 estimated dead, millions of livestock lost.

1977-1978 WESTERN UNITED STATES: Severe drought compromises agriculture.

1981-1986 AFRICA: Drought in 22 countries, including Angola, Botswana, Burkina Faso, Chad, Ethiopia, Kenya, Mali, Mauritania, Mozambique, Namibia, Niger, Somalia, South Africa, Sudan, Zambia, and Zimbabwe, results in 120 million people affected, several million forced to migrate, significant loss of life and of livestock.

1986-1988 MIDWEST: Many farmers are driven out of business by a drought.

1986-1992 SOUTHERN CALIFORNIA: Drought brings increased water prices, loss of water for agricultural production, water rationing.

1998 MIDWEST: Drought destroys crops in the southern part of the Midwest.

1999 LARGE PART OF UNITED STATES: Major drought strikes the Southeast, the Atlantic coast, and New England; billions of dollars in damage to crops.

Beg. 2006 WESTERN UNITED STATES: Drought-induced wildfires cause significant property and environmental damage in both urban and rural areas.

agement, which includes such things as economic disincentives and legal restrictions on water use. Second is water-supply management, which might involve enlarging existing water-storage reservoirs, arranging for long-distance transport of water supplies from unaffected areas, desalinating seawater, increasing the depths of bores and wells, planning for the emergency use of lakes, and recycling used water. The third category is mitigation of drought impact, which includes insurance to handle economic losses from drought, gradual reduction in livestock numbers, loans to farmers and ranchers to provide supplemental feed for livestock, and subsidies for well drilling or stock slaughter.

ENVIRONMENTAL AND OTHER IMPACTS

The damage caused by drought includes direct losses, such as livestock and human deaths owing to water shortage, and indirect losses, such as reduction in agricultural production and hydroelectric power generation, the occurrence of forest fires, and increases in pests and diseases in crops and animals. The patterns of damage from drought disasters are familiar ones. In poorer countries, drought disasters tend to cause high loss of life and comparatively lower economic losses. In developed industrial societies, loss of life is generally low but economic damages can be very high.

Globally, overall growth has been seen in the damage caused by droughts as populations have grown and economies have developed, leading to greater accumulations of property and wealth to be damaged. The impacts of drought have also increased as some populations have experienced growing vulnerability as the result of increased pressure on limited water resources. More humid, less drought-prone regions are often most vulnerable. It has been suggested that human-caused global climate change may have an effect on the frequency of droughts and their spatial patterns of occurrence.

LINKS WITH DESERTIFICATION

Drought is part of the climate pattern of semiarid areas, as rainfall in such places is highly variable. However, in certain regions, such as the Sahel in Africa, droughts display remarkable spatial coherence—for example, they often occur over the whole sub-Saharan belt (Sahel) simultaneously—which can ultimately lead to desertification. Desertification is the effect of progressive desiccation and land degradation in arid environments associated with human use of the land. Land degradation manifests itself as destruction of biological potential resulting from the loss of soil or damaging changes within the soil. The Sahelian zone of Africa is the world's casebook on drought and desertification.

In societies of the developing countries of the Sahel, the effects of drought are repetitious because of climatic variability but are aggravated by rapidly increasing population. A large part of the population of the Sahel engages in subsistence farming and herding, and the amount of rainfall determines the carrying capacity of the land. Desertification sets in when the numbers of people and livestock exceed the carrying capacity. In precolonial Africa, the numbers of grazing livestock and the size of the area cultivated increased with rainfall. As a dry cycle began, overgrazing and overcultivation resulted in severe environmental stress and degradation. Carrying capacity decreased and population declined as people either died or moved away. In modern times, in contrast, when carrying capacity begins to decline, aid in the forms of medicine, food, and shelter suppresses the death rate, illness, and the need for out-migration. The continued presence of large numbers of people reduces the carrying capacity of the land still further. A larger and more resilient population remains, providing an elevated base from which increased growth can continue when rainfall temporarily increases again.

Stress on the land results in the destruction of plant cover beyond the minimum required to prevent severe soil erosion. The more fertile topsoil is removed, less soil remains, and therefore less moisture is retained. The soil dries out and soil temperatures increase, further speeding drying and collapse of the system. The added effects of human-induced feedback processes accentuate and perpetuate drought

phases. For example, overgrazing leads to loss of vegetation, which means more windblown dust that blocks incoming solar radiation. Less surface heating means less uplift of air by convection and therefore less rain. Scientists have suggested that other, more complex interactions may also be at work, but the key is that human-caused environmental degradation affects feedback processes that ultimately modify the vulnerability and susceptibility of the population to drought and desertification.

C. R. de Freitas

FURTHER READING

De Freitas, C. R. "The Hazard Potential of Drought for the Population of the Sahel." In *Population and Disaster,* edited by John I. Clarke et al. Oxford: Blackwell, 1989.
Iglesias, Ana, et al., eds. *Coping with Drought Risk in Agriculture and Water Supply Systems: Drought Management and Policy Development in the Mediterranean.* New York: Springer, 2009.
Wilhite, Donald A., ed. *Drought: A Global Assessment.* 2 vols. London: Routledge, 2000.
_____. *Drought and Water Crisis: Science, Technology, and Management Issues.* New York: Taylor & Francis, 2005.

SEE ALSO: Africa; Airborne particulates; Carrying capacity; Cloud seeding; Deforestation; Desertification; Erosion and erosion control; Grazing and grasslands; Overgrazing of livestock; Soil conservation.

Dubos, René

CATEGORIES: Activism and advocacy; ecology and ecosystems
IDENTIFICATION: French-born American bacteriologist and environmental writer
BORN: February 20, 1901; Saint-Brice, France
DIED: February 20, 1982; New York, New York
SIGNIFICANCE: Through his writings, Dubos encouraged exploration of the manner in which humans interact with the environment.

René Dubos was born into a family of poor French shopkeepers. He excelled as a student in the Collège Chaptal in Paris and decided to pursue a career as a scientist. Later, inspired by the work of Louis

Pasteur, about whom he wrote two books, Dubos determined to study bacteriology. In 1924 he settled in the United States, and he eventually became a U.S. citizen. From 1924 to 1927 Dubos undertook study and research at Rutgers University in New Jersey and then moved to Rockefeller University in New York City. Except for a brief period at Harvard (from 1942 to 1944), he remained associated with Rockefeller University until he became director of Environmental Studies at the State University of New York College at Purchase in 1971.

Dubos first achieved international recognition through his discovery of what would become the

Bacteriologist and environmental writer René Dubos. (AP/Wide World Photos)

first commercially marketed antibiotic, tyrothricin, in 1939. The development of tyrothricin awakened medical researchers to the enormous possibilities for soil-produced antibacterial agents.

During the 1950's Dubos developed an interest in ecology, and over the next twenty-five years, he became one of the world's most outspoken and heralded environmentalists. Aided by his second wife, Letha Jean Porter, a former research associate, Dubos wrote prolifically about environmental issues and the way humans ought to relate to the natural world. He emphasized that the quality of human life (mental, physical, and spiritual) depends primarily on the establishment of a symbiotic relationship between humankind and nature. He rejected the claims made by more extreme environmentalists that human life is destined to become extinct on a planet filled with chemical waste.

Dubos's book *The Unseen World* (1962) gained him international recognition as an influential commentator on the environment. There followed a series of books over the next thirty years, including *So Human an Animal* (1968), *A God Within* (1972), and *The Wooing of Earth* (1980). In 1969 he was cowinner of the Pulitzer Prize in general nonfiction for *So Human an Animal*. In the 1960's and 1970's Dubos achieved a sort of celebrity status and was often featured in major newsmagazines and newspapers.

Dubos's most important contribution to the environmental dialogue was a report published in 1972 as

Only One Earth: The Care and Maintenance of a Small Planet. Maurice Strong, general secretary for the United Nations Conference on the Human Environment, commissioned Dubos to chair an international committee of experts that would prepare a commentary on the condition of the earth for the United Nations Conference on the Human Environment, held in Stockholm, Sweden, in 1972. Dubos then chose well-known British environmentalist Barbara Ward to assist in preparing and writing the final report. Much of the Ward-Dubos report focused on the need to maintain a balance in the earth's ecosystems—a balance that was in jeopardy. Although many of the experts consulted were displeased that *Only One Earth* did not predict doom, the report inspired the United Nations to focus attention on worldwide environmental problems. As a consequence, the Stockholm Conference established the United Nations Environment Programme to observe environmental issues and coordinate information about them on a worldwide basis.

Ronald K. Huch

FURTHER READING

Cooper, Jill E. "A Brief Romance with Magic Bullets: René Dubos at the Dawn of the Antibiotic Era." In *Silent Victories: The History and Practice of Public Health in Twentieth-Century America*, edited by John W. Ward and Christian Warren. New York: Oxford University Press, 2007.

Moberg, Carol L. *René Dubos, Friend of the Good Earth: Microbiologist, Medical Scientist, Environmentalist.* Washington, D.C.: ASM Press, 2005.

SEE ALSO: Spaceship Earth metaphor; United Nations Conference on the Human Environment; United Nations Environment Programme.

Ducktown, Tennessee

CATEGORIES: Places; land and land use; resources and resource management

IDENTIFICATION: Town in southeastern Tennessee

SIGNIFICANCE: Copper mining operations that began in Ducktown, Tennessee, during the late nineteenth century led to the eradication of plant life and subsequent soil erosion in the region. The transformation of forestland into an artificial desert demonstrated the hazards of extracting natural resources without consideration for its effect on the surrounding area.

In 1843 a gold rush led prospectors to the southeastern corner of Tennessee, where, instead of gold, they found copper. This discovery led to the founding of several mining companies in 1850. The Copper Basin, an area of approximately 19,400 hectares (48,000 acres) surrounding the area where the city of Ducktown is located, was the only place in the eastern United States to produce significant amounts of copper, and by 1902 Tennessee was the sixth-largest copper-producing state in the nation.

After the copper ore was mined, it was put through a process called roasting to separate the copper from the rest of the ore, which included zinc, iron, and sulfur. Firewood was harvested from the surrounding forests and placed in heaps in roofed sheds with open sides. The ore was piled onto the heaps of firewood, which were then lit. These heaps, which covered vast areas of ground, were then allowed to roast for three months.

The smoke arising from the ores during the roasting was highly sulfuric. In addition to hindering vision, the smoke contained sulfuric gas, sulfuric dust, and sulfuric acid that descended onto the earth. The extent of the pollution was first acknowledged around

1895, when individual citizens filed lawsuits for damage to vegetation. The state of Georgia filed similar lawsuits in 1904 and 1905. The heap method was soon abandoned. Instead, raw ore was smelted in furnaces without being roasted first.

The ore-processing companies also built taller smokestacks to take advantage of higher air currents that would disperse the smoke, but that only spread the sulfuric vapors over a larger area. It was not until 1911 that the problem of sulfuric pollution was controlled to some extent by a process that captured the smoke for the valuable sulfuric acid that could be derived from it. By that time, the years of sulfuric pollution and systematic deforestation had turned the Ducktown area into the only desert east of the Mississippi River. Erosion washed away the topsoil, leaving a bare, rocky terrain that, at its greatest extent, covered about 130 square kilometers (50 square miles). Copper mining continued until 1986, when it became impossible for local producers to compete with the price of imported copper.

During the 1930's copper mining companies worked with the Tennessee Valley Authority (TVA) to reforest the area. More than fourteen million trees were planted over the next several decades, and grasses were also planted. Little resulted from the plantings until the 1970's, when improved methods of fertilization and soil churning increased the plants' chances of survival. By the 1990's only about 405 hectares (1,000 acres) of land appeared to be bare, with large gullies and widely spaced trees. The restoration of the forest was not totally accepted by local residents, however, who believed that a small portion of the desert should be saved as an example of severe environmental degradation.

Rose Secrest

FURTHER READING

Davis, Donald Edward. *Where There Are Mountains: An Environmental History of the Southern Appalachians.* Athens: University of Georgia Press, 2000.

Pipkin, Bernard W., et al. *Geology and the Environment.* 5th ed. Belmont, Calif.: Thomson Brooks/Cole, 2008.

SEE ALSO: Ashio, Japan, copper mine; Desertification; Erosion and erosion control; Mine reclamation; Reclamation Act.

Dust Bowl

CATEGORIES: Disasters; land and land use; resources and resource management

THE EVENT: Environmental disaster marked by huge dust storms in the southern region of the Great Plains of the United States

DATE: 1930's

SIGNIFICANCE: The Dust Bowl revealed the damage that mechanized agriculture could cause if not accompanied by a program of soil management.

Droughts periodically occur in the Great Plains of the United States. During such periods, winds pick up loose soil and create dust storms, especially during the spring months. Settlers reported numerous examples of this natural phenomenon during the nineteenth century. During the twentieth century, new agricultural practices and overgrazing by cattle speeded soil erosion in the region. Tractors and other machines allowed farmers to plow larger areas for planting wheat. In the process, they destroyed the natural grasses, the root systems of which had stabilized the soil. Because the wheat replaced the grasses, most farmers remained unaware that they were contributing to a coming catastrophe.

In 1931 a severe drought struck the Great Plains; it centered on the Texas and Oklahoma panhandles, northeastern New Mexico, eastern Colorado, and southwestern Kansas. The wheat crop withered in the fields, and its root systems were no longer able to support the soil. As the drought continued, soil particles that normally clustered together separated into a fine dust. When the winds blew in early 1932, they lifted the dust into the air, marking the beginning of the environmental disaster that a newspaper reporter later dubbed the Dust Bowl.

Although their number and severity increased, dust storms remained an issue of local and regional concern for the first two years. However, as the drought continued into 1934, the storms grew so immense that they caused damage in areas far from the

A farmer's son sits on a sand dune near his home in Liberal, Kansas, during the 1930's. (Library of Congress)

The Dust Bowl

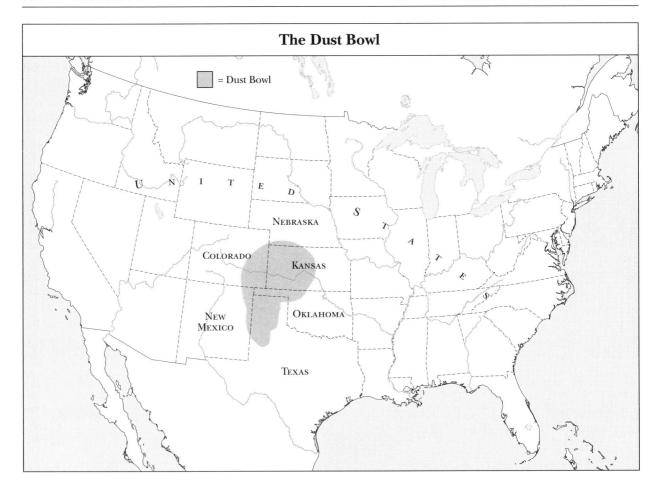

= Dust Bowl

plains. A storm that emanated from Montana and Wyoming in May, 1934, deposited an estimated twelve million tons of dust on Chicago, Illinois. Ships that were some 480 kilometers (300 miles) offshore in the Atlantic Ocean reported that dust from the same storm landed on their decks. Incidents such as these provoked national concern over the growing crisis on the plains.

Scientists identified two types of dust storms: those caused by winds from the southwest and those resulting from air masses moving from the north. While no less damaging, the more frequent southwest storms tended to be milder than the terrifying northern storms, which came to be known as "black blizzards." Huge walls of dust, sometimes more than 1.6 kilometers (1 mile) high, rolled across the plains at 100 kilometers per hour (62 miles per hour) or faster, driving frightened birds before them. The sun would disappear, it would become as dark as night, and frightened people would huddle in their homes, their windows

often taped shut. On occasion, people stranded outside during these severe storms suffocated. Some black blizzards lasted less than one hour; others reportedly continued for longer than three days.

Most historians argue that the Dust Bowl was one of the worst ecological disasters in the United States, one that could have been mitigated had farmers practiced soil conservation in the years before drought struck. Instead, farms were ruined, causing some 3.5 million people to abandon the land. Many of them moved into small towns on the plains, while others journeyed to California in search of opportunity. Cattle and wildlife choked to death. Human respiratory illnesses increased markedly during the Dust Bowl era, and a number of people died from an ailment known as dust pneumonia. Anecdotal evidence indicates that many people grew depressed as the dust storms continued year after year.

The mid-1930's marked the peak of the Dust Bowl, with seventy-two storms that reduced visibility to less than 1.6 kilometers (1 mile) reported in 1937. The return of the rain in the late 1930's eased the crisis, and by 1941 the disaster was over. However, by that time ecologists and farmers had begun to undertake soil conservation measures in response to the crisis. The U.S. government provided expertise and financial support for many of these efforts. Farmers practiced listing, a plowing process that makes deep furrows to capture the soil and prevent it from blowing. Alternating strips of planted wheat with dense, drought-resistant feed crops such as sorghum slowed erosion by blocking wind and retaining moisture, which prevented the soil from separating into dust. On lands not farmed, natural grasses were planted to prevent erosion. The government also sponsored the Shelterbelt Project, a program that used rows of trees to form windbreaks. Millions of trees were planted throughout the Great Plains, with more than 4,828 kilometers (3,000 miles) of shelterbelts created in Kansas alone.

Despite the experiences of the 1930's, once the drought ended many farmers returned to the farming practices that had damaged their fields. Soil conservation experts worried that the region would suffer a return of Dust Bowl conditions when the rains stopped. Their predictions came to pass in 1952, when another drought led to a series of dust storms, including several storms with wind gusts clocked at 129 kilometers (80 miles) per hour. That drought ended in 1957, but in accord with a twenty-year cycle, the region again faced a shortage of rainfall in the

early 1970's. At that time some analysts confidently predicted that dust storms such as those seen in the 1930's were a thing of the past. They claimed that irrigation with aquifer water from deep wells would prevent soil erosion. However, shrewd observers pointed out that the fate of the region was now tied to a resource, aquifer water, that would become increasingly precious in the coming years. The possibility that the Great Plains could again witness a devastating ecological catastrophe like that of the 1930's remains.

Thomas Clarkin

FURTHER READING

Bonnifield, Paul. *The Dust Bowl: Men, Dirt, and Depression.* Albuquerque: University of New Mexico Press, 1979.

Cunfer, Geoff. *On the Great Plains: Agriculture and Environment.* College Station: Texas A&M University Press, 2005.

Hurt, R. Douglas. *The Dust Bowl: An Agricultural and Social History.* Chicago: Burnham, 1981.

Lookingbill, Brad D. *Dust Bowl, USA: Depression America and the Ecological Imagination, 1929-1941.* Athens: Ohio University Press, 2001.

Worster, Donald. *Dust Bowl: The Southern Plains in the 1930s.* 25th anniversary ed. New York: Oxford University Press, 2004.

SEE ALSO: Droughts; Erosion and erosion control; Soil conservation; Strip farming.

E

Earth Day

CATEGORIES: Activism and advocacy; ecology and ecosystems

THE EVENT: Annual observance intended to promote public concern for environmental problems

DATE: Inaugurated on April 22, 1970

SIGNIFICANCE: In many ways the first Earth Day in 1970 marked the birth of the environmental movement, as twenty million Americans either engaged in demonstrations or gathered to hear speeches. Although participation was disappointing during the 1970's, a growing number of grassroots organizations celebrated Earth Day throughout the 1980's, and in the intervening decades participation in Earth Day activities has continued to increase.

In June, 1969, Democratic U.S. senator Gaylord Nelson of Wisconsin, having observed how anti-Vietnam War "teach-ins" had influenced public opinion, conceived of the idea of a large teach-in to educate the general public about the importance of environmental issues. He suggested that such an event should be planned for April 22, a day when many states commemorated Arbor Day and a day that would not conflict with the timing of final exams on college campuses. Recognizing the potential of the idea, a small group of concerned citizens founded the organization Environmental Action to sponsor the event. They were able to raise the modest sum of $125,000, and a dynamic young law student, Denis Hayes, was put in charge of publicizing and coordinating activities. Senator Nelson and Republican congressman Paul N. McCloskey of California were named official cochairs of the event.

Numerous historical factors contributed to the great success of the first Earth Day. By the late 1960's a growing number of organizations were helping to sensitize the public to environmental problems, and an unprecedented number of publications on environmental themes were being produced by prominent writers, such as Rachel Carson, Walter Udall,

Lynn White, and Paul R. Ehrlich. Of even greater significance, Americans in locations throughout the country were witnessing the harmful effects of environmental damage. In 1968 and 1969 members of Congress reflected public opinion as they considered nearly 140 bills related to environmental issues. Almost four months before Earth Day, President Richard Nixon signed the National Environmental Policy Act of 1969, which required analysis and review of public projects. For many people, especially the young, the impressive achievements of the Civil Rights movement represented a model for reform based on a moral appeal. In addition, the spirit of youthful rebellion embodied in the antiwar movement was inspiring parallel movements throughout American society. In short, Senator Nelson could not have chosen a more auspicious context for the launching of his idea.

Conservative business and political leaders were not enthusiastic about the idea of Earth Day. Although President Nixon's press secretary announced the administration's support for the day, the president took no active role in any of the events. While some members of the administration suspected that the observance was a means for advancing the agenda of liberal Democrats, Nixon's secretary of the interior, Walter Hickel, urged Nixon to proclaim a national holiday and become an active participant. Hickel later wrote that he gave "marching orders" to Interior Department personnel to visit college campuses and that fifteen hundred employees of the department did so. The White House was embarrassed when the press reported that Controller General James Bentley had spent sixteen hundred dollars in public funds to send telegram warnings of a possible left-wing plot after he observed that Earth Day fell on the birthday of Soviet revolutionary leader Vladimir Ilich Lenin. Bentley was forced to apologize and pay for the telegrams with his own money.

APRIL 22, 1970

In all measurable ways, the first Earth Day was a huge success. In New York City, Chicago, and Philadelphia large crowds gathered to hear speeches by

politicians, poets, ecologists, and other concerned citizens. Some fifteen hundred college campuses, as well
as ten thousand elementary and secondary schools,
scheduled programs of one kind or another. The National Education Association estimated that about ten
million schoolchildren participated in some kind of
environmental activity for the day. Also, approximately two thousand communities planned environmental ceremonies of one kind or another.

An impressively diverse array of activities were conducted throughout the United States, and the atmosphere at many events was euphoric and theatrical. In
Washington, D.C., about ten thousand young people
attended a rock concert in front of the Washington
Monument. The University of Wisconsin held fifty-
eight separate programs. To dramatize air-pollution
problems caused by internal combustion engines,
several universities held enthusiastic automo-
bile-wrecking events called "wreck-ins." Some localities also held "bike-ins." In New York City,
Fifth Avenue was closed to motor vehicle traffic
for two hours. Many idealistic people helped
proenvironmental efforts. At the University of
Washington, four hundred people planted trees
and shrubs during a "plant-in" in an abandoned
area near the campus. In Ohio, one thousand
students from Cleveland State University gathered litter and loaded it into garbage trucks. In
hundreds of communities, groups of Boy Scouts
and Girl Scouts held cleanup campaigns and
picked up litter.

The first Earth Day was an occasion for numerous speeches, including many by the best-known
spokespersons for the environmental movement.
Barry Commoner, Paul Ehrlich, Ralph Nader,
and the aging René Dubos were among the speakers in greatest demand. Both houses of Congress
were adjourned for the day, and many politicians
were seen at various rallies. Senator Nelson spoke
on nine university campuses in Wisconsin, California, and Colorado. Senator Thomas McIntyre,
who delivered fourteen speeches in his home
state of New Hampshire, set the record for the
greatest number of speeches given by one person
on that first Earth Day.

No major acts of violence marred the celebrations, but scattered incidents of militancy did occur. At Boston's Logan Airport, thirteen demonstrators were arrested for blocking traffic during
a demonstration to protest a proposed expansion of the airport. In Washington, D.C., about twenty-
five hundred demonstrators assembled before the offices of the Department of the Interior to protest the
approval of oil leases. Students at the University of
California at Berkeley conducted a sit-in to register
their disapproval of the presence on campus of job re-
cruiters from Ford Motor Company, while at the University of Texas, twenty-six students were arrested for
perching in trees to try to prevent the trees' destruction.

EARTH DAY 1990

In 1971 Earth Day was expanded into Earth Week,
but the expanded observance was not successful. For
a few years Earth Day attracted limited interest. By the
mid-1980's, however, the celebration of Earth Day was
regaining popularity, as environmentalists viewed the

*A woman dressed as the Statue of Liberty rides a float full of trash during a
parade held in observance of the first Earth Day in 1970. (AP/Wide
World Photos)*

celebrations as a repudiation of President Ronald Reagan's conservative environmental policies. Leaders of the environmental movement wisely decided to concentrate their efforts on commemorating the twentieth anniversary of Earth Day, and Denis Hayes was chosen as the chairperson of the occasion.

On April 22, 1990, an estimated 200 million people in 140 countries participated in Earth Day. Organizers claimed that this was the largest grassroots demonstration in history. For this occasion, Hayes had the assistance of a large coalition of environmentalist and other socially conscious groups, and together they raised and spent about $4 million. The day was celebrated with marches, rallies, parades, concerts, and a large assortment of activities on all continents. Although the largest demonstrations were held in the developed industrial countries, scattered events also took place within the poorer and less developed regions of the world.

In Boston, a crowd of 200,000 people turned out; in New York City's Central Park, the various rallies attracted an estimated 750,000 participants; in Washington, D.C., some 125,000 people participated in a demonstration on the National Mall; in St. Louis, an estimated 10,000 people planted 10,000 trees on the banks of the Mississippi River. Throughout the day, speeches were given by Hayes, Gaylord Nelson, Morris Udall, Barry Commoner, Bruce Babbitt, Senator Edward Muskie, and countless others. In a prime-time national television program called *The Earth Day Special*, aired by the American Broadcasting Company (ABC), Bette Midler played an abused Mother Earth who collapsed as a result of global warming, deforestation, and toxic poisoning. Although the events of the day were almost uniformly peaceful, eco-guerrilla groups attracted headlines by destroying oil-exploration gear and pouring sand in the fuel tanks of logging machinery.

At a time of controversy over issues such as protection of the northern spotted owl, President George H. W. Bush and his administration appeared distrustful of Earth Day 1990. However, President Bush, who had earlier referred to himself as "the environmental president," addressed several crowds via a telephone hookup. Although speakers at Earth Day events often criticized the Bush administration's policies, they tended to spend more time attacking the earlier policies of former president Ronald Reagan. In comparison with those of twenty years before, the speeches of Earth Day 1990 were more somber and realistic, as it had become clear that environmental problems would not be quickly solved in a painless manner.

Earth Day 1990 was truly a global festival. In Brazil, a concert by Paul McCartney paid special attention to the environment. In West Germany, Green organizations sponsored the ceremonial planting of trees. Thailand's top rock band, the Carabano, held a concert with the theme "We Love the Forest." In Hong Kong a daylong educational entertainment featured singers, mimes, and exhibits of "green" consumer goods. Other activities included a roadway "lie-down" by five thousand protesters against car fumes in Italy, an 800-kilometer (500-mile) human chain across France, a trash-cleanup campaign on Mount Everest in Nepal, and a "flyby" of three thousand kites made by schoolchildren in Tours, France.

IMPACT OF EARTH DAY

By 1990 it appeared clear that yearly celebrations of Earth Day had become a mainstream institution. The day's silver anniversary in 1995 attracted considerable interest, but organizers decided to concentrate greater efforts on the thirtieth anniversary in the year 2000. With the help of Internet organization, some five thousand environmental groups working in 184 countries around the world joined together to celebrate Earth Day 2000.

The 1970 celebration of Earth Day tended to be a predominantly white, middle-class affair. Many African Americans were suspicious that the day would detract from issues of racial and economic justice. Although such views did not completely disappear, they tended to decline with time. In 1969 polls indicated that only 33 percent of African Americans wanted the government to pay more attention to environmental issues; by 1976, approximately 58 percent of African Americans expressed the same viewpoint. By the time of Earth Day 1990, African American leaders had more evidence that pollution tends to be especially severe in areas where poor and marginalized people live, and thus they were able to use the day to publicize the issue of environmental racism.

Although celebrations of Earth Day are primarily important as a reflection of public opinion at a particular time, there is evidence that such celebrations have some impact on public attitudes and that they can solidify the commitment of environmental organizations. In a 1965 Gallup Poll, only 17 percent of the responding Americans said they considered the reduction of air and water pollution to be one of the most pressing problems demanding governmental action. Immediately after Earth Day 1970, the figure was 53 percent, but by 1980 it had fallen to 24 percent. Ironically, the environmental policies of President Ronald Reagan appeared to encourage the popularity of Earth Day during the 1980's, and the success of Earth Day 1990 is partially explained by a survey of the time in which 80 percent of Americans said they would support more strenuous environmental efforts, regardless of costs.

Without doubt, there are always faddish, trendy elements in the celebrations and speeches of Earth Day. At times the rhetoric seems rather excessive, and environmentalists are often offended when some of the nation's worst polluters attempt to exploit the day, to use exposure at events as a form of commercial advertising. The important point, however, is that Earth Day promotes education and reflection about serious problems that are experienced by people in their daily lives.

Thomas T. Lewis

FURTHER READING

Dunlap, Riley E., and Angela G. Mertig, eds. *American Environmentalism: The U.S. Environmental Movement, 1970-1990.* Philadelphia: Taylor & Francis, 1992.

Gottlieb, Robert. "The Sixties Rebellion: The Search for a New Politics." In *Forcing the Spring: The Transformation of the American Environmental Movement.* Rev. ed. Washington, D.C.: Island Press, 2005.

National Staff of Environmental Action. *Earth Day: The Beginning—A Guide for Survival.* New York: Bantam Books, 1970.

Nelson, Gaylord, with Susan Campbell and Paul Wozniak. *Beyond Earth Day: Fulfilling the Promise.* Madison: University of Wisconsin Press, 2002.

Scheffer, Victor. *The Shaping of Environmentalism in America.* Seattle: University of Washington Press, 1991.

Shabecoff, Philip. *A Fierce Green Fire: The American Environmental Movement.* Rev. ed. Washington, D.C.: Island Press, 2003.

Switzer, Jacqueline Vaughn. *Green Backlash: The History and Politics of Environmental Opposition in the U.S.* Boulder, Colo.: Lynne Rienner, 1997.

SEE ALSO: Animal rights movement; Earth Summit; Environmental education; Environmental justice and environmental racism; Green movement and Green parties; Public opinion and the environment; Sun Day.

Earth First!

CATEGORIES: Organizations and agencies; activism and advocacy; preservation and wilderness issues
IDENTIFICATION: Radical environmental movement
DATE: Founded in 1980
SIGNIFICANCE: Earth First! gained notoriety during the 1980's for employing unorthodox and controversial means to protest the abuse of wilderness areas. The movement's use of ecologically motivated sabotage and highly visible civil disobedience actions, such as tree sitting to prevent logging, has drawn public attention to environmental causes at the same time it has drawn criticism from some environmentalists.

Earth First! was founded in the United States in 1980 by five men who were concerned about what they perceived to be a lack of passion and commitment within the old-line conservation organizations. The five founders were Howie Wolke, a Friends of the Earth representative; Ron Kezar, a National Park Service worker; Bart Koehler, a Wilderness Society representative; Mike Roselle, a Yippie activist; and Dave Foreman, a lobbyist for the Wilderness Society who had once been a Barry Goldwater Republican. Other like-minded persons quickly joined the cause.

A movement rather than a formally structured organization, Earth First! has gained notoriety and sparked controversy through its environmental activism. The movement is committed to naturalist John Muir's wilderness preservation ideals as well as to the principles of deep ecology. The philosophy of Earth First! is expressed in the slogan "No compromise in defense of Mother Earth." Members of the movement view private and public developers as their enemies, and they see the U.S. Forest Service as being too willing to capitulate to developers. The movement is also highly critical of mainstream conservation organizations, which Earth First! accuses of having become passive and bureaucratized.

Earth First!ers use various strategies to inhibit the destruction of wilderness, from letter-writing campaigns, petitions, boycotts, and legal actions to demonstrations, nonviolent civil disobedience, and ecologically motivated sabotage (also called ecotage or monkeywrenching). The movement entered the media spotlight in 1981 when Earth First!ers unfurled a

300-foot-long piece of black polyethylene depicting a "crack" on the face of the Glen Canyon Dam, located just south of the Utah-Arizona border, in symbolic opposition to the dam and its recently created reservoir, Lake Powell. Their methods led the media to dub Earth First!ers a real-life Monkey Wrench Gang, for the titular heroes of Edward Abbey's 1975 novel who sabotage construction equipment in the American Southwest to stop environmental destruction.

Black plastic quickly gave way to monkeywrenching: Earth First!ers pulled up survey stakes, blocked logging roads, disabled construction equipment, and spiked old-growth trees (a controversial tactic in which metal or ceramic spikes are pounded into trees to ruin their commercial use and thwart logging equipment). Such actions have been widely condemned and criticized. Tree spiking, in particular, has the potential to harm loggers and sawmill workers by shattering saw blades that contact spikes; it was made illegal in the United States in 1988. Foreman published a detailed how-to book on ecotage, *Ecodefense: A Field Guide to Monkeywrenching* (1985), in which he—like the movement itself—claimed to neither condemn nor condone the use of such tactics.

In 1989, Foreman and three Arizona Earth First!ers were arrested in an undercover sting operation conducted by the Federal Bureau of Investigation (FBI); they were charged with planning to sabotage power lines. Foreman was released on probation, but the others served prison sentences—a fact that caused resentment against Foreman among some Earth First!ers. In the decade since the movement's inception, a rift had developed within the movement between conservationists such as Foreman and social justice activists such as Roselle. Earth First!'s anti-hierarchical lack of structure and leadership had attracted a politically anarchic following, and the movement's focus had broadened from wilderness preservation to a host of leftist causes. In 1990 Foreman and others dissatisfied with what Earth First! had become left the movement.

That year marked Earth First!'s Redwood Summer—a conscious attempt to emulate the Civil Rights movement's 1964 Freedom Summer—when a number of protesters faced off against the lumber industry in California's redwood forests. The protesters demonstrated a more communal, countercultural, and leftist spirit, with greater involvement by women, an increasing emphasis on social justice, and attempts to appeal to the loggers as a class against the corporations. One of the organizers, Judi Bari, was seriously injured by a bomb that had been placed in her automobile by an unknown assailant. Earth First! continued its confrontational campaign to save the redwoods through the decade, with Julia "Butterfly" Hill's two-year-long occupation of an old-growth redwood tree in 1997-1999 gaining international media attention. In 1998 fellow redwood-defending Earth First!er David "Gypsy" Chain lost his life when he was struck by a tree that a logger refused to stop cutting.

Ethics of an Eco-Warrior

In the following excerpt from his memoir Confessions of an Eco-Warrior *(1991), environmentalist Dave Foreman conveys the sense of passion and frustration that ultimately led to his founding of the grassroots organization* Earth First!

We, as human beings, as members of industrial civilization, have no divine mandate to pave, conquer, control, develop, or use every square inch of this planet. As Edward Abbey . . . said, we have a right to be here, yes, but not everywhere, all at once.

The preservation of wilderness is not simply a question of balancing competing special-interest groups, arriving at a proper mix of uses on our public lands, and resolving conflicts between different outdoor recreation preferences. It is an ethical and moral matter. A religious mandate. Human beings have stepped beyond the bounds; we are destroying the very process of life. . . .

The crisis we now face calls for *passion*. When I worked as a conservation lobbyist in Washington, D.C., I was told to put my heart in a safe deposit box and replace my brain with a pocket calculator. I was told to be rational, not emotional, to use facts and figures, to quote economists and scientists. I would lose credibility, I was told, if I let my emotions show.

But, damn it, I am an animal. A living being of flesh and blood, storm and fury. The oceans of the Earth course through my veins, the winds of the sky fill my lungs, the very bedrock of the planet makes my bones. I am alive! I am not a machine, a mindless automaton, a cog in the industrial world, some New Age android. When a chain saw slices into the heartwood of a two-thousand-year-old Coast Redwood, it's slicing into my guts. When a bulldozer rips through the Amazon rain forest, it's ripping into my side. When a Japanese whaler fires an exploding harpoon into a great whale, my heart is blown to smithereens. I am the land, the land is me.

Earth First! groups sprang up in the United Kingdom during the early 1990's, as did a radical spin-off group called the Earth Liberation Front (ELF). As Earth First! began to work to distance itself from criminal activities, focusing instead on nonviolent civil disobedience, ELF embraced economic sabotage as a means to save a dying planet. ELF, which went on to become an international presence, regards extreme methods such as arson and bombing as legitimate weapons in the war against profit-motivated environmental destruction. Housing developments, chain stores, construction sites, and automotive dealerships selling sport utility vehicles are favorite ELF targets. In 2002 the FBI listed ELF as America's largest and most active domestic terrorist group. Spokespersons for ELF maintain that the group advocates violent action only against inanimate objects, but it does not shy away from harassment and threats against developers and corporate heads—"the real terrorists," according to ELF.

By 2010 the Web site of the *Earth First! Journal* listed Earth First! groups in Canada, Iceland, the United Kingdom, Ireland, Belgium, the Netherlands, Germany, Italy, the Czech Republic, Poland, Russia, Nigeria, the Philippines, and Australia, in addition to groups in twenty-seven U.S. states.

Eugene Larson
Updated by Karen N. Kähler

FURTHER READING

Foreman, Dave, and Bill Haywood, eds. *Ecodefense: A Field Guide to Monkeywrenching*. 3d ed. Chico, Calif.: Abbzug Press, 2002.

Liddick, Donald R. *Eco-Terrorism: Radical Environmental and Animal Liberation Movements*. Westport, Conn.: Praeger, 2006.

Scarce, Rik. *Eco-Warriors: Understanding the Radical Environmental Movement*. Updated ed. Walnut Creek, Calif.: Left Coast Press, 2006.

Wall, Derek. *Earth First! and the Anti-Roads Movement: Radical Environmentalism and Comparative Social Movements*. New York: Routledge, 2002.

SEE ALSO: Abbey, Edward; Biocentrism; Deep ecology; Ecocentrism; Ecotage; Ecoterrorism; Foreman, Dave; Glen Canyon Dam; Monkeywrenching.

Earth resources satellites

CATEGORY: Resources and resource management
DEFINITION: Unmanned scientific satellites that collect information about the earth and its resources and relay it to land-based scientific stations
SIGNIFICANCE: Earth resources satellites provide scientists with systematic, repetitive measurements of the surface conditions of the earth, including atmospheric and oceanic conditions. The data the satellites collect help scientists understand phenomena such as climate change and how events such as deforestation, coastal erosion, and pollution affect the earth as a whole.

The first earth resources satellite was launched on July 23, 1972, from Vandenberg Air Force Base in California. Earth Resources Technology Satellite 1, which became better known as Landsat 1, was placed in an orbit in which it passed over the North and South Poles to provide scientists with information about those areas. Since that first launch, earth resources satellites have provided scientists with a wealth of data about the earth. They accurately measure and transmit values for atmospheric water vapor, cloud cover, ocean currents and temperatures, ozone distribution, surface altitude and temperatures, wave height, wind speed and direction, and a variety of other atmospheric, geological, and oceanic information. By tracking the fluctuations in these measurements, scientists can study how various events affect the earth's surface and climate. They can also use this information to observe and monitor the earth's limited resources, document the spread of pollution and changes in climate, and better understand how the earth's limited resources can be used and conserved in the best ways possible.

Earth resources satellites also document events such as earthquakes, fires, floods, oil spills, and volcanic eruptions, giving scientists a clearer picture of how these events interact across the planet. For example, scientists came to a better understanding of the El Niño weather phenomenon by using satellite-gathered data to track the ocean's surface currents, temperatures, and winds through several El Niño cycles. During the 1991 Persian Gulf War, images from these satellites were used to document and monitor the damage caused to Kuwait's oil fields during the Iraqi occupation.

Workers inspect the Landsat 7 earth resources satellite before it is launched into orbit. (NASA)

Information gathered by earth resources satellites may also be used to locate mineral deposits and monitor the condition of forests, fisheries, and farms. Agricultural uses include finding underutilized areas of both sea and land and determining whether plots of land can be irrigated to produce new farming areas or whether aquaculture could be a fitting use for particular areas in the sea. Satellite-gathered information has also been used to pinpoint the origins of crop diseases and to monitor the spread of such diseases.

History

During the 1960's weather satellites and photographs from high-flying aircraft provided scientists with some information about the earth's surface. These information sources were relatively inefficient, however; the black-and-white images from weather satellites did not provide significant detail, and it was impossible to take enough aerial photographs to provide sufficient information (it was estimated that

some thirty thousand photographs would need to be taken to cover the United States alone). William T. Pecora, who worked for the U.S. Geological Survey, became interested in developing a type of orbiting platform that would carry scientific recording devices and provide better information. This type of equipment was already being developed for space exploration and for possible military applications, and he saw its potential for providing detailed maps of the earth. At the same time, Archibald B. Park, who worked in the U.S. Department of Agriculture, was interested in the same type of information; he believed that if the world's crops and forests could be mapped and surveyed, farmers and ecologists could use the information to help them better manage land resources.

For about ten years, what became the Landsat program was managed by a private-sector company (the Earth Observation Satellite Company, a partnership between Hughes Aircraft Company and the RCA Corporation), but throughout most of the history of these satellites, the U.S. government has managed Landsat through a joint program of the National Aeronautics and Space Administration (NASA) and the U.S. Geological Survey. Other countries also have earth resources satellites circling the globe, either through their own national programs or joint ventures; these include Australia, Brazil, China, Canada, Japan, and various European nations.

The first five U.S.-based earth resources satellites were launched in 1972, 1975, 1978, 1982, and 1984. The sixth satellite, launched in 1993, failed to achieve orbit, but a seventh, known as Landsat 7, was successfully launched in 1999. All of the satellites have carried various types of cameras, including infrared cameras. The older models collected photographic images of the earth's surface; the later models (1982, 1984, and 1999) also carried additional sensors and a scanning radiometer capable of providing scientists with high-resolution images of the earth's surface.

Marianne M. Madsen

Further Reading

Bukata, Robert P. *Satellite Monitoring of Inland and Coastal Water Quality: Retrospection, Introspection, Future Directions.* Boca Raton, Fla.: CRC Press, 2005.

China Remote Sensing Satellite Ground Station. *The Majestic Earth: A Selection of Earth Resources Satellite Images.* Hong Kong: Science Press, 1996.

Chuvieco, Emilio, ed. *Earth Observation of Global Change: The Role of Satellite Remote Sensing in Monitoring the Global Environment.* New York: Springer, 2008.

Kramer, Herbert J. *Observation of the Earth and Its Environment: Survey of Missions and Sensors.* New York: Springer, 2002.

Mack, Pamela E. *Viewing the Earth: The Social Construction of the Landsat Satellite System.* Cambridge, Mass.: MIT Press, 1990.

Maul, G. A. *Introduction to Satellite Oceanography.* Dordrecht, Netherlands: Martinus Nijhoff, 1985.

Short, Nicholas M. *Mission to Earth: Landsat Views the World.* Washington, D.C.: National Aeronautics and Space Administration, 1976.

SEE ALSO: Beaches; Deforestation; Desertification; El Niño and La Niña; Floodplains; Glacial melting; Logging and clear-cutting; Oil spills; Rain forests; Renewable resources; Resource depletion.

Earth-sheltered construction

CATEGORIES: Energy and energy use; resources and resource management

DEFINITION: Construction of homes or other buildings mostly underground

SIGNIFICANCE: Compared with traditional above-ground structures, earth-sheltered buildings have several environment-related advantages, in particular the fact that they require significantly lower amounts of energy for heating and cooling.

Human beings have been using earth-sheltered construction techniques since they first began creating shelters for themselves as protection against the elements. In the past earth-sheltered buildings were often carved into hillsides or constructed from turf, but modern versions are usually made of concrete. Earth-sheltered homes have been found in all cultures and all parts of the world, from Iceland to China to the American Southwest.

During the 1970's "back to the land" movement, building earth-sheltered homes became popular with

An earth-sheltered house. (©iStockphoto.com)

many people who wanted to live in harmony with nature. Malcolm Wells, often referred to as the "father of earth-sheltered construction," was an architect who believed that underground architecture was a promising and overlooked way to build without destroying the land, a "silent, green alternative to the asphalt society." Wells credited his interest in such building to an underground housing exhibit he saw at the 1964 New York World's Fair and to a trip he took to Taliesin West, the Arizona home of architect Frank Lloyd Wright, where an underground theater stayed cool despite soaring outdoor temperatures.

An earth-sheltered building is best located on a well-drained hillside; correct excavation and preparation of the building site are critical. It is not recommended that earth-sheltered buildings be constructed in areas of permafrost, areas with high water tables, or areas with high underground radon concentrations.

Two basic types of earth-sheltered buildings are common: underground and bermed. In underground construction, the structure is created and then covered with earth (at a depth of anywhere from 15 centimeters to 2.7 meters, or 6 inches to 9 feet). In a bermed building, the structure is banked with earth surrounding one or more outside walls. Both types generally have earth-covered roofs. The choice between underground or bermed construction depends on the vegetation, climate, soil, and drainage of the building site area.

An earth-sheltered home may cost 10 to 20 percent more to construct than a typical aboveground dwelling, but often low maintenance costs and energy savings offset these higher costs somewhat. The earth surrounding an earth-sheltered building serves as insulation, so that the interior maintains a constant temperature. Proponents of earth-sheltered construction have estimated that families living in earth-sheltered homes can save up to 80 percent in heating and air-conditioning costs in comparison with these costs for a traditional aboveground house. The insulation of the earth on earth-sheltered homes also keeps water lines from freezing, makes these buildings safer than traditional houses during storms, and provides soundproofing. Proponents of earth-sheltered construction also note that the outer parts of these structures are nearly maintenance-free (they require no painting, for instance) and that, compared with aboveground buildings, earth-sheltered structures have fewer problems with break-ins (as earth-sheltered buildings have fewer entrance points) and with

damage from fire (as they are usually built of concrete).

Problems can arise in earth-sheltered buildings if they are not properly constructed. Potential problems include poor air circulation, flooding, condensation, and water seepage. Earth-sheltered homes may also not have very good resale value, as the numbers of people seeking out such homes remain somewhat limited.

Marianne M. Madsen

FURTHER READING

Roy, Rob. *Earth-Sheltered Houses: How to Build an Affordable Underground Home.* Gabriola Island, B.C.: New Society, 2006.

Underground Space Center, University of Minnesota. *Earth Sheltered Housing Design: Guidelines, Examples, and References.* New York: Van Nostrand Reinhold, 1979.

Wells, Malcolm. *The Earth-Sheltered House: An Architect's Sketchbook.* Rev. ed. White River Junction, Vt.: Chelsea Green, 2009.

SEE ALSO: Back-to-the-land movement; Energy conservation; Green buildings.

Earth Summit

CATEGORIES: Ecology and ecosystems; resources and resource management

IDENTIFICATION: Conference convened by the United Nations in Rio de Janeiro, Brazil, to address environmental issues at the global level

DATES: June 3-14, 1992

SIGNIFICANCE: Despite disagreements between the nations of the Northern and Southern Hemispheres, the Earth Summit yielded important results, including the signing of the Convention on Climate Change and the Convention on Biological Diversity, the endorsement of the Rio Declaration and the Statement of Forest Principles, the adoption of Agenda 21, and the creation of the U.N. Commission on Sustainable Development.

The Earth Summit held in Rio de Janeiro in 1992 was a follow-up to the Stockholm Conference of 1972. The United Nations General Assembly passed a resolution on December 20, 1988, that the U.N. Conference on Environment and Development (UNCED), or Earth Summit, would convene in 1992. Considerable changes had taken place over the years after the Stockholm Conference: Many countries had negotiated treaties and agreements to manage the earth's resources and preserve the environment. Serious degradation of the environment was still occurring, however, and international debates continued, particularly between the countries of the Northern and Southern Hemispheres.

The first preparatory meeting for the Earth Summit was held in 1990. The decision to convene the conference was based on the belief that many environmental issues needed to be addressed at the global level; sustainable development was the core concept around which the attendees' deliberations would revolve.

It was evident at that first meeting in 1990 that the same issues between nations of the Northern Hemisphere and those of the Southern Hemisphere that had been prevalent at Stockholm twenty years before still remained. The Group of 77, representing the developing nations in the South, repeated demands for preferential, noncommercial, and concessional technology transfer from the advanced states in the North. They also wanted additional financial resources made available to the less developed states. The Northern nations were divided on these issues. Some, particularly the Nordic countries, felt that there would never be agreement on environmental issues unless Northern nations demonstrated willingness to make substantial economic concessions to the South. They wanted a working group established to consider the South's demands for technology transfer and to discuss resource issues.

The United States and the European Community were opposed to formalizing the demands of the South. In fact, the U.S. delegation was working under orders from the White House to keep such issues off the summit agenda. The United States had its way when a working group was established and given a mandate to discuss only legal and institutional issues. The only forum in which the South could bring up its issues would be the plenary session at Rio. In response, the Southern nations changed their strategy, making their participation in any agreements to be reached at Rio conditional on recognition of their demands.

At the second meeting of the Preparatory Committee, the Malaysian representatives presented a declaration, supported by the other developing countries, stating that Malaysia would negotiate the issue of forests only if the North would pay for the South's participation in any agreement that was reached. The declaration also demanded that those nations with advanced economies reduce their energy consumption and sign an agreement concerning climate change.

THE CONFERENCE

Delegations representing 178 countries, with more than 110 heads of state, met in Rio de Janeiro from June 3 to June 14, 1992, for the Earth Summit. With representatives of nongovernmental organizations (NGOs), which met in a parallel Global Forum, and other observers, the Rio meeting involved some twenty-five thousand people. The conference had been convened to address urgent problems related to environmental protection and socioeconomic development. The U.N. General Assembly had added the word "development" to the official conference name to include the concerns of developing countries that still, as they had at Stockholm, believed that economic growth was of greater importance to them than protection of the environment.

The developed countries came to Rio with an agenda different from that of the developing countries. The developed nations were concerned with specific environmental problems, such as ozone depletion, global warming, deforestation, and acid rain. The developing countries were focused on the negative impacts of the economic policies of the North on poorer countries whose economies were slow to improve or, in some cases, were experiencing negative growth patterns.

A possible treaty on forests became one of the major areas of disagreement between developed and developing countries. At a Preparatory Committee meeting held in March, 1991, the United States had tried to get approval for negotiations on a forest treaty outside of UNCED. Instead, the committee banned any outside negotiations and created an ad hoc group within the conference to study the topic of forests. Although all the world's forests have suffered from deforestation, the greatest depletion of forests has taken place in Africa, with Asia also experiencing significant losses. The countries of the South insisted, however, that the issue not focus on their forests alone.

AGREEMENTS REACHED

Notwithstanding the difficulties of getting the participant nations to negotiate regarding issues that they considered intrusive into their own internal affairs, the

Fellow world leaders and diplomats applaud as Brazilian president Fernando Collor de Mello signs the United Nations Framework Convention on Climate Change at the Earth Summit in June, 1992. (AP/Wide World Photos)

Earth Summit yielded significant results. The Convention on Climate Change and the Convention on Biological Diversity were both signed at Rio. In addition, the Rio Declaration and the Forest Principles were endorsed, and Agenda 21 was adopted. The U.N. Commission on Sustainable Development (CSD) was created to monitor the results of the summit.

Agenda 21 is a three-hundred-page plan for achieving sustainable development in the twenty-first century. Since 1992 more than 1,800 cities and towns worldwide have created their own "local Agenda 21," and 150 countries have established national advisory councils to promote dialogue among government officials, environmentalists, and businesspeople on sustainable development policies for their own countries. In addition, countries have been able to reach agreement on new treaties on desertification and high-seas fishing.

A range of environmental topics are included in Agenda 21 under the following classifications: agriculture and rural development, biotechnology, business and industry issues, capacity building in developed countries, changing consumption patterns, children's issues, combating poverty, resources, deforestation, desertification, hazardous wastes, human health, human habitat, and NGO participation. Agenda 21 entered into force on June 13, 1992, when the participating governments signed it. It proposes comprehensive strategies and programs to counteract environmental degradation and promote sustainable development. Its strategies are intended to create a balance between the environment and development, and emphasize the importance of participation by all countries.

The U.N. Framework Convention on Climate Change entered into force on March 21, 1994. The main objective of this convention is to achieve stabilization of the production of greenhouse gases. It sets out principles through which countries can acquire an understanding of global warming, including such issues of importance to the South as the sharing of research and development and technology transfer.

The U.N. Convention on Biological Diversity entered into force on December 29, 1993. The convention addresses conservation of biological species; use of genetic resources; technology transfer; identification and monitoring of problems; research, training, and education; and impact assessment.

The Rio Declaration on Environment and Development entered into force on June 13, 1992, at the time of adoption by the participating states. Its purpose is to establish cooperation among member states so that they can reach agreements on laws and principles promoting sustainable development. It creates a balance between the sovereign right of states and states' responsibilities to the rest of the international community. The declaration reaffirms and updates Principle 21 of the Stockholm Declaration and itself contains twenty-seven principles. It addresses the environmental impacts of development, poverty, ecosystem protection, sharing of scientific ideas, public participation and access to information, implementation of legislation, economic policies of states, internationalization of environmental costs and the "polluter pays principle," notification of pollution incidents and environmental impact statements, and indigenous cultures.

The Statement of Forest Principles is a compromise that was reached by the environmental ministers who represented the nations taking part in the Earth Summit. Statements of principles, considered "soft law," are not binding on the states and allow them to find some areas of agreement without committing themselves. While acknowledging the right of states to exploit their own forest resources, the Statement of Forest Principles encourages nations to manage such resources without causing long-term damage to them.

The Commission on Sustainable Development is a high-level group created to monitor and report on implementation of the agreements reached at the summit. It has been referred to as a watchdog over the actions of the states in carrying out their obligations under the agreements. The CSD reports to the U.N. General Assembly through the U.N. Economic and Social Council (ECOSOC). It is also intended to serve as an ongoing forum for negotiating further global policies on the environment and development.

The conference delegates agreed to a proposal that called for a treaty to be negotiated on the topic of desertification. They also agreed on a mechanism for financing programs discussed at Rio, including the projects included in Agenda 21. Developed countries were urged to meet a target of providing 0.7 percent of their gross national product to help developing countries implement sustainable development programs. It was decided that this funding would be distributed by the Global Environment Facility (GEF), a multinational organization.

PROGRESS REVIEWS

The governmental leaders who participated in the Earth Summit agreed to a five-year review of progress, to be held in 1997. The review took place in a special session of the U.N. General Assembly that was held in New York City from June 23 to June 27, 1997. The session was devoted to examining how well countries, international organizations, and various sectors of civil society had responded to the challenges given them by the Earth Summit.

The objectives of the 1997 session were to revitalize and energize commitments to sustainable development, recognize failures and determine the reasons for them, recognize achievements and identify actions to build on them, define priorities beyond 1997, and raise the profile of issues that had not been addressed sufficiently at Rio. While recognizing the progress that had been made, the state participants concluded that the global environment was continuing to deteriorate. The more than fifty heads of state who attended did agree to further action on freshwater, energy, and transportation issues, but few concrete commitments were made. The five-year review process continued with the 2002 World Summit on Sustainable Development, which was held in Johannesburg, South Africa; the United States did not participate in this conference, which also produced few concrete commitments.

Colleen M. Driscoll

FURTHER READING

Adede, A. *International Environmental Law from Stockholm 1972 to Rio de Janeiro 1992: An Overview of Past Lessons and Future Challenges.* Maryland Heights, Mo.: Elsevier Science, 1992.

Andersen, Stephen O., and K. Madhava Sarma. *Protecting the Ozone Layer: The United Nations History.* Sterling, Va.: Earthscan, 2002.

Kane, Hal, and Linda Starke. *Time for Change: A New Approach to Environment and Development.* Washington, D.C.: Island Press, 1992.

Porter, Gareth, Janet Welsh Brown, and Pamela

Chasek. "The Emergence of Global Environmental Politics." In *Global Environmental Politics*. Boulder, Colo.: Westview Press, 2000.

Speth, James Gustave, and Peter M. Haas. "From Stockholm to Johannesburg: First Attempt at Global Environmental Governance." In *Global Environmental Governance*. Washington, D.C.: Island Press, 2006.

U.N. Conference on Environment and Development. *The Global Partnership for Environment and Development: A Guide to Agenda 21.* New York: United Nations, 1993.

SEE ALSO: Agenda 21; Global Environment Facility; Sustainable development; United Nations Conference on the Human Environment; United Nations Environment Programme; World Summit on Sustainable Development.

Eat local movement

CATEGORIES: Agriculture and food; philosophy and ethics

IDENTIFICATION: Movement devoted to building locally based, self-reliant food economies in which sustainable food production, processing, distribution, and consumption are integrated

SIGNIFICANCE: Proponents of the eat local movement assert that by buying and eating only locally grown or produced foods, people can enhance social and economic health while at the same time providing benefits to the environment, such as by supporting small farmers who employ organic and sustainable agricultural practices.

One goal of the eat local movement is to reduce the carbon footprint of food by reducing the amount of energy it takes to get food from the field to the plate. According to proponents of the movement (often called locavores), the food in an average American meal travels at least 3,200 kilometers (2,000 miles) from its sources to the consumer. Those who endorse the practice of eating locally encourage people to eat only foods that are grown or produced near where they live. This, locavores assert, serves to keep people in better touch with the earth and the seasons and provides a connection between farm and table, reduces the carbon footprint of the food, sustains lo-

cal economies, and encourages consumption of high-quality foods. Proponents of the movement generally seek out food that is grown in ways that are healthy, that respect the environment and those who work in the food-growing industry, and that are humane to animals. Although no single definition of "local" has been established, many proponents of the eat local movement suggest that people should buy and eat only those foods that are grown or produced within a radius of roughly 160 kilometers (100 miles) of where they live.

"Local food" includes food grown in one's own garden and food produced on local farms or by local community-sponsored agriculture (CSA) groups. In addition to reducing the expenditure of energy on shipping and other means of transporting food, the eat local movement is concerned with sustainable agriculture; buying and eating locally grown foods supports local small farms, many of which raise crops, maintain dairy cattle, and raise livestock for food according to the principles of organic farming, without using chemical fertilizers or pesticides and without the hormones or antibiotics often used by large food producers. Locavores point to fresher, better-tasting, lower-cost produce with higher amounts of vitamins and other healthy compounds that are lost when fruits and vegetables are picked green for shipping or are handled extensively before they reach the consumer. Locavores also note that cutting out the middlemen—the large food suppliers—means that local farmers receive a larger percentage of each dollar spent on food.

Often, eat local and CSA groups offer classes or other kinds of instruction on buying local foods in bulk during their peak season and preserving them for later use, such as by canning or drying; many also provide recipes that focus on the locally grown foods in their own areas. Eat local and CSA groups also generally encourage members and others to patronize restaurants that feature local foods that are in season.

Detractors of the eat local movement have noted that although eating locally may save energy by cutting down on the transportation of food, the overall environmental costs of growing food can remain quite high. Is it more environmentally friendly, for example, to grow fruits and vegetables outdoors in a sunny environment and then ship them to a store in another area than it is to use electricity and other energy sources to grow the same kinds of produce in greenhouses in less naturally sunny environments and then sell them locally? Debate on such issues is ongoing.

Marianne M. Madsen

FURTHER READING

Bendrick, Lou. *Eat Where You Live: How to Find and Enjoy Fantastic Local and Sustainable Food No Matter Where You Live.* Seattle: Skipstone Press, 2008.

Nabhan, Gary Paul. *Coming Home to Eat: The Pleasures and Politics of Local Food.* New York: W. W. Norton, 2009.

Pollan, Michael. *The Omnivore's Dilemma: A Natural History of Four Meals.* New York: Penguin Press, 2007.

SEE ALSO: Back-to-the-land movement; Community gardens; Organic gardening and farming; Sustainable agriculture.

Eaubonne, Françoise d'

CATEGORY: Philosophy and ethics

IDENTIFICATION: French novelist, poet, essayist, and journalist

BORN: March 12, 1920; Paris, France

DIED: August 3, 2005; Paris, France

SIGNIFICANCE: D'Eaubonne, who coined the term "ecofeminism," influenced thinking about the relationship between ecology and feminist theory, both of which are built on the concept of the interconnectedness of living and nonliving beings.

Although she is well known in the French-speaking world as the author of more than fifty volumes of fiction, nonfiction, essays, and verse, Françoise d'Eaubonne is not very familiar to most English-speaking readers. Of her dozens of books, only a handful have been translated into English. Her significance to the environmental movement stems from her early advocacy of ecofeminism, which would become a powerful philosophical segment of the movement.

The daughter of an insurance executive and a professor, d'Eaubonne formed radical feminist opinions as a young adult. She began publishing fiction and poetry in the 1950's, and she was cofounder, in 1971, of a revolutionary homosexual group, the Front homosexuel d'action révolutionnaire. She had published more than a dozen works by the time of her 1974 book *Le Féminisme ou la mort* (feminism or death), in which

she coined the term *éco-féminisme*, or "ecofeminism," to denote the connection between ecology and feminism, as well as the potential for women to effect reforms promoting the well-being of the human species and the earth. In 1979 she elaborated on her ideas in a work titled *Ecologie-Féminisme.*

According to d'Eaubonne and others who share her views, ecofeminism is both a theory and a sociopolitical movement. As a sociopolitical movement, ecofeminism recognizes the historical connection between women and nature. Ecofeminists believe that the human condition can be transformed through the nurturing of an environment in which people live at peace with one another, with nonhuman species, and with the natural environment, and they maintain that such a transformation is necessary if life on earth is to survive.

Ecofeminists believe that feminism and ecology need each other. Feminism cannot liberate women from oppression if it does not work to liberate the environment from human exploitation, because an ecology-based science that fails to recognize the oppression of peoples becomes irrelevant and useless. According to ecofeminists, women especially have an important role in saving the environment.

Ecofeminist thinkers have forced a reexamination of the idea of development, particularly in developing nations, where women have traditionally been the managers of resources and providers of food for their people. This reevaluation of development has focused on sustainability rather than on resource extraction and commodity production for profit maximization. Ecofeminists represent a wide range of disciplines, including the natural, social, and behavioral sciences, as well as philosophy, theology, and literature.

Critics assert that the woman-nature connection claimed by ecofeminists defines women primarily as biological beings, in contrast to feminism's longtime struggle to overcome the view that women are primarily childbearers and nurturers. While ecofeminists do not deny this, they tend to ignore the fact that many women around the world have assumed roles in what have historically been traditionally male professions and occupations.

Some question the validity of ecofeminism as a theory because of its references to mysticism, the Goddess tradition, magic, and witchcraft. Ecofeminists especially romanticize prehistoric cultures and the simple lifestyles they entailed. Essentially, critics claim that ecofeminism cannot really be called a theory; rather, it is more of an ideology built on a belief in the universal ideal of the life-giving, caring, nurturing woman.

Alexander Scott

FURTHER READING

Gates, Barbara T. "A Root of Ecofeminism: *Ecofémin-isme.*" *Interdisciplinary Studies in Literature and Environment* 3, no. 1 (1996):7-16.

Kheel, Marti. *Nature Ethics: An Ecofeminist Perspective.* Lanham, Md.: Rowman & Littlefield, 2008.

Warren, Karen J. *Ecofeminist Philosophy: A Western Perspective on What It Is and Why It Matters.* Lanham, Md.: Rowman & Littlefield, 2000.

SEE ALSO: Ecofeminism; Environmental justice and environmental racism; Sustainable development.

Echo Park Dam opposition

CATEGORIES: Activism and advocacy; preservation and wilderness issues

THE EVENT: Environmentalists' efforts to halt the building of a dam in the Echo Park river bottom in Utah's Dinosaur National Monument

DATES: 1952-1956

SIGNIFICANCE: The Sierra Club's successful attempt to stop construction of the proposed Echo Park Dam in Dinosaur National Monument during the 1950's signaled the emergence of the environmental movement as a powerful political force in the United States.

The U.S. Bureau of Reclamation first suggested building a high dam at the Echo Park site, located 3.2 kilometers (2 miles) below the confluence of the

Steamboat Rock in Echo Park, part of Dinosaur National Monument, photographed in 1957 after the Sierra Club's victory in preserving the area. (AP/Wide World Photos)

Green and Yampa rivers on the border of Utah and Colorado, during the 1930's. No formal request was made to the U.S. Congress for authorization of the project until 1950, however. The terms of the Organic Act of 1916, passed following the controversy surrounding construction of the Hetch Hetchy Dam in Yosemite National Park in California, prohibited such a project, but administrators within the bureau believed legislators would be willing to make an exception for Echo Park. Subsequent events proved them wrong.

At the time that the Bureau of Reclamation asked Congress for permission to build a high dam within the boundaries of Dinosaur National Monument, few people expected any significant opposition. In the years following the Great Depression, both the U.S. government and the general public saw dam development as good for the economy and thus good for the country. True, noted photographer Ansel Adams had helped mobilize opposition to hydroelectric development on the Kings River in California a decade earlier, leading to the creation of Kings Canyon National Park, but the Kings River area was home to giant sequoia trees—its scenic wilderness value was obvious. Dinosaur National Monument, in contrast, appeared barren. As long as the Echo Park Dam would not inundate the dinosaur fossil quarries, advocates of wilderness preservation initially voiced few objections. According to David Brower, executive director of the Sierra Club at the time, even members of his organization described the monument as being nothing but sagebrush.

This changed in 1952 following a Sierra Club member's trip through Dinosaur National Monument. The home movie footage he shot of the canyons within the monument persuaded Brower and others to take a closer look. In 1953 the Sierra Club began organizing rafting trips along the Green River through Dinosaur National Monument. As more people traveled through the spectacular river canyons, opposition to dam construction within the boundaries of the monument grew. Other wilderness preservation and conservation groups, such as the Wilderness Society and the Izaak Walton League, along with prominent writers and politicians, joined with the Sierra Club in fighting the Echo Park Dam proposal.

In 1956 this coalition of preservationists and conservationists was successful: The Bureau of Reclamation dropped its plans for the Echo Park Dam. The victory for the environmental preservationists proved bittersweet, however. In exchange for the cancellation of the plans for Echo Park, the environmentalists agreed not to fight the Bureau of Reclamation's plan to build Glen Canyon Dam on the Colorado River, a decision Brower later regretted. Still, by preventing construction of the Echo Park Dam, the environmentalists reaffirmed the important principle that no industrial development should ever take place within a national park.

Nancy Farm Männikkö

FURTHER READING

Lowry, William Robert. *Dam Politics: Restoring America's Rivers.* Washington, D.C.: Georgetown University Press, 2003.
Palmer, Tim. "The Beginnings of River Protection." In *Endangered Rivers and the Conservation Movement.* 2d ed. Lanham, Md.: Rowman & Littlefield, 2004.

SEE ALSO: Brower, David; Dams and reservoirs; Glen Canyon Dam; Hetch Hetchy Dam; Sierra Club.

Ecocentrism

CATEGORY: Philosophy and ethics
DEFINITION: The view that the natural world is morally important and should be valued independent of present or future human interests
SIGNIFICANCE: Ecocentrism emerged from a concern that human interests, culture, and history have limited and shaped human understanding of ethics and of what is valuable, thereby obscuring the importance of the earth's ecosystem.

Sometimes called deep ecological ethics or dark green ethics, ecocentrism is the view that the natural world is morally important and should be valued independent of present or future human interests; its importance should be reflected in human ethics and in human treatment of the earth. The natural world includes the organisms of the earth as well as the earth itself and its elements and ecosystems. Ecocentrism thus differs from anthropocentrism, which claims that only human interests and values ultimately matter or that they matter more than any other interests and values, and biocentrism, which extends moral importance only to certain animals or living organisms in addition to humans. Formulation of the ecocentric ethic is usually credited to Aldo Leopold; it was later developed by Arne Naess and George Sessions.

Ecocentrism calls for a new ethical outlook based on the recognition that human life has emerged from and is dependent on the ecosystem. Humans are one constituent part of a larger whole; they are, along with other inhabitants, equal citizens of the earth, not the earth's masters. Such egalitarianism insists that human interests do not have automatic priority over the natural world. In some situations of conflict between human and nonhuman interests, priority should be granted to the nonhuman interests, as when commercial or economic development is restricted for the sake of habitat or species preservation.

Demands for equality between humans and nonhumans are often based on claims that the natural world has intrinsic value, but intrinsic value has been understood in different ways. One approach recognizes that organisms such as plants have identifiable interests, or ends, that should be respected alongside those of sentient beings. Another claims fundamental value for the existence and preservation of all species. The value of the biotic community is also identified with the robust, highly integrated functioning of the ecosystem and features such as integrity, stability, and beauty.

Some ecocentric thinkers worry, however, that an appeal to objective and abstract properties has too much in common with the kinds of reductive and scientific analyses of the world that have contributed to environmental degradation. An alternative strategy appeals for the rediscovery of a sense of reverence toward nature in its abundance and vitality. It points to the human experience of nature as an endless series of sensuous particulars, stories and emotional encounters with places, that imply that the value of nature can never be fully articulated.

Ecocentrism is also practical. It denies that environmental ethics should be based on enlightened human self-interest and efficient management. Human understanding of the earth's innumerable elements and systems, as well as humanity's effects on them, is limited. The appropriate response is skepticism toward adopting technological solutions to environmental problems (technocentrism). Ecocentrism favors familiar lower-tech and environmentally friendly means of addressing ecological problems. Further, some go beyond practical concern for the environment and seek reform of the social structures and economic systems, including capitalism, that have caused neglect of the ecosystem. Others emphasize that the goals of ecocentrism should be secured through piecemeal and local action and argument rather than by appeal to big facts.

Critics of ecocentrism accuse it of misanthropy and authoritarianism; they assert that it uses claims about the earth's intrinsic value dogmatically to secure priority for nature over individual people and communities. Some versions are also accused of mysticism and impracticality, insofar as they call for identification with a greater self or whole or an attitude of reverence toward something not fully understood.

Andrew Lambert

FURTHER READING

Boylan, Michael, ed. *Environmental Ethics.* Upper Saddle River, N.J.: Prentice Hall, 2001.

Curry, Patrick. *Ecological Ethics: An Introduction.* Malden, Mass.: Polity Press, 2006.

Sessions, George, ed. *Deep Ecology for the Twenty-first Century.* Boston: Shambhala, 1995.

SEE ALSO: Anthropocentrism; Biocentrism; Deep ecology; Leopold, Aldo; Naess, Arne.

Eco-fashion

CATEGORY: Philosophy and ethics

DEFINITION: Philosophy in the world of clothing design that is concerned with the use of sustainable materials produced in a socially responsible manner

SIGNIFICANCE: Eco-fashion designers seek to reduce the negative environmental impacts associated with traditional methods of textile and clothing production, which include air and water pollution.

The fashion industry's impacts on the environment began during the Industrial Revolution with the large-scale production of fabrics under conditions that included little regulation of pollution and little concern with production by-products. Since the advances in agricultural science of the Green Revolution in the mid-twentieth century, the growing of fibers used in textiles has relied increasingly on pesticides and synthetic fertilizers, a practice that has led to negative ecological impacts. Both revolutions made

A showing of eco-fashion designs presented in Los Angeles in March, 2008. (WireImage/Getty Images)

many kinds of textiles more widely available around the world. With the introduction of ready-to-wear clothing during the 1960's, consumer demand for product diversity and seasonal fashions rose dramatically. This demand led to practices that came to dominate the world market, causing widespread environmental and social harms.

Eco-fashion developed in the late twentieth century as an alternative design approach involving the use of organic, vintage, recycled, locally based, and natural materials to bring consumers ecologically and socially sustainable choices in clothing. Those who engage in eco-fashion practices focus on minimizing the use of hazardous chemicals and the production of waste by-products, maximizing efficiency in their use of energy and water, and establishing fair wages and production standards that are healthy for workers.

Eco-fashion design is evolving constantly to reduce the environmental and social impacts of textile production. Initially it targeted negative preexisting methods, such as the toxicity of pretreatment, dying, finishing, drying, and laundry processes in which

emissions from formaldehyde, acids, and volatile organic compounds pollute the air and salts, surfactants, heavy metals, toxic chemicals, biocides, detergents, emulsifiers, and dispersants create aquatic toxicity. The manufacture and transportation of these products also require large amounts of electricity, principally generated from carbon dioxide-emitting fossil fuels. Scrap waste from garment assembly and unsold garments from retail are increasingly directed to landfills. Given that factory workers suffer from such methods owing to their exposure to chemicals, fiber dust, and polluted water, designers who support eco-fashion include healthy workplace standards and sufficient wages within their definition of eco-fashion.

The concept of sustainable fashion first appeared with the introduction of organic cotton as an alternative to conventionally grown crops in the United States. Brands such as Esprit and Vanity Fair began working to create large-scale sustainable clothing lines during the late 1980's, widening the market and developing environmentally sound practices. During the 1990's the Organic Trade Association (OTA) formed and worked with the Organic Fiber Council to adopt the Organic Fiber Certification Standard. In 1996 council member and environmental business leader Marci Zaroff coined and trademarked the term "ECOFashion" to brand and identify this market further. In the same year, the Patagonia apparel company emerged as a leader of eco-fashion, committing to the use of organic cotton in all of its cotton clothing items. The Organic Exchange was created in 2002 to build a global community of farmers, manufacturers, brands, and retailers committed to producing organic fibers with sustainable practices. Organizations from four nations—the United Kingdom, the United States, Germany, and Japan—worked together to create the Global Organic Textile Standard (GOTS), a revised version of which was published in 2008; the GOTS provides a basis for the certification of fibers as organic.

With growing market demand for environmentally conscious products of all kinds, greenwashing (deceptive claims that products are green or eco-friendly) on the part of companies that do not fully adhere to eco-fashion principles has made some observers skeptical of the industry. Additionally, the lack of transparency in clothing manufacture supply chains poses difficulties; disconnects often occur between designers and the agents who fill designer requests, and between designers and their manufacturing and packaging sites. Eco-fashion designers must research their supply chains carefully, and sometimes they must select more expensive and slower-paced alternative production sources to ensure that the principles of eco-fashion are not violated. For these reasons, eco-fashion companies may often be at a competitive disadvantage vis-à-vis conventional production lines. Eco-fashion companies, however, seek to reduce their long-term costs through augmentation of product longevity and versatility, closed-loop recycling, and participation in rental programs; all of these strategies show promise in making eco-fashion increasingly competitive.

Elizabeth A. Barthelmes and Brian J. Gareau

FURTHER READING

Allwood, Julian M., et al. *Well Dressed? The Present and Future Sustainability of Clothing and Textiles in the United Kingdom.* Cambridge, England: University of Cambridge, Institute for Manufacturing, 2006.

Fletcher, Kate. *Sustainable Fashion and Textiles: Design Journeys.* Sterling, Va.: Earthscan, 2008.

Humphrey, Liz, and Nick Robins. *Sustaining the Rag Trade.* London: International Institute for Environment and Development, 2000.

SEE ALSO: Green Revolution; Greenwashing; Industrial Revolution; Intensive farming; Organic gardening and farming; Recycling; Volatile organic compounds.

Ecofeminism

CATEGORY: Philosophy and ethics

DEFINITION: Philosophy that bridges the issues of feminism and environmentalism with the understanding that gender discrimination and environmental degradation are related manifestations of systematic oppression

SIGNIFICANCE: Ecofeminism is an important movement within environmental philosophy, environmental activism, and environmental justice that addresses harms against nature and the patriarchal oppression of women, the feminization of nature and the naturalization of women, and the logic that connects all hierarchical relationships.

A theoretical philosophy and an activist stance, ecofeminism understands that all oppression is linked by a shared logic and that historical, theoretical, and practical relationships exist between gender discrimination and environmental degradation. Ecofeminists assert that in order to address environmental harm, human beings need to attend to power-laden gender relationships; in order to address gender inequity, humans need to understand the logic that enacts hierarchical relationships. Ecofeminists argue that the feminine has long been associated with the natural, the body, and emotion, symbolized in metaphors such as the nurturing image of Mother Nature. Alternately, the masculine is tied to traits such as rationality and civilized (nonnatural) progress. Shared logic perpetuates these dualisms—female/male, nature/culture, body/mind—which are overlaid with corresponding value judgments: Rationality, male traits, and culture are good; expressions of the body, female traits, and nature are bad. Ecofeminism seeks to understand and address these dualisms.

Activist ecofeminism emerged during the 1970's with women-led groups who performed acts of civil disobedience to protest harms against nature such as nuclear proliferation, deforestation, and widespread pollution. These gatherings provoked reaction from academic feminists, who explored the theoretical relationships among multiple forms of degradation. Early ecofeminism was tied closely to the spirituality of nature- and female-centered religions, including paganism and Native American mythologies. Essentialist arguments for the "special" relationship between women and nature also permeated early scholarship.

One essentialist argument asserts that some female quality—the ability to bear children, to nurture life—imbues women with an innate sensitivity for and connection to environmental issues, locating female identity in biology. A second essentialist argument suggests that metaphysical gender-specific essences exist separate from biology and social constructions of gender. Another essentialism posits female identity ahistorically and thus assumes the oppressed "female" experience across time is a universal experience, regardless of class, race, ethnicity, or sexuality, which can serve to privilege the dominant female voice. Other essentialisms claim that particular ethnic and racial groups have an innate closeness with nature, often associated with cultural worldview, or imagine nature itself as fixed and unchanging.

While some ecofeminists argue that generalizations bring groups together in unity for a cause, and thus value essentialism's activist purpose, the more common rhetoric is antiessentialist: Gender differences are shaped by culture and result from lived experience, not biology. Many scholars worry that collapsing the differences among women enacts the same dichotomies that ecofeminists strive to overcome, though with reversed value associations. Some believe that female identities are far more flexible and fluid than essentialist arguments describe; limited descriptions in turn limit the available prescriptions for social change. Scholars such as Donna Haraway believe that essentialist rhetoric wrongly focuses on mystical connections with nature and an oversimplified understanding of women rather than on the experiences of actual women. Ecofeminism has evolved to address these essentialist concerns, and discussion

has come to center on material issues of justice related to the conditions that cause women to bear more severely the burdens of environmental harm, on the promise—based on women's social roles—of empowering women to address ecological disaster, on the logic that enables all forms of discrimination, and on the historical roots of hierarchical relationships.

Karen Warren has written about this "logic of domination" that founds hierarchical relationships. She argues for a critical ecofeminism that understands the impacts of anthropocentric rationality in the context of history, culture, and social structure. Other ecofeminists, including Carolyn Merchant, also trace the connected exploitations of women and nature to the impacts of reductionism—adopted by early scientific and religious institutions—on Western thought. In order to address the dichotomies perpetuated by this worldview, they explain, human beings need to address the logic that supports divisive relationships.

Ecofeminism in the early twenty-first century is more academic than activist, though it does drive small movements across the globe. Intellectually, it has splintered into several branches, including spiritual, essentialist, critical, transformative, radical, and materialist ecofeminism. Although ecofeminism has not launched the grand social change originally imagined by early thinkers, the shared concerns of ecological destruction and gender oppression continue to permeate environmental discourse and action.

Lissy Goralnik

FURTHER READING

Kheel, Marti. *Nature Ethics: An Ecofeminist Perspective.* Lanham, Md.: Rowman & Littlefield, 2008.

Sturgeon, Noël. *Ecofeminist Natures: Race, Gender, Feminist Theory, and Political Action.* New York: Routledge, 1997.

Warren, Karen J. *Ecofeminist Philosophy: A Western Perspective on What It Is and Why It Matters.* Lanham, Md.: Rowman & Littlefield, 2000.

SEE ALSO: Anthropocentrism; Antinuclear movement; Chipko Andolan movement; Eaubonne, Françoise d'; Environmental ethics; Environmental justice and environmental racism; Environmentalism.

Ecological economics

CATEGORY: Resources and resource management

DEFINITION: Interdisciplinary research field pursuing human and environmental well-being under the premise of recognizing the human economy as a subsystem that functions within the planet's ecological system

SIGNIFICANCE: Ecological economics seeks to replace the mainstream economics paradigm of economic welfare and continuous monetary growth with a model of intra- and intergenerational, holistic, sustainable well-being for both humankind and the environment.

By utilizing transdisciplinary approaches, ecological economics investigates and tries to understand the complex interdependence of society, human economy, and ecological environment. Ecological economics acknowledges the fact that the interdependence of these systems is structured in such a way that the economic sector is part of the human society, which itself is embedded in the planet's ecosystem. The existence of both, the economic sector and the society, are impossible without the ecological system.

In the Western context, precursors of ecological economics can be traced to ancient Greece: Plato drafted a model of a fairly sustainable economy in his dialogue *Nomoi* (*Laws*); also, according to Stoic philosophy, men should live according to nature. Kenneth E. Boulding's seminal essay "The Economics of the Coming Spaceship Earth" (1966) cleared the ground for ecological economics, employing the concept of Spaceship Earth, which emphasizes that the planet—like a spaceship—is an (almost) closed system (insolation, the reception of solar energy by the earth, is an exception, along with other examples of radiation that reach the planet or are emitted by it). Herman E. Daly and Robert Costanza, among others, are considered the main initiators of both the International Society for Ecological Economics (founded in 1989) and the society's journal, *Ecological Economics*, which began publishing in 1989.

TRANSDISCIPLINARY APPROACHES TO SUSTAINABLE WELL-BEING

Although having its point of departure at the intersection of economics and natural (especially ecological) sciences, ecological economics' transdisciplinary approach utilizes research findings from various other academic fields, such as philosophy—especially (environmental) ethics—history, political science, sociology, psychology, anthropology, biology, chemistry, geology, climatology, and physics. This approach is deemed essential, since mainstream economics has various epistemic shortcomings owing to its lack of multidisciplinarity and its instrumental view of the ecosystem (in mainstream economics, ecosystemic goods and services are considered economic resources). In addition, such an approach is vital for addressing the well-being of individuals, the human society as a whole, and the planet's ecological systems in a sustainable way, which in turn means that well-being is important not only for the present time but also for future generations and future ecosystems.

ECOSYSTEMIC DIMENSIONS AND POLICY INSTRUMENTS

Ecological economics observes that continuing overpopulation beyond the earth's carrying capacity, unsustainable resource exploitation, overconsumption, and contamination beyond the sink function (the environment's assimilative ability to absorb pollutants) will increase entropy (ecosystemic disorganization) and reduce biodiversity as well as the aggregated quality of life on the planet. Particularly irreversible actions—such as the extinction of species and the overexploitation of nonrenewable resources—have to be avoided. Therefore, the principle of uncertainty and the precautionary principle play important roles in the planning of economic actions and the implementation of policy instruments; it is, for example, extremely difficult to predict very accurately the long-term impacts of nuclear waste on the ecosystem. Hence, to facilitate intra- and intergenerational justice or equity and individual, social, and ecological sustainable well-being (instead of short-term economic welfare for a few), particular policy instruments are recommended.

Mainstream economics is associated with certain market failures—for example, inefficient allocation of services and goods such as clean water and air in some locations. In ecological economics, the prices of resources, products, and services have to speak the ecological truth. The price of an airplane ticket should reflect the monetarized impacts of, for example, air and noise pollution. The approach of calculating and charging ecologically correct prices is also known as the internalization of external effects. To limit emissions, governments, communities, unions of countries, and certain nonstate entities may issue pollution permits that producers can buy and trade. Ecological taxing and incentive systems are other measures that have been discussed.

Such measures are, however, not sufficient to prevent unsustainable external effects on the ecosystem. Concrete suggestions by ecological economists therefore include a decrease in the world population growth rate, strong limitations on the exploitation of nonrenewable resources, and sustainable use of renewable resources (that is, use of these resources in such a way that they can regenerate). Other recommendations include, but are not limited to, more extensive as well as stricter recycling and upcycling, a paradigm shift from conventional to organic farming, and a revaluation of economic mainstream values such as monetary wealth and economic growth.

Roman Meinhold

FURTHER READING

Boulding, Kenneth E. "The Economics of the Coming Spaceship Earth." 1966. In *Valuing the Earth: Economics, Ecology, Ethics*, edited by Herman E. Daly and Kenneth N. Townsend. Cambridge, Mass.: MIT Press, 1993.

Common, Michael S., and Sigrid Stagl. *Ecological Economics: An Introduction*. New York: Cambridge University Press, 2005.

Costanza, Robert, ed. *Ecological Economics: The Science and Management of Sustainability*. New York: Columbia University Press, 1991.

_____, et al. *An Introduction to Ecological Economics*. Boca Raton, Fla.: St. Lucie Press, 1997.

Daly, Herman E. *Ecological Economics and Sustainable Development*. Cheltenham, England: Edward Elgar, 2008.

Daly, Herman E., and Joshua C. Farley, eds. *Ecological Economics: Principles and Applications*. 2d ed. Washington, D.C.: Island Press, 2010.

Edwards-Jones, Gareth, Ben Davis, and Salman Hussain. *Ecological Economics: An Introduction*. Malden, Mass.: Blackwell, 2004.

Eriksson, Ralf, and Jan Otto Andersson. *Elements of Ecological Economics*. New York: Routledge, 2010.

Zografos, Christos, and Richard Howarth, eds. *Deliberative Ecological Economics: Ecological Economics and Human Well-Being*. New York: Oxford University Press, 2008.

SEE ALSO: Carrying capacity; Ceres; Ecosystem services; Environmental economics; Externalities; Free market environmentalism; Intergenerational justice; Natural capital; Polluter pays principle; Pollution permit trading; Precautionary principle; Renewable resources; Schumacher, E. F.; Sustainable development.

Ecological footprint

CATEGORIES: Ecology and ecosystems; resources and resource management

DEFINITION: Measure used to quantify and assess the impact of human activities on ecosystems

SIGNIFICANCE: Although inconsistency in the ways in which ecological footprints are calculated has resulted in some mistrust of the resulting figures, the concept of the ecological footprint provides a valuable educational tool regarding environmental sustainability.

Nature provides for the needs of humans worldwide, but individuals are not always aware of how many natural resources they are using and how their consumption of those resources affects the environment at large. The concept of the ecological footprint (EF) emerged during the early 1990's as the favored measure of human beings' demands on nature. It was first articulated by Mathis Wackernagel and William Rees of the University of British Columbia, who originally used the term "appropriated carrying capacity" but later adopted "ecological footprint" because it was more easily understood.

Measuring the EF of a person, group, or other entity essentially consists of comparing the entity's demands on nature with the earth's capacity to regenerate the resources used and to provide services. This measurement takes into consideration how much (biologically productive) land and water are (or would be) required to produce the resources the entity consumes and to absorb and render harmless its corresponding wastes, given current knowledge and available technologies. The concept of EF can be used to estimate the use of resources by a population, a person, a region, a city, a country, a business or a sector of the economy, an organization or institution, or even a particular lifestyle. It makes possible comparisons of resource use among different countries and the calculation of per-capita EF measurements.

An ecological footprint can be calculated in many different ways, but the analysis usually requires computing a large amount of data and involves several steps. The data most commonly included in the calculation concern carbon, food, housing, goods and services, waste, and recycling.

The many differences among the methodologies and accounting procedures used in calculating ecological footprints have led some observers to question the reliability of EF measurements and even the concept's usefulness. For this reason, in 2006 the organization Global Footprint Network began efforts to devise a set of common standards, which it subsequently developed further and improved in partnership with the research and policy organization Redefining Progress.

Although critics have increasingly focused on the limitations of the EF measure—particularly its accounting and calculation procedures—and have called for its continued improvement, the concept of EF is still widely considered to be useful. It can help people to understand the concept of environmental sustainability, educating them about carrying capacity and overconsumption and perhaps leading to the eventual alteration of world trends in consumption behaviors, as it becomes clear to increasing numbers of people that the lifestyles pursued by many in the developed world are unsustainable.

Nader N. Chokr

FURTHER READING

Chambers, Nicky, Craig Simmons, and Mathis Wackernagel. *Sharing Nature's Interest: Ecological Footprints as an Indicator of Sustainability.* 2000. Reprint. Sterling, Va.: Earthscan, 2007.

Fiala, Nathan. "Measuring Sustainability: Why the Ecological Footprint Is Bad Economics and Bad Environmental Science." *Ecological Economics* 67, no. 4 (2008): 519-525.

Rees, William, and Mathis Wackernagel. *Our Ecological Footprint: Reducing Human Impact on the Earth.* Gabriola Island, B.C.: New Society, 1996.

Venetoulis, Jason. "Redefining the Footprint (Footprint 2.0)." In *Sustainable Development: Principles, Frameworks, and Case Studies*, edited by Okechukwu Ukaga, Chris Maser, and Mike Reichenbach. Boca Raton, Fla.: CRC Press, 2010.

Venetoulis, Jason, and John Talberth. *Ecological Footprint of Nations: 2005 Update.* Oakland, Calif.: Redefining Progress, 2005.

SEE ALSO: Accounting for nature; Carbon footprint; Carrying capacity; Ecological economics; Environmental education; Environmental ethics; Environmental impact assessments and statements; Life-cycle assessment; Resource depletion; Sustainable development.

Ecology as a concept

CATEGORIES: Ecology and ecosystems; philosophy and ethics

DEFINITION: Interdisciplinary scientific study of the interactions among organisms and their environments

SIGNIFICANCE: The concept of ecology covers many broad areas of importance, including life processes that help to explain adaptation, which remains a central unifying principle of ecology; the broad distribution and abundance of organisms supported by radiation from the sun; the successful development of robust ecosystems; and the abundance and distribution of biodiversity across the environment.

The global sum of ecosystems is known as the biosphere, and this is where ecological theory has been used to explain regulatory phenomena at the planetary scale. One of the best-known holistic theories to explain this is the Gaia hypothesis, as developed by James Lovelock. Lovelock's thesis, like Aldo Leopold's seminal "land ethic," affirms that the biosphere together with its atmospheric environment forms a single entity or natural system.

Ecology crosses between many disciplines, including ecophysiology, the study of how physiological functions of organisms influence the way they interact with the environment; ecomechanics, which uses physics and engineering principles to examine the interactions of organisms with their environment and with other species; behavioral ecology, which examines the role of behavior in enabling a species to adapt to its environment; community ecology, or synecology, which explores the interrelationships among species within ecological communities; and ecosystem ecology, which is concerned with the flows of energy and matter through the components of the ecosystem.

LIGHT VERSUS DEEP ECOLOGY

Ecology that tends to take a "managerial" approach to the environment, with the status quo almost implicitly accepted, is sometimes characterized as light ecology, in contrast with deep ecology, which is considered to be more progressive if somewhat abstract and idealistic. The founding father of deep ecology, Arne Naess, considered it as having fundamental ethical implications and going beyond the transformation of technology and politics to a transformation of humanity.

Contradictory positions between light and dark ecology can initially be reconciled through a focus on the seminal writings of the environmental guru Aldo Leopold. He wrote of how the land ethic rests on a single unifying premise: "that the individual is a member of a community of interdependent parts." Leopold's vision served to enlarge the boundaries of community to include soils, water, plants, and animals. Especially since Leopold's writings were rediscovered during the 1960's, his land ethic thesis has become a central tenet of environmental thinking, and the symbiotic relationship he proposes between human beings and nature has remained the dominant orthodoxy across much ecological thinking.

Philosophical theories of the value of nature and ecology that cut across these deep and shallow divisions can be grouped into three broad categories: anthropocentrism, inherentism, and ecocentrism. Anthropocentrism recognizes nature and ecology in all its manifestations primarily as a resource that simply contributes to human value and can be used in whatever way is beneficial to human beings, as masters of their ecological environment. Inherentism recognizes that the very concept of value is innately human; thus this philosophical notion remains at odds with the tenets of deep ecology. Ecocentrism, in contrast, aims to challenge all ideas that suggest that human beings are somehow exempt from the natural processes and natural laws within which all other animals have to live.

It may be stating the obvious to observe that modern societies are based on anthropocentric principles, but unless this view is challenged, ecocentrists assert, ecological and environmental problems will not be solved. Ecocentrists argue that societies have to move toward a more earth-centered approach, which puts nature first. They point out the hidden social processes underlying developed societies' comfortable lifestyles and the price paid in damage to the natural

environment as a result. Some go so far as to suggest, as Philip W. Sutton has, that the Western world in particular is suffering from the "disease of over-consumption." Furthermore, ecocentrists use the well-known theories of German sociologist Ulrich Beck regarding risk management, together with the Gaia hypothesis and other holistic models, to help frame current ecological thinking about the pressing environmental dangers facing the planet, most notably global warming.

Pat Brereton

FURTHER READING

Knight, Richard L., and Suzanne Riedel, eds. *Aldo Leopold and the Ecological Conscience.* New York: Oxford University Press, 2002.

Leopold, Aldo. *A Sand County Almanac, and Sketches Here and There.* 1949. Reprint. New York: Oxford University Press, 1987.

Porritt, Jonathon. *Seeing Green: The Politics of Ecology Explained.* New York: Blackwell, 1984.

Sutton, Philip W. *The Environment: A Sociological Introduction.* Cambridge, England: Polity Press, 2007.

Taylor, Paul. *Respect for Nature: A Theory of Environmental Ethics.* Princeton, N.J.: Princeton University Press, 1986.

SEE ALSO: Anthropocentrism; Cultural ecology; Deep ecology; Ecocentrism; Leopold, Aldo; Naess, Arne; Social ecology.

Ecology in history

CATEGORY: Ecology and ecosystems

DEFINITION: Development of the interdisciplinary scientific study of the interactions among organisms and their environments

SIGNIFICANCE: Ecology, the science that studies the relationships among organisms and their biotic and abiotic environments, emerged as a discipline in the late nineteenth century. It gained prominence in the latter half of the twentieth century as general interest in and awareness of environmental issues increased.

The study of ecological topics arose in ancient Greece, but these studies were part of a catchall science called natural history. The earliest attempt to organize an ecological science separate from natural history was made by Carolus Linnaeus in his essay

Oeconomia Naturae (1749; *The Economy of Nature*, 1749), which focused on the balance of nature and the environments in which various natural communities exist. Although the essay was well known, the eighteenth century was dominated by biological exploration of the world, and Linnaeus's science did not develop.

EARLY ECOLOGICAL STUDIES

The study of fossils led some naturalists to conclude that many species known only as fossils must have become extinct. However, Jean-Baptiste Lamarck argued in his *Philosophie zoologique* (1809; *Zoological Philosophy*, 1914) that fossils represent the early stages of species that evolved into different, still-living species. In order to refute this claim, geologist Charles Lyell mastered the science of biogeography and used it to argue that species do become extinct and that competition from other species seems to be the main cause. In his book *On the Origin of Species by Means of Natural Selection: Or, The Preservation of Favoured Races in the Struggle for Life* (1859) English naturalist Charles Darwin blends his own research with the influence of Linnaeus and Lyell in order to argue that some species do become extinct, but existing species have evolved from earlier ones. Lamarck had underrated and Lyell had overrated the importance of competition in nature.

Although Darwin's book was an important step toward ecological science, Darwin and his colleagues mainly studied evolution rather than ecology. However, German evolutionist Ernst Haeckel realized the need for an ecological science and coined the name *oecologie* in 1866. It would be another three decades before steps were actually taken to organize this science. Virtually all of the early ecologists were specialists in the study of particular groups of organisms, and it was not until the late 1930's that some efforts were made to write textbooks covering all aspects of ecology. Since the 1890's most individual ecologists have viewed themselves as plant ecologists, animal ecologists, marine biologists, or limnologists.

Nevertheless, general ecological societies were established. The first was the British Ecological Society, which was founded in 1913 and began publishing the *Journal of Ecology* in the same year. Two years later, ecologists in the United States and Canada founded the Ecological Society of America, which launched the journal *Ecology* in 1920. The British Ecological Society and the Ecological Society of America have been the leading organizations in ecology ever since, though

other national and regional societies have also been established. More specialized societies and journals also began appearing; for example, the Limnological Society of America was established in 1936 and expanded in 1948 into the American Society of Limnology and Oceanography. It publishes the journal *Limnology and Oceanography*.

Although Great Britain and Western Europe were active in establishing ecological sciences, it was difficult for their trained ecologists to obtain full-time employment that utilized their expertise. European universities were mostly venerable institutions with fixed budgets; they already had as many faculty positions as they could afford, and these were all allocated to the older arts and sciences. Governments employed few, if any, ecologists. The situation was more favorable in the United States, Canada, and Australia, where universities were still growing. In the United States, the universities that became important for ecological research and the training of new ecologists were mostly in the Midwest. The reason was that most of the country's eastern universities were similar to European ones in being well established with scientists in traditional fields.

ECOLOGY AFTER 1950

Ecological research in the United States was not well funded until after World War II. With the advent of the Cold War, science was suddenly considered important for national welfare. In 1950 the U.S. Congress established the National Science Foundation, and ecologists were able to make the case for their research along with the other sciences. The Atomic Energy Commission had already begun to fund ecological research efforts by 1947, and under its patronage the Oak Ridge National Laboratory and the University of Georgia gradually became important centers for radiation ecology research. (In 1966, five years after its formation, the Institute of Radiation Ecology at the University of Georgia would become simply the Institute of Ecology. In 2007, it would be renamed the Eugene P. Odum School of Ecology in memory of the University of Georgia professor widely regarded as the father of modern ecology.)

Another important source of research funds was the International Biological Program (IBP), which, though international in scope, depended on national research funds. Officially established in 1964, it began operations in 1967 after an extended planning phase. Even though no new funding sources were created after the IBP ended in 1974, its existence meant that more research money flowed to ecologists than in previous years.

Ecologists learned to think big. Computers became available for ecological research shortly before the IBP got under way, and so computers and the IBP became linked in ecologists' imaginations. The first Earth Day in 1970 helped awaken Americans to the environmental crisis, and they expected ecologists to advise on environmental policy. The IBP encouraged a variety of studies, but in the United States, studies of biomes (large-scale environments) and ecosystems were most prominent. The IBP-funded biome studies were grouped under the headings of desert, eastern deciduous forest, western coniferous forest, grassland, and tundra (a proposed tropical forest program was never funded). Even after the IBP ended, a number of the biome studies continued at reduced levels.

Ecosystem studies were also large in scale, at least in comparison with many previous ecological studies, though smaller in scope than biome studies. The goal of ecosystem studies was to gain a total understanding of how individual ecosystems—such as a lake, a river valley, or a forest—work. IBP funds enabled research students to collect data and to use computers to process the data. However, ecologists could not agree on what data to collect, how to compute outcomes, and how to interpret the results. Therefore, thinking big did not always produce impressive results.

PLANT AND ANIMAL ECOLOGIES

Because ecology is enormous in scope, the discipline was bound to experience growing pains. It arose at the same time as the science of genetics, but because genetics is a cohesive science, it reached maturity much sooner than ecology. Ecology can be subdivided in a wide variety of ways, and any collection of ecology textbooks shows how diversely it is organized by different ecologists. Nevertheless, self-identified professional subgroups tend to produce their own coherent findings.

Plant ecology progressed more rapidly than other subgroups and has retained its prominence. In the early nineteenth century, German naturalist Alexander von Humboldt's many publications on plant geography in relation to climate and topography were a powerful stimulus to other botanists. By the early twentieth century, however, the idea of plant communities was the main focus for plant ecologists. Henry C. Cowles began his studies at the University of Chi-

cago in geology but switched to botany and studied plant communities on the Indiana dunes of Lake Michigan. He received his doctorate in 1898 and stayed at that university as a plant ecologist. He trained others in the study of community succession.

Frederic E. Clements received his doctorate in botany in the same year from the University of Nebraska. He carried the concept of plant community succession to an extreme by taking literally the analogy between the growth and maturation of an organism and that of a plant community. His numerous studies were funded by the Carnegie Institute in Washington, D.C., and even ecologists who disagreed with his theoretical extremes found his data useful. Henry A. Gleason was skeptical; his studies indicated that plant species that have similar environmental needs compete with each other and do not form cohesive communities. Although Gleason first expressed his views in 1917, Clements and his disciples held the day until 1947, when Gleason's individualistic concept received the support of three leading ecologists. Debates over plant succession and the reality of communities helped increase the sophistication of plant ecologists and prepared them for later studies on biomes, ecosystems, and the degradation of vegetation by pollution, logging, and agriculture.

Animal ecology emerged from zoology. A good illustration of the transition is the career of Stephen A. Forbes, professor of zoology and entomology at the University of Illinois and head of the State Laboratory of Natural History. His responsibilities focused his attention on the practical uses of zoology for agriculture and for fish and wildlife management; he also had a theoretical interest in both evolution and ecology. He brought together these various interests in his 1887 essay "The Lake as a Microcosm."

One important aspect of the early history of animal ecology was the attempt to understand and describe the growth or decline of animal populations mathematically. Mathematics is a universal language, and the fluctuation of animal populations is a universal problem. Therefore, this aspect of ecology developed globally rather than regionally. It was also possible to use the same mathematical methods to study population changes from the standpoints of ecology, evolution, and genetics. This situation promoted a lively exchange and rapid progress in the development of population ecology in the United States, Great Britain, Australia, Italy, and the Soviet Union. The great challenge was to develop equations that could help predict the pattern of population fluctuations. It turned out to be easier to develop mathematical models than to understand or predict the fluctuations of real populations. Nevertheless, these efforts eventually paid off in the ability of fish and wildlife biologists to gauge the level of harvesting that could maintain stable populations versus the level that would cause a population to decline.

LIMNOLOGY AND MARINE ECOLOGY

Limnology, the scientific study of bodies of fresh water, is important for managing freshwater fisheries and water quality. The Swiss zoologist François A. Forel coined the term and also published the first textbook on the subject in 1901. He taught zoology at the Académie de Lausanne and devoted his life's research to understanding Lake Geneva's characteristics and its plants and animals. In the United States in the early twentieth century, the University of Wisconsin became the leading center for limnological research and the training of limnologists. There, zoologist Edward Birge and fellow faculty member Chancey Juday pioneered North American limnology with their extensive field studies. The university, which has retained its preeminence in the field, established its Center for Limnology in 1982.

Marine ecology is viewed as a branch of either ecology or oceanography. Early studies were made either from shore or close to shore because of the great expense of committing oceangoing vessels to research. The first important research institute was the Statione Zoologica at Naples, Italy, founded in 1874. Its successes soon inspired the founding of others in Europe, the United States, and other countries. Karl Möbius, a German zoologist who studied oyster beds, was an important pioneer of the community concept in ecology. Great Britain dominated the seas during the nineteenth century and made the first substantial commitment to deep-sea research by equipping the HMS *Challenger* as an oceangoing laboratory that sailed the world's seas from 1872 to 1876. Its scientists collected so many specimens and such quantities of data that they called upon marine scientists in other countries to help them write the fifty large volumes of reports (1885-1895). The development of new technologies and the funding of new institutions and ships in the nineteenth century enabled marine ecologists to monitor the world's marine fisheries and other resources and provide advice on harvesting marine species.

The twentieth century brought advances in deep-sea exploration technology that allowed marine ecologists to gain a much more profound understanding of marine ecosystems. Pioneering work in oceanic acoustic research during World War I was followed by the development of acoustic sounding devices and other deep-marine electronic oceanographic instruments. Naturalist William Beebe and engineer Otis Barton ushered in the era of manned deep-sea exploration in the 1930's with their bathysphere dives. The next decade saw the advent of deep-ocean camera systems. Further advances included the first successful missions of untethered research submersibles in the 1950's and the unmanned remotely operated oceanographic systems that emerged in the 1960's. A landmark discovery came in 1977 when a manned submersible expedition explored the distinctive hydrothermal vent ecosystems along the mid-ocean ridge on the ocean floor near the Galápagos Islands.

SPACE-BASED ECOLOGICAL OBSERVATION

Since the late twentieth century satellite-based monitoring capabilities have provided researchers with invaluable tools for assessing and monitoring ecosystems while enabling them to increase their understanding of the earth as a collection of integrated systems. Remote-sensing instrumentation aboard orbital platforms provides scientists with both localized details and an overall global view of terrestrial and marine ecosystems. Regular, repeated collection of data from a particular area affords a means for monitoring how that area is changing over time. Because the data are collected in digital form, they are comparatively easy to integrate with other data. Satellite coverage includes remote areas that would be difficult or impossible to monitor by other means.

Since the United States launched the first satellite in the Landsat series in 1972, a host of environmental satellites equipped with a variety of remote-sensing instruments have been deployed by a number of nations. Notable among these is the Earth Observing System (EOS) series of satellites. A project of the U.S. National Aeronautics and Space Administration (NASA), EOS comprises several orbital missions to gather data on various factors influencing the earth's climate system and their interactions. The first EOS satellite, Terra (originally called EOS AM-1), was launched in 1999.

Optical instruments that detect different wavelengths within the ultraviolet, visible, and infrared spectral regions, along with radars, light detection and ranging instruments (lidars), and more, provide researchers with insights into a wide range of earth systems. Remote sensing via satellite can be used to monitor and study surface temperatures, snowfield conditions, soil moisture, the contours of the surfaces and floors of the oceans, ocean temperatures, air chemistry, air quality, water quality, sedimentation and organic matter in bodies of water, thermal pollution, weather patterns, urban influences on local climate, global climate change, alterations in land cover and land use, human activity and infrastructure, terrestrial and marine animal population size and distribution, habitat loss, plant health and response to stressors, crop yields, deforestation, coral reef conditions, algal blooms, droughts, floods, fires, major oil spills, disease outbreaks and pest infestations, and encroachment of invasive species.

In combination with ground-based ecological studies, satellite-based research facilitates investigations exploring how individual ecosystems contribute to larger ecologies and how pervasive factors such as an increase in global temperature would affect those individual ecosystems. At a time when scientists are striving to understand the global impacts of human activity, the ability to view ecosystems and their functions within a worldwide context is critical.

Frank N. Egerton
Updated by Karen N. Kähler

FURTHER READING

Golly, Frank B. *A History of the Ecosystem Concept in Ecology.* New Haven, Conn.: Yale University Press, 1993.

Molles, Manuel Carl. *Ecology: Concepts and Applications.* 5th ed. Boston: McGraw-Hill, 2009.

Morris, Christopher. "Milestones in Ecology." In *The Princeton Guide to Ecology,* edited by Simon A. Levin. Princeton, N.J.: Princeton University Press, 2009.

National Aeronautics and Space Administration. *Earth Science Reference Handbook: A Guide to NASA's Earth Science Program and Earth Observing Satellite Missions.* Washington, D.C.: Author, 2006.

National Research Council Space Studies Board. *Earth Science and Applications from Space: National Imperatives for the Next Decade and Beyond.* Washington, D.C.: National Academies Press, 2007.

Odum, Eugene P., and Gary W. Barrett. *Fundamentals of Ecology.* 5th ed. Belmont, Calif.: Thomson Brooks/Cole, 2005.

Raffaelli, David G., and Christopher L. J. Frid, eds. *Ecosystem Ecology: A New Synthesis.* New York: Cambridge University Press, 2010.

Real, Leslie A., and James H. Brown, eds. *Foundations of Ecology: Classic Papers with Commentaries.* Chicago: University of Chicago Press, 1991.

SEE ALSO: Balance of nature; Biomes; Darwin, Charles; Earth resources satellites; Ecology as a concept; Ecosystems; International Biological Program.

Ecosystem services

CATEGORIES: Ecology and ecosystems; resources and resource management

DEFINITION: Natural resources and processes of the environment that are beneficial to humankind

SIGNIFICANCE: The world's ecosystems provide a unique set of services for the global population. Increasing awareness of the threat of environmental degradation to the continued operation of many of these services has stirred governments, nongovernmental organizations, and individuals to take steps to protect them.

The processes of natural ecosystems are rarely experienced directly by the individuals who use the resources those ecosystems provide. Ecosystem services are complex functions of ecosystems that support every aspect of life on the planet. Examples of such services are pollination, food webs, and the nitrogen and carbon cycles. As human beings have become increasingly aware of the relationship between human activities and environmental degradation, many have also realized that these services are endangered.

The Millennium Ecosystem Assessment, which was initiated by the United Nations in 2001, groups ecosystem services into the categories of provisioning, regulating, supporting, and cultural. Services in the provisioning group include the items collected from given ecosystems, such as crops, fish, water, pharmaceuticals, and energy in the form of tidal power or biomass. Regulating ecosystem services are the benefits humans gain from constant environmental processes; examples include carbon sequestration, waste detoxification, and pollination. Supporting ecosystem services are those services that are necessary for the sustaining of other ecosystem services. These include nutrient cycling—that is, the cycling of the key elements phosphorus, nitrogen, sulfur, carbon, iron, and silicon, all of which have been significantly altered by human activity.

Cultural ecosystem services are the most difficult to value; these are the nontangible benefits that humans gain from the environment, in such forms as recreation, spiritual connection, and aesthetic appreciation. A common method used to determine the value of cultural ecosystem services is contingent valuation, which involves conducting surveys in which respondents are asked about the values they attach to these nonmarket resources.

The foundation for the concept of ecosystem services came from the work of environmentalists William Vogt and Aldo Leopold during the late 1940's. Both expressed a deep appreciation for and understanding of the connection between humans and the environment. The first significant study in the valuation of ecosystem services was conducted by ecological economist Robert Costanza and his colleagues; their report on their work, "The Value of the World's Ecosystem Services and Natural Capital," was published in the journal *Nature* in 1997. These researchers estimated that ecosystems provide at least US$33 trillion worth of services annually; in comparison, they noted that the then-current global gross national product was approximately US$18 trillion.

In the late twentieth century some developing nations began to seek payment from developed nations in exchange for taking steps to maintain or improve the developing nations' local ecosystems. Generally, such arrangements involve organizations or nations providing funds to specific areas of the world to protect the services provided by their ecosystems, whether for direct benefit or for global benefit. An example of this is Brazil seeking payment for preserving the Amazon rain forest.

Some critics of attempts to attach monetary values to ecosystem services have noted that approaching the worth of such services by examining their value for humans propagates an anthropocentric (that is, human-centered) view of the world. Many environmentalists argue that the financial value of the environment should not be the foundation for its worth; rather, the environment has intrinsic value.

Kara Kaminski

FURTHER READING

National Research Council. *Valuing Ecosystem Services: Toward Better Environmental Decision-Making.* Washington, D.C.: National Academies Press, 2005.

Ninan, K. N. *Conserving and Valuing Ecosystem Services and Biodiversity: Economic, Institutional, and Social Challenges.* Sterling, Va.: Earthscan, 2008.

Vogt, William. *Road to Survival.* New York: W. Sloane Associates, 1948.

SEE ALSO: Accounting for nature; Anthropocentrism; Bees and other pollinators; Conservation biology; Debt-for-nature swaps; Ecological economics; Ecosystems; Environmental economics; Leopold, Aldo.

Ecosystems

CATEGORY: Ecology and ecosystems

DEFINITION: Specific areas of the earth and the interactions among organisms and the physical-chemical environments present in those areas

SIGNIFICANCE: The ecosystem concept has proven to be a useful tool for dissecting environmental problems. Ecosystem science is being used to define and address the impacts of environmental changes caused by humans.

Ecologists view ecosystems as basic units of the biosphere, much as biologists consider cells as the basic units of an organism, self-organized and self-regulating entities within which energy flows and resources are cycled in a coordinated, interdependent manner to sustain life. Disruptions and perturbations to, or within, a unit's organization or processes may reduce the quality of life or cause its demise.

Ecosystem boundaries are arbitrary; they are usually defined by the research or management questions being asked. An entire ocean can be viewed as an ecosystem, as can a single tree, a rotting log, or a drop of pond water. Those systems with tangible boundaries—such as forests, grasslands, ponds, lakes, watersheds, seas, and oceans—that separate them from contiguous ecosystems are especially useful to ecosystem research.

ECOSYSTEM RESEARCH

The word "ecosystem" was coined in 1935 by English ecologist Arthur G. Tansley. The ecosystem concept was first put to use in American research by limnologist Raymond L. Lindeman in the classic study he conducted on Cedar Bog Lake, Minnesota, which resulted in his article "The Trophic Dynamic Aspect of Ecology" (1942). Lindeman's study, along with the publication of Eugene P. Odum's *Fundamentals of Ecology* (1953), converted the ecosystem notion into a guiding paradigm for ecological studies, thus making it a concept of theoretical and applied significance.

Ecologists study ecosystems as integrated components through which energy flows and resources cycle. Although ecosystems can be divided into many components, the four fundamental ones are abiotic (nonliving) resources, producers, consumers, and decomposers. The ultimate sources of energy come from outside the boundaries of the ecosystem (solar energy or chemothermo energy from deep-ocean hydrothermal vent systems). Since this energy is captured and transformed into chemical energy by producers and translocated through all biological systems through consumers and decomposers, all organisms are considered as potential sources of energy. As the first law of thermodynamics states, energy cannot be created or destroyed; thus the energy follows the tenets of the second law of thermodynamics upon reaching the end of its flow and enters the state of entropy. In this state energy is completely randomized and divested of its ability to do work.

Abiotic resources (water, carbon dioxide, nitrogen, oxygen, and other inorganic nutrients and minerals) primarily come from within the boundaries of the ecosystem. From these, producers utilizing energy synthesize the biomolecules, which are transformed, upgraded, and degraded as they cycle through the living systems that comprise the various components. The destiny of these bioresources is to be degraded to their original abiotic forms and recycled.

Environmental research conducted using the ecosystem approach is a major endeavor. It requires amassing large amounts of qualitative and quantitative data relevant to the structure and function of each component. The researchers then integrate these data among and between the components in an attempt to determine linkages and relationships. Such work is attractive to many researchers because it takes a holistic approach to the examination of complex systems. The ecosystem approach to research involves the use of systems information theory, predictive models, and computer applications and simulations. As ecosystem ecologist Frank B. Golly

states in his book *A History of the Ecosystem Concept in Ecology* (1993), the ecosystem approach to the study of ecosystems is "machine theory applied to nature."

ECOSYSTEM RESEARCH PROJECTS

Initially, ecosystem ecologists used the principles of Tansley, Lindeman, and Odum to determine and describe the flow of energy and resources through organisms and their environment. The objectives of this research were to answer the fundamental academic questions that plagued ecologists, such as those concerning controls on ecosystem productivity: What are the connections between animal and plant productivity? How are energy and nutrients transformed and cycled in ecosystems?

Once fundamental insights were obtained, computer-model-driven theories were constructed that linked and integrated the components to provide an understanding of the biochemophysical dynamics that govern ecosystems. Researchers could then examine the responses of ecosystem components by altering or manipulating parameters within the simulation models. Early development of the ecosystem concept culminated, during the 1960's, in the definition of the approach of ecosystem studies. Ecosystem projects were funded primarily under the umbrella of the International Biological Program (IBP). Other funding came from the Atomic Energy Commission and the National Science Foundation. The intention of the IBP was to integrate data collected by teams of scientists at research sites that were considered typical of wide regions. Although the IBP was international in scope, studies in the United States received the greatest portion of the funds—approximately $45 million during the life of the IBP (1964-1974).

Five major IBP ecosystem studies involving grasslands, tundra, deserts, coniferous forests, and deciduous forests were undertaken. The Grasslands Project, directed by George Van Dyne, received the largest portion of IBP funding, primarily because of the dynamism of the director. The Grasslands Project set the research stage for, and gave direction to, the other four endeavors. The research effort was so extensive in scope, however, that the objectives of the IBP were not totally realized. Because of the large number of scientists involved and their cultural, visionary, and motivational differences, little coherence in results was obtained even within individual projects.

Other problems also plagued the early development of the ecosystem concept. Evolutionary biolo-

Ecosystem Energetics

The diverse forms of life in all ecosystems are powered by a single energy source, sunlight, which enters the biosphere through a process called photosynthesis. Organisms such as plants and algae capture a small fraction of sunlight's energy, storing it as chemical energy in sugar or other complex organic molecules. The energy then passes through an ecosystem via different feeding (trophic) levels. Each time the energy is passed on, a portion of it is lost as heat. Thus, energy needs constant replenishment from the original source, the Sun.

How does energy flow through communities? It flows through ecosystem energetics. Of the energy that reaches Earth, much is either reflected or absorbed as heat by the atmosphere and Earth's surface, leaving only about 1 percent to power all life. Of this 1 percent, green plants capture 3 percent or less. All life on this planet is therefore supported by less than 0.03 percent of the energy reaching Earth from the Sun. Photosynthetic organisms are called autotrophs, or producers, because they produce food for themselves. Directly or indirectly, they produce food for nearly all other forms of life as well. Organisms that cannot photosynthesize are called heterotrophs, or consumers, because they must acquire energy prepackaged in the molecules of the bodies of other organisms.

There are three basic principles that govern ecosystem energetics. First, the amount of life an ecosystem can support is defined by the energy captured by the producers. This energy, made available to consumers over a given period, is called net primary productivity. It is usually measured in units of energy stored per unit area in a given period, or measured as biomass, the dry weight of the total organic material added to the ecosystem per unit area in a given time span. The net primary productivity is influenced by a variety of environmental factors, including the amount of sunlight, the availability of water, the amount of nutrients available to producers, and the temperature. Among all environmental variables, the most limiting variable is the one that determines net primary productivity, for instance, water in the desert or light in the deep ocean.

Second, within the community, energy is passed from one feeding level to another. Energy flow moves from producers to primary consumers to secondary and tertiary consumers. Primary consumers are normally herbivores that directly consume producers. Secondary and tertiary consumers are typically carnivores that eat meat of other consumers. Certain consumers may occupy more than one feeding level. As energy is passed through feeding levels via food chains or food webs, the transfer is never efficient. This is the third basic principle of ecosystem energetics. Each time the energy is transferred to the next feeding level, the bulk of it is lost as heat. On average, only 10 percent of the energy is transferred from one feeding level to the next. In other words, the higher the feeding level an organism occupies, the less energy is available to it. This so-called 10 percent law or energy pyramid puts a cap on how much life a particular ecosystem can sustain.

Ming Y. Zheng

gists faulted the concept for it inability to link with the evolutionary theory. This link could not be established because of the lack of fundamental knowledge of individuals, species, populations, and behavior. A more pervasive concern, voiced by environmentalists and scientists alike, was that little of the information obtained from ecosystem simulation models could be applied to the solution of existing environmental problems.

An unconventional project partially funded by the IBP was an examination of the Hubbard Brook watershed ecosystem—located in New Hampshire and studied by F. Herbert Bormann and Gene E. Likens—that redirected the research approach for studying ecosystems from the IBP computer-model-driven theory to more conventional scientific methods of study. Under the Hubbard Brook approach, an ecosystem phenomenon is observed and noted. A pattern for the phenomenon's behavior is then established for observation, and questions are posed about the behavior. Hypotheses are developed to allow experimentation and manipulation in an attempt to explain the observed behavior. This approach requires detailed scrutiny of the ecosystem's subsystems and their linkages. Since each ecosystem functions as a unique entity, this approach has greater utility than do other approaches. The end results provide insights specific to the activities observed within particular ecosystems. Explanations for these observed behaviors can then be made in terms of biological, chemical, or physical principles.

Utility of the Ecosystem Concept

Publicity from the massive ecosystem projects and the publication of Rachel Carson's *Silent Spring* (1962) helped stimulate the environmental movement of the 1960's. The public began to realize that human activity was destroying some of the bioecological matrices that sustain life. By the end of the 1960's, the applicability of the IBP approach to ecosystem research was proving to be purely academic; the approach provided few solutions to the problems plaguing the environment. Scientists realized that, because of the lack of fundamental knowledge about many of the systems and their links and because of the technological shortcomings that existed, researchers could not simply divide ecosystems into three to five components and analyze them by computer simulation.

The more applied approach taken in the Hubbard Brook project, however, showed that the ecosystem approach to environmental studies could be successful if the principles of the scientific method were used. The Hubbard Brook study area and the protocols used to study it were clearly defined. This ecosystem allowed hypotheses to be generated and experimentally tested. Applying the scientific method to the study of ecosystems had practical utility for the management of natural resources and for testing some possible solutions to environmental problems. When perturbations such as diseases, parasites, fire, deforestation, and urban and rural centers disrupt ecosystems from within, this approach helps answer problems and defines potential mitigation and management plans. Similarly, external causative agents within atmospheric areas, drainage flows, or watersheds can also be considered.

The principles and research approach of the ripening ecosystem concept are being used to define and attack the impacts of environmental changes caused by humans. Such problems as human population growth, apportioning of resources, toxification of the biosphere, loss of biodiversity, global warming, acid rain, atmospheric ozone depletion, land-use changes, and eutrophication are being holistically examined. Management programs related to woodlands (such as the approach known as new forestry) and urban and rural centers (such as the Urban to Rural Gradient Ecology, or URGE, program) have found the ecosystem perspective useful, as have other governmental agencies investigating water and land use, fisheries, endangered species, and exotic species introductions.

Ecosystems are also viewed as systems that provide the services necessary to sustain life on earth. Most people either take these services for granted or do not realize that such natural processes exist. Ecosystem research has identified seventeen naturally occurring services, including water purification, regulation, and supply, as well as atmospheric gas regulation and pollination. An article by Robert Costanza and others, "The Value of the World's Ecosystem Services and Natural Capital" (published in *Nature*, May 15, 1997), places a monetary cost to humanity should these services, for some disastrous reason, need to be maintained by human technology. The amount is staggering, ranging from $16 trillion to $54 trillion per year and averaging $33 trillion. Humanity could not afford this; the global gross national product is only about $18-20 trillion.

Academically, ecosystem science has been shown to be a tool useful for dissecting environmental problems, but this has not been effectively demonstrated to the public and private sectors, especially decision makers and policy makers at governmental levels. Additionally, the idea that healthy ecosystems provide socioeconomic benefits and services is controversial. It has been suggested that, to bridge this gap between academia and the public, ecosystem scientists need to be "bilingual"; that is, they need to be able both to speak their scientific language and to translate it so that the nonscientist can understand.

Richard F. Modlin

FURTHER READING

Coleman, David C. *Big Ecology: The Emergence of Ecosystem Science.* Berkeley: University of California Press, 2010.

Daily, Gretchen C., ed. *Nature's Services: Societal Dependence on Natural Ecosystems.* Washington, D.C.: Island Press, 1997.

Dodson, Stanley I., et al. *Ecology.* New York: Oxford University Press, 1998.

Golly, Frank B. *A History of the Ecosystem Concept in Ecology.* New Haven, Conn.: Yale University Press, 1993.

Odum, Eugene P., and Gary W. Barrett. *Fundamentals of Ecology.* 5th ed. Belmont, Calif.: Thomson Brooks/Cole, 2005.

Raffaelli, David G., and Christopher L. J. Frid, eds. *Ecosystem Ecology: A New Synthesis.* New York: Cambridge University Press, 2010.

Vogt, Kristiina A., et al. *Ecosystems: Balancing Science with Management.* New York: Springer, 1997.

SEE ALSO: Balance of nature; Biodiversity; Biosphere; Biosphere reserves; Ecology as a concept; Ecosystem services; Tansley, Arthur G.

Ecotage

CATEGORIES: Activism and advocacy; philosophy and
    ethics
DEFINITION: Sabotage tactics used by radical
    environmentalists to stop projects they perceive
    as ecologically destructive
SIGNIFICANCE: Many mainstream environmentalists
    believe that the actions of those who engage in so-
    called ecotage have alienated some people who
    would otherwise support the environmentalist
    cause, but others assert that such extreme acts are
    sometimes necessary to call attention to the need
    to defend nature.

In 1972 the group Environmental Action
published the handbook *Ecotage!*, which
compiled ideas about how to sabotage environmentally destructive projects. Edward Abbey's
novel *The Monkey Wrench Gang* (1975), featuring a small group of ecoguerrillas who destroy
construction equipment to stop development
in the southwestern United States, inspired
Dave Foreman and others to start the radical
environmental movement Earth First! In 1985
Foreman published *Ecodefense: A Field Guide to
Monkeywrenching*, a manual of ecotage methods
and information on related issues such as
safety and security. In the early 1990's Foreman
wrote that "those willing to commit ecotage are
needed today as never before." Several other
environmental advocates, such as Greenwar International, have also promoted ecotage.

The early proponents of ecotage, known as
ecoteurs, were dismayed by industrial development of wilderness areas that the government
refused to protect. Frustrated that civil disobedience did not achieve their goals, ecoteurs decided to preserve the environment by illegally
damaging machinery used to degrade wilderness areas. Such militant acts of destruction often impeded future development efforts and
reduced industrialists' profits.

Many ecoteurs, including Foreman, had belonged to mainstream environmental groups
during the 1960's and 1970's but had become
disillusioned by the dominance of conservative
political leaders in these groups during the
1980's. Ecoteurs are often critical of environmentalists in such organizations as the Sierra
Club, asserting that they are passive, ignore opportunities to preserve the wilderness, and appease industrialists and governmental agencies by sacrificing the
environment. Ecoteurs denounce environmentalists
with anthropocentric attitudes who view the environment as a source of production and resources to fulfill
human needs.

Most ecoteurs consider themselves a symbiotic part
of the environment and justify their bold, destructive
conduct as acts of self-defense. "It is time to act hero-
ically and admittedly illegally in defense of the wild,"
Foreman stated, "to put a monkeywrench into the
gears of the machinery destroying natural diversity."
He stressed that ecotage "is sabotage, not terrorism,
because it's about property destruction. It's saying,

*Greenpeace activists obstruct logging in an old-growth forest near Peters-
burg, Alaska.* (©Les Stone/Zuma/CORBIS)

I'm operating as part of the wilderness, defending myself."

Sometimes calling themselves "ecowarriors," ecoteurs strategically select their targets for ecotage in their efforts to disrupt environmentally harmful development. Many have focused on obstructing logging and strip-mining operations. Common monkeywrenching procedures include tree spiking, in which large nails are driven into trees in old-growth forests to deter loggers. Other tactics include pouring sand into the gas tanks of trucks and bulldozers, as well as slashing tires. Ecoteurs have also pulled up survey stakes, cut power lines, blockaded roads, stolen machinery parts, and ruined tools. Some ecoteurs have damaged offices belonging to the U.S. Forest Service to protest logging in national forests, and others have sprayed baby seals with dye so their fur would be unusable for coats. The Federal Bureau of Investigation (FBI) and industrial leaders have offered rewards for the arrest and conviction of ecoteurs.

During the 1990's, many of the people in the Earth First! movement came to believe that ecotage often does more to ostracize people than to protect the environment, and they reassessed their protest tactics. Columns in the *Earth First! Journal* during that period discussed individuals' ecotage efforts, but Earth First!ers disagreed about the role of ecotage in the environmental movement. While some believed all ecotage should be ceased, others argued that only certain acts of sabotage should be curtailed—those that might result in injuries to loggers or miners. Although some ecoteurs began to question their tactics in the 1990's, ecotage continued into the twenty-first century.

Elizabeth D. Schafer

FURTHER READING

Foreman, Dave, and Bill Haywood, eds. *Ecodefense: A Field Guide to Monkeywrenching*. 3d ed. Chico, Calif.: Abbzug Press, 2002.

Scarce, Rik. *Eco-Warriors: Understanding the Radical Environmental Movement*. Updated ed. Walnut Creek, Calif.: Left Coast Press, 2006.

Sterba, James P., ed. *Earth Ethics: Introductory Readings on Animal Rights and Environmental Ethics*. 2d ed. Upper Saddle River, N.J.: Prentice Hall, 2000.

SEE ALSO: Abbey, Edward; Earth First!; Ecoterrorism; Environmental ethics; Foreman, Dave; Monkeywrenching; Sea Shepherd Conservation Society; Seal hunting.

Ecoterrorism

CATEGORIES: Activism and advocacy; philosophy and ethics

DEFINITION: Clandestine activities conducted by radical environmentalists with the intent of disrupting environmental damage or preventing cruelty to animals

SIGNIFICANCE: The extreme activities of some radical environmentalists have caused considerable controversy, with critics noting that ecoterrorists have damaged the reputations of many hardworking environmentalist groups around the world.

Ecoterrorism began in the United States in the 1970's, peaked in the early 1990's, and waned later in that decade, although acts of ecoterrorism continued into the twenty-first century. Other terms, such as "ecological sabotage," "monkeywrenching," "ecotage," and "decommissioning," roughly connote the same concept, but with much less pejorative implications. "Monkeywrenching" was coined by Edward Abbey in his novel *The Monkey Wrench Gang* (1975). "Ecotage" refers to acts of sabotage for environmental ends. The radical environmentalist group Earth First! probably engaged in more ecoterrorist activities during the peak period than any other group, including People for the Ethical Treatment of Animals (PETA), the Sea Shepherd Conservation Society, and the Animal Liberation Front (ALF). In 1985 Dave Foreman, cofounder of Earth First!, published *Ecodefense: A Field Guide to Monkeywrenching*, which describes a variety of radical environmental activities and ways to carry them out.

Lorenz Otto Lutherer and Margaret Sheffield Simon chronicle the history of ecoterrorism conducted by animal rights groups in *Targeted: The Anatomy of an Animal Rights Attack* (1992). They describe numerous cases of break-ins, vandalism, thefts of animals and equipment, and threats of violence against researchers and businesspeople. Animal rights activists who engage in these activities, these authors maintain, are a menace to society, scientific and economic progress, and democracy itself.

The types of activities that initially led critics to characterize particular environmental activists as ecoterrorists can be divided into three categories. The first involves legal protests that are still clear cases of monkeywrenching. Examples are sit-ins in front of offices, laboratories, factories, bulldozers, and even

Time Line of Ecoterror Incidents

YEAR	INCIDENT
1998	Arson of a U.S. Department of Agriculture animal damage control building near Olympia, Washington, resulting in $2 million in damages
1998	Arson of a Vail, Colorado, ski resort, resulting in $12 million in damages
1999	Arson at a genetic-engineering research office at Michigan State University
2003	Arson at a HUMMER dealership in West Covina, California, destroying 125 SUVs and resulting in an estimated $1 million in damages
2005	Arson of five townhouses under construction in Hagerstown, Maryland
2008	Arson at the Street of Dreams housing development in Woodinville, Washington, resulting in $12 million in damages

locomotives, sometimes accentuated by activists' chaining themselves to gates or trees. Activists have also often engaged in a process called tree spiking, in which metal spikes are driven into trees to prevent logging with chain saws.

The activities in the second category are characterized by their illegal nature, such as decommissioning machinery by pouring sand, sugar, or water into gas tanks and damaging motor vehicles by smashing distributors and spark plugs. Most break-ins at animal research facilities are included in this category. The third category consists of more daring but potentially hazardous operations, such as the ramming of whaling ships by Sea Shepherd activists.

Although many environmentalists oppose acts of violence in the name of their cause, others view monkeywrenching activists as "the conscience of the environmental movement." However, most agree that the term "ecoterrorism" more aptly applies to those who plunder the earth and its atmosphere in the name of capitalism and progress. Some environmental activists compare themselves to Resistance fighters of World War II; they see environmental damage as equivalent to the Holocaust and other war crimes, meriting drastic reprisals. In general, however, environmental activists believe that their actions ought to be nonviolent, and most who participate in ecotage

are aggressive in taking preventive measures. For example, to avoid injuries, antilogging activists inform loggers and mill workers about trees that contain spikes by sending letters, telephoning, or marking trees with paint.

Opinions regarding the moral implications of ecoterrorism vary. Many environmental activists, whether or not they approve, agree that it works. However, critics and even some activists feel that some radical activists have damaged the reputations of many hardworking, peace-loving environmentalists around the world.

Chogollah Maroufi

FURTHER READING

Foreman, Dave, and Bill Haywood, eds. *Ecodefense: A Field Guide to Monkeywrenching.* 3d ed. Chico, Calif.: Abbzug Press, 2002.

Lutherer, Lorenz Otto, and Margaret Sheffield Simon. *Targeted: The Anatomy of an Animal Rights Attack.* Norman: University of Oklahoma Press, 1992.

Scarce, Rik. *Eco-Warriors: Understanding the Radical Environmental Movement.* Updated ed. Walnut Creek, Calif.: Left Coast Press, 2006.

SEE ALSO: Earth First!; Ecotage; Foreman, Dave; Monkeywrenching; Operation Backfire.

Ecotopia

CATEGORY: Philosophy and ethics

IDENTIFICATION: Novel by Ernest Callenbach describing an ecologically ideal world

DATE: Published in 1975

SIGNIFICANCE: *Ecotopia* presents a wide variety of environmental issues and illustrates how a society might address them ideally and practically. A number of the measures described in the novel have been implemented in some parts of the world, whereas the practicality and efficiency of others remain topics of debate among environmentalists, politicians, and economists.

The term "ecotopia" is derived from the Greek words *oikos* (household, home) and *topos* (place); it is also a shortening of the term "ecological utopia." In the novel *Ecotopia: The Notebooks and Reports of William Weston*, Ecotopia is a society (the geography of which comprises the former states of Washington and

Oregon as well as Northern California) where ecology guides all human and political attitudes and actions. Ecologically compatible high technology exists alongside the postmaterial(istic) lifestyles and attitudes of Ecotopia's citizens.

The environmental paradigms of Ecotopia include intergenerational justice, sustainability, sufficiency economics, prices of goods that reflect their real costs (that speak the "ecological truth"), anticonsumerism, slow reduction of population, and strict environmental laws. The ecocentric worldview gives preference to the quality of life, not to the economic paradigm of growth. The novel presents highly detailed descriptions of the many ecologically friendly practices and programs carried out in Ecotopia, including all possible types of waste recycling; the use of plant-derived, biodegradable durable plastics; the use of renewable energy sources, such as solar, wind, tidal, and geothermal energy and biogas; the provision of a wide variety of public transport opportunities, both high tech and simple, such as free public bicycles; organic farming; and reforestation.

Roman Meinhold

SEE ALSO: Alternative energy sources; Balance of nature; Ecocentrism; Eco-fashion; Electric vehicles; Environmental ethics; Nature writing; Organic gardening and farming; Recycling; Renewable resources.

Ecotourism

CATEGORY: Preservation and wilderness issues
DEFINITION: Environmentally, socially, and culturally responsible recreational travel intended to preserve ecosystems and improve the well-being of local populations
SIGNIFICANCE: Supporters claim that ecotourism dollars help save endangered wilderness areas that might otherwise be subject to indiscriminate exploitation of natural resources. Detractors note that even the most conscientious ecotourism can contribute to the destruction of fragile ecosystems and cultures, and poor ecotourism practices can wreak even more havoc.

Improvements in travel after World War II, especially the development of jet aircraft, dramatically increased the numbers of tourists in all areas of the globe. With this trend came increased interest in visiting exotic locations to enjoy unspoiled landscapes, view unusual wildlife, and participate in recreational adventures. According to the United Nations World Tourism Organization (UNWTO), by the late twentieth century tourism had become a main income source and the top export category for many developing countries.

The rise in tourism as a leisure activity brought economic benefits such as development and employment opportunities to many pristine areas, but it was sometimes accompanied by negative social, cultural, and environmental impacts. Local communities and lifestyles were sometimes displaced, and ecosystems were altered with the building of hotels, roads, and other amenities for guests. The growing numbers of tourists threatened the very vistas and animals that lured visitors in the first place.

Despite these problems, environmentalists recognize tourism as a means to benefit preservation efforts. While developing countries have the option of exploiting their natural resources to provide revenue, preserving those resources can provide an ongoing alternative source of income—tourist dollars—that gives governments an incentive to protect wilderness areas. Coupled with a "no-impact" ethic, ecotourism is seen as a method of saving ecosystems that are quickly disappearing.

Ideally, ecotourism operations employ practices that have minimal negative impacts on the environment and local cultures. Tours focus on natural destinations and rotate the routes they travel and the sites they visit. Participants gain an understanding of their surroundings and how human activity—including their own—affects the ecosystem. Local communities and indigenous populations are involved in managing ecotourism and reap economic benefits from it. The revenues produced by ecotourism are used to help preserve the natural environment.

Ecotourism proponents point to the regions that have successfully used ecotourism to preserve environments and support local communities. Ecotourism in the Ecuadoran rain forest has staved off oil exploration and provides income to native peoples in the area. A former director of a mountain gorilla project in Africa credits ecotourism with the survival of mountain gorillas and their habitats; gorilla ecotourism has also provided significant revenue for local communities. In Costa Rica, the market demand for pristine wilderness has led to the establishment of national parks and protected areas over more than 20

Principles of Ecotourism

The International Ecotourism Society, the oldest international nonprofit organization devoted to the promotion of ecotourism as a tool for conservation, states that "ecotourism is about uniting conservation, communities, and sustainable travel" and advises that "those who implement and participate in ecotourism activities" should follow these principles:

- Minimize impact.
- Build environmental and cultural awareness and respect.
- Provide positive experiences for both visitors and hosts.
- Provide direct financial benefits for conservation.
- Provide financial benefits and empowerment for local people.
- Raise sensitivity to host countries' political, environmental, and social climate.

percent of the nation's territory. In Kenya, hundreds of millions of annual tourist dollars provide a powerful incentive to ensure the survival of the country's elephant and rhinoceros populations.

Ecotourism is a growth industry. According to a 2001 report by the Worldwatch Institute, ecotourism grew 20 to 34 percent every year beginning during the 1990's. The UNWTO has reported that in 2004 ecotourism and nature tourism were expanding three times faster than the international tourism industry as a whole.

NEGATIVE IMPACTS

Ecotourism is not without its drawbacks. Observers in Costa Rica, for example, have noted that although some national parks are large, most visitors want to see specific sites, which leads to overcrowding, trail erosion, and pollution at those sites. Also, scientists have noted changes in the behavioral patterns of local wildlife that appear to be linked to human activity. In Africa, the proximity of ecotourist groups to mountain gorillas puts the great apes at risk from human infectious diseases such as measles, polio, influenza, and tuberculosis.

Growth in ecotourism also promotes development outside protected areas, with attendant environmental degradation. In addition, not all of the people who participate in ecotourism activities have a deep understanding of the no-impact philosophy and a full appreciation of its importance; some of these people contribute to negative impacts through their actions in sensitive areas.

To complicate matters, some purported ecotourism is little more than greenwashed tourism. The bur-

geoning popularity of ecotourism has led to a proliferation of companies offering purported ecotours that actually fail to employ sustainable practices. In the absence of regulation or even consensus on what constitutes ecotourism, some operators sell their products as ecotours despite the fact that they do not meet the standards of the term as it is usually understood. One Costa Rican tourism project touted as an ecodevelopment included environmentally unfriendly amenities such as a shopping center and a golf course.

Studies indicate that local communities often do not benefit from activities in their surrounding areas touted as ecotourism. In many countries, foreign interests own tourist facilities and recreational sites, thus ensuring that profits flow out of the local area. Environmentally insensitive tourism can displace native populations into marginal lands or drive them from a subsistence lifestyle into poverty-wage service jobs. In Nepal, local families earn little money while serving as porters for tourists. In areas where locals do profit, problems can still arise. Some communities in Costa Rica, for example, have moved from a subsistence to a market economy, a transition that belies the ethic of maintaining the integrity of local cultures.

Critics maintain that the concept of ecotourism is inherently flawed. They argue that ecotourists merely pave the way for mass tourists, people who demand the comforts of home while they visit remote areas. Moreover, the developing nations that offer ecotourist attractions are often the least able to invest the funds necessary to counter the negative impacts of tourism. Only a small percentage of tourist dollars may go toward the management of natural resources.

Opponents of ecotourism assert that it is merely a variant of tourism that will inevitably despoil the very areas it is intended to protect. The deluge of tourists visiting the Galápagos Islands, for example, has overwhelmed the Ecuadoran government's ability to manage them. The annual number of visitors to the islands surpassed the government's target limit of 25,000 people decades ago; by 2005, the number of visitors per year had swelled to more than 121,000. Economic development to accommodate the tourist traffic has caused appreciable damage to the fragile island environment. Environmental advocates recommend that potential ecotourists carefully review the literature of any organization that offers ecotours to be sure that its practices and philosophy are in keeping with the goals of environmental and cultural preservation.

EMERGING STANDARDS

Interest in ecotourism's role in sustainable development and concerns regarding the detrimental effects of ecotourism's mismanagement led to the first World Ecotourism Summit, held in Quebec, Canada. A joint initiative of the UNWTO and the United Nations Environment Programme (UNEP), the summit was held in 2002, designated by the United Nations as the International Year of Ecotourism. The summit laid the groundwork for the Global Sustainable Tourism Criteria (GSTC), introduced in 2008 by the United Nations Foundation, the UNWTO, UNEP, and the Rainforest Alliance. The first international criteria for sustainable tourism practices, these voluntary standards are based on four key elements of sustainable tourism: effective sustainability planning, maximum social and economic benefits for local communities, minimum negative impacts on cultural heritage, and minimum negative impacts on the environment. The criteria are meant not only for ecotourism but also to guide the tourism industry in general toward sustainable practices.

In 2010 the Tourism Sustainability Council, a global membership body, began developing an accreditation program for the world's existing ecotourism certification bodies to bring ecotourism businesses into compliance with universal standards. The program as proposed will use measurable indicators—such as electricity and energy consumption per serviced area, freshwater consumption and waste production per guest per night, and the quality of water discharged from on-site wastewater treatment facilities—to distinguish true ecotourism businesses from greenwashed enterprises.

Thomas Clarkin
Updated by Karen N. Kähler

FURTHER READING

Fennell, David A. *Ecotourism.* 3d ed. New York: Routledge, 2008.
France, Lesley. *The Earthscan Reader in Sustainable*

Ecotourists participate in bird-watching in Panama's Soberanía National Park. (AP/Wide World Photos)

Tourism. 1997. Reprint. Sterling, Va.: Earthscan, 2002.

Honey, Martha. *Ecotourism and Sustainable Development: Who Owns Paradise?* 2d ed. Washington, D.C.: Island Press, 2008.

McLaren, Deborah. *Rethinking Tourism and Ecotravel.* 2d ed. Bloomfield, Conn.: Kumarian Press, 2003.

Patterson, Carol. *The Business of Ecotourism: The Complete Guide for Nature and Culture-Based Tourism Operators.* 3d ed. Victoria, B.C.: Trafford, 2007.

Schellhorn, Matthias. "Development for Whom? Social Justice and the Business of Ecotourism." *Journal of Sustainable Tourism* 48, no. 1 (2010): 115-136.

Weaver, David B. *Ecotourism.* 2d ed. Milton, Qld.: John Wiley & Sons Australia, 2008.

_____, ed. *The Encyclopedia of Ecotourism.* Oxford, England: CABI, 2001.

SEE ALSO: Environmental economics; Green marketing; Greenwashing; Mountain gorillas; National parks; Nature reserves; Sustainable development.

Ehrlich, Paul R.

CATEGORIES: Activism and advocacy; population issues

IDENTIFICATION: American biologist and environmental philosopher

BORN: May 29, 1932; Philadelphia, Pennsylvania

SIGNIFICANCE: Ehrlich has published several books that have been influential in raising awareness and promoting action concerning such problems as the dangers of overpopulation and the possible effects of nuclear war.

Paul Ralph Ehrlich, the son of William and Ruth Ehrlich, displayed an early interest in nature study. Following high school, he enrolled at the University of Pennsylvania, from which he graduated in 1953 with a zoology degree. Ehrlich conducted his graduate work at the University of Kansas, earning his M.A. in 1955 and his Ph.D. in 1957. His doctoral research was in the field of entomology, and his first published book dealt with identification of butterflies. He married Anne Howland in 1954; they had one child, a daughter.

After he received his doctorate in 1957, Ehrlich worked as a research associate on various studies. He joined the faculty of Stanford University's Biology Department in 1959 and became a full professor there in 1966. During this time his interest in ecology and conservation developed more fully. He was promoted to the position of Bing Professor of Population Studies at Stanford in 1976.

Ehrlich is best known to the general public for his vigorous support of conservation, including what some consider radical ideas for preserving the earth's resources. Foremost among the topics he has addressed is the potentially devastating effect that increased human consumption could have on the environment. When his book *The Population Bomb* first appeared in 1968, it was widely circulated and caused much discussion regarding worldwide population growth. In the book Ehrlich maintained that increased population, coupled with decreased food production, would, in the following few decades, result in billions of deaths from starvation. This prediction, however, did not come true—although population growth is increasing, food production, with the help of modern agricultural technology, has increased at an even faster rate. Ehrlich suggested in *The Population Bomb* that the government should place limitations on the number of children a couple could have and went so far as to suggest forced vasectomies for men in overpopulated countries with high birthrates.

Ehrlich's suggestions were not limited to population control. He criticized developed countries, especially the United States, for unwise and excessive consumption of natural resources. He predicted that as resources became depleted, inflation would follow. Developing countries, unable to afford even basic necessities, would suffer the most. He urged the U.S. government to pass legislation mandating limited consumption of natural fuels, proper treatment and disposal of wastes, and extensive conservation of fish and wildlife areas.

In his writings, Ehrlich has also predicted increases in air pollution, ozone depletion, species extinctions, and numbers of deaths caused by acquired immunodeficiency syndrome (AIDS); decreases in food production because of poor farming practices; and growing disparity between rich and poor unless corrective measures were taken. His predictions have increased public awareness of environmental problems, and this increased awareness has had some influence on efforts by government officials and grassroots organizations to address these concerns.

In the years after *The Population Bomb* appeared, Ehrlich wrote several other books—many coauthored with his wife—detailing his concerns for conservation and ecological restraint. Among these are *The Population Explosion* (1990), *Healing the Planet* (1991), *Betrayal of Science and Reason* (1996), and *The Dominant Animal: Human Evolution and the Environment* (2008).

Gordon A. Parker

FURTHER READING

De Steiguer, J. E. "Paul Ehrlich and *The Population Bomb*." In *The Origins of Modern Environmental Thought.* Tucson: University of Arizona Press, 2006.

Ehrlich, Paul R., and Anne H. Ehrlich. *One with Nineveh: Politics, Consumption, and the Human Future.* Washington, D.C.: Island Press, 2004.

Simmons, Ian G. "Paul Ehrlich, 1932-   ." In *Fifty Key Thinkers on the Environment,* edited by Joy A. Palmer. New York: Routledge, 2001.

SEE ALSO: Antinuclear movement; Population-control movement; Population growth.

El Niño and La Niña

CATEGORY: Weather and climate

DEFINITION: El Niño is a quasi-periodic abnormal warming of surface waters in the central and eastern tropical Pacific Ocean; La Niña is a quasi-periodic abnormal cooling of surface waters in the central and eastern tropical Pacific Ocean

SIGNIFICANCE: The El Niño-Southern Oscillation climate pattern has implications for weather around the world. El Niño and La Niña events affect seasonal temperatures and precipitation patterns in many different regions and are associated with extreme weather such as heavy rainfall, floods, and droughts.

Ecuadoran and Peruvian fishermen gave the name El Niño to a warm, southward-flowing ocean current that would occur off the west coasts of their countries every year around Christmastime (*El Niño* is Spanish for "the little boy" or, more specifically, "the Christ child"). Originally used to describe a brief, localized, annual phenomenon, the term later became associated with unusually strong ocean warming in this area that occurred every few years, disrupting lo-

cal fish and bird populations. "Anti-El Niño" ocean cooling events were called either La Niña (the little girl) or El Viejo (the old man).

The term "El Niño" has come to be associated with the large warm-water anomalies covering extensive portions of the tropical Pacific Ocean off the coast of Latin America that persist for many months, and with the related weather effects noted around the globe. It has been known for many years that when warm water appears off the coast of Peru, atmospheric pressure drops over the eastern Pacific and rises over Australia and the Indian Ocean. Because of this relationship, major El Niño events are usually associated with other global weather phenomena, including drought in Africa and Australia and the failure of the Indian monsoon. Scientists know the atmospheric component of this global pattern as the Southern Oscillation. The coupled global oceanic-atmospheric system is called El Niño-Southern Oscillation (ENSO). El Niño is sometimes called an ENSO warm event or the warm phase of ENSO; similarly, La Niña may be referred to as an ENSO cold event or the cold phase of ENSO.

Typically, El Niño and La Niña events occur every three to five years, although the interval may vary from two to seven years. El Niño events tend to last nine to twelve months but have been known to last as long as two, three, or even four years. The typical La Niña event lasts one to three years. Both El Niño and La Niña conditions typically develop during the months of March through June, reach peak intensity during the months of December through April, and diminish during the months of May through July.

CAUSES AND PREDICTION

When an El Niño condition develops in the eastern Pacific, the sea surface temperature and rainfall in the eastern tropical Pacific are at their seasonal peaks. Major El Niño occurrences are closely tied to global weather patterns and the circulation of currents in the Pacific Ocean. Variations in Indian monsoon circulation sometimes precede variations in the Southern Oscillation, indicating that there is a possible feedback mechanism linking these phenomena. The period of the Southern Oscillation is irregular, with a return period of about three to four years, so about two El Niños occur per decade. The amplitude of the Southern Oscillation is highly irregular. If some global atmospheric perturbation contributes to the amplitude of the Southern Oscillation while an El Niño is developing, a major El Niño might be ex-

Years in Which El Niño and La Niña Were Observed, 1900-2007

EL NIÑO YEARS (warm water in eastern Pacific)	LA NIÑA YEARS (cold water in eastern Pacific)
1902, 1905, 1911, 1914, 1918, 1923, 1925, 1930, 1932, 1939, 1941, 1951, 1953, 1957, 1965, 1969, 1972, 1976, 1982, 1986, 1991, 1994, 1997, 2002, 2004, 2006, 2009	1904, 1908, 1910, 1916, 1924, 1928, 1938, 1950, 1955, 1964, 1970, 1973, 1975, 1988, 1995, 1998, 2007, 2010

Note: Many El Niños begin in one calendar year and end during the following calendar year. Only the beginning year is listed.

pected to occur. However, if a global perturbation subtracts from the amplitude of the Southern Oscillation, the El Niño might be weak.

The point at which scientists decide that a major El Niño condition is occurring has been historically contentious. A network of buoys has been established to augment satellite monitoring of sea surface temperatures in the Pacific Ocean. This network played a key role in the early detection of the 1997-1998 El Niño. When sea surface temperatures reach 3 to 5 degrees Celsius (5.4 to 9.0 degrees Fahrenheit) above normal for the season in the eastern equatorial Pacific, scientists can be fairly certain that an El Niño is occurring. The classic El Niño begins off the coast of Peru, slowly propagating westward. Since the 1990's, however, a new type of El Niño has been observed in which the maximum ocean warming occurs instead in the central-equatorial Pacific. Known variously as a central Pacific El Niño, dateline El Niño, warm-pool El Niño, or El Niño Modoki, this type of event occurred in 1991-1992, 1994-1995, 2002-2003, 2004-2005, and 2009-2010.

During a major El Niño, the normal westerly trade winds subside, and the height of the eastern Pacific sea surface rises. This is coupled with a decline in the height of the western Pacific sea surface, which sometimes causes normally submerged coral reefs in the western Pacific to appear above the ocean surface.

Since historical recording of El Niño events began, long-term variations in their strength have been observed. The 1920's and 1930's experienced only weak El Niño events. In contrast, the El Niños of the 1980's and 1990's were generally moderate to strong. El Niño conditions result from a complex interplay of atmospheric and oceanic forces, and the reasons for the waxing and waning in strength of El Niño events over periods of decades are not understood.

Some scientists have noted that unusual El Niños have followed volcanic eruptions. A major volcanic eruption causes the formation of a stratospheric aerosol layer, which may lead to an increase in solar radiation being reflected back into space. The 1951 El Niño followed the eruption of Mount Lamington in Papua New Guinea; the 1982 El Niño followed the eruptions of El Chichon in Mexico and Galunggung in Indonesia; the 1991 El Niño followed the eruption of Mount Pinatubo in the Philippines. However, while 1951 was a weak El Niño year, both 1982 and 1991 were strong. Although it is tempting to focus on a single parameter such as sea surface temperatures or the period of the Southern Oscillation as a predictor of El Niño, history shows that many factors, both atmospheric and oceanic, contribute to the development of a strong El Niño event.

ENVIRONMENTAL CONSEQUENCES

When unusually warm water appears off the coast of Peru, the local anchovy fishing industry falters. Sportfishing off Baja California, California, and Oregon, in contrast, enjoys a boom, as marlin and other highly prized fish usually found in more tropical southern waters move north. Other marine ecosystems around the globe may also be affected by El Niño conditions. Factors such as increased sea surface temperature, decreased sea level, and salinity changes related to high rainfall affect the algae that protect coral reefs, causing the coral to bleach and die. This, in turn, has negative impacts on fish and plant life within reef ecosystems. The 1997-1998 El Niño event was marked by coral reef bleaching in the Indo-Pacific region, the Caribbean, and the Florida Keys. Extensive coral bleaching also occurred around the globe in association with the 2009-2010 El Niño.

During El Niño episodes, the intertropical convergence zone, a band of major tropical convection circling the globe, moves southward. This southward shift in precipitation patterns causes torrential rains in some places that are normally dry and dry conditions in places that are usually wet. In the Galápagos Islands, El Niño brings much higher than normal precipitation in March, April, and May. During major El Niños, Peru and Ecuador experience torrential rains and flooding. In Guayaquil, Ecuador, El Niño was blamed for causing more than 3 meters (9.8 feet) of rain between October, 1982, and January, 1983. During the 1997-1998 El Niño, severe flooding occurred in Ecuador along rivers where rain forests had been cleared to establish shrimp farms. By contrast, the moderate El Niño of 2002-2003 had little effect on the weather of these two countries.

During many major El Niño events, countries bordering the western Pacific and Indian oceans experience droughts. In El Niño years India often receives lower-than-normal rainfall, while Sri Lanka's rainfall tends to be unusually high. During the 1982-1983 El Niño, Indonesia and Australia were stricken by drought. Curiously, the Australian drought associated with the strong 1997-1998 El Niño was not as severe as the one that accompanied the moderate 2002-2003 El Niño. Early in the 1990's southern Africa experienced

its worst drought of the twentieth century, probably worsened by the El Niño that began in late 1991. El Niño years in Japan are associated with mild winters, cool summers, and lengthy rainy seasons.

Many diverse ecological, environmental, and economic events throughout the world are often attributed to El Niño occurrences. Sometimes these events may indeed be related to El Niño, but some occurrences can be attributed to other factors. Sometimes conflicting claims are made about the effects of El Niños. Just as there is no consistent relationship between the failure of the Indian monsoon and El Niño years, other claims about El Niño and weather may hold up during a statistically significant number of years but not in all years.

Although the 1997 El Niño was credited with causing the unusually mild winter of 1997-1998 in the eastern United States, the 1976 El Niño was blamed for causing extreme cold in the same region in December, 1976, and January, 1977. The Sonoran Desert of Arizona and California tends to be wet in El Niño years. The Florida drought of 1998 was attributed to

Demolished homes in Pacifica, California, teeter on the edge of a cliff that was washed away by 1997-1998 El Niño storms. (AP/Wide World Photos)

El Niño, although the warm-water anomaly off Peru had virtually disappeared by July, when forest fires were plaguing Florida. Because Texas often experiences more precipitation during the growing season after an El Niño, farmers planted crops requiring more moisture than usual in 1998. The summer of 1998 was unusually dry in Texas and Oklahoma, however, leaving many of the affected farmers to conclude that El Niño-based forecasts might be less than reliable. The failure of the Soviet harvest in 1972 was attributed by some to El Niño. In El Niño years, Moscow frequently experiences very cold winters; December, 1997, fit into this pattern.

La Niña

As the warm-water anomalies in the eastern Pacific fade, scientists know that El Niño is ending. When a large body of colder-than-average water establishes itself off the coast of Peru, along with a strong ridge of high pressure, meteorologists announce that a La Niña event is occurring. A La Niña may follow closely on the heels of an El Niño, as was the case with the 2010-2011 La Niña, or occur after a year or two of neutral conditions. During a La Niña, the Pacific sea surface height in the eastern Pacific is measurably lower than when an El Niño is occurring. The westerly trade winds are also stronger than normal. During a La Niña, the average temperature of the tropical troposphere may be 1 degree Celsius (1.8 degrees Fahrenheit) lower than during an El Niño. La Niñas tend to be much more variable in strength than El Niños.

During periods when the waters of the eastern Pacific have been observed to be anomalously cold, the Pacific Northwest tends to be wetter and cooler than normal, especially during winter. During a La Niña year, winter temperatures in the southern and southeastern United States are often warmer than normal. In the United States, some link La Niña years to very hot summers; the summer of 1988, which was very hot and dry, is the prototype summer for this weather phenomenon. Tropical cyclones are more common on the northern Australian coast during La Niña events. Widespread flooding in eastern Australia tends to occur early in the calendar year (late summer) during La Niñas. La Niña conditions also tend to be accompanied by more damaging Atlantic hurricanes, while a reduction in hurricane activity is typically associated with El Niño.

Anita Baker-Blocker
Updated by Karen N. Kähler

FURTHER READING

Changnon, Stanley Alcide, ed. *El Niño, 1997-1998: The Climate Event of the Century.* New York: Oxford University Press, 2000.

Clarke, Allan J. *An Introduction to the Dynamics of El Niño and the Southern Oscillation.* San Diego, Calif.: Academic Press, 2008.

D'Aleo, Joseph S., and Pamela G. Grube. *The Oryx Resource Guide to El Niño and La Niña.* Westport, Conn: Oryx Press, 2002.

Glantz, Michael H. *Currents of Change: Impacts of El Niño and La Niña on Climate and Society.* 2d ed. New York: Cambridge University Press, 2001.

_____. *La Niña and Its Impacts: Facts and Speculation.* Tokyo: United Nations University Press, 2002.

Philander, S. George. *Our Affair with El Niño: How We Transformed an Enchanting Peruvian Current into a Global Climate Hazard.* Princeton, N.J.: Princeton University Press, 2006.

Rosenzweig, Cynthia, and Daniel Hillel. *Climate Variability and the Global Harvest: Impacts of El Niño and Other Oscillations on Agroecosystems.* New York: Oxford University Press, 2008.

SEE ALSO: Climate change and oceans; Climate models; Climatology; Droughts; National Oceanic and Atmospheric Administration; Ocean currents; Thermohaline circulation.

Electric vehicles

CATEGORY: Energy and energy use
DEFINITION: Vehicles that use electricity to provide some or all of their power for propulsion
SIGNIFICANCE: The replacement of fossil-fuel-powered vehicles with electric vehicles offers several benefits to the environment. Electric vehicles do not contribute to air pollution with tailpipe emissions, and overall they use less energy than gasoline-powered vehicles because of their higher efficiencies.

Nearly every major automaker in the world has an active program to develop, manufacture, and sell electric vehicles. Electric vehicles include automobiles, trucks, buses, and motorbikes. The vehicles may be divided into types based on their power sources; they include battery electric vehicles, hybrid electric vehicles, and fuel cell-powered vehicles. Electric vehi-

An electric car, produced by the Shanghai Automotive Industry Corporation, is recharged at the Beijing International Automotive Exhibition in 2010. (AP/Wide World Photos)

cles deliver instant torque, smooth acceleration, quiet operation, and lower maintenance costs than their internal combustion-powered counterparts.

A battery electric vehicle uses a rechargeable battery to power a motor controller that in turn powers one or more electric motors to propel the vehicle at different speeds depending on the position of the accelerator pedal. Neither a transmission nor an internal combustion engine is needed. A hybrid electric vehicle uses two different energy sources, an onboard internal combustion engine linked to an electric motor or generator to drive the vehicle and to recharge the batteries. Hybrid electric vehicles operate almost twice as efficiently as traditional internal combustion vehicles. In a fuel cell-powered vehicle, a fuel cell, rather than batteries, provides the electric energy. Hydrogen is the fuel used most often in such fuel cells, but other fuels—such as methanol, natural gas, and petroleum—may be used.

The replacement of significant numbers of gasoline-powered vehicles with electric vehicles would re-

sult in large reductions in the vehicle emissions that contribute approximately two-thirds of all air pollution in most cities. The amount of reduction in a given area would depend on the local power plants used to generate the electricity for battery charging and the plants that produce hydrogen for fuel cell vehicles. In areas where power is generated using solar, wind, hydro, nuclear, or carbon-capture technologies, overall pollution would be decreased much more than in areas that depend on fossil-fuel power plants, particularly those relying on coal. Battery electric vehicles themselves produce no greenhouse gas emissions in their operation, and if their batteries are charged from an electrical grid that is powered by clean sources, the battery-charging process also produces no greenhouse gases. It has been estimated that electric vehicles can reduce greenhouse gas emissions by about 20 percent compared with gasoline-powered internal combustion vehicles in areas where coal-burning plants supply all the electrical power. In areas that use cleaner electrical-grid technologies, such as

Arizona and California, greenhouse emissions could be reduced by more than 70 percent.

The limitations of electric vehicles include the cost and availability of batteries with high energy densities, short charge time, and long life. Lithium-ion batteries have demonstrated the energy densities necessary to deliver driving ranges of approximately 250 kilometers (155 miles) at speeds of up to 130 kilometers (80 miles) per hour, with recharge times of less than four hours with charging systems using 208-volt, 40-amp power supplies. The overall efficiency of electric vehicles is affected by battery charging and discharging efficiencies, which depend on the efficiency of the local electricity-generating grid. On average, battery electric vehicles have been shown to be three times more efficient than internal combustion vehicles.

The aims of future advancements in battery and charger technologies are to achieve charging times that are roughly equivalent to the amount of time it takes to fill the gas tank of a gasoline-powered vehicle and to produce such charging at a lower cost per kilometer than filling a gas tank. Also, increases in production of lithium-ion or comparable batteries are expected to reduce the prices of the batteries significantly in the future, which should reduce the cost of electric vehicles.

Alvin K. Benson

FURTHER READING

Caputo, Richard. *Hitting the Wall: A Vision of a Secure Energy Future.* San Rafael, Calif.: Morgan & Claypool, 2009.

Erjavec, Jack. *Hybrid, Electric, and Fuel-Cell Vehicles.* Clifton Park, N.J.: Thomson Delmar Learning, 2007.

Fuhs, Allen E. *Hybrid Vehicles and the Future of Personal Transportation.* Boca Raton, Fla.: CRC Press, 2008.

SEE ALSO: Air pollution; Alternatively fueled vehicles; Automobile emissions; Energy conservation; Fossil fuels; Hybrid vehicles; Internal combustion engines; Photovoltaic cells.

Electronic waste

CATEGORIES: Waste and waste management; pollutants and toxins

DEFINITION: Electronic equipment or parts of equipment discarded when broken or obsolete

SIGNIFICANCE: As consumers replace outdated or broken electronic equipment or appliances, the old items are discarded, and many end up in landfills. These discarded electronic devices often contain hazardous materials that can leach into the environment.

Electronic waste, or e-waste, is generated when consumers discard broken or obsolete electronic equipment or parts of such equipment. Industrial waste that is produced during the manufacture of electronic equipment is also sometimes referred to as e-waste. Electronic waste raises environmental concerns because many electronic devices contain numerous heavy elements, such as lead and mercury, and other toxic elements, such as cadmium and arsenic. Additionally, some devices, such as smoke detectors, can include radioactive elements. Some electronic devices also contain plastics that have been treated with fire-retardant chemicals, many of which can be toxic when released into the environment.

Unlike the disposal of many other consumer products containing hazardous materials, the disposal of electronic devices did not receive regulatory attention until the early twenty-first century. When these devices were first introduced, they existed in limited quantity and were expensive. If such a device began to malfunction or stopped working, it would be repaired. By the late twentieth century, however, electronic devices such as computers, microwave ovens, televisions, and cell phones had become common and comparatively inexpensive. The cost of repairing a broken device was often comparable to the cost of simply replacing it.

The pace of technology, too, was such that newer devices often had more desirable features than older ones, and many consumers elected to replace broken electronics rather than repair them. Some manufacturers began to construct electronic devices that were nonserviceable—that is, they could not be repaired—so consumers would have to buy new ones when the old ones no longer worked. Manufacturers began to make some devices in which batteries could not be changed when they died; some sold devices to con-

The EPA's Goal for Electronic Waste

The U.S. Environmental Protection Agency has stated that its overall goal in regard to electronic waste is "to promote greater electronics product stewardship":

Product stewardship means that all who make, distribute, use, and dispose of products share responsibility for reducing the environmental impact of those products. EPA intends to work towards this goal in three ways:

1. Foster a life-cycle approach to product stewardship, including environmentally conscious design, manufacturing, and toxics reduction for new electronic products.

2. Increase reuse and recycling of used electronics.

3. Ensure that management of old electronics is safe and environmentally sound.

sumers with the knowledge that the devices would not function with later generations of the technology. Such planned obsolescence exacerbated the problem of electronic waste.

At first, most electronic waste simply ended up in landfills. However, by the 1990's, public awareness of the environmental hazards posed by electronic waste began to rise. Grassroots movements put pressure on governments and industry to curb the landfilling of electronic waste. By 2010, twenty U.S. states had passed laws designed to reduce electronic waste, generally by establishing recycling centers devoted to electronics or by requiring retailers or manufacturers to implement take-back programs for used electronics.

Electronic waste is difficult and expensive to recycle. Many electronic devices have materials in them that are valuable if recovered—such as lead, copper, and small amounts of gold, platinum, and silver—but recovering these materials is a labor-intensive process. For this reason, much of the electronic waste collected for recycling in the United States is shipped to developing countries, where unskilled and poorly paid laborers break apart the devices to get to the commercially useful materials. Often this work is done without the kind of oversight and regulation required by the environmental laws of industrialized nations, creating environmental hazards in those developing countries.

International efforts to limit the environmental damage done by electronic waste have led some devel-

oping countries to limit imports of e-waste or to regulate its disposal, leaving developed nations with fewer options for disposing of their e-waste. Manufacturers and governments have sought ways to raise revenues to address the expensive process of recycling electronic waste. To handle the cost of electronic waste disposal, some companies charge fees to accept old electronics. California has legislated disposal fees that consumers must pay when they purchase new electronic devices.

Raymond D. Benge, Jr.

FURTHER READING

Consumer Reports. "Where to Recycle Electronics, Free." June, 2009, 11.

Hester, Ronald E., and Roy M. Harrison, eds. *Electronic Waste Management*. Cambridge, England: RSC, 2009.

Jozefowicz, Chris. "Waste Woes." *Current Health*, January 2, 2010, 24-27.

United Nations Environment Programme. *Recycling: From E-Waste to Resources*. Berlin: Oktoberdruck, 2009.

SEE ALSO: Environmental law, U.S.; Heavy metals; Landfills; Planned obsolescence; Recycling; Solid waste management policy.

Elephants, African

CATEGORY: Animals and endangered species

DEFINITION: Largest living land mammals, native to Africa

SIGNIFICANCE: Owing primarily to habitat destruction and hunting, both legal and illegal, the population of African elephants has been reduced steadily over time. Although populations in protected areas have achieved some stability, poaching and progressive desertification continue to threaten the species.

Elephants of Africa's savanna regions belong to the taxonomic group *Loxodonta africana*. Debate continues over whether the lesser-known African bush elephant is a separate species (as *Loxodonta cyclotis*) or subspecies. Some biologists postulate that an extinct elephant species may have also previously existed in the Atlas Mountains of North Africa (*Loxodonta africana pharaonensis*), but if so it disappeared by the

A female African elephant and her calf at the Chobe River in Botswana. (©Steve Allen/Dreamstime.com)

end of the Roman Empire owing to hunting and habitat change.

Elephants are the largest living land mammals known, with male African elephants often standing 3.7 meters (12 feet) tall at the shoulder and weighing up to 6 tons. Both male and female African elephants have tusks. They can live to sixty to eighty years in the wild, possibly longer in captivity. Female elephants are highly social, living in groups led by the eldest female, to whom they are generally related, with the adult males usually living alone on the group perimeter, although bachelor herds also exist for periods of time.

Although the primary habitat of the largest group of African elephants is the great savanna, elephants have historically ranged across the entire African continent, south of the great Sahara. They were at one time ubiquitous in this geographic region, but African elephant populations have shrunk greatly in the past two centuries because of human activity, especially habitat loss—resulting from land mismanagement and climatic aridization—along with hunting and poaching, primarily for their ivory tusks. Esti-

mated around 1980 as numbering up to 1.3 million (quite a few experts believe this number was inflated), by 2010 they numbered around 400,000. Where protected in nature preserves, this population is relatively stable, but outside these conservation areas African elephants are still considered to be an endangered species ("threatened" or "vulnerable" status). Habitat loss for African elephants owing to desertification and deforestation is also considered to be highly problematic for wild populations, especially if desertification continues at the rates seen since the late twentieth century.

Trade in ivory has been mostly banned since 1989 by the Convention on International Trade in Endangered Species (CITES), backed by the United Nations, but illicit trade continues in politically unstable parts of Africa. Some ivory exportation is allowed, although only a fraction of the revenue generated appears to have ever benefited Africans, according to Oxford University studies conducted in 2000. One of the most egregious single incidents of deliberate elephant poaching took place in southeastern Chad in

2006 just outside Zakouma National Park, where more than one hundred elephants were slaughtered for ivory. It is estimated that 96 percent of the 300,000 Chadian elephant population was lost between 1970 and 2010, when numbers were down to approximately 10,000 elephants in Zakouma National Park.

Individuals and organizations such as Sir Richard Leakey and Sir Iain Douglas-Hamilton of Save the Elephants, both in Kenya, and J. Michael Fay of the Wildlife Conservation Society are strong advocates for African elephant preservation. Because elephants are migratory, following water and plant seasons, they can cover a territory of up to 3,600 square kilometers (about 1,400 square miles), which is problematic when they leave protected areas where hunting is forbidden.

Patrick Norman Hunt

FURTHER READING

Bradshaw, G. A. *Elephants on the Edge: What Animals Teach Us About Humanity.* New Haven, Conn.: Yale University Press, 2009.

Chadwick, Douglas H. *The Fate of the Elephant.* San Francisco: Sierra Club Books, 1992.

Christo, Cyril. *Walking Thunder: In the Footsteps of the African Elephant.* London: Merrell, 2009.

Meredith, Martin. "Ivory Wars." In *Elephant Destiny: Biography of an Endangered Species in Africa.* New York: PublicAffairs, 2001.

Walker, John Frederick. *Ivory's Ghosts: The White Gold of History and the Fate of Elephants.* New York: Grove/Atlantic, 2009.

SEE ALSO: Africa; Convention on International Trade in Endangered Species; Ecotourism; Hunting; Poaching.

Eminent domain

CATEGORY: Land and land use

DEFINITION: The power of government to take property for public use with just compensation

SIGNIFICANCE: Proponents of environmental preservation have often favored governments' use of the power of eminent domain when such action can prevent property owners from using their land in ways that would be detrimental to the environment.

Eminent domain is an inherent attribute of sovereignty that permits a government to seize property and air, water, or land rights in property for a public use without the owner's consent as long as the government provides monetary compensation. The eminent domain or takings clause of the Fifth Amendment to the U.S. Constitution limits the federal government's eminent domain power, and the Fourteenth Amendment's due process clause applies the Fifth Amendment's limitations to state governments. Land-use restrictions and public-use issues have dominated the debate over the meaning of the takings clause.

LAND-USE REGULATIONS

The takings clause requires government to pay compensation if it takes private property, but not if it regulates the property to further its police power goals of promoting public health, safety, or welfare. In Pennsylvania Coal Co. v. Mahon (1922), the U.S. Supreme Court limited the state's police power when it held that a statute that prohibited subsurface mining under houses and towns to promote public health and safety was a taking because it drastically reduced the value of the company's property in coal.

The Supreme Court did not develop a takings test until *Penn Central Transportation Co. v. City of New York* (1978). In *Penn Central,* the Court balanced two factors to hold that a historic preservation ordinance did not take the company's property in Grand Central Station: The ordinance created a generally applicable program intended to produce a general public benefit, and the ordinance did not interfere with the company's past use and a reasonable return on its investment. In rare instances the Court has recognized that a land-use regulation is a categorical taking. In *Lucas v. South Carolina Coastal Council* (1992) it held that a state statute protecting fragile coastal areas from erosion was a taking because it prohibited the owner from building on land purchased prior to enactment of the statute. In *Tahoe-Sierra Preservation Council, Inc. v. Tahoe Regional Planning Agency* (2002), the Court limited *Lucas* and held that a regional agency's moratorium on land development was not a total temporary taking. Instead, the Court used the *Penn Central* test to uphold the moratorium, finding that it had been used to create a comprehensive land-use plan for the fragile Tahoe basin and had not interfered with land values, which were expected to increase in spite of the moratorium.

ACCESS MANDATES AND LAND-USE RESTRICTIONS

With access mandates, however, the Supreme Court has closely scrutinized government decisions that require owners to open their lands to the public. In *Kaiser Aetna v. United States* (1979) the Court rejected the federal government's argument that a private pond, opened by its owners to the ocean, had become a navigable waterway and held that granting the public access would be a taking. When access mandates are tied to land-use restrictions and owners are required to dedicate a portion of their property for a public use in exchange for a land-use right, the Court has required an even closer fit between the government's regulations and its objectives. In *Nollan v. California Coastal Commission* (1987) the Court held that an owner of beachfront property could not be required to grant public access to the beach adjacent to the ocean in exchange for a permit to rebuild his house, because the easement would not substantially advance the state's interest in reducing obstacles to viewing the beach.

Dolan v. City of Tigard (1994) enhanced the Court's rigorous review by adding a second requirement. The city could not require an owner in exchange for a permit to expand her business to dedicate a portion of her property as a bicycle path because the city failed to provide quantitative evidence of a rough proportionality between the bike path dedication and the increased traffic generated by the business expansion.

PUBLIC USE

The takings clause limits taking of private property for public uses and forbids giving the property to another person. The U.S. Supreme Court has found that three categories of public uses justify a taking. In the first, the government takes title to property and builds a school, road, or park. In the second, the government takes private property and conveys it to a private business, such as a railroad or utility, but requires it to provide a public service on reasonable and equal terms. Third, the government identifies a public purpose, such as improving an economically depressed community, takes property, and conveys it to another private party for a private use.

The Supreme Court's creation of the third and most controversial category of public-use takings began with *Berman v. Parker* (1954), which upheld an economic development plan to condemn a Washington, D.C., slum and use part of the land to construct public facilities, including schools, and to sell or lease the remainder to private parties to construct low-cost housing. The Court extended public-use takings in *Hawaii Housing Authority v. Midkiff* (1984) by holding that the state's Land Reform Act, which condemned large private landholdings and sold single-family lots to tenants on those lands, was a reasonable means to achieve the public purpose of eliminating land oligarchy and opening the housing market. In *Kelo v. City of New London* (2005), the Court extended *Berman* and *Midkiff* by deferring to the city's judgment that its plan, which condemned owner-occupied homes in a nonblighted area, served the public purpose of creating jobs, generating tax revenue, and revitalizing the city center. In sum, the Court's deferential review of public land-use regulations and broad interpretation of public use have permitted governments substantial discretion in the use of their eminent domain power and made it difficult to distinguish that power from the police power.

William Crawford Green

FURTHER READING

Dana, David A., and Thomas W. Merrill. *Property: Takings.* New York: Foundation Press, 2002.

Ely, James W. *The Guardian of Every Other Right.* New York: Oxford University Press, 1992.

Epstein, Richard A. *Supreme Neglect: How to Revive Constitutional Protection for Private Property.* New York: Oxford University Press, 2008.

SEE ALSO: Land-use planning; Property rights movement; Urban planning; Urban sprawl; Zoning.

Enclosure movement

CATEGORIES: Land and land use; agriculture and food

DEFINITION: Movement to convert agricultural lands from the medieval system of open fields and lands held in common to fenced or hedged fields and pastures

SIGNIFICANCE: The European enclosure movement, along with improvements in agricultural techniques, dramatically changed land use.

Under the medieval feudal system of agriculture in Europe, typically three large, open fields without fences or hedges surrounded each village. Each field was left fallow for one year out of three. The fields were subdivided into long, narrow strips, with each farmer's allotted strips scattered over the open fields rather than adjoining one another. Farmers were allowed to graze set numbers of animals on pastureland held in common. During certain seasons—especially after harvest—the arable and hay lands were opened up to grazing by the livestock of the whole community. Woodlands and other uncultivated lands were also held in common.

In England during the thirteenth century, farmers' allotments of arable land began to be consolidated into fields enclosed by hedgerows or fences. The ancient common lands were divided among the farmers and enclosed. In the process, large areas of forestland and uncultivated land were converted into plowland. English enclosure reached its height during the late eighteenth and early nineteenth centuries and was largely complete by the mid-nineteenth century. The process occurred somewhat later in Continental Europe. Enclosure drove many farmers off the land, leading to social unrest.

Agricultural efficiency increased within enclosed fields. Farmers could rotate crops and pasture scientifically, without regard to what their neighbors did. The resulting increase in soil fertility helped remove the need for fallow fields. Farmers could raise livestock more easily with herds in enclosures and could grow fodder without having it eaten by livestock that belonged to others. With enclosed land to grow fodder, farmers could maintain livestock through the winters. The manure from the growing herds also improved soil fertility.

Over time, as the hedgerows naturally accumulated plant species, they became important wildlife habitats, compensating somewhat for the conversion of woodlands into fields and pastureland during enclosure. More than eight hundred kinds of plants have been found in British hedgerows, including such woody perennials as blackthorn, hawthorn, oak, beech, ash, hazel, roses, crabapple, and holly. Most of Great Britain's woodland birds and small mammals use hedgerows at some time during their lives. For many species, hedgerows are the only remaining habitat.

The landscape of rectangular, hedged fields largely persisted in Britain until the 1950's, as fields remained small and were regularly rotated between crops and pasture. However, many farmers began selling off their livestock and turning to crops, such as wheat, cultivated with large equipment that required broad expanses of open land. In the process, many of the hedgerows, some of great antiquity, were destroyed. This loss transformed the appearance of the countryside and was detrimental to wildlife.

Another major threat to hedgerows has been neglect. Because the strict maintenance that hedgerows require costs more than many farmers are willing to pay, the British government started a program in 1989 to pay subsidies to farmers for planting and maintaining hedgerows.

Jane F. Hill

FURTHER READING

Dixon-Gough, Robert W., and Peter C. Bloch. *The Role of the State and Individual in Sustainable Land Management.* Burlington, Vt.: Ashgate, 2006.

Kain, Roger J. P., John Chapman, and Richard R. Oliver. "The Enclosure Movement in England and Wales." In *The Enclosure Maps of England and Wales, 1595-1918.* New York: Cambridge University Press, 2004.

SEE ALSO: Grazing and grasslands; Land-use policy; Range management; Sustainable agriculture.

Endangered Species Act

CATEGORIES: Treaties, laws, and court cases; animals and endangered species

THE LAW: U.S. federal law designed to protect species threatened with extinction

DATE: Enacted on December 28, 1973

SIGNIFICANCE: The Endangered Species Act is an important part of the movement toward the protection of biodiversity in the United States. The legislation has been amended frequently since 1973, but its fundamental purpose—the preservation of species—has remained the same.

The U.S. Congress first demonstrated concern for the conservation of species in the Lacey Act of 1900, which prohibited the transportation in interstate commerce of any fish or wildlife taken in violation of national, state, or foreign laws. Following the extinction of the passenger pigeon, the Migratory Bird Treaty Act of 1918 authorized the secretary of the interior to adopt regulations for the protection of migratory birds.

In the Endangered Species Preservation Act of 1966, Congress declared that the preservation of species was a national policy. The statute authorized the interior secretary to identify native fish and wildlife threatened with extinction and to purchase land for the protection and restoration of such species. The Endangered Species Conservation Act of 1969 further empowered the secretary to list species threatened with "worldwide extinction" and prohibited the importation of any listed species into the United States. The only species eligible for the list were those threatened with complete extinction. Although the 1966 and 1969 statutes did not include any penalties for destroying species on the list, the legislation was the most comprehensive of its kind to have been enacted by any nation.

In legislative hearings in 1973, it was reported that species were being lost at the rate of about one per year and that the pace of disappearance seemed to be accelerating, with potential damage to the total ecosystem. The majority of the members of Congress concluded that it was necessary to stop a further decline in biodiversity, and they passed the Endangered Species Act (ESA), which President Richard Nixon signed into law on December 28, 1973.

The ESA outlines a process for the listing of protected species, authorizes appropriate regulations, and provides for state subsidies and funding for habitat acquisition. The act provides that any species of wild animals or plants may receive federal protection whenever the species has been listed as "endangered" or "threatened." The statute defines "endangered" to mean that the species is currently in danger of becoming extinct within a significant geographical region. The term "threatened" means that the species probably will become endangered within the near future.

The definition of a "species" includes any subspecies or any distinct population that interbreeds within a specific region. Species found only in other parts of the world are eligible for inclusion on the U.S. list. The only creatures not eligible for inclusion are those insects that are determined to pose an extreme risk to human welfare.

The ESA makes it a federal offense to take, buy, sell, or transport any portion of a threatened or endangered species. Listed animals, however, may be taken in defense of human life, and Alaska Natives are allowed to use listed animals for subsistence purposes. Additional exemptions may be granted for special cases involving economic hardship, scientific research, or projects aimed at the propagation of species. Individuals may be fined ten thousand dollars for each violation of the law committed knowingly and one thousand for a violation committed unknowingly. Harsher criminal penalties are available in extreme cases.

The ESA assigned most enforcement and regulatory powers to the heads of two executive departments. The secretary of commerce, through the National Marine Fisheries Service (NMFS), has responsibility for threatened and endangered marine species. The secretary of the interior exercises formal responsibility for the protection of other species, but the secretary delegates most of the work to

President Bill Clinton announces the 1999 removal of the bald eagle from the endangered species list, an act that highlighted the success of the Endangered Species Act. (AP/Wide World Photos)

Nixon on the Endangered Species Act

President Richard M. Nixon made the following statement upon signing the Endangered Species Act of 1973 into law:

At a time when Americans are more concerned than ever with conserving our natural resources, this legislation provides the Federal Government with needed authority to protect an irreplaceable part of our national heritage—threatened wildlife.

This important measure grants the Government both the authority to make early identification of endangered species and the means to act quickly and thoroughly to save them from extinction. It also puts into effect the Convention on International Trade in Endangered Species of Wild Fauna and Flora signed in Washington on March 3, 1973.

Nothing is more priceless and more worthy of preservation than the rich array of animal life with which our country has been blessed. It is a many-faceted treasure, of value to scholars, scientists, and nature lovers alike, and it forms a vital part of the heritage we all share as Americans. I congratulate the 93d Congress for taking this important step toward protecting a heritage which we hold in trust to countless future generations of our fellow citizens. Their lives will be richer, and America will be more beautiful in the years ahead, thanks to the measure that I have the pleasure of signing into law today.

the U.S. Fish and Wildlife Service (FWS), which is assisted by the Office of Endangered Species (OES).

In order to benefit from the ESA, species must be officially designated as either endangered or threatened. The courts have consistently ruled that the ESA cannot be used to protect an unlisted species. Species may be proposed for listing by the NMFS, the FWS, private organizations, or citizens. Species are listed only after comprehensive investigations have been conducted; open hearings are also held, and opportunities are offered for public involvement in these decisions.

The first list of endangered species, published in 1967, included 72 species. By 1976 the list had grown to 634 species. As of 1995, 1,526 species of plants and animals were listed, including more than 500 that were foreign, and almost 4,000 candidate species were awaiting a listing determination. By 2010, a total of 1901 plants and animals were listed. Although the FWS is required to prepare a recovery plan for each listed species, only a few have recovered sufficiently to be taken off the list.

The act requires that critical habitat for threatened or endangered species be designated whenever possible. All federal agencies have special obligations to determine whether their projects or actions jeopardize the continued existence of a species. Following the U.S. Supreme Court's controversial ruling in *Tennessee Valley Authority v. Hill*, Congress passed the ESA amendments of 1978, which allow consideration for economic factors in the designation of critical habitat. Especially controversial is the section of the act requiring the FWS to formulate and enforce regulations on private lands that provide habitat for listed species. The government must compensate owners in those rare cases when regulations eliminate almost all productive and economic uses of their property, but not when landowners continue to have partial productive use of their land.

Many people who live in rural regions of the United States, particularly in the West, have been highly critical of the ESA, charging that it causes significant job losses as it protects minor subspecies, such as the northern spotted owl. In 1995-1996, a conservative coalition of Republican members of Congress tried to pass a bill that would have weakened the ESA. The controversy that ensued, however, demonstrated that the existing law enjoyed considerable support, and the proposed bill was never passed. Most experts agree that the economic impact of the ESA on the national economy is minimal, but the act does cause hardship for small landowners in some instances. Many environmentalists would support revisions of the law that would give less emphasis to particular species and place more concern on the need for sufficient habitat to support healthy biodiversity, but others fear that such complexity would make the law ineffective.

Thomas T. Lewis

FURTHER READING

Baur, Donald C., and William Robert Irvin. *The Endangered Species Act: Law, Policy, and Perspectives*. Chicago: ABA Publishing, 2002.

Burgess, Bonnie B. *Fate of the Wild: The Endangered Species Act and the Future of Biodiversity*. Athens: University of Georgia Press, 2001.

Goble, Dale D., J. Michael Scott, and Frank W. Davis, eds. *Renewing the Conservation Promise*. Vol. 1 in *The Endangered Species Act at Thirty*. Washington, D.C.: Island Press, 2005.

Kohm, Kathryn, ed. *Balancing on the Brink of Extinction:*

The Endangered Species Act and Lessons for the Future. Washington, D.C.: Island Press, 1991.

Mann, Charles, and Mark Plummer. *Noah's Choice: The Future of Endangered Species.* New York: Knopf, 1995.

Noss, Reed, Michael O'Connell, and Dennis Murphy. *Habitat Conservation Under the Endangered Species Act.* Washington, D.C.: Island Press, 1997.

Regenstein, Lewis. *The Politics of Extinction.* New York: Macmillan, 1975.

Rohlf, Daniel. *The Endangered Species Act: Protection and Implementation.* Stanford, Calif.: Stanford Environmental Law Society, 1989.

Scott, J. Michael, Dale D. Goble, and Frank W. Davis, eds. *Conserving Biodiversity in Human-Dominated Landscapes.* Vol. 2 in *The Endangered Species Act at Thirty.* Washington, D.C.: Island Press, 2006.

SEE ALSO: Biodiversity; Convention on International Trade in Endangered Species; Endangered Species Act; Endangered species and species protection policy; Extinctions and species loss; *Tennessee Valley Authority v. Hill.*

Endangered species and species protection policy

CATEGORIES: Animals and endangered species; forests and plants

DEFINITIONS: Plant and animal species whose numbers are so reduced that the species are in danger of becoming extinct if protection is not provided, and high-level governmental plans of action to support the survival and recovery of endangered and threatened species

SIGNIFICANCE: Natural causes as well as pollution, habitat fragmentation and destruction, and other environmental stresses imposed by human activity can drive species toward extinction. Once a population's size declines past a certain point, various factors will eventually wipe out the population entirely. Implementing protective policies can save a declining species from extinction and, ideally, enable it to recover and thrive.

Extinction of a species does not occur in a vacuum. Causes, typically environmental, are many and often complex. Likewise, because of the many intricate, interconnected relationships existing within ecosystems, the loss of any member may have a ripple effect, eventually having profound negative results. For example, the extinction of a single insect, bird, or bat species may result in the extinction of one or more plant species dependent on the animal species for pollination. If the plant is a critical item in the diet of certain animals, those too may be adversely affected.

Paul R. Erhlich and Anne H. Ehrlich introduce their book *Extinction: The Causes and Consequences of the Disappearance of Species* (1981) by referring to fictitious "rivet poppers"—workers whose job it is to remove rivets from the wings of airplanes. The expectation is that many rivets could be removed without the wings falling off. By analogy, the Ehrlichs consider many world leaders—politicians, bureaucrats, industrialists, engineers, religious leaders, and even some scientists—to be rivet poppers. Through their policies and practices, these leaders espouse programs that will, by design or neglect, result in the loss of endangered species. Ecosystems, by their nature, are somewhat redundant: They are likely to continue to function even after the loss of several species. Ecologists refer to this capacity as "resistance." Ecosystems also possess resilience, or the ability to recover after disturbances, including those in which species are lost. However, just as one would not wish to fly in an airplane from which even a few rivets have been removed, it seems only prudent to take reasonable steps to prevent endangered species from becoming extinct.

Extinction is the conclusion of a long, gradual process typically involving a considerable span of time. When a species undergoes a drastic reduction in the extent of its range, accompanied by a reduction in the number of individuals, it may be designated as a rare species. As this trend continues, the species is likely to be considered threatened prior to being recognized as endangered.

FACTORS CONTRIBUTING TO SPECIES LOSS

Whether because of their intrinsic nature or environmental conditions, some species are naturally more predisposed to becoming endangered or extinct than others. As one would expect, species with smaller numbers of individuals are more vulnerable than those with more. Each species has a critical population size. Once the numbers fall below that size, the species is especially subject to extinction. Natural populations undergo year-to-year fluctuations in numbers; therefore, a small population will "crash" more readily than a large one.

Endangered and Threatened Species, 2008

	MAM-MALS	BIRDS	REP-TILES	AMPHIB-IANS	FISHES	SNAILS	CLAMS	CRUSTA-CEANS	INSECTS	ARACH-NIDS	PLANTS
Total listings	**357**	**275**	**119**	**32**	**151**	**76**	**72**	**22**	**61**	**12**	**747**
Endangered species	325	254	79	21	85	65	64	19	51	12	599
United States	69	75	13	13	74	64	62	19	47	12	598
Other countries	256	179	66	8	11	1	2	—	4	—	1
Threatened species	32	21	40	11	66	11	8	3	10	—	148
United States	12	15	24	10	65	11	8	3	10	—	146
Other countries	20	6	16	1	1	—	—	—	—	—	2

Source: Data from U.S. Department of Commerce, *Statistical Abstract of the United States, 2009,* 2009.

Note: Numbers reflect species listed by U.S. government as "threatened" or "endangered"; actual worldwide totals of species that could be considered threatened or endangered are unknown but are believed to be higher.

Several categories of animal and plant species are at high risk of becoming endangered or extinct. Among these are species restricted to special habitats. Most such animal or plant species, by becoming tolerant of an unusual situation, lose their ability to compete in a more general one. One example is island species: If threatened by humans, predators, competing exotic species, or diseases, native island species cannot easily escape. A disproportionate number of animals native to islands have become extinct. Large species with low reproductive rates are also at risk. Large species require more space than do smaller species; therefore, the number of large specimens occupying a given area is lower than the number of smaller ones. Also, most large species, whether whales or trees, are likely to reproduce less often than smaller ones. Even when large species are protected, it is difficult for them to increase their numbers.

Neotropical migratory birds such as warblers, orioles, and tanagers winter in tropical Central or South America or the Caribbean and breed in eastern North America. Their migratory pattern is advantageous in that they can take advantage of the availability of summer food in the north while escaping harsh conditions in winter. Migration, however, is a process that is fraught with danger. As the tropical forests in which they spend the winter are destroyed and the temperate forests in which they breed are fragmented, neotropical migrants may be threatened; thus, they are subject to double jeopardy.

Among the other at-risk species are those at the end of long food chains. Animals such as hawks, owls, and various cat species suffer when any of the links in their food chain are affected. Also, they may be more subject to damage by toxic substances such as environmentally persistent pesticides because of chemical amplification along the food chain. Finally, species of economic value are also in a precarious situation. Many animals have been hunted to extinction; an often-cited example is the passenger pigeon. Plants used medicinally, such as ginseng, have been subjected to overcollecting. Regulatory protections are in place to control the harvesting of American ginseng, which has been dug in eastern North America for centuries.

CONSERVATION AND MANAGEMENT

In order to preserve biodiversity and not lose species that are important to the health and existence of an ecosystem, wildlife conservation and management practices must be put into effect. There are three basic approaches to wildlife conservation and management: the species approach, the ecosystem approach, and the wildlife management approach.

The species approach involves giving endangered species legal protection, protecting and managing their habitats, propagating species in captivity, and reintroducing species into safe habitats. In 1903 President Theodore Roosevelt established the first wildlife refuge in the United States. The refuge, located on Pelican Island on the east coast of Florida, was developed to protect the brown pelican, which was in decline. (Only in 2009 was the species officially declared to be out of danger.) Since then the National Wildlife Refuge System has grown to more than 550 refuges and other units. Habitats in the United States are also protected through the national park and forest systems and the National Wilderness Preservation System. In addition to the government, private conservation organizations such as the National Audubon Society, the Sierra Club, and the Nature Conservancy have been of tremendous value in acquiring and protecting sensitive landscapes.

According to statistics reported by the International Union for Conservation of Nature and Natural Resources (IUCN), as of 2000 about thirty thousand protected areas had been established around the world. These areas, which include strict nature reserves and wilderness areas, national parks, natural monuments, habitat and species management areas, protected landscapes and seascapes, and managed resource protected areas, represent more than 13,250,000 square kilometers (5,115,854 square miles) of the planet's land surface.

Other forms of the species approach to saving diversity include gene banks, botanical gardens, and zoos. The seeds of many endangered plant species are preserved in climatically controlled environments. The organization Botanic Gardens Conservation International estimates that more than eighty thousand plant species are in cultivation in the world's botanic gardens. Many botanical gardens, such as Kew Gardens in England, are repositories for plant species that are endangered or have even ceased to exist in the wild. Some of these plants are reintroduced into native habitats after being cultivated for decades in these gardens or in seed banks.

Egg pulling and captive breeding are two methods that zoos and animal research centers use for preserving endangered animal species. Egg pulling involves collecting eggs from endangered species in the wild and hatching the eggs in zoos or research centers, as was done with California condors beginning in 1983. Endangered species still in the wild are sometimes captured and put into research centers to breed in a controlled environment. When the captive populations become large enough, some of the individuals are reintroduced into protected habitats. The Arabian oryx, a large antelope species that originated in the Middle East, was hunted to extinction in the wild; however, the species survived thanks to captive breeding programs that began in San Diego, Los Angeles, and Phoenix zoos. The oryx has since been reintroduced into its native habitats, although the captive population outnumbers the population in the wild.

The second approach to saving biodiversity is the ecosystem approach, which emphasizes preserving balanced populations of species within their native habitats. It involves establishing legally protected wilderness areas and wildlife reserves. An important part of making sure that the habitat is safe is to eliminate all alien species from the area. The Minnesota Zoo has formed a partnership with other organizations to help protect certain animals in their native habitats, notably the desert black rhino and Hartmann's mountain zebra. Instead of moving these animals to Minnesota, the zoo supports conservation efforts to study and protect these animals in their native habitats in Namibia, Africa.

The third approach to preserving biodiversity is the wildlife management approach. When it is decided which species or group of species will be managed in a given area, a management plan is put into effect. Steps in the plan include investigating and determining the kinds of cover, food, water, and space the targeted species requires. Action is then taken to grow the plants that provide the needed cover and food for the species.

HUNTING AND INTERNATIONAL COOPERATION

Legal and illegal commercial hunting has led to the extinction or near extinction of many animal species. Despite policies to regulate hunting, poaching remains a lucrative business, particularly in underdeveloped countries. Some threatened and endangered species are killed for their hides, horns, or other ornamental or medicinal parts, while others are captured and smuggled alive, as there is a market for exotic pets and decorative plants.

Species, primarily game species, are managed through the establishment of laws that regulate hunting and hunting quotas. Hunters are required to have licenses and to use only certain types of hunting equipment and are permitted to hunt only during certain months of the year. Limits are set on the size, number, and sex of animals that can be hunted in a given game refuge.

Management plans and international treaties have been developed to protect migrating game species, such as waterfowl. In North America, waterfowl such as ducks, geese, and swans nest in Canada during the summer and migrate to the United States and Central America in the fall and winter. The United States, Canada, and Mexico have signed agreements to protect these waterfowl from overhunting and habitat destruction.

Some waterfowl refuges in the United States include human-built nesting sites, ponds, and nesting islands. The U.S. National Wildlife Refuge System includes thirty-eight wetland management districts administering more than twenty-six thousand waterfowl production areas, which contribute to the protection of migratory birds. In 1986, amid concerns regarding record lows in waterfowl populations, the United States and Canada entered into an agreement (later joined by Mexico) to attempt to restore the continental waterfowl population and associated habitats. By the end of 2009, $4.5 billion had been invested to protect, restore, and enhance 6.3 million hectares (15.7 million acres) of waterfowl habitat in North America.

In July, 1975, a wide-reaching treaty to protect endangered species, the Convention on International Trade in Endangered Species of Wild Fauna and Flora (CITES) went into force. This agreement, which by 2010 had been agreed to by 175 national governments, extends varying protections to more than thirty thousand plant and animal species. Under CITES, some endangered and threatened species cannot be commercially traded, either alive or as products. Others can be traded, but only by persons who obtain the proper export licenses. One of the best-known results of the CITES agreement is the 1989 ban (subsequently weakened) on the international trade in ivory. The ban was enacted to halt the decline of the African elephant, which had dwindled in population from 2.5 million animals in 1950 to approximately 350,000 at the treaty's inception.

In 1980 the World Wildlife Fund (now the World Wide Fund for Nature), the United Nations Environment Programme, and IUCN developed a world conservation strategy. The plan, which was expanded in 1991, seeks to preserve biological diversity, combine wildlife conservation and sustainable development, encourage rehabilitation of degraded ecosystems, and monitor sustainability of ecosystems. Fifty countries have established national conservation programs in response to this plan.

U.S. Laws

Important U.S. laws that control imports and exports of endangered wildlife and wildlife products began with the Lacey Act of 1900, passed in response to an egret population decline resulting from the commercial value of their feathers as decoration. The Lacey Act prohibited transporting live or dead wild animals or their parts across state borders without a federal permit. Later came the Endangered Species Preservation Act of 1966, then the Endangered Species Act (ESA) of 1973, which has been amended several times. The ESA was unique in that, where previous wildlife regulations had focused primarily on game animals, the ESA program focused on identification of all endangered species and populations in order to save biodiversity, regardless of the species' usefulness to humans. The act classifies endangered species as those that are in immediate danger of extinction and threatened species as those that are likely to become endangered in a given habitat in the future. Some species are classified as being locally threatened even though they can be found in fairly large numbers in some parts of their former habitats. The U.S. Fish and Wildlife Service is required to prepare a recovery plan for each species that the ESA lists as officially endangered. In 2010 the ESA listings for endangered and threatened U.S. species included 578 animals and 793 plants.

The ESA provides that a listed species cannot be harassed, harmed, pursued, hunted, shot, trapped, killed, captured, or collected, either on purpose or by accident. It further prohibits importing or exporting endangered species, as well as possessing, selling, transporting, or offering to sell any endangered species. Violators face fines and imprisonment. In 1995 the U.S. Supreme Court ruled to extend further protection for endangered species by ruling that habitat essential for species survival must be protected, whether on public or private land.

Critics view the Endangered Species Act as a major stumbling block to economic progress. A classic example is the delay that occurred in the late 1970's in the construction of Tellico Dam in the mountains of eastern Tennessee because of the presence of the snail darter, a small endangered fish. The dam was ultimately completed, and other populations of the fish were unexpectedly found elsewhere, resulting in the snail darter's being removed from the endangered list. Some people who criticize the ESA assert that business and public interests should take precedence over the protection of wildlife, especially plants and animals of no obvious value to human beings.

Some of the ESA's recovery plans have proved successful. Bald eagles, which numbered only 800 in 1970, were able to rebound to a population of almost 9,800 by 2006, largely because of the U.S. ban on the pesticide dichloro-diphenyl-trichloroethane (DDT). The American alligator was listed as an endangered species in 1967 after its population declined because of habitat destruction and high demand for alligator meat and products made from alligator hides. Because of ESA protection, the alligator was reestablished in its southern range; it was removed from the endangered list in 1987. Most of the ESA success stories have involved such "charismatic megafauna." Less glamorous endangered species, such as fungi, wildflowers, liverworts, mosses, and insects, receive less attention even though their roles in ecosystems may be more important. In spite of ESA protection, a number of species remain critically endangered, largely because their standing was so precarious by the time they were listed as protected. ESA listed status has not saved some species from going extinct.

OTHER PROTECTIVE MEASURES

The loss of aquatic species has generally attracted less attention than the extinction of land species. With the realization of the importance of healthy freshwater and marine species, however, governments have begun establishing marine preserves. Fishing, construction, tourism, pollution, and other human disturbances are closely regulated and restricted in these areas. The National Marine Sanctuary Program, which was developed in 1972 in the United States, has established fourteen marine protected areas. The most extensive of these, the Papahānaumokuākea Marine National Monument in the northwest Hawaiian Islands, is the single largest conservation area in the United States and the world's largest fully protected marine area. It includes both marine and terrestrial habitats.

International measures to protect species from destruction or exploitation include the 1946 International Convention for the Regulation of Whaling. Overwhaling worldwide caused a huge decline in whales, from an estimated 4.4 million in 1900 to approximately 1 million by the end of the twentieth century. Overharvesting of whales affected almost every whale species of commercial value. In 1946 the International Whaling Commission (IWC) was established to set annual whaling quotas to prevent commercial overharvesting and the extinction of whales. However, many whaling countries ignored the suggested quotas. In 1971 the United States stopped all commercial whaling and banned imports of all whale products. In 1974 the IWC began to regulate whaling ac-

cording to the principle of maximum sustainable yield. When a given species of whale—such as the right whale, bowhead whale, or blue whale—fell below the optimal population for such a yield, the IWC issued a ban on hunting that species.

Other international agreements to protect endangered species include the Convention on the Conservation of Migratory Species of Wild Animals (Bonn Convention), which entered into force in 1983, and the Convention on Biological Diversity (CBD), which entered into force in 1993. The Bonn Convention is the only international treaty that focuses on the conservation of terrestrial, marine, and avian migratory species, their habitats, and their migration routes. The CBD, which opened for signature at the 1992 Earth Summit in Rio de Janeiro, Brazil, concerns the significance of biodiversity for future generations, the sovereignty of each nation over its resources, and each nation's need and right to conserve and protect its own biodiversity. The treaty details a plan that directs industrialized countries to help fund projects for the protection of biodiversity within developing countries; it also stresses the right of national governments to decide who may have access to their resources. The treaty provides for a sharing of technologies, particularly biotechnologies that have been developed from plants originating in developing countries, thereby giving developing countries substantial benefits from any technologies that are developed based on theses countries' genetic resources. Agenda 21, a comprehensive international plan of action adopted at the Earth Summit, addresses the need for nations to "promote the rehabilitation and restoration of damaged ecosystems and the recovery of threatened and endangered species."

Thomas E. Hemmerly, Toby Stewart, and Dion Stewart
Updated by Karen N. Kähler

FURTHER READING

Bräutigam, Amie, and Martin Jenkins. *The Red Book: The Extinction Crisis Face to Face.* Gland, Switzerland: IUCN, 2001.

Chiras, Daniel D. "Preserving Biological Diversity." In *Environmental Science.* 8th ed. Sudbury, Mass.: Jones and Bartlett, 2010.

Groom, Martha J., Gary K. Meffe, and Carl Ronald Carroll. *Principles of Conservation Biology.* 3d ed. Sunderland, Mass.: Sinauer, 2006.

Mackay, Richard. *The Atlas of Endangered Species.* Rev. 3d ed. Sterling, Va.: Earthscan, 2009.

McNeely, Jeffrey A., et al. *Conserving the World's Biological Diversity.* Washington, D.C.: Island Press, 1990.

Vié, Jean-Christophe, Craig Hilton-Taylor, and Simon N. Stuart, eds. *Wildlife in a Changing World: An Analysis of the 2008 IUCN Red List of Threatened Species.* Gland, Switzerland: IUCN, 2009.

Wagner, Viqi, ed. *Endangered Species: Opposing Viewpoints.* Detroit: Greenhaven Press, 2008.

SEE ALSO: American alligator; Biodiversity; Captive breeding; Convention on Biological Diversity; Convention on International Trade in Endangered Species; Endangered Species Act; Extinctions and species loss; Habitat destruction; Hunting; International Convention for the Regulation of Whaling; International Union for Conservation of Nature.

Energy conservation

CATEGORIES: Energy and energy use; resources and resource management

DEFINITION: Avoidance of wasteful use of energy through decreased consumption or improved efficiency of technology or services

SIGNIFICANCE: The generation of energy often involves the use of nonrenewable resources, such as fossil fuels, the burning of which contributes to environmental degradation. Energy conservation helps to reduce the negative environmental impacts of energy generation and preserves nonrenewable resources for the future.

Many common daily practices require some use of energy resources, including fuels used to generate electricity, fuels burned for heat, and fuels burned to power vehicles. As populations around the world continue to grow and developing countries become increasingly industrialized, the amounts of energy used are also increasing. Environmentalists assert that energy conservation is a necessary approach to addressing the problems of limited natural resources, global pollution, and climate change. Conserving energy can involve both changing behaviors to reduce the amount of energy used and changing technologies or services to alternatives that require less energy to begin with.

NONRENEWABLE AND RENEWABLE RESOURCES

Energy can be derived from renewable resources, which can easily replenish or renew themselves over time, or from nonrenewable resources, the supplies of which are limited and cannot be replenished. The majority of the energy used in the world comes from such nonrenewable resources as petroleum, natural gas, coal, and uranium. These sources are fixed in supply, and burning them contributes to the presence of greenhouse gases in the atmosphere, which has been linked to global climate change. Increased awareness that the supplies of nonrenewable energy sources are limited and that their use has negative environmental impacts has led governments, environmental advocates, and others to promote a movement toward energy conservation and energy efficiency, including increasing the use of renewable and nonpolluting resources. Governments, energy utility companies, and others offer financial incentives for the reduction of energy consumption in the forms of rebates and tax credits.

The use of energy derived from renewable resources, such as hydropower, solar power, wind power, and geothermal power, does not generally contribute to global pollution and climate change. The burning of some kinds of fuels derived from renewable sources, such as biomass, can contribute to pollution, however. Among the drawbacks to the use of some

Top Energy-Consuming Countries, 2005

	TOTAL (QUADRILLION BTUS)	PER CAPITA (MILLION BTUS)
Canada	14.3	436.2
China	67.1	51.4
France	11.4	181.5
Germany	14.5	176.0
India	16.2	14.8
Japan	22.6	177.0
Russia	30.3	212.2
United Kingdom	10.0	165.7
United States	100.7	340.0
World	462.8	71.8

Source: Data from U.S. Energy Information Administration, *International Energy Annual, 2005.*

Note: Values are in British thermal units (Btus). Totals are in quadrillions; hence U.S. consumption of 100.7 is 100,700,000,000,000,000. Per capita consumption is in millions; hence U.S. per capita consumption of 340.0 is 340,000,000.

kinds of renewable energy sources are habitat destruction and noise pollution; in addition, these sources can be subject to variability in the amount of power they can generate.

BEHAVIORAL CHANGES AND ENERGY RETROFITS

Individuals, particularly in industrialized nations, can take many different steps to reduce the amounts of energy they use both in their homes and in their workplaces. In addition to the positive environmental impacts of energy conservation, the practice can help home owners and building managers to reduce utility costs. One simple way to practice energy conservation is by reducing the use of household appliances and other devices that run on electricity or that heat water or air using gas; these include chargers for electronic devices, dishwashers, washing machines and dryers, and water heaters. Behaviors such as turning off lights, unplugging appliances when not in use, taking shorter showers, and reducing the use of air-conditioning or heating can have dramatic effects in energy savings.

Modifying homes and other buildings with energy-efficiency retrofits can also help to conserve energy. Such retrofits include improved insulation and the installation of energy-efficient windows and doors to reduce the amount of energy needed for heating and cooling. The use of water heaters can be reduced with the installation of low-flow showerheads and faucets as well as high-efficiency dishwashers and washing machines.

Courtney A. Smith

FURTHER READING

Chiras, Daniel D. "Foundations of a Sustainable Energy System: Conservation and Renewable Energy." In *Environmental Science.* 8th ed. Sudbury, Mass.: Jones and Bartlett, 2010.

Elliott, David. *Energy, Society, and Environment.* 2d ed. New York: Routledge, 2003.

Horn, Miriam, and Fredd Krupp. *Earth: The Sequel— The Race to Reinvent Energy and Stop Global Warming.* New York: W. W. Norton, 2009.

Krigger, John, and Chris Dorsi. *The Homeowner's Handbook to Energy Efficiency: A Guide to Big and Small Improvements.* Helena, Mont.: Saturn Resource Management, 2008.

Sawhill, John C., and Richard Cotton, eds. *Energy Conservation: Successes and Failures.* Washington, D.C.: Brookings Institution Press, 1986.

SEE ALSO: Air-conditioning; Air pollution; Alternative energy sources; Alternative fuels; Compact fluorescent lightbulbs; Conservation movement; Conservation policy; Geothermal energy; Renewable energy; Solar energy; Wind energy.

Energy-efficiency labeling

CATEGORY: Energy and energy use

DEFINITION: Consumer product labeling systems that provide information on the amounts of energy used or conserved by particular products

SIGNIFICANCE: Energy-efficiency labeling enables consumers to take energy consumption into account when they make purchases, allowing them to see how an energy-efficient option might be more expensive in the short run but less expensive in the long run because of reduced energy consumption over time.

Energy rating systems were created in an effort to encourage the conservation of energy and thereby minimize the pollution created and resources consumed when fuel is burned for power. Primary energy sources include coal, petroleum and petroleum products, natural gas, water, uranium, wind, sunlight, and geothermal energy. Coal, petroleum and petroleum products, and natural gas are fossil fuels, which are considered nonrenewable energy sources because what is consumed takes millions of years to replace. Fossil fuels are burned to power motor vehicles, heat homes, and generate electrical power.

In the United States, every major home appliance sold carries an EnergyGuide label. A requirement of the Federal Trade Commission, this label shows how much it will cost the consumer to use the appliance for one year. The estimated cost is based on an average cost per kilowatt-hour in the appliance retailer's local area. (Electrical power is measured and billed by kilowatt-hours. A watt-hour is a unit of energy supplied steadily through an electric circuit for 1 hour; 1 kilowatt-hour is equal to 1,000 watt-hours.) One purpose of EnergyGuide labeling is to encourage manufacturers to make products that use the lowest amounts of electricity possible or that use electricity as efficiently as possible. EnergyGuide and similar labels allow consumers to compare the operation costs of appliances, so that they can make informed choices;

although a particularly energy-efficient appliance may cost more than one that is less efficient, consumers may be more likely to buy the product that saves energy because of its lower operating costs over time. Using less electricity not only saves the consumer money but also helps to reduce pollution and slow the depletion of natural resources.

Energy-efficiency labeling that appears on electronic equipment shows how energy-efficient computers and other devices consume less electricity, even when turned off, than standard equipment. (When plugged into electrical outlets, many electronic devices and other appliances draw standby power even when turned off. Audiovisual equipment in standby mode accounts for about 60 percent of a home's "leaking electricity.") Energy-efficiency labeling systems identify equipment that goes into "sleep" mode, or draws less energy, when not in use.

Energy-efficiency ratings systems have expanded to encompass many kinds of equipment used in homes, including air conditioners and windows. Newly built homes are often rated on their overall energy efficiency. The amount of energy needed to heat or cool a building depends in large part on how well insulated the building is; the windows and the quality of their installation can also be a big factor, as poorly installed or poor-quality windows can allow heated or cooled air to escape or allow hot or cold outside air to enter the building. Many of the products used to improve the energy efficiency of existing buildings, such as windows and insulation, receive energy-efficiency ratings.

In 1992, the year of the United Nations Earth Summit, a number of energy-efficiency labeling programs were launched in an effort to lower the amount of electrical power consumed and the amount of pollution generated by electricity plants that burn fossil fuels. These programs include Energy Star, TCO certification, and the EU Energy Label. In the United States, the Environmental Protection Agency and the Department of Energy jointly established the Energy Star program, and Energy Star labeling subsequently became an international standard employed in Canada, the European Union, Australia, New Zealand, Japan, and Taiwan. It helps consumers make energy-efficient choices in household appliances, new homes, and home-improvement projects. The program claims to have saved consumers in the United States $17 bil-

Logo of the Energy Star program, one of the most widely known energy-efficiency labeling programs in the United States. (Courtesy, EPA)

lion on their utility bills and 30 million cars' worth of greenhouse gas emissions in 2009 alone.

The Swedish Confederation of Professional Employees, a national trade union center, created TCO certification. This standard, which applies to computer monitors and other office equipment marketed in the European Union, takes into account ergonomics, emissions, and recyclability as well as energy consumption. The European Union itself launched the EU Energy Label, which covers a wide range of energy-consuming devices, from lightbulbs to major home appliances and automobiles. In 1998, China established its Certification Center for Energy Conservation Products, which issues a voluntary energy-efficiency endorsement label for major home appliances and consumer electronics.

Lisa A. Wroble
Updated by Karen N. Kähler

FURTHER READING

International Energy Agency and Organisation for Economic Co-operation and Development. *Cool Appliances: Policy Strategies for Energy-Efficient Homes.* Paris: OECD, 2003.

_____. *Energy Labels and Standards: Energy Efficiency Policy Profiles.* Paris: OECD, 2000.

McLean-Conner, Penni. *Energy Efficiency: Principles and Practices.* Tulsa, Okla.: PennWell, 2009.

U.S. Environmental Protection Agency. *Energy Star and Other Climate Protection Programs: 2008 Annual Report.* Washington, D.C.: Author, 2009.

Wiel, Stephen, and James E. McMahon. *Energy-Efficiency Labels and Standards: A Guidebook for Appliances, Equipment, and Lighting.* 2d ed. Washington, D.C.: Collaborative Labeling and Appliance Standards Program, 2005.

Williams, Wendy. *Eco-Labelling Technology: For You and the Planet.* Stockholm: TCO Development, 2008.

SEE ALSO: Alternative energy sources; Carbon footprint; Compact fluorescent lightbulbs; Department of Energy, U.S.; Energy conservation; Environmental Protection Agency; Fossil fuels; Green buildings; Green marketing.

Energy policy

CATEGORY: Energy and energy use

DEFINITION: High-level governmental plan of action pertaining to issues of energy supply, demand, and utilization

SIGNIFICANCE: Until the 1970's the United States made no concerted effort to develop a comprehensive national energy policy. After the 1973 oil crisis created a new view of energy as a distinct and important policy arena, successive presidential administrations undertook various initiatives, and Congress enacted several broad laws addressing aspects of energy production and consumption. No enduring, coherent national policy has resulted, however.

Energy production and consumption are strongly linked to a nation's environmental quality, economic well-being, and security. Energy use is one of the greatest sources of environmental degradation. Acid rain, along with most other ecosystem and health-related air pollution, originates with power plant and vehicle fossil-fuel combustion, as do carbon dioxide emissions, which are the most important contributors to global warming. Energy extraction, transportation, and utilization facilities create land-use and conservation impacts and also use and pollute water. Nuclear power plants generate thousands of tons of long-lived radioactive wastes.

Historically, economic growth has been accompanied by rising energy use. Beginning during the mid-1970's, however, experience in the industrialized nations demonstrated that the two factors were not necessarily tightly coupled and that using energy more efficiently could significantly slow energy growth rates while still permitting vigorous economic expansion. For less developed nations, the connection remains firmer until the transition has been made to an industrialized economy.

The national security implications of energy have become increasingly important as more countries have been forced to look beyond their borders for necessary energy supplies. U.S. reliance on imported oil was first underscored by the 1973 oil embargo imposed by the Organization of Petroleum Exporting Countries (OPEC) in response to U.S. support for Israel in the 1973 Arab-Israeli War. It was soon highlighted twice more: by the 1979-1980 "second oil shock" precipitated by the Iranian revolution and unexpected collapse of Iran's substantial oil production, and by the 1990 Iraqi invasion of Kuwait—a major oil producer—and perceived threat to the vast oil fields of Saudi Arabia, which led to the Persian Gulf War. In the early twenty-first century, the wars in Afghanistan and Iraq affected the production and price of oil.

EARLY HISTORY

Before the 1970's no serious effort had been undertaken in the United States to craft a national energy policy. The principal reason was that despite transient imbalances, cheap and abundant energy supplies were readily available. Unrelated policy initiatives focused on individual fuels and industries, driven largely by the different characteristics of each and by a desire to keep energy prices low and supplies ample.

Coal was the dominant fuel in the United States through the 1940's, having eclipsed fuelwood during the mid-1880's. Because the coal industry was large, dispersed, and competitive, there was little incentive for federal regulation. However, coal's share of total American energy consumption began to decline as oil and gas production rose after World War I.

Oil's share of total energy increased particularly rapidly. By 1950 petroleum overtook coal as the largest U.S. energy source. A complex system of tax subsidies, import quotas, and other mechanisms arose to protect the domestic oil industry. Natural gas use also increased sharply after World War II as a consequence of gradual improvements in pipeline technology, wartime pipeline construction subsidies, and a rise in demand during the war. In 1938, the Federal Power Commission was given authority to regulate aspects of interstate commerce in gas; states had exercised some regulatory authority since the late nineteenth century. Federal price regulation continued to be a thorny issue up through the mid-1980's, when most gas prices were finally deregulated.

During the early 1950's the federal government began to push strongly for the development of commercial nuclear power and provided large research, development, and demonstration (RD&D) subsidies in a unique effort to stimulate the nuclear industry. No other fuel or technology received such promotional assistance from government. Nuclear power is also unique in that most state regulation is preempted by federal law. Despite government subsidies, the development of nuclear power stalled in the marketplace during the 1970's for reasons related to costs, safety concerns, public resistance, and an intractable radioactive waste disposal problem. The U.S. Nuclear Regulatory Commission issued no construction permits for new nuclear power plants between 1973 and the end of the century. The first reactor construction application of the twenty-first century was not submitted until 2007.

Meanwhile, the electric utility industry continued to grow, with electricity consumption rising by more than 1,100 percent during the latter half of the twentieth century. This led to an increase in total fuel consumption, because when fuels are burned to generate electricity (as opposed to being used directly, such as when natural gas is used for home heating), two of every three units of fuel are unavoidably lost as waste heat during the conversion process. The monopoly market power of electric utilities was recognized early, and both federal and state regulation was imposed. During the 1990's political pressure mounted to deregulate the utility industry and open up electricity generation to competition. By 2000 twenty-four U.S. states had passed deregulation laws; over the next decade, however, deregulation-related problems with price and supply led eight of these states to postpone, suspend, or repeal deregulation.

Prior to the 1970's technologies and initiatives aimed at increasing the efficiency of energy utilization (often popularly referred to as "energy conservation") and developing renewable energy sources such as solar, wind, geothermal, and biomass energy re-

ceived little attention and no significant federal funding. Conservation and renewables began to receive more attention and funding in the wake of the 1973 energy crisis.

1973 OIL EMBARGO

The 1973 oil embargo drove up oil prices in the United States and, combined with other complex factors, created localized shortages. Suddenly energy, never before regarded as a distinct federal policy arena, was thrust to the top rank of the policy agenda in the United States. The elements of national policy remained deeply embedded in the unique industries, arrangements, and regulatory regimes that had co-evolved with the three fossil fuels (oil, natural gas, and coal), nuclear power, hydroelectricity, and electric utilities, but for the first time there was also strong interest in promoting technologies to utilize renewable energy sources and increase the efficiency of energy use through a combination of federal regulation, research funds, tax subsidies, and public education. Federal RD&D funds for these alternative energy sources rose sharply during the 1970's in response to concerns regarding the energy crisis.

Between 1973 and 1983 several comprehensive, groundbreaking energy policy studies were produced by groups in the private, nonprofit, and government sectors. These served to focus government and public attention on the importance of the issue. Not surprisingly, the sudden rise in concern about energy came during the same period that the modern environmental movement arose. Beginning during the late 1960's, a growing recognition of the connection between energy and environmental issues was accompanied by frequent conflicts between energy and environmental policy goals. For example, the National Environmental Policy Act of 1969 affected many energy projects and industries, while new laws regarding coal mining and the 1970 amendments to the Clean Air Act tended to discourage coal combustion. During the early 1970's a wide assortment of new energy-related initiatives were adopted, but many—such as oil price controls, efforts to jump-start a domestic synthetic fuels industry, and a nuclear breeder reactor program—were soon abandoned.

Funding for energy RD&D peaked in 1980, only to drop dramatically during President Ronald Reagan's administration (1981-1989), which opposed support for them. Market forces were trusted to resolve imbalances in energy supply and demand, and in those

Cold War years weapons development and construction were prioritized over energy research. Energy RD&D spending was increased during President George H. W. Bush's administration (1989-1993), and renewables and energy efficiency received particular attention in the early years of President Bill Clinton's administration (1993-2001). Cost-sharing RD&D programs involving public-private partnerships became more common, and in 1999 government spending on energy research reached a twenty-six-year low. Midway through George W. Bush's presidency (2001-2009) the total government RD&D investment in energy was only 2 percent, compared to 10 percent in 1980. The year 2007 saw an increase in funding for basic energy sciences. In 2009, during the first year of President Barack Obama's administration, funding was provided for the Advanced Research Projects Agency-Energy (ARPA-E), created for potential paradigm-shifting research. The 2011 national budget decreased research funding for fossil-fuel and nuclear energy projects while increasing RD&D moneys for projects concerned with energy efficiency, renewables, and the country's electricity distribution grid. With that fiscal year, energy became the third-largest recipient of federal RD&D funding, behind defense and health.

FEDERAL ADMINISTRATIVE STRUCTURE

Until the 1970's responsibilities for energy matters were scattered among various federal agencies, and there was little coordination from either a regulatory or a policy perspective. After the 1973 oil embargo, efforts were made to improve the situation, but there was no encompassing centralization of function until the Department of Energy (DOE) was established in 1977. The creation of this federal department reflected the elevation of energy at the national policy level and marked the first time that energy had been afforded cabinet-rank status. Even then, several regulatory commissions remained independent, and some other federal departments retained important responsibilities.

At the state level, public-service commissions exercised long-established jurisdiction over electric utility rates, and other state agencies controlled land-use decisions that could affect power plant construction. Many states established energy agencies, and some began to formulate explicit energy policies and fund energy research.

Following the 1973 oil embargo several administrations drafted national energy plans. The first, Project

Independence, was launched during President Richard Nixon's administration (1969-1974). It proposed national energy self-sufficiency by the 1980's—a goal soon recognized as politically attractive but unrealistic. President Jimmy Carter's administration (1977-1981) produced the 1977 National Energy Plan, which emphasized short-term strategies to reduce oil imports, reduce total energy demand through increased energy efficiency, raise coal production, and increase the use of renewable energy. Carter's proposals regarding renewable energy sources and increased efficiency marked the first time that the federal government had proposed a serious, tangible commitment to these alternatives.

During the Reagan administration no comprehensive energy policy document was produced, although the administration strongly favored fossil and nuclear fuels and opposed funding for energy efficiency and renewables. George H. W. Bush's administration put forth a 1991 National Energy Strategy that emphasized oil, gas, and nuclear power production but offered little support for energy efficiency and renewables. The response of the environmental community was sharply critical, but many of the proposals were included in legislation enacted the following year. The Clinton administration followed with energy plans in 1994 and 1998, as well as the Partnership for a New Generation of Vehicles, launched in 1993, and the Climate Change Technology Initiative, begun in 1998. However, an opposition-led Congress was generally unreceptive, the administration did not place high priority on its energy proposals, and no major new legislation or initiatives were adopted during the Clinton years.

FEDERAL LEGISLATION

Federal legislation sometimes marks national recognition of the importance of a hitherto ignored policy area. Prior to the 1970's, energy legislation was directed at individual fuels and energy sectors. During the 1970's, however, Congress began adopting broad-scope energy laws that, for the first time, addressed disparate energy matters in single pieces of legislation. Some contained new research and regulatory initiatives and addressed issues or technologies that had never before received serious federal attention or support. Among the most important were the Energy Reorganization Act of 1974, which established the short-lived Energy Research and Development Administration (superseded in 1977 by the DOE) and

Carter's Energy Address to the Nation

Our Nation's energy problem is very serious—and it's getting worse. We're wasting too much energy, we're buying far too much oil from foreign countries, and we are not producing enough oil, gas, or coal in the United States.

In order to control energy price, production, and distribution, the Federal bureaucracy and redtape have become so complicated, it is almost unbelievable. Energy prices are high, and they're going higher, no matter what we do.

The use of coal and solar energy, which are in plentiful supply, is lagging far behind our great potential. The recent accident at the Three Mile Island nuclear power plant in Pennsylvania has demonstrated dramatically that we have other energy problems. . . .

Federal Government price controls now hold down our own production, and they encourage waste and increasing dependence on foreign oil. Present law requires that these Federal Government controls on oil be removed by September 1981, and the law gives me the authority at the end of next month to carry out this decontrol process.

In order to minimize sudden economic shock, I've decided that phased decontrol of oil prices will begin on June 1 and continue at a fairly uniform rate over the next 28 months. The immediate effect of this action will be to increase production of oil and gas in our own country.

As Government controls end, prices will go up on oil which has already been discovered, and unless we tax the oil companies, they will reap huge and undeserved windfall profits. We must, therefore, impose a windfall profits tax on the oil companies to capture part of this money for the American people. This tax money will go into an energy security fund and will be used to protect low income families from energy price increases, to build a more efficient mass transportation system, and to put American genius to work solving our long-range energy problems. . . .

We are dangerously dependent on uncertain and expensive sources of foreign oil. Since the 1973 embargo, oil production in the United States has actually dropped. Our imports have been growing. Just a few foreign countries control the amount of oil that's produced and the price that we must pay. . . .

This growing dependence has left us dangerously exposed to sudden price rises and interruptions in supply. In 1973 and 1974, shipment of oil was embargoed, and the price quadrupled almost overnight. In the last few months, the upheaval in Iran again cut world supplies of oil, and the OPEC cartel prices leaped up again. . . .

There is no single answer. We must produce more. We must conserve more. And now we must join together in a great national effort to use American technology to give us energy security in the years ahead.

The most effective action we can take to encourage both conservation and production here at home is to stop rewarding those who import foreign oil and to stop encouraging waste by holding the price of American oil down far below its replacement or its true value.

This is a painful step, and I'll give it to you straight: Each of us will have to use less oil and pay more for it. But this is a necessary step.

companion legislation that funded a program of energy RD&D that included a renewables and efficiency component. The 1975 Energy Policy and Conservation Act mandated automotive fuel-efficiency standards, authorized a variety of energy-efficiency programs and standards, and established the Strategic Petroleum Reserve as a hedge against future supply disruptions. The 1976 Energy Conservation and Production Act funded state conservation programs and authorized a federal program to weatherize low-income housing.

President Carter signed several pieces of legislation constituting the National Energy Act of 1978. These included the National Energy Conservation Policy Act, which expanded weatherization programs, authorized utility residential conservation programs and an energy grants program for schools and hospitals, and also authorized efficiency standards for household appliances. The 1978 Public Utilities Regulatory Policies Act (PURPA) paved the way for the tremendous expansion of nonutility, independent electric power producers that occurred during the deregulation of the 1990's. The 1978 Powerplant and Industrial Fuel Use Act (PIFUA) barred the use of oil or natural gas fuel in new electric generating plants or major industrial facilities (this act was repealed in 1987). To set an example for alternative energy use, President Carter had solar panels placed on the White House, but they were removed during Reagan's administration. Carter's energy policies are widely regarded as effective in reducing oil consumption and encouraging greater use of alternative sources of energy.

No major energy legislation was passed during the Reagan administration. Congress enacted the 1992 National Energy Policy Act at the end of the following Bush administration. The most wide-ranging legislation adopted since the late 1970's, the law included provisions to help independent power producers, notably by opening access to utilities' transmission systems to nonutility electricity generators. Other important changes included easing licensing requirements for new nuclear power plants and restricting public access to the process, initiatives intended to foster energy-efficiency improvements and renewable energy resources, and a variety of measures related to alternative fuels, energy-related taxes, coal development, and other matters. No major energy legislation was enacted during the Clinton administration.

The first significant energy legislation of the twenty-first century was the Energy Policy Act of 2005, which provided tax incentives for clean energy as well as subsidies and mandates for ethanol use. Other major energy legislation passed during the George W. Bush era includes the Energy Independence and Security Act of 2007, an omnibus energy policy law intended to increase energy efficiency and the availabil-

ity of renewable energy. Its provisions included raising standards for vehicle fuel efficiency, establishing energy-efficiency standards for lighting and commercial and residential appliances, and encouraging the energy efficiency of commercial, residential, and federal buildings. Some oil and gas tax incentives were repealed to offset the cost of the revised fuel economy standards. The economic stimulus bill of fall 2008 included federal tax incentives for energy efficiency and renewables.

A similar stimulus bill was passed in the winter of 2009 after Obama took office. Another piece of energy-related legislation that year was the Supplemental Appropriations Act, which included the so-called Cash for Clunkers program, which was intended to encourage U.S. residents to trade in less fuel-efficient vehicles for newer, cleaner-operating, more efficient models. The Obama administration also accelerated the timetable for implementing the new vehicle fuel-

At the end of his first week in office in 2009, President Barack Obama signs an executive order dealing with climate change and energy independence. Behind him are Transportation Secretary Ray LaHood and Environmental Protection Agency Administrator Lisa Jackson. (Chuck Kennedy/MCT/LANDOV)

efficiency standards set by the previous administration and added the goal of reducing tailpipe greenhouse gas emissions. In 2010 attempts to pass comprehensive energy and climate change legislation foundered for lack of bipartisan support.

U.S. CONSUMPTION TRENDS

As oil prices rose and fell after the events of 1973 and 1979 and other issues began to capture the public's attention, energy began to drop down the list of pressing policy concerns. The 1990 Iraqi invasion of Kuwait and the subsequent Persian Gulf War briefly raised oil prices, as well as the visibility of energy policy and energy security. Subsequently, however, friction among OPEC cartel members and increased oil production by non-OPEC countries resulted in unexpectedly abundant oil supplies and low oil prices mid-decade. After a few years' recovery, average real gasoline prices in the United States fell to record low levels in 1998. In response, public and governmental attention again shifted elsewhere. Periodic peaks such as the record-breaking high oil prices of 2008 did not shake petroleum from its place as the nation's dominant energy source. In 2009 petroleum accounted for approximately 35 percent of the total energy consumed in the United States and 94 percent of the transportation sector's total energy use. Transportation consumed 71 percent of all petroleum used in the nation.

U.S. oil production peaked in 1970; since 1993 imports have exceeded production. In 2009 the United States produced 5.31 million barrels of crude oil per day; by contrast, the nation's imports were 9.06 million barrels per day. Crude oil imports peaked in 2005 at 10.13 million barrels per day. Although non-OPEC countries have supplied more than half of U.S. imports since 1992, OPEC oil still accounted for almost 41 percent of those imports in 2009.

Total U.S. energy consumption remained remarkably stable between 1973 and 1990 in response to substantial increases in the efficiency of energy use. A key measure of energy efficiency, energy intensity (the amount of energy used per dollar of gross domestic product), declined by 28 percent during the period but leveled off after 1986, largely in response to falling energy prices. Energy consumption between 1992 and 2000 rose to record levels in each successive year as lower energy prices and vigorous economic growth led industries and consumers to give less consideration to energy efficiency in planning and purchases.

With the exceptions of 2001 and 2006, consumption continued to rise in the early twenty-first century, reaching an all-time high in 2007. Record high oil prices in 2008—double what they had been in 2007—caused a subsequent consumption decline before they subsided. In 2010 energy consumption in the United States was twice what it was in 1963 and 40 percent greater than it was during the low point in 1975 that followed the oil crisis.

For decades, energy experts have expressed concern about the economic, environmental, and national security implications of energy supply and consumption patterns in the United States. Since the 1990's environment-related concerns have been exacerbated by the growing international recognition of the probable impacts of global climate change. The ability of most of the world's governments to meet carbon emission targets set forth in the international climate agreements first made in the 1990's hinged on changes in energy technologies and consumption patterns that have yet to be implemented to a sufficient extent.

CONTINUING PROBLEMS

The United States still faces several fundamental energy policy problems. A heavy reliance on imported oil has continued for decades. No real effort has been made to stem rising transportation-sector oil consumption caused by increased total annual miles driven and the widespread use of sport utility vehicles, minivans, and light trucks as personal transportation, a trend that took hold during the 1990's when gasoline was cheap. It was not until model year 2005 that improved fuel-efficiency standards were required for light trucks, and heavier SUVs and vans were not subject to any fuel-efficiency standards before model year 2011.

Large quantities of coal are still used to produce electricity. Attempts to establish meaningful policies to curb fossil-fuel use and accompanying greenhouse gas emissions meet with legislative gridlock; meanwhile, the United States remains one of the world's top emitters of such gases. (Long the largest emitter, the United States was surpassed by China in 2006.) Another problem is the long-standing failure of the United States to include the substantial externalized social and environmental costs of energy extraction, utilization, and security in the price of energy.

Some observers argue that such problems are not serious and that aggressive responses to them would be premature and economically devastating. Environmentalists contend that the economic, social, and environmental costs of continuing the status quo will be far more serious than the economic costs of taking steps to mitigate these problems.

Several comprehensive studies by government and nonprofit groups have concluded that an aggressive national commitment to energy efficiency and renewable energy sources could, over time, result in the displacement of significant fractions of fossil and nuclear fuels and a sharp reduction in energy-consumption growth rates, with no severe economic penalty and sizable environmental benefits. For example, analyses by the DOE and others have found that energy savings in the United States resulting from appliance and equipment standards alone totaled approximately 1.2 quadrillion British thermal units in 2000, equivalent to about 1.3 percent of the nation's overall electricity use. This energy savings represented a $9 billion reduction in consumer energy bills. According to a 2002 report from the National Academy of Sciences, vehicle fuel economy standards saved the country 2.8 million barrels of oil per day in 2000, or 15 percent of oil consumption for that year.

Among the innovative energy policies favored by environmental advocates are the imposition of meaningful taxes on carbon fuels, including additional gasoline taxes, and the abolition of large and long-standing tax and research subsidies for the fossil-fuel and nuclear industries. Another strategy would involve removing market barriers and providing ample, stable RD&D funds and tax incentives for projects concerning energy efficiency and renewable energy, including alternative transportation fuels. Environmentalists also support stringent fuel economy standards for all vehicles; tighter energy-efficiency standards for appliances, motors, and buildings; and management of the deregulation of the electric utility industry in such a way that renewable energy and energy-efficiency investments, markets, and potential are protected. Environmentalists acknowledge that the strong political opposition to many of these measures makes it unlikely that effective change will be adopted without committed federal leadership and a shift in national sentiments about the importance of energy.

ENERGY POLICIES AROUND THE WORLD

Like the United States, most industrialized nations developed explicit energy policies after the oil price and supply dislocation that occurred in 1973. In every country, energy receives more or less policy attention depending on the availability and cost of domestic and imported fuels, competition from other pressing policy and social concerns, and the degree of industrial development.

For many years, European nations pursued markedly different energy strategies, but during the 1990's the increasing economic integration of Europe led to efforts to forge a common European Union (EU) energy policy. In 2008 EU leaders adopted a far-reaching comprehensive reform of European energy policy to ensure reliable energy supplies while reducing greenhouse gas emissions and becoming a world leader in low-carbon and renewable energy technologies. Japan developed its reliance on nuclear power because no significant domestic fossil-fuel resources were available to power the tremendous postwar growth of the Japanese economy. The country remains completely dependent on imports for its oil needs, and this lack of resources has led Japan to become a world leader in energy efficiency. Continued high consumption of coal in China, the United States, India, South Africa, Japan, and Russia has raised concerns about associated greenhouse gas emissions. China, while more focused on securing energy resources for the future than on curbing greenhouse gas emissions, has the distinction of being the world's top producer of renewable energy.

Most developing countries face pressing social and economic problems that eclipse energy and environmental considerations, even though the latter often aggravate the former. For example, the cost of imported energy—usually oil—constitutes a significant drain on foreign exchange for many developing nations. At the same time, scarce capital and technical expertise make it difficult for these nations to develop alternative energy resources, technologies, and infrastructure without foreign investment and assistance.

Phillip A. Greenberg
Updated by Karen N. Kähler

FURTHER READING

Chiras, Daniel D. "Foundations of a Sustainable Energy System: Conservation and Renewable Energy." In *Environmental Science*. 8th ed. Sudbury, Mass.: Jones and Bartlett, 2010.

Davis, David Howard. *American Environmental Politics.* Belmont, Calif.: Wadsworth, 1998.

Elliott, David. *Energy, Society, and Environment.* 2d ed. New York: Routledge, 2003.

Gerrard, Michael. *Global Climate Change and U.S. Law.* Chicago: American Bar Association, 2007.

Horowitz, Daniel. *Jimmy Carter and the Energy Crisis of the 1970s: A Brief History with Documents.* Boston: Bedford/St. Martin's Press, 2005.

Kash, Don E., and Robert W. Rycroft. *U.S. Energy Policy: Crisis and Complacency.* Norman: University of Oklahoma Press, 1984.

McCartney, Laton. *The Teapot Dome Scandal: How Big Oil Bought the Hardy White House and Tried to Steal the Country.* New York: Random House, 2009.

Sandalow, David. *Freedom from Oil: How the Next President Can End the United States' Oil Addiction.* Columbus, Ohio: McGraw-Hill Professional, 2008.

Temples, James R. "The Politics of Nuclear Power: A Subgovernment in Transition." *Political Science Quarterly* 95, no. 2 (Summer, 1980): 239-260.

SEE ALSO: Cash for Clunkers; Coal; Coal-fired power plants; Corporate average fuel economy standards; Department of Energy, U.S.; Energy Policy and Conservation Act; Fossil fuels; Nuclear power; Oil crises and oil embargoes; Renewable energy.

Energy Policy and Conservation Act

CATEGORIES: Treaties, laws, and court cases; energy and energy use; resources and resource management

THE LAW: U.S. federal law intended to address American demands for energy while simultaneously promoting conservation of energy resources

DATE: Enacted on December 22, 1975

SIGNIFICANCE: The Energy Policy and Conservation Act was the first serious attempt by the federal government to address energy independence. Many of the programs the act established, including the Strategic Petroleum Reserve, corporate average fuel economy standards for automobiles, and efficiency standards for appliances, remain focal points for debate in the twenty-first century.

The first call for the United States to set aside a supply of petroleum for emergencies came in 1944, from Secretary of the Interior Harold Ickes. Over the next three decades, the need was reconsidered, but no action was taken until an embargo by members of the Organization of Petroleum Exporting Countries (OPEC) limited the supply of oil coming to the United States in 1973. The result was a dramatic increase in the prices of oil and of gasoline, long lines and rationing at gas stations, declining stock prices, calls to reduce energy consumption immediately, and even, in some cities, bans on outdoor Christmas lights. This energy crisis spurred the federal government to take action to ensure that the United States could face future embargoes, or other emergencies, without panic.

The most important part of the Energy Policy and Conservation Act of 1975 (EPCA) was the establishment of the Strategic Petroleum Reserve (SPR), a supply of petroleum stored in salt domes in the Gulf of Mexico. Planned eventually to be a supply of 1 billion barrels (42 billion gallons), the full capacity of the SPR is 727 million barrels (30.5 billion gallons), roughly the amount in the SPR inventory as of 2010. Oil can be removed from the SPR only by order of the president of the United States. Because of limits imposed by the technology that draws the petroleum from the reserve, only about 4.4 million barrels (184.8 million gallons) can be drawn each day in an emergency—about one-fourth of the average daily petroleum consumption of the United States. The SPR has been tapped only rarely since its founding, including during the first Gulf War in 1991-1992 and after Hurricanes Rita and Katrina, when oil processing was slowed.

The far-reaching EPCA addressed energy use and conservation in other ways as well. It called on the Department of Energy to create efficiency standards for home and commercial appliances such as air conditioners, refrigerators, dishwashers, water heaters, and clothes washers and dryers. It created the first corporate average fuel economy (CAFE) standards for auto manufacturers, calling for a doubling in the fuel efficiency of cars and light trucks to 27.5 miles per gallon by 1985. To move the country away from reliance on oil and natural gas, the act encouraged power-generating plants to shift to burning coal. It directed the Federal Trade Commission to encourage the recycling of oil and the safe disposal of used oil and oil products, and it created a complex set of price and import controls for petroleum. Finally, the EPCA included requirements for increased energy conservation in federal buildings and established the State Energy Conservation Program, which helped to fund energy offices and energy management plans for individual states.

Cynthia A. Bily

FURTHER READING

Bamberger, Robert, and Robert L. Pirog. *Strategic Petroleum Reserve*. Hauppauge, N.Y.: Nova Science, 2006.

Gerrard, Michael. *Global Climate Change and U.S. Law*. Chicago: American Bar Association, 2007.

Sandalow, David. *Freedom from Oil: How the Next President Can End the United States' Oil Addiction*. Columbus, Ohio: McGraw-Hill Professional, 2008.

SEE ALSO: Air-conditioning; Alternative energy sources; Corporate average fuel economy standards; Department of Energy, U.S.; Energy conservation; Energy-efficiency labeling; Oil crises and oil embargoes.

Environment Canada

CATEGORIES: Organizations and agencies; human health and the environment; atmosphere and air pollution; water and water pollution

IDENTIFICATION: Department of the Canadian federal government responsible for environmental policies and programs

DATE: Established on June 11, 1971

SIGNIFICANCE: Environment Canada serves as a model for best practices in monitoring, public education, and action in protection of the environment. The department works to develop, implement, and enforce policies that can prevent future damage to the environment and repair damage already done.

Environment Canada (officially the Department of the Environment) was created in 1971 through the combination of various existing elements of the Canadian federal government, including the Meteorological Service (established in 1871) and the Wildlife Service (established in 1947). The first five services established by the new department were the Atmospheric Environment Service, the Environmental Protection Service, the Fisheries Service, the Water Management Service, and the Land, Forest, and Wildlife Service. In 1979, organizational changes were made, and the Fisheries Service left Environment Canada to form the Department of Fisheries and Oceans.

Environment Canada's stated mandate includes the following elements: preservation and enhancement of the quality of the natural environment, including water, air, soil, flora, and fauna; conservation of Canada's renewable resources; conservation and protection of Canada's water resources; the forecasting of weather and environmental changes; enforcement of rules relating to boundary waters; and coordination of environmental policies and programs for the Canadian federal government. The various services and programs of Environment Canada are concerned with protecting, conserving, and enhancing the environment by shaping how Canadians think about the environment, developing and supporting partnerships, and establishing economic incentives for industries and individuals to make sound environmental decisions.

The department uses scientific research and technologies to track and manage wildlife populations, improve understanding of ecosystems and support the recovery of degraded ecosystems, and assess environmental risk. The department also supports policy and legislative action aimed at promoting environmental health and sustainable practices. It is a participant in several United Nations organizations, including the United Nations Environment Programme, the World Meteorological Organization, and the Commission on Sustainable Development, and is involved in joint efforts with international organizations such as the Arctic Council and the Inter-American Institute for Global Change. In addition, the department works with industries, businesses, farmers, municipalities, customs officials, hunters, the U.S. Environmental Protection Agency, and the Royal Canadian Mounted Police to enforce legislation related to areas of environmental concern, including the import and export of harmful substances, the protection of migratory birds and endangered species, and the protection and trade of both domestic and internationally shared waters.

Wendy C. Hamblet

FURTHER READING

Biggs, David, et al. *Life in 2030: Exploring a Sustainable Future for Canada*. Vancouver: University of British Columbia Press, 1996.

Boyd, Susan C., Dorothy E. Chunn, and Robert Menzies, eds. *Toxic Criminology: Environment, Law, and the State in Canada*. Halifax, N.S.: Fernwood, 2002.

Fafard, Patrick C., and Kathryn Harrison, eds. *Managing the Environmental Union: Intergovernmental Relations and Environment Policy in Canada*. Kingston,

Ont.: School of Policy Studies, Queen's University, 2000.

Holland, Kenneth M., F. L. Morton, and Brian Galligan, eds. *Federalism and the Environment: Environmental Policymaking in Australia, Canada, and the United States.* Westport, Conn.: Greenwood Press, 1996.

SEE ALSO: Convention on Long-Range Transboundary Air Pollution; Experimental Lakes Area; Great Lakes International Joint Commission; Green Plan; Lake Erie; Montreal Protocol; National parks; St. Lawrence Seaway; Sudbury, Ontario, emissions.

Environmental determinism

CATEGORY: Philosophy and ethics

DEFINITION: The theory that features of the environment ultimately determine human culture, character, and societal development

SIGNIFICANCE: Environmental determinists hold that human activity, culture, and character are ultimately determined by environmental factors. From this point of view, the environment, rather than human activity, is the dominant party in the relationship between humans and environment.

The notion of environmental determinism achieved considerable prominence during the late nineteenth and early twentieth centuries, especially through the work of geographer Ellen Churchill Semple. The "environment" in this context is understood to be the natural physical features of a setting—the geographical features, such as climate and landform, as opposed to factors directly or indirectly the result of human activity (such as social and economic factors). According to environmental determinists, these environmental factors determine the course of human character, action, and cultural development. Semple, for example, wrote that the climate of northern Europe causes the development of an "energetic, provident, serious, thoughtful rather than emotional, cautious rather than impulsive" character in the peoples resident there; in contrast, tropical climates lead to the development of laziness because the warm weather and ready availability of food make survival easy and do not encourage the development of a keen work ethic.

Because of this tendency toward categorization of peoples according to environmental ancestry, environmental determinism became associated with racist and imperialist attitudes. It is important to note, however, that environmental determinism is not a normative moral theory (that is, it is not a theory about how human beings "ought" to behave). It simply attempts to explain differences in human development through differences in environmental ancestry. It passes no judgment on the moral merits or otherwise of any particular character trait or culture and does not in itself justify discrimination or imperialism.

Environmental determinism has been criticized for being overly simplistic and for not providing sufficient support for its central claim. It seems implausible that one set of factors should have an ultimate determinative influence on human development, and environmental determinists have been criticized for providing very little supporting evidence for such a strong claim. Such criticism has led some environmental determinists to concede that other factors play some role, but they still believe that environmental factors are dominant.

Environmental determinism is often thought to deny human free will. The natural environment into which a person is born is not a matter of that person's choosing, and so if environment determines individuals' outlooks, values, and thereby their decisions, then humans are not free agents and not morally responsible for their behavior—they are just victims of their circumstances.

Some of these concerns are misplaced, however. Environmental determinism does not deny that human beings are rational deliberators who weigh and assess options and then make decisions based on those deliberations. Further, environmental determinism need not deny that an individual's decisions play a crucial role in determining the course of that person's character development. The most that an environmental determinist is committed to is that the course of an individual's deliberations and thus the role of those deliberations in the development of the individual's character will have been ultimately determined by environmental factors. That is quite different from the claim that human beings' deliberations are pointless and inert. They are one part of the mechanism by which the environment determines the course of human development. Whether this still challenges the status of humans as fully free, morally responsible agents turns on the finer details of what free will is taken to involve.

Gerald K. Harrison

FURTHER READING

Sauer, Carl. *Carl Sauer on Culture and Landscape: Readings and Commentaries.* Edited by William M. Denevan and Kent Mathewson. Baton Rouge: Louisiana State University Press, 2009.

Semple, Ellen Churchill. *Influences of Geographic Environment, on the Basis of Ratzel's System of Anthropo-Geography.* 1911. Reprint. Ann Arbor, Mich.: University Microfilms International, 1993.

Sutton, Mark Q., and E. N. Anderson. *Introduction to Cultural Ecology.* 2d ed. Lanham, Md.: AltaMira Press, 2010.

SEE ALSO: Anthropocentrism; Biocentrism; Ecocentrism; Environmentalism.

Environmental economics

CATEGORY: Resources and resource management

DEFINITION: The study of the relationship between the economy and the environment

SIGNIFICANCE: Environmental economics—which focuses on such areas of concern as the allocation of costs associated with pollutants, the allocation of natural resources, and efforts to place monetary values on resources for which there are no markets—provides insights into the ways in which the environment and the economy affect each other.

The concept of environmental economics grew out of the awareness of environmental issues that impinged on the social consciousness beginning in the 1960's and 1970's. Visions of a "silent spring," polluted rivers, and smog-filled cities posed questions about whether a free market economy efficiently allocates resources. A fourfold increase in the price of oil, long lines at gas stations, and the new view of Earth from space as a small blue sphere against a black void prompted debate about whether sufficient nonrenewable resources exist to sustain economic growth. Deforestation, species extinction, and global warming raised doubts about whether markets adequately value environmental resources, prompting economists to inquire into how to value resources for which there are no markets.

Traditional economic theory largely ignores the relationship between economics and the environment. The assumption of economic rationality depicts firms as profit maximizers and consumers as pleasure maximizers. The invisible hand of the market conveys the view that voluntary exchange promotes economic harmony. Firms and consumers in pursuit of their self-interest unintentionally promote the interest of all. Nature is reduced to an input into the production process, providing both renewable and nonrenewable resources. Pollution is treated as an aberration, an example of market failure requiring some form of government intervention to restore the harmony of the market.

An alternative view, expressed by Herman E. Daly and John B. Cobb in their book *For the Common Good: Redirecting the Economy Toward Community, the Environment, and a Sustainable Future* (1989), conceives of the economy and nature as interdependent. Emphasis is placed on the concept of extended rationality; that is, individuals find it in their self-interest to protect the environment and care for future generations. The economy and nature are viewed in terms of coevolutionary processes, each affecting the other. This view focuses on creating institutions to channel self-interest in environmentally sensitive ways.

POLLUTION

Despite highly restrictive assumptions, economists most often use the perfectly competitive model to evaluate environmental policy. Under perfect competition, no single agent has the power to influence price, resulting in an equilibrium price that efficiently allocates resources. Efficiency means that the benefit of producing one more unit equals the cost of producing that unit. Social welfare (happiness of the individuals in society) is maximized because each unit produced prior to equilibrium yields more benefit than it costs.

For example, consider a firm engaged in the production of copper by the ton. As the consumption of copper rises, the benefits obtained from an additional ton decline; conversely, as production rises, the costs to the firm of producing an additional ton rise. The intersection of these supply and demand factors results in an equilibrium that sets a price for the copper and also determines the amount of copper that the firm should produce.

Suppose, however, that the firm also produces as a by-product pollution that injures others. In this case, the free market price would not reflect the external cost imposed on others or on society at large resulting from the pollution. Pollution is a type of externality, a cost involuntarily imposed on one party as a result of the activities of others. If corrections were made to the supply-and-demand equilibrium to account for the external cost, the price of the copper would rise, and production would fall. Note that correcting for external costs does not eliminate pollution; it merely requires that producers and consumers consider the external costs in their decisions.

The existence of externalities implies that the free market misallocates resources: Too much is produced for a price that does not include the external costs. Social welfare is not maximized, because costs to the firm plus external costs of the last units produced exceed the benefit of the last units.

DEALING WITH EXTERNALITIES

Several options are available to government in correcting for externalities: standards, taxes and subsidies, property rights, and marketable permits. In the past, environmental laws generally imposed fixed standards to which all businesses must conform. The simplicity of standards from a policy point of view has made them widely used. However, standards have been criticized for their coercive element, their failure to consider local circumstances, and their apparent arbitrariness.

In the 1930's, Arthur Pigou, one of the first economists to address externalities, recommended that government impose a tax equal to the external cost. Critics, however, cite difficulty in measuring the external costs, measuring the amount of pollution, determining the location to measure the pollution, and so on.

Ronald H. Coase, in a classic article titled "The Problem of Social Cost," advocates a free market solution. Coase recommends assigning property rights for the air or water, leading the affected parties to bargain over who pays the costs associated with pollution. If the injured party owns the resource, he or she charges the polluter an amount not less than the damages for using the resource. If the polluter owns the resource, the injured party would bribe the polluter not to pollute. The injured party would not pay an amount exceeding the damages created by pollution, and the polluter would not accept an amount less than the profits forgone. To maintain a lack of bias, Coase advocates the doctrine of ethical neutrality. In the absence of property rights, the polluter is no more responsible for pollution than is the injured party. Who pays depends on which method reduces the transaction costs (costs of identifying the party or parties harmed and

transacting the compensation). In most cases, minimizing transaction costs requires the injured party or parties to bribe the polluters not to pollute.

Market permits were adopted by the Clean Air Act amendments of 1990. The government sells permits that allow purchasers to emit limited amounts of pollution. Advocates argue that such permits provide polluters with incentive not to pollute. If a firm finds that the cost of installing pollution-control equipment is less than the cost of a permit, the firm reduces its pollution. If the cost of the equipment exceeds the price of the permit, the firm buys the permit. Critics, however, charge that the option to buy such permits implies that government endorses pollution.

RENEWABLE AND NONRENEWABLE RESOURCES

Natural resources may be categorized as either renewable or nonrenewable. Renewable resources are those that may be replenished, such as forests. Nonrenewable resources cannot be replenished.

The conditions for sustaining the environment are easy to identify in theory but difficult to achieve in practice. First, the harvest of natural resources must be less than the growth rates of those resources. Deforestation and the depletion of fisheries, for example, indicate that in many instances harvest rates exceed growth rates. Second, nonrenewable resources are, by definition, nonrenewable. The World Resources Institute estimates that fossil fuels provide 90 percent of the commercial energy used in the world. At current rates of use, the world's reserves of coal are estimated to last approximately five hundred years, and reserves of oil are estimated to last less than one hundred years. Ultimately, if economic growth is to be sustained, renewable resources must be substituted for nonrenewable resources. Third, emissions must not exceed nature's ability to absorb wastes. Markets can be used to provide incentives to encourage people to reduce, reuse, and recycle. Innovative businesses have reduced pollution by altering production processes and input mixes, thereby also reducing their costs.

Garrett Hardin addresses resource depletion in a classic article titled "The Tragedy of the Commons" (1968). "The commons" refers to a resource owned by no one and available to everyone. Assuming rationality, individuals exploit the resource as long as the benefit exceeds the cost. The result is that self-interest leads individuals to destroy the resource. Hardin's recommendation is to privatize the resource.

Economists are divided on whether the market alone is sufficient to drive the transition from nonrenewable to renewable resources. Conservative economists assert that the market works. As oil production slows, the price of oil rises, providing incentives to entrepreneurs to find alternatives. This assumes that if markets are allowed to work, new technologies will be developed, making inputs infinitely substitutable.

Markets, however, rarely allocate resources in ways that preserve the environment. The interest rate, for example, reflects society's preference between allocating resources for present use versus future use. The higher the rate of interest, the more society discounts the use of the resources in the future. If a firm finds that the market interest rate exceeds the rate of increase in the price of a resource, such as oil, then the profit-motivated firm will sell its resource and invest the proceeds.

The market reflects the preferences of those who have the "dollar votes." Nature and habitats, while important, do not vote. The services that nature provides in the forms of recycling wastes, maintaining the climate, providing oxygen, absorbing carbon dioxide, and providing aesthetic pleasure are not reflected in market values. Hence what is most profitable is not necessarily consistent with environmental preservation.

VALUING NATURAL RESOURCES

Markets allocate resources based on their price or value. How does one place a value on something for which there is no market? The answer better enables policy makers to allocate resources among competing uses, such as whether to use public lands for recreation or oil drilling. Placing a market value on natural resources generally means transforming them into commodities: a forest transformed into lumber, for example, or the mining of gold in Yellowstone. There are exceptions: The Nature Conservancy uses a strategy called debt-for-nature swaps, in which land in developing nations is purchased for preservation, thus reducing the nations' debt and protecting rain forests. In many cases, however, such as the administration of public lands, market solutions are unavailable.

In addressing the value of a resource, economists have developed three classifications for value: user value, option value, and existence value. User value is the value to the individual in using the resource; this is reflected in the value of the resource to hikers, recreationists, and skiers. Option value is the value of having the option to develop the resource at some future point. Existence value is the value of bequeathing the environmental resources to future generations, to habitat, and so on.

To determine the value of a resource, economists have employed a number of approaches. The most widely used are known as willingness to pay and willingness to accept. Willingness to pay asks individuals how much they would be willing to pay to enjoy environmental benefits. Willingness to accept asks how much individuals would be willing to accept in order to incur some loss. The difficulties with all approaches, however, reveal that there is no objective way to determine the value of a resource.

John P. Watkins

FURTHER READING

Baumol, William J., and Wallace E. Oates. *The Theory of Environmental Policy.* New York: Cambridge University Press, 1998.

Callan, Scott J., and Janet M. Thomas. *Environmental Economics and Management: Theory, Policy, and Applications.* 4th ed. Mason, Ohio: Thomson South-Western, 2007.

Coase, Ronald H. "The Problem of Social Cost." *Journal of Law and Economics* 3 (October, 1960): 1-44.

Cropper, Maureen L., and Wallace E. Oates. "Environmental Economics: A Survey." *Journal of Economic Literature* 30, no. 2 (1992): 675-740.

Daly, Herman E., and John B. Cobb, Jr. *For the Common Good: Redirecting the Economy Toward Community, the Environment, and a Sustainable Future.* 2d ed. Boston: Beacon Press, 1994.

Hanley, Nick, and Colin J. Roberts, eds. *Issues in Environmental Economics.* Malden, Mass.: Blackwell, 2002.

Hardin, Garrett. "The Tragedy of the Commons." *Science* 162 (December 13, 1968): 1243-1248.

Jaeger, William K. *Environmental Economics for Tree Huggers and Other Skeptics.* Washington, D.C.: Island Press, 2005.

Tietenberg, Tom, and Lynne Lewis. *Environmental Economics and Policy.* 6th ed. Boston: Addison-Wesley, 2010.

SEE ALSO: Accounting for nature; Benefit-cost analysis; Contingent valuation; Debt-for-nature swaps; Ecological economics; Externalities; Pollution permit trading.

Environmental education

CATEGORIES: Philosophy and ethics; ecology and ecosystems

DEFINITION: Organized efforts to teach how natural environments function and particularly how human beings can manage their behavior and ecosystems in order to live sustainably

SIGNIFICANCE: Many organizations and government agencies concerned with environmental issues have increasingly focused their efforts on educating the public about those issues, disseminating information to various audiences as a way to encourage behavior change and to gain support for their programs.

Early writers such as the eighteenth century philosopher Jean-Jacques Rousseau stressed the importance of education concerning the environment, which they nonetheless believed could not simply be taught or learned secondhand from books. The celebration of environmental education underpinned much of the early Romantic movement, and the works of writers such as William Wordsworth and John Ruskin are still recognized as representative of human beings' desire for a deep appreciation of landscape alongside a deep concern for all the flora and fauna across the planet.

HISTORY

An urge to protect the environment and educate people on the intrinsic benefits of nature can be seen in the development of pressure groups as far back as 1854, with the organization of an environmental protection group in France, for example. This global phenomenon is further illustrated by the Lake District Defence Society in Great Britain, which was founded in 1883; the Audubon Society in the United States, the beginnings of which go back to 1886; and the Wildlife Preservation Society of Australia, which started in 1907. In more recent times, a new type of environmental education, described by some as conservation education, emerged in the 1930's as a result of the Great Depression and the resultant Dust Bowl in the United States, which upset the balance of nature and the fertility of the soil for much-needed food production.

The modern environmental education movement, which came to prominence during the late 1960's and early 1970's, has built on these and other developments, spearheaded by global organizations such as Greenpeace and Friends of the Earth alongside the development of Green political parties, which have helped promote a wide range of environmental issues and educational debates. Furthermore, events such as the Civil Rights movement and the fallout from the Vietnam War in the United States—in addition to the long Cold War, which raised fears of radiation from possible nuclear attacks—helped create global fears around environmental catastrophe, coupled with the urge to protect the fragile planet. The notion of stewardship of the environment was further augmented by the publication in 1962 of Rachel Carson's famous American study *Silent Spring*, which dramatized the harmful effects of the pesticide dichloro-diphenyltrichloroethane (DDT). Carson effectively expressed the "death-producing effects" of chemicals for Americans in particular and helped to make the question of life on earth a public, ethical, and political issue.

EARTH DAY AND BEYOND

Some observers have suggested that the first Earth Day on April 22, 1970, finally paved the way for the modern environmental movement, especially in the United States. In 1971, the National Association for Environmental Education (later renamed the North American Association for Environmental Education) was created to promote environmental education programs generally by providing resources for teachers. Internationally, environmental education gained recognition when the United Nations Conference on the Human Environment, held in Stockholm, Sweden, in 1972, declared it an important tool for addressing global environmental problems.

Within primary schools, environmental education has been developed extensively; at this level, the inculcation of notions of civic engagement and pride in local environments appears to have widespread political, cultural, and even religious acceptance. In the early twenty-first century, a trend within environmental education was to move away from an approach of preaching and activism to one that would enable students to make informed decisions and take actions to promote an ecologically sustainable future based on their own experiences as well as on data provided.

Sociologists, anthropologists, and historians over time have found that people's attitudes toward the natural world are variable. Not all societies exhibit the same attitudes as those that had become common by the late twentieth century in the United States. The less frightened people are of natural forces and processes, the more they are able to appreciate these as sources of beauty, and the more they appreciate natural environments as beautiful, the closer they feel themselves to be to their own "natural" selves. Many observers consider Green politics, as an approach to environmental education, to be the single most significant international movement since the birth of socialism.

Environmental philosopher Paul Taylor has effectively consolidated environmental ethics as an essential part of any educational agenda to include the following: an ultimate moral attitude of "respect for nature," a belief system that he calls "the biocentric outlook," and a set of rules of duty that express "the attitude of respect." The central tenet of environmental education is embodied by the notion of harmony with nature together with a recognition of finite resources. Everything else is either peripheral to or at best ancillary to these all-inclusive affirmations. There is often little agreement, however, among the large rainbow of "Green supporters" on the specific means, especially the priorities and time frames, for achieving these ends.

Social philosopher André Gorz warns in *Ecology as Politics* that environmentalism is continually being "commandeered" by the dominant groups in Western society for their own ends. According to Gorz and others, the forces of capitalism are very capable of adapting an "environmental conscience" to meet the needs of the dominant culture. Those who are involved in environmental education must tease out and face up to such contradictions and ambiguities as they address many contentious issues, including global warming, nature management, and population control. It is very difficult to get consensus on these issues in a world where the developed West has gained the most from a long Industrial Revolution, in contrast to the developing nations, which want to acquire a similar level of wealth creation but find themselves facing restrictions related to protection of the environment.

Pat Brereton

Eleven-year-old Sarah Howe holds a bullfrog during an educational presentation held on Earth Day, 2007, in Eau Claire, Wisconsin. Since the inception of Earth Day in 1970, a major purpose of the event has been environmental education. (AP/Wide World Photos)

FURTHER READING

Gorz, André. *Ecology as Politics*. Translated by Patsy Vigderman and Jonathan Cloud. 1980. Reprint. London: Pluto Press, 1987.

Gruen, Lori, and Dale Jamieson, eds. *Reflecting on Nature: Readings in Environmental Philosophy*. New York: Oxford University Press, 1994.

Johnson, Edward, and Michael Mappin, eds. *Environmental Education and Advocacy: Changing Perspectives of Ecology and Education*. New York: Cambridge University Press, 2005.

Sutton, Philip W. *The Environment: A Sociological Introduction*. Cambridge, England: Polity Press, 2007.

Taylor, Paul. *Respect for Nature: A Theory of Environmental Ethics*. Princeton, N.J.: Princeton University Press, 1986.

SEE ALSO: Earth Day; Environmental ethics; Environmentalism; Friends of the Earth International; Green movement and Green parties; Greenpeace; United Nations Conference on the Human Environment.

Environmental engineering

CATEGORY: Resources and resource management

DEFINITION: Branch of engineering concerned with protecting the environment from the negative impacts of human activity

SIGNIFICANCE: Environmental engineering measures are employed to preserve or restore environmental quality, conserve resources, and protect human health. Environmental engineering encompasses a wide range of concerns, including but not limited to the control and prevention of air and water pollution, solid waste management, industrial waste minimization, contamination remediation, wastewater treatment, transportation system planning, and energy conservation.

For millennia, civil engineering has dealt with problems involving the design, construction, and upkeep of buildings, roads, bridges, dams, canals, and other works. Environmental engineering, a subdiscipline of civil engineering, deals with the environmental consequences of human activity. Long concerned almost exclusively with sanitation issues, environmental engineering has evolved into a complex, multidisciplinary field.

The theory and practice of modern environmental engineering owe much to R. Buckminster Fuller, a twentieth-century inventor, engineer, and architect who advocated doing more with less. His "Dymaxion House" prototype built in the 1940's typified his environment-conscious sensibilities. The innovative modular living structure featured water-saving bathroom fixtures, graywater recycling, natural heating and cooling, energy-efficient design, and low-cost construction that made a minimal impact on the land. Fuller's pioneering work, along with the many advances made in the science of ecology in the twentieth century and the growing awareness of the need for sustainable development, contributed to the expansion of environmental engineering beyond the realms of water supply and sewage management.

INTEGRATING BUILT AND NATURAL ENVIRONMENTS

The environment as a whole is large and complicated. To reduce the level of complexity in analysis, it is helpful to invent relatively simple ways to look at the whole and its constituent parts. This simplification is sometimes called a model for systems analysis. An example of such a model is one in which a researcher considers an environmental system as consisting of distinct elements that interact with one another. Air, water, soil, waste products, and hazardous products may be considered constituent parts of a whole. It is important to understand how each part of the large system behaves. The quality of each system, system interactions, and the impact of human activities on each system are of enormous importance to the well-being of the whole.

Although different in their ultimate goals, environmental science and environmental engineering are intimately related. Environmental science provides a means for understanding how the natural environment is constituted, how it sustains itself, and how it is affected by human activities. Based on this understanding, rigorous criteria can be developed to help maintain high standards in the quality of the environment. In turn, such criteria can be used to help establish realistic standards that will guide the development and utilization of new technologies.

Technology has three aspects: The first aspect is helpful to human life and the environment, while the second is detrimental to both of them. The third aspect of technology can help solve many environmental problems, even those that are created by technology itself. For this last aspect to be realized, scientists, engineers, technicians, social scientists, and health professionals must combine their talents to gain an understanding of how human activities affect the environment and what can be done to reduce or eliminate the damage that has been done to it.

Human civilization continues to require increasing amounts of fuel for its machines, chemicals for its industrial plants, fertilizers and pesticides for its farms, products to satisfy consumer demand, and raw materials from which to manufacture those products. Industry, agriculture, population centers, and more generate a staggering variety of wastes that must be dealt with safely. The essence of environmental engineering is finding practical ways to help solve the many environmental problems presented by human activity in an efficient and affordable manner.

Environmental engineering is a broad and multidisciplinary subject. Generally, academic preparation in this field emphasizes basic engineering and scientific principles, as well as the design and application of environmental engineering methods and processes. Thus, at the graduate level, the field attracts scientists, engineers, and students from a variety of engineering and science backgrounds.

CONTEXTS

Environmental engineering operates in three broad contexts: natural systems, engineered systems, and design. Environmental engineering in natural systems is concerned with the chemistry and biology of air, water and soil quality, water and air pollution, limnology (the scientific study of bodies of fresh water), global atmospheric change, the fate and transport of pollutants in the natural environment, hazardous substances, and risk analysis. Environmental engineering in engineered systems concentrates on systems made by humans, such as water supply and treatment processes, wastewater treatment processes, processes for air-pollution control, groundwater remediation, and solid and hazardous waste management. Environmental engineering design is the application of physical, chemical, and biological operations and processes in natural and engineered systems. It must take into account five sets of requirements: legal, public health and sanitary, socioeconomic, aesthetic and sociocultural, and engineering.

The use of water in the home provides a good example of environmental engineering design. Running water in every home is an expected convenience in many countries. After water has been used, however, it must somehow be disposed of. The safe disposal of domestic wastewater is necessary in order to protect the health of individuals, families, and communities. This requires the design of treatment, recycling, or disposal systems that will dispose of domestic wastewater in such a way as not to violate any laws, regulations, or ordinances regarding water pollution and sewage disposal (legal requirement). Supplies of drinking water and of waters that are used for recreation cannot be contaminated by wastewater and must be protected from possible carriers of public health hazards such as animals and insects (public health and sanitary requirement). In addition, treatment systems cannot exceed the cost that the community can bear (socioeconomic requirement) and cannot cause a nuisance such as excessive noise, unpleasant odor, or unsightly appearance (aesthetic and sociocultural requirement). Finally, sewage should be transported, discharged, treated, and disposed of in specified ways. Typically, this is achieved through the development of a system that takes into account the four interconnected stages of waste treatment: the wastewater source, its collection, its treatment, and its disposal or reuse (engineering requirements).

Similarly, environmental engineering measures for controlling air emissions from an industrial plant must be designed with several factors in mind: What are the regulatory criteria for airborne concentrations of the chemical compounds present in the plant emissions (legal requirement)? At what concentrations would these compounds cause illness or death within the neighboring communities, and might long-term exposure to lower concentrations result in a deterioration of health over time (public health and sanitary requirement)? Can the plant keep emissions at safe and compliant levels while carrying out cost-effective operations, and, if not, what would be the impact on the community's economy if the plant were to downsize or close altogether (socioeconomic requirement)? Would emissions produce offensive odors, obscure visibility, be unsightly, stain or erode buildings, or damage trees (aesthetic and sociocultural requirement)? Are technologies available, or can new ones be developed, for cost-effective emissions control, and could a change in management practices also reduce emissions (engineering requirement)? Striking a balance among all these requirements is often challenging, especially when economic considerations or technological limitations come into conflict with goals for protecting human or ecosystem health.

Awareness of the threat that human activities pose to the shared environment is translated into pressures that are brought to bear on developers of technology, those who sell technology, and those who consume it. This awareness is typically converted into legislation and regulations to which people and corporations must adhere. Legislation such as the U.S. Clean Air Act and Clean Water Act makes pollution prevention into national environmental policy. Regulatory requirements present the environmental engineer with challenges as well as opportunities.

EXPANSION OF THE FIELD

Concerns for the quality of the environment have added new dimensions to old fields of study in engineering and science. For example, traditional hydraulics engineering has developed the subspecialty of environmental hydraulics, which is the study of how problems connected with human activities relate to the quality and motion of fluid in rivers, lakes, estuaries, coastal waters, and air. Water resource agencies have traditionally been concerned merely with ensuring adequate supplies of water, but the importance of water quality has increased substantially along with environmental concerns. Particular emphasis is now placed on the studies of pollutant mixing in bodies of water, modeling flow and contaminant transport in surface and subsurface environments, creating methods for control of turbidity in urban runoff, and designing and building outfalls for the discharge of wastewater and thermal effluent. Furthermore, studying these environmental concerns has become increasingly sophisticated, requiring the integration of many new technologies: complex computer models, electronic laboratory and field instrumentation systems, and data acquisition and transmission for real-time control.

Groundwater sources are continuously threatened by excessive pumpage and contamination. Stringent federal regulations on groundwater quality and quantity management have become necessary, and they have created a demand for technically proficient labor resources with strong multidisciplinary backgrounds in areas such as hydrology, hydrogeology, geochemistry, optimization, computational techniques, and software engineering.

In the field of transportation, the movement of people, goods, and services provides for an improved quality of life and strong economic activities. Efficient transportation systems thus form integral building blocks of a developed society. However, the physical construction of pathways and sites, the operation of transportation facilities and vehicles, and the travel behaviors associated with the vehicles themselves have negative impacts on the environment. Transportation therefore requires not only the efficient design and operation of systems and vehicles but also the careful linkage among travel behavior, urban life, and environmental quality and policy.

As environmental engineering has evolved, it has become increasingly focused on sustainability and holistic approaches in design. The field has expanded to encompass such wide-ranging concerns as energy-efficient appliances and vehicles, ecological landscape design, green buildings, reclamation of usable materials from waste streams, sustainable manufacturing, urban planning, and biomimetics (the use of biological forms and functions as they occur in nature as models for the design and engineering of structures, materials, and processes).

The United Kingdom's Hockerton Housing Project, a five-unit residential development competed in 1998, is an excellent example of holistic environmental design and engineering. The project was designed in a way that would minimize pollution, energy use, and resource consumption. Solar and wind power installations generate electricity, and a passive solar system provides heating. The roofs and rear walls of homes within the project are covered in soil rather than tiles or slates, a design strategy that helps the structures remain at a comfortable temperature regardless of the weather. Rainwater is collected from the roofs, filtered, purified, and used as drinking water. Rainwater falling elsewhere on the estate is collected and transferred to a reservoir, where it is retained for nondrinking use, such as washing laundry; it passes through filters before reaching the homes. Household organic wastes are processed in compost bins. Wastewater from homes is initially transferred to a septic tank; after several days during which solids settle out, the effluent enters a biological treatment system, which relies on reeds and bacteria. The treated effluent, which flows into an on-site lake, meets bathing water standards. Private yards provide garden space for each residence, in addition to an organic vegetable garden, an orchard, and an apiary that are for communal use.

Josué Njock Libii
Updated by Karen N. Kähler

Further Reading

Davis, Mackenzie Leo, and David A. Cornwell. *Introduction to Environmental Engineering.* 4th ed. Boston: McGraw-Hill Higher Education, 2008.

Maczulak, Anne E. *Environmental Engineering: Designing a Sustainable Future.* New York: Facts On File, 2010.

Mihelcic, James R., and Julie Beth Zimmerman, eds. *Environmental Engineering: Fundamentals, Sustainability, Design.* Hoboken, N.J.: John Wiley & Sons, 2010.

Nathanson, Jerry A. *Basic Environmental Technology.* 5th ed. Upper Saddle River, N.J.: Prentice Hall, 2007.

Nazaroff, William W., and Lisa Alvarez-Cohen. *Environmental Engineering Science.* Hoboken, N.J.: John Wiley & Sons, 2001.

Nemerow, Nelson L., et al. *Environmental Engineering.* 6th ed. Hoboken, N.J.: John Wiley & Sons, 2009.

Pfafflin, James R., ed. *Encyclopedia of Environmental Science and Engineering.* 5th ed. Boca Raton, Fla.: CRC Press, 2006.

Vesilind, P. Aarne, Susan M. Morgan, and Lauren G. Heine. *Introduction to Environmental Engineering*. 3d ed. Stamford, Conn.: Cengage Learning, 2010.

SEE ALSO: Best available technologies; Earth-sheltered construction; Energy conservation; Green buildings; Planned communities; Rainwater harvesting; Sewage treatment and disposal; Urban planning; Waste management; Water treatment.

Environmental ethics

CATEGORY: Philosophy and ethics

DEFINITION: Field of inquiry concerned with evaluating the ethical responsibilities humans have for the natural world

SIGNIFICANCE: The issue of human beings' relationship to the natural world—including humans' responsibility for stewardship of the land and natural resources, and for protection or preservation of plant and animal life—has long been a part of philosophical inquiry. Points of view in the field of environmental ethics continue to evolve.

The ethical responsibilities that humans have for the natural world, including natural resources, have been examined from many, often conflicting, perspectives, including anthropocentrism, individualism, ecocentrism, and ecofeminism. Each perspective has strengths and weaknesses.

Anthropocentrism is a human-centered philosophy that holds that moral values should be limited to humans and should not be extended to other creatures or to nature as a whole. A justification for this perspective is that moral relationships are sets of reciprocal rules followed by humans in their mutual relationships. Nonhumans are excluded from moral relationships because they lack comprehension of these rules. Some anthropocentrists argue that, from an evolutionary perspective, successful species should not work for the net good of other species; species that have done so in the past have become extinct.

Some anthropocentrists oppose restrictions on the use of natural resources because such restrictions may have negative impacts—for example, the loss of jobs or the loss of products beneficial to humans. Others stress that the natural world is a critical life-support system for humans and advocate effective environmental controls so that it will maintain its full value for present and future generations. This anthropocentric regard for the environment is based on the practical value of the natural world for meeting human needs rather than on a belief that the natural world has intrinsic value.

Those who hold an individualist philosophy believe that humans should extend moral concern to individual animals of certain species. Individualists include advocates of animal liberation and animal rights. Individualists accept that all humans have intrinsic value; they also argue that because some animals share morally relevant qualities valued in humans, these animals should be extended moral concern. Animal liberationists define the capacity for pleasure and pain (sentience) as the morally relevant feature to be considered. Animal rightists value more complex qualities, including desires, consciousness, a sense of the future, intentionality, and memories; they commonly associate these qualities with most mammals. Individualists generally are not concerned with the use of natural resources unless that use involves a direct threat to individuals of a species deserving moral concern, as through hunting or trapping.

Ecocentrism is based on the belief that the natural world has intrinsic value; it includes both the land ethic and deep ecology perspectives. Land ethic advocates believe that moral concern should be extended to the natural world, including natural units such as ecosystems, watersheds, and bioregions. Land ethic advocates emphasize respect (rather than rights) for the natural world. Ecocentrists may justify a land ethic by noting that all living creatures have a common origin and history on the planet and are ecologically connected and interdependent. The notions of common origin and history, as well as interdependence, are viewed as analogous to the human concept of family. Ecocentrists view humans as members of a large family comprising all of nature. Because family relationships entail not only privileges but also responsibilities for the well-being of the other family members and their environment, it follows that humans have responsibility for the natural world.

Impact on land health is an important criterion by which natural resource use is assessed in a land ethic. Characteristics of land health include the occurrence of natural ecological functioning, good soil fertility, absence of erosion, and having all the original species properly represented at a site (biodiversity). From a land ethic perspective, natural resource use should minimize long-term impacts on land health or should even enhance land health.

Deep ecology is often viewed as an ecosophy—an ecological wisdom that calls for a deep questioning of lifestyles and attitudes. Some guidelines regularly cited by deep ecologists include living lives that are simple in means but rich in ends, honoring and empathizing with all life-forms, and maximizing the diversity of human and nonhuman life.

Ecofeminists believe that many environmental problems are tied to human beings' desire to dominate nature, and this desire is closely linked with the problem of the domination of women and other groups in society. Ecofeminists believe that these problems would decline with a transformation in societal attitudes from dualistic, hierarchical, and patriarchal thinking to an emphasis on enrichment of underlying relationships and greater focus on egalitarian, empathetic, and nonviolent attitudes. Ecofeminism emphasizes less intrusive and more gentle use of natural resources.

Many Westerners have reexamined established cultural and religious perspectives for inspiration and insights in developing an environmental ethic. Native American cultures are often viewed as a source of moral insights on the human relationship to the environment. While it is difficult to generalize, given the many diverse Native American cultures, several perspectives appear common to many Native American groups: a strong sense of identity with a specific geographic feature, such as a river or a mountain; the notion that all of the world is enspirited and has being, life, and self-consciousness; and a strong sense of kinship with the natural world. Such Native American views are commonly associated with reduced environmental impacts and harmonious relationships with the natural world.

Judaism, Christianity, and Islam share common traditions; each contains elements upon which scholars have drawn for insights into environmental responsibility. Some scholars emphasize portions of the biblical book of Genesis, where the world is seen as God's creation for the free use and enjoyment of humans. Subjugation and use of nature are acceptable, but the land also must be appreciated and protected as belonging to God. Others emphasize the special role of humans as caretakers or advocate close relationships to the natural world, as exemplified by Saint Francis of Assisi. Attitudes toward the natural world and the use of natural resources vary widely among different groups of Jews, Christians, and Muslims. Some Eastern philosophies, such as Daoism and Buddhism, contain insights for environmental ethics. Both encourage a caring behavior toward nature.

Richard G. Botzler

FURTHER READING

Armstrong, Susan J., and Richard G. Botzler, eds. *Environmental Ethics: Divergence and Convergence.* 3d ed. New York: McGraw-Hill, 2003.

DesJardins, Joseph R. *Environmental Ethics: An Introduction to Environmental Philosophy.* 4th ed. Belmont, Calif.: Thomson/Wadsworth, 2006.

Keller, David R., ed. *Environmental Ethics: The Big Questions.* Malden, Mass.: Blackwell, 2010.

Pojman, Louis P., and Paul Pojman, eds. *Environmental Ethics: Readings in Theory and Application.* 5th ed. Belmont, Calif.: Thomson Wadsworth, 2008.

Sterba, James P., ed. *Earth Ethics: Introductory Readings on Animal Rights and Environmental Ethics.* 2d ed. Upper Saddle River, N.J.: Prentice Hall, 2000.

SEE ALSO: Animal rights; Animal rights movement; Anthropocentrism; Antienvironmentalism; Deep ecology; Ecocentrism; Ecofeminism; Speciesism.

Environmental health

CATEGORIES: Human health and the environment; pollutants and toxins

DEFINITION: Major discipline in the public health field that applies scientific study to environmental agents having a detrimental effect on the health and well-being of human populations

SIGNIFICANCE: The field of environmental health involves protecting the general public from disease vectors and contaminated air, water, and food; ensuring safe handling and disposal of nonhazardous and hazardous wastes; and reducing risks from contaminated surroundings. Since World War II, the field has been broadened to include noise pollution, radiological health and safety, and the impact of environmental disasters on large populations.

In the United States the responsibilities for ensuring the environmental health and safety of the population are shared by the National Institute of Environmental Health Sciences (NIEHS), the U.S. Public

Health Service (PHS) and its Centers for Disease Control and Prevention (CDC), the Environmental Protection Agency (EPA), the Nuclear Regulatory Commission (NRC), the U.S. Department of Agriculture (USDA), the Food and Drug Administration (FDA), the Federal Emergency Management Agency (FEMA), and numerous other government agencies. Each U.S. state has mechanisms for environmental health education, enforcement, and oversight. Substantial responsibilities fall on local public health inspectors, sanitarians, coroners, animal-control officers, and a host of elected and appointed officials at the city and county levels.

Within the United States corporate enthusiasm has grown for environmental audits. These audits validate corporate compliance with federal, state, and local environmental laws and regulations and help clearly define and publicize policies and procedures within the corporations. An audit enables a corporation to recognize environmental risks, bring them under control, and adjust resources and personnel needed to complete environmental work.

WATER TREATMENT

Most cities in industrialized countries have municipal water treatment plants that strive to keep drinking-water supplies free from pathogenic organisms and harmful substances. In addition, many communities fluoridate their drinking water to prevent tooth decay, a practice the CDC hails as one of the great public health achievements of the twentieth century. By 2006 nearly 70 percent of the people residing in the United States who were receiving their water from public water systems had fluoridated water.

During the 1990's a number of large U.S. cities experienced epidemics resulting from inadequate water treatment. Some of the most persistent problems occurred in Milwaukee, Wisconsin, where agricultural runoff resulted in heavy loading of *Cryptosporidium*, a genus of parasitic protozoan, which water treatment failed to destroy. A diarrheal disease called cryptosporidiosis affected more than 403,000 people and caused an estimated 69 deaths, mostly among people with acquired immunodeficiency syndrome (AIDS). Microorganisms such as *Cryptosporidium* tend to cause more serious infections in members of the population with damaged immune systems, such as people with AIDS and chemotherapy patients.

Several cities with aging water distribution systems have suffered epidemics related to contamination that occurred after the water left the treatment plant but before it reached users. Water leaving treatment plants has usually been disinfected using a chlorination or ozonation process, leaving a residual disinfectant that prevents the growth of pathogens. However, when water pressure is low at the fringes of the distribution system, the residual may be insufficient to prevent the growth of pathogenic microorganisms. In such a situation, immunocompromised persons must either use bottled water or boil their drinking water to minimize the risk of serious infection.

The goals of sewage treatment are the elimination of pathogenic organisms and the reduction of the amount of organic materials in the wastes discharged into the environment. Because of the potential for the cultural eutrophication of lakes by nutrients in treated wastewater, most communities discharge treated wastewater into rivers or oceans. Sewage normally contains a number of heavy metals—including mercury, lead, copper, and iron—which become concentrated in sewage sludge. Sludge must be disposed of properly; most is dewatered and landfilled.

ARTHROPOD- AND ANIMAL-CONTROL PROGRAMS

Insects and other arthropods may serve as vectors in the transmission of disease; environmental health programs are crucial in controlling these threats. Effective vaccines are available for some diseases, such as yellow fever, and mass vaccination programs are recommended for high-risk populations. In areas of the world where malaria, yellow fever, dengue fever, filariasis, viral encephalitis, and other mosquito-transmitted diseases pose public health threats, environmental health officials seek to eliminate mosquito breeding grounds and control mosquitoes near population centers using pesticides.

Prairie dogs, rats, mice, rabbits, and deer harbor a number of infectious diseases that may be transmitted to humans via fleas and ticks. The most important bacterial pathogens include plague, tularemia, and Lyme disease. Rickettsia, small microorganisms similar to bacteria, cause a number of infectious diseases, including epidemic typhus and Rocky Mountain spotted fever. Anywhere from 250 to 1,200 cases of Rocky Mountain spotted fever are reported in the United States each year; despite the disease's name, more than half of the cases occur in the south Atlantic region of the country. Global climate change has the potential to alter or expand the geographic areas affected by arthropod-transmitted diseases. Some researchers believe that such an expansion has already begun to occur.

Wild and feral animals harbor microorganisms that can produce disease in domestic animals and humans. Veterinary public health efforts seek to limit the spread of zoonosis (disease that can be transmitted from wild or domesticated animals to humans), and environmental health efforts are directed at preventing the spread of disease from animals to humans. Bovine tuberculosis (TB) in deer and buffalo, which may be transmitted to cattle and then to humans, is a national problem. Cattle herds are routinely tested for TB in the United States. Cattle ranchers near Yellowstone National Park complain that they must shoot buffalo that stray onto their property to prevent their cattle herds from becoming infected. Each year people who hunt deer, moose, and elk are advised to have their kills tested for TB.

Periodic epidemics of rabies among raccoons are a continuing problem. People often will try to help an obviously sick animal, not realizing the risk to themselves. Vaccination of all pet dogs and cats is an important step in controlling the spread of rabies. Wild animals that bite humans are usually euthanized and examined for evidence of rabies; if the test is positive, or for some reason is not performed, rabies vaccine is usually administered to the bite victim.

Rodents, especially rats and mice, carry bacteria such as *Salmonella typhimurium*. Environmental health programs designed to control rodent populations in urban areas help prevent epidemics caused by these bacteria. A 1993 outbreak in the American Southwest of an unfamiliar and deadly respiratory disease, eventually dubbed hantavirus pulmonary syndrome, led to increased efforts to control the wild rodent populations that spread the illness. Related hantaviruses were subsequently discovered elsewhere in the United States, Canada, and South America.

FOOD SAFETY

The FDA is responsible for approving food additives; substances that are poisonous or carcinogenic may not be added to foods. The pesticides used on fresh produce must also be approved. The CDC is responsible for identifying the strains and origins of microorganisms that cause nationwide food-borne epidemics.

Food poisoning can be broken down into two categories: noninfective and infective. Noninfective food poisoning is caused by contamination with a toxic substance such as an insecticide or a bacterial toxin such as botulism. An example of noninfective food poisoning is that caused by melamine, a chemical component of certain plastics. Melamine was used in China as a food adulterant because, in laboratory tests, it could make products seem to have a higher protein content than they actually did. The use of melamine in pet, fish, and livestock feed was discovered in 2007 after an outbreak of kidney failures among pets in the United States. A massive product recall followed. The following year, hundreds of thousands of infants in the People's Republic of China suffered kidney damage and six died from melamine-tainted milk and infant formula. China responded by imposing more stringent regulations, implementing more rigorous food inspections, trying and imprisoning several people responsible for the adulteration practice, and executing two of the offenders.

Infective food poisoning is caused by a pathogenic organism, most commonly a bacterium, virus, or parasite, that infects people who consume contaminated food. Outbreaks have involved a variety of products, ranging from alfalfa sprouts and breakfast cereal to fast-food hamburgers and ice cream. In mid-1996, Guatemalan-grown raspberries that were contaminated with the intestinal parasite cyclospora sickened approximately 1,500 people in the United States and Canada. In late 1996 and early 1997, frozen strawberries grown in Mexico and processed in California were contaminated with the hepatitis A virus, causing a food-borne epidemic in several states; more than 200 schoolchildren in Michigan contracted the disease. Notable incidents of infective food poisoning in the twenty-first century have included widespread salmonella cases in 2006-2007 and 2008-2009 that were traced to contaminated peanut butter and outbreaks of salmonella in 2010 that were traced to contaminated eggs.

In the United States a number of epidemics have been caused by food and beverages contaminated with pathogenic strains of the bacterium *Escherichia coli* (*E. coli*). The best-known outbreaks have involved hamburger; such outbreaks have been referred to as "hamburger disease" in Canada. Despite efforts by various government agencies and the news media to educate the public about the necessity of thoroughly cooking hamburger, epidemics repeatedly forced recalls during the 1990's that involved thousands of tons of meat. Other notable outbreaks of *E. coli* in the United States have involved unpasteurized apple

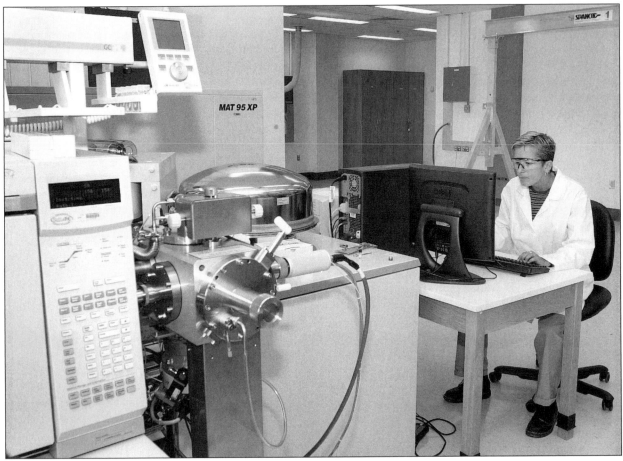

A chemist with the Centers for Disease Control and Prevention measures polychlorinated biphenyls (PCBs) in a sample at the CDC's Environmental Health Laboratory in Chamblee, Georgia. The laboratory's employees develop and apply advanced methods for measuring toxic substances in body fluids and tissues. (AP/Wide World Photos)

juice (1996), alfalfa sprouts (1997), contaminated drinking water (1999), spinach (2006), onions (2006), and lettuce (2010). Efforts are under way to develop a vaccine against *E. coli* infection.

Meat safety issues during late 1990's centered on the communicability of bovine spongiform encephalopathy (BSE), or mad cow disease, a neurodegenerative prion disease that entered the European food chain as the result of British cattle being given a commercial feed made from animal remains that included the carcasses of sheep killed by a prion disease called scrapie. A number of deaths caused by the human form of BSE, classed as new-variant Creutzfeldt-Jakob disease (nvCJD), caused Great Britain to ban the sale of beef brains, marrow, and spinal cord products. In 1989 the United States banned the import of British beef because of the BSE outbreak, and Canada banned British beef in 1990. To prevent a similar out-

break of BSE in the United States, authorities banned the use of ruminant remains in animal feed. Although bans have served to minimize the number of cases among cattle since the disease peaked in 1993, BSE is no longer confined to the United Kingdom. By June, 2010, more than two hundred people worldwide had died from nvCJD, most of them residents of the United Kingdom or Europe.

RADIATION

Radioisotopes are widely used in medical treatment, by industries, and by governments throughout the world. Wastes generated by the mining and purification of radionuclides are a worldwide problem. Within the United States, many tons of radioactive waste in tanks at various government laboratories require constant monitoring. Within the United States the greatest exposure to radioactive substances is from radon, an inert gas that is produced from the decay of radium in soil, rocks, and building materials.

Following a number of accidents at nuclear reactors, including those at Three Mile Island in Pennsylvania in 1979 and Chernobyl in the Soviet Union in 1986, most people began to recognize that living near a nuclear plant carries some risk. The NRC monitors nuclear power plants to ensure that they are functioning correctly and that all safety equipment is in place to prevent unplanned releases of radioactive materials. Before a nuclear power plant can receive a license to operate, an evacuation plan for the local community must be in place.

NOISE AND ACCIDENTS

Many environments in industrialized societies experience sound levels that approach the threshold of pain. It is difficult and expensive to control noise pollution, whether it takes the form of background levels resulting from community sources or intrusive noise from aircraft and car alarms. Many communities try to minimize noise complaints by limiting construction to daylight hours and banning airplane traffic between certain hours of the day. In 1982 about 1,100 noise-control programs existed at various governmental levels in the United States; by 1990 fewer than 20 programs were left—the rest had fallen victim to budget cuts. Although the federal Noise Control Act of 1972 and the Quiet Communities Act of 1978 remain in effect, they are not funded, and the primary responsibility for noise regulation has been handed off to state and local governments.

Preventing accidental injury and death and protecting the environment through implementation of safety programs have long been mainstays of environmental health. Accidents are a leading cause of death for Americans. In 2006, according to the CDC, unintentional injuries claimed 121,599 lives, making accidents the fifth leading cause of death in the nation. Automobile accidents accounted for 45,316 of those deaths, with about 24 percent of them occurring among persons ages fifteen to twenty-four. Efforts to reduce injury and death in automobile accidents include mandatory driver education in many states and mandatory use of seat belts in most states; auto manufacturers have been required to install driver and passenger front air bags and side bars or side air bags to protect against broadside impacts.

Gunshot wounds are also a significant cause of injury and death in the United States. Unintentional

discharge of firearms killed 642 people in 2006, roughly 38 percent of them ages five to twenty-four. Efforts to reduce this problem have centered on making guns more difficult to buy; however, studies indicate that many young people have access to guns in the home.

Continuing public education on accident prevention is essential in industrial societies, where many individuals have access to hazardous substances and the large-scale transport of potentially hazardous substances is routine. This responsibility usually rests with police and fire departments on the local level.

LAND CONTAMINATION AND AIR POLLUTION

Appropriate land use has emerged as a major concern in the United States. Multiple epidemiological studies have found that living close to landfills heightens the risk of neural tube birth defects such as spina bifida. The U.S. National Environmental Protection Act of 1969 requires the preparation of environmental impact statements on planned federal actions that will affect the human environment. Among the federal legislation that helps prevent land contamination in the United States are the Toxic Substances Control Act (1976); the Resource Conservation and Recovery Act (1976); the Comprehensive Environmental Response, Compensation, and Liability Act (1980), also referred to as Superfund; the Hazardous and Solid Waste Amendments (1984); the Superfund Amendments and Reauthorization Act (1986); and the Pollution Prevention Act (1990). Despite such legislation, however, abandoned hazardous waste sites continue to be discovered next to schools and playgrounds, where heavy metals contaminate dust and soil with which the children come into contact. Remedial actions to mitigate health threats from contaminated land require extensive funding to be effective.

The Air Quality Act of 1967 and subsequent Clean Air Act amendments of 1970 and 1990 yielded significant improvements in air quality in most urban areas in the United States. A 2002 report published in the *Journal of the American Medical Association* attributed a reduction in carbon monoxide-related deaths between 1968 and 1998 to the regulation of automobile emissions in the United States; according to the report, over the thirty-year period some 11,700 lives were saved. The EPA monitors and reports on selected air pollutants (carbon monoxide, nitrogen dioxide, ozone, lead, sulfur dioxide, and minute respirable particles) nationwide.

Poor air quality and other environmental problems tend to affect disadvantaged populations disproportionately. In the United States and elsewhere in the developed world, poorer inner-city residents are more likely than more well-off persons to be exposed to hazardous substances in their homes and workplaces. As a result, they tend to suffer from higher rates of asthma, cancer, infant mortality, and other maladies. Similarly, poorer countries often bear a disproportionate share of the world's environmental ills. Industries often "export" practices deemed environmentally unacceptable in the United States and other developed nations, moving their operations to developing countries where environmental regulations are weaker or poorly enforced. As a result, the residents of these poorer nations suffer the environmental health consequences.

Anita Baker-Blocker
Updated by Karen N. Kähler

FURTHER READING

Brown, Phil. *Toxic Exposures: Contested Illnesses and the Environmental Health Movement.* New York: Columbia University Press, 2007.

Conant, Jeff, and Pam Fadem. *A Community Guide to Environmental Health.* Berkeley, Calif.: Hesperian Foundation, 2008.

Friis, Robert H. *Essentials of Environmental Health.* Sudbury, Mass.: Jones and Bartlett, 2007.

Frumkin, Howard, ed. *Environmental Health: From Global to Local.* 2d ed. Hoboken, N.J.: John Wiley & Sons, 2010.

Leiss, William, and Douglas A. Powell. *Mad Cows and Mother's Milk: The Perils of Poor Risk Communication.* 2d ed. Montreal: McGill-Queen's University Press, 2004.

Maxwell, Nancy Irwin. *Understanding Environmental Health: How We Live in the World.* Sudbury, Mass.: Jones and Bartlett, 2009.

Moeller, Dade W. *Environmental Health.* 3d ed. Cambridge, Mass.: Harvard University Press, 2005.

Yassi, Annalee, et al. *Basic Environmental Health.* New York: Oxford University Press, 2001.

SEE ALSO: Air pollution; Birth defects, environmental; Cultural eutrophication; Drinking water; Environmental illnesses; Environmental justice and environmental racism; Environmental Protection Agency; Land pollution; Waste management; Water pollution; Water treatment.

Environmental illnesses

CATEGORIES: Human health and the environment; pollutants and toxins

DEFINITION: Ailments caused by exposure to chemical agents, radiation, physical hazards, and nature's reactions to invasions by humankind

SIGNIFICANCE: With growing awareness of the illnesses that can be caused by exposure to certain elements in the environment have come the passage of legislation and increasing support for regulations designed to protect human health. Workplace-related environmental illnesses have come under particular scrutiny in the United States.

Environmental illnesses—a category of ailments that includes occupational illnesses—are noninfectious and infectious diseases caused by environmental exposures, in addition to injuries caused by physical hazards considered beyond the immediate control of the individual. Nonoccupational environmental diseases identified by the U.S. Department of Health and Human Services include asthma, heatstroke, hypothermia, heavy metal poisoning, pesticide poisoning, carbon monoxide poisoning, acute chemical poisoning, and methemoglobinemia.

Physicians in ancient Egypt noted environmental conditions that had negative impacts on health, and some historians believe that lead poisoning was a strong contributor to the fall of the Roman Empire in 476 C.E. Awareness of environmental illnesses intensified during the Industrial Revolution with the realization that some diseases were strongly associated with specific occupational settings. Some early examples include silicosis, a lung disease contracted by large numbers of industrial workers, miners, and potters who were exposed to silica dust, and a delayed form of bone disease in laborers working within manufacturing plants that contained phosphorus.

Most industrialized countries had implemented early forms of environmental protection laws by the 1920's, but the increased use of caustic chemicals and radioactive materials made research involving ecology (scientific study of how living organisms are affected by their environment) increasingly complex. The ecology of infection involves interactions among the climate (as shown by the seasonal increases in influenza and pneumonia); contaminated air, water, and food; and nature itself, with many serious diseases such as tuberculosis, cholera, malaria, and typhoid fever significantly decreasing in incidence upon implementation of appropriate changes within the environment.

Environmental illnesses can affect every organ and system of the body in both mild and severe forms, with diagnosis made more difficult when specific exposures cannot be identified or when symptoms of the illness are delayed. The onset of some disease symptoms occurs immediately, but many symptoms do not appear until long after exposure; some forms of cancer, for example, have latency periods exceeding thirty years. Epidemiological studies of exposed populations are complicated by the fact that clinical features are often nonspecific. Furthermore, many illnesses can be enhanced by both the environment and personal habits such as smoking and medication abuse.

Chemical Agents

Environmental hazards that influence health and disease processes include natural stressors such as heat, cold, altitude, relative humidity, and wind speed. Unnatural environmental illnesses are created by humans rather than by nature and are generally caused by one of three factors: chemical agents, radiation, or human-made physical hazards. Exposure routes include direct or indirect contact with toxins and contaminated air, water, and food. Risk is greatly increased when multiple toxic agents act together, as illustrated by the increased risk of lung cancer in asbestos workers who also smoke or inhale secondhand smoke. Toxic waste dumps pose considerable environmental risks, as they can expose people to multiple hazardous chemicals simultaneously. Thousands of hazardous chemicals have been introduced into the environment with advances in industry; common inorganic examples include dichloro-diphenyl-trichloroethane (DDT), vinyl chloride, and polychlorinated biphenyls (PCBs), and common organic examples include asbestos, mercury, lead, and arsenic.

A young woman carries her son past the Patarcocha lagoon, which is used as a dump for sewage and other waste in the mining town of Cerro de Pasco, Peru. Soil, homes, and water in the town have been found to be saturated with toxic levels of lead. Many environmental illnesses are associated with exposure to contaminants in air, soil, and water. (AP/Wide World Photos)

The pesticide DDT is the most widely referenced example of the danger of introducing synthetic compounds into the environment before long-term effects have been researched. Used for years following World War II, DDT nearly eliminated malaria worldwide. The compound is not easily biodegradable, however, and persists in the environment for years. The use of DDT was eventually banned in nearly all developed countries following detection of the chemical in essentially every living organism tested. Many other agricultural pesticides are designed to deter or eliminate weeds, insects, fungi, or rodents that pose a threat to crops. When these toxins drift with the wind or are absorbed into the crops they are designed to protect, illnesses such as cancer and birth defects can result, with their extent related to dosage and exposure duration. Many chemicals and chemical combinations have the potential to produce delayed forms of cancer. For example, exposure to asbestos may lead to lung cancer and mesothelioma, vinyl chloride may cause liver cancer, and benzene may cause leukemia. The expression of diseases that come from chemical agents and radiation depends on agent entry into the body, metabolic processes within the body, routes by which the body attempts to excrete the substance, and medical treatments.

Airborne pollutants have a much greater influence on the body during physical exertion than during rest because the increased rate and depth of breathing during exertion exposes more particulate matter to the delicate tissues of the lungs. Physical work requires a transition from nose breathing to mouth breathing, with inhaled air thus bypassing the body's natural air-purifying system in the nasal hairs and mucous membranes, which are generally very effective at removing pollutants at low ventilation rates. Tiny particulates are more dangerous than larger particles because they are not trapped in the upper respiratory tract, they attach solidly to alveoli in the lungs, and they cannot be effectively exhaled.

Ozone is extremely toxic to humans, causing lung irritation, chest pain, bronchospasm, headaches, and nausea. Long periods of breathing ozone combined with hydrocarbons, aerosols, and sulfur and nitrogen dioxide may be a contributing factor to allergies, asthma, emphysema, bronchitis, and lung cancer. Temperature inversions in cities located at high altitudes and in basins surrounded by mountains that block winds and trap pollutants produce a greenhouse effect. Temperature inversions cause a reversal of the normal atmospheric temperature gradient that heats the harmful chemicals, thus enhancing their effects on the body. For example, the strong eastern winds blowing toward Denver, Colorado, trap a brown cloud of pollutants against the Rocky Mountains, requiring the daily broadcasting of air-quality reports for sensitive persons, such as senior citizens and cardiorespiratory patients, who are often advised to stay indoors when the air quality is poor.

RADIATION, PHYSICAL HAZARDS, AND NATURE

Ever since the 1945 atomic bomb attack on Hiroshima, Japan, scientists have become increasingly concerned about the health effects of radioactive pollution. Even small-scale testing of nuclear weapons directly affects the environment, a realization that led the United States, Great Britain, and the Soviet Union to sign the Limited Test Ban Treaty in 1963. Both ionizing and nonionizing radiation can cause acute and chronic health problems such as chromosome damage, with workers continually exposed to radioactive metals and X rays being most susceptible. The 1986 Chernobyl nuclear plant malfunction in the Soviet Union, among the worst peacetime nuclear disasters in history, caused cancer, birth defects, and skin disease among those exposed to radiation. The disposal of nuclear wastes also poses health concerns, given that many radioactive substances have a half-life of more than ten thousand years.

The predominant source of physical hazards that cause environmental illnesses are human-made environments that increase the incidence of traumatic injuries and create noise pollution. Accidents occurring in unsafe work surroundings account for a large proportion of preventable injuries. Noise in the workplace can cause hearing loss, the most prevalent occupational impairment, which can progress to permanent deafness. Health problems related to noise pollution have increased over time among musicians and their audiences, as well as in urban environments where the constant din of traffic and construction contribute to illnesses such as headaches, depression, and insomnia.

Nature can also cause illnesses as it responds to ecological imbalances introduced by human beings; examples are rabies, giardiasis, Rocky Mountain spotted fever, Lyme disease, and the diseases caused by hantaviruses. Hantaviruses do not cause obvious illness in their rodent hosts, but their effects are transmitted to humans who inhale dust or mist containing dried traces of the urine or feces of infected animals. Hantaviruses are distant cousins of the fearsome ebola virus, and hantavirus outbreaks must be handled much like outbreaks of hepatitis. Rabies is transmitted to humans through bites or scratches containing the saliva of rabid animals such as infected dogs or bats. The disease attacks the nervous system. Giardiasis is a nonbacterial intestinal illness caused by a parasite found in untreated or improperly treated surface water taken from streams and lakes. Symptoms of infection include diarrhea, nausea, reduced appetite, abdominal cramps, bloated stomach, and fatigue.

The symptoms of Rocky Mountain spotted fever, an infection caused by a dog tick, are fever, headache, rash, and nausea. As infection progresses, the original red spots may change in appearance to look more like bruises or bloody patches under the skin. Lyme disease was classified following a mysterious juvenile arthritis outbreak and has since accounted for more than 90 percent of vector-borne illnesses in North America. Spread exclusively through bites from infected ticks, its early stages are marked by fatigue, malaise, chills, fever, headaches, muscle and joint pain, swollen lymph nodes, and skin rashes. Later stages may include arthritis, nervous system abnormalities, and heart conduction disturbances.

AGENCIES AND LEGISLATION

Agencies that have federal authority to investigate environmental issues related to disease in the United States include the Department of Labor, under which fall the Environmental Protection Agency (EPA) and the Occupational Safety and Health Administration (OSHA), and the Department of Health and Human Services, under which fall the Food and Drug Administration (FDA), the National Institutes of Health (NIH), the Centers for Disease Control and Prevention (CDC), and the Health Resources and Services Administration (HRSA). The National Institute of Occupational Safety and Health (NIOSH) conducts ongoing research to identify hazards and develop safety standards, and many large companies now employ industrial health advisers.

International coordination regarding environmental and occupational health concerns is provided by the World Health Organization (WHO), founded in 1942 as an agency of the United Nations. WHO is extremely active in developing countries as industrialization, poverty, and population growth continue to increase. Its broad scope of activities includes control-

ling widespread disease such as malaria and tuberculosis, establishing purified water supplies and sanitation systems, and providing health education and health planning assistance.

During the 1960's, the U.S. Congress increasingly took up legislation intended to regulate workplace practices and sources of pollution that could lead to environmental illnesses. Federal laws that remain the most relevant include the Occupational Safety and Health Act of 1970, the Federal Environmental Pesticide Control Act of 1972, the Toxic Substances Control Act of 1976, the Resource Conservation and Recovery Act of 1976, and the Comprehensive Environmental Response, Compensation, and Liability Act of 1980 (CERCLA; also known as Superfund). In 1985 several "right to know" laws went into effect; these laws require the managers of manufacturing plants to supply employees with health and safety information regarding toxic materials.

Daniel G. Graetzer

FURTHER READING

Barrett, Stephen, and Ronald E. Gots. *Chemical Sensitivity: The Truth About Environmental Illness.* Amherst, N.Y.: Prometheus Books, 1998.

Brown, Phil. *Toxic Exposures: Contested Illnesses and the Environmental Health Movement.* New York: Columbia University Press, 2007.

Gittleman, Ann Louise. *How to Stay Young and Healthy in a Toxic World.* New York: McGraw-Hill, 1998.

Kroll-Smith, J. Stephen, and H. Hugh Floyd. *Bodies in Protest: Environmental Illness and the Struggle over Medical Knowledge.* 1997. Reprint. New York: New York University Press, 2000.

Nash, Linda. *Inescapable Ecologies: A History of Environment, Disease, and Knowledge.* Berkeley: University of California Press, 2006.

Vig, Norman J., and Michael E. Kraft. *Environmental Policy in the 1990's.* Washington, D.C.: CQ Press, 1990.

Wargo, John. *Our Children's Toxic Legacy: How Science and Law Fail to Protect Us from Pesticides.* New Haven, Conn.: Yale University Press, 1998.

SEE ALSO: Air pollution; Airborne particulates; Asbestos; Cancer clusters; Carcinogens; Dichloro-diphenyl-trichloroethane; Environmental health; Lead; Limited Test Ban Treaty; Right-to-know legislation; Sick building syndrome; Silicosis; Smog; Water pollution.

Environmental impact assessments and statements

CATEGORY: Land and land use

DEFINITION: Evaluations of the likely impacts of proposed or existing human activities on given environments, and the reports of the findings in documents for public review

SIGNIFICANCE: Implementation of legal requirements that environmental impact assessments be performed and statements of the results be made public have resulted in changes in construction, resource extraction, and land-use planning that take into account the larger environment.

Various kinds of human activities may trigger environmental impact assessments (EIAs); these include construction, resource extraction, and land-management policy implementation. An entity planning a development can choose to do an EIA, but formal EIAs are mandated responses to specific legislation. Under some legislation, an initial scoping process is done to determine whether a more lengthy process, the EIA, is required. The results of a legally mandated or regulatory EIA are documented in an environmental impact statement (EIS).

The mandating of EIAs has become increasingly common in the United States and in other nations because of intensified public pressure on legislators and policy makers to improve resource management and conservation. The U.S. National Environmental Policy Act of 1969 (NEPA) helped usher in the era of environmental assessment for government decision making. By 1998 more than one hundred countries had established EIA processes. Ideally, administration of an EIA program promotes government and corporate accountability for environmental alterations and institutionalizes systematic, science-based policy analysis.

The environmental movement, spurred by Rachel Carson's book *Silent Spring* (1962), influenced U.S. legislators to reconsider the lack of a national policy for the environment. NEPA, signed into law on January 1, 1970, heralded a new role for citizens to participate in reviews of government decisions. The EIA process produces a draft environmental impact statement (DEIS) and a final environmental impact statement (FEIS) for public comment. However, once the FEIS has been accepted, NEPA has been satisfied, but this in itself does not constitute approval or denial of a proposed project. By focusing on the process rather than the end result (as most environmental permit programs do), the EIA reflects a compromise between environmental and political interests. The goal is to ensure that a suitable EIS or "finding of no significant impact" (FONSI) is prepared. A well-planned project should be able to withstand the public scrutiny. Other laws and permitting processes may be required before a proponent can actually go ahead with a proposed development or action, but these processes can build upon or use the data gathered in the EIA.

In the United States, NEPA marked a change for federal agencies because it added environmental accountability to every agency's mission, along with a specific method to carry out environmental reviews. NEPA not only provided a common thread among agencies but also comprehensively linked various categories or media in which environmental impacts occur. In the EIA process, impacts ranging from archaeological resource depletion to air pollution are examined. Social impacts resulting from proposed actions are as much a part of the EIA as are issues related to water resources, noise, solid waste disposal, and other common types of environmental impacts. Aesthetics has also proved an important if initially nebulous category of impact, although a significant body of literature has arisen to treat the need to quantify impacts normally considered subjective. Socioeconomic values of environmental resources (for life support, amenities, and raw materials) are also used in evaluating trade-offs among alternatives. Formal EIAs in most countries generally include consideration of similar wide arrays of environmental impacts.

The broad categories of impacts are intended to reflect the interconnectedness of environmental settings and allow the interplay of social, economic, political, and environmental issues. The breadth also extends to the type of projects that require environmental assessment. In the United States, projects that are subject to the EIA process include any undertaken by a federal agency; any involving a federal license, permit, or funding; and any taking place on federal property. Approximately two dozen U.S. states have their own equivalent assessment requirements. Nations that use the EIA process are more readily able to participate in global trade, qualify for funding, and meet the increasing international demand for and appreciation of environmental quality.

An EIA assesses more than the proposed action; it also looks at the impacts of legitimate alternatives to the action, as well as the impacts of not doing the project. The EIA process includes comparing the costs and benefits of the alternatives and the various impacts. The EIS documents the impacts, costs, and benefits for the project and its alternatives. Under NEPA-type legislation, a DEIS is circulated, and public comments are solicited either in writing or at hearings. The FEIS is issued after consideration of public and other agency input. The courts provide a forum for class-action suits and other assessment-related disputes. The majority of challenges have been based on allegations of either failure to prepare an EIS or failure to consider fully the proper alternatives. From a public policy perspective, as well as from the perspective of peer-reviewed science, it is the public nature of the EIS that determines the success of the EIA process.

The EIA ideally is part of the planning process rather than an afterthought for a project that has already commenced; however, the EIA must be conducted late enough in the planning stages that a sufficient description of the project exists to allow assessment of the impacts. One response to this problem is the strategic environmental assessment (SEA). The forecasting of impacts, especially for SEAs, cannot simply be done through direct observation. In conducting SEAs, teams of professionals use physical models, mathematical models, qualitative models, checklists, and expert opinions.

Starting with a project description, an EIS proceeds to identification of associated or expected direct and indirect impacts. Next, the existing environmental conditions are described. Then relevant laws and regulations are examined for standards and applicability. Specific environmental impacts are predicted, and their significance is evaluated. The final step is description of the incorporation of the EIA's results into the project to reduce or mitigate the impacts; this includes information on monitoring, reporting, and responding to postconstruction impacts. EIA and EIS notifications appear in legal sections of major newspapers, on government and corporate Internet sites, and at agency offices. EIS's are generally available from the preparers upon request or can be viewed by concerned citizens at various public locations.

Robert M. Sanford and Hubert B. Stroud

FURTHER READING
Cantor, Larry W. *Environmental Impact Assessment.* New York: McGraw-Hill, 1996.

Garb, Yaakov, Miriam Manon, and Deike Peters. "Environmental Impact Assessment: Between Bureaucratic Process and Social Learning." In *Handbook of Public Policy Analysis: Theory, Politics, and Methods,* edited by Frank Fischer, Gerald J. Miller, and Mara S. Sidney. Boca Raton, Fla.: CRC Press, 2007.
Glasson, John, Riki Therivel, and Andrew Chadwick. *Introduction to Environmental Impact Assessment.* 3d ed. New York: Routledge, 2005.
Rogers, Peter P., Kazi F. Jalal, and John A. Boyd. "Environmental Assessment." In *An Introduction to Sustainable Development.* Sterling, Va.: Earthscan, 2008.

SEE ALSO: Environmental law, international; Environmental law, U.S.; Land-use planning; Land-use policy; National Environmental Policy Act.

Environmental justice and environmental racism

CATEGORY: Philosophy and ethics
DEFINITIONS: Environmental justice is the fair treatment of all individuals in terms of issues related to the environment, particularly environmental hazards; environmental racism involves patterns of racial prejudice with regard to such issues
SIGNIFICANCE: The issues of environmental justice and environmental racism have become increasingly important elements in debates regarding industrial and government environmental practices that have impacts on low-income and other disadvantaged communities. Although the goals of a clean environment for all and an end to racist practices are attractive to many people, achieving these goals can be difficult, because human beings tend to desire both justice and manufactured goods that are linked to environmental degradation.

During the 1960's most Americans involved in environmental activities were white and members of the upper or middle classes. Issues such as conservation and preservation of natural resources were of little interest to members of low-income and minority groups, who were often more concerned with civil rights and the improvement of economic conditions. In the 1970's and 1980's, however, as concerns mounted about lead poisoning, the dangers of haz-

Benjamin Chavis, left, executive director of the National Association for the Advancement of Colored People (NAACP), speaks with U.S. vice president Al Gore during the Summit on Environmental Justice, held in December, 1993, in Washington, D.C. At the summit, Gore announced that President Bill Clinton's administration planned to issue a presidential directive to respond to what Gore called a crisis of disproportionate pollution in minority communities. (AP/Wide World Photos)

ardous waste dumps, and the effects of soil and water pollution, minority leaders began to take notice. Researchers found that garbage dumps and other contaminated sites were disproportionately located near communities with higher-than-average percentages of minority group residents. A 1983 study of landfill and incinerator sites in Houston, Texas, for example, showed that these facilities were usually found near African American neighborhoods. This and other studies led to a grassroots movement during the 1980's to address the problem known as environmental racism.

Activists contend that racism continues to prompt governments to issue permits for waste facilities in low-income and minority areas. In some cases, communities have welcomed such facilities as sources of employment. In addition, members of these communities are less likely to possess the knowledge and the resources to oppose regulatory decisions. Through public protests and political pressure, activists seek to bring national attention to the problem, with the hope of influencing the decisions of local and national policy makers.

Grassroots organizations can point to some successes. Residents of one California community suc-

cessfully pressured their town council to implement a program to screen for environmental lead poisoning. When citizens in Halifax County, Virginia, learned that the federal government was considering their community as the location for a nuclear waste depository, they formed a group to fight the proposal. More than fourteen hundred residents, both African American and white, voiced their opposition at a public meeting, and shortly thereafter, government officials dropped the county as a potential depository site. In 1986 the residents of Revilletown, Louisiana, received cash settlements and relocation assistance from a nearby chemical manufacturer after they sued the company for damages to health and property caused by emissions from the chemical plant. These and other victories indicate the potential power of grassroots movements that seek environmental justice.

Such successes do not mean that these movements are without their critics, however. Some argue that the concept of environmental justice is so broad and vague that it cannot serve as a guide for policy makers. Moreover, they maintain that the available evidence regarding environmental racism is flawed. They contend that studies have failed to determine whether harmful facilities have been located in already existing minority communities or the communities coalesced around the facilities.

Proponents of environmental justice reject these arguments as further evidence of injustice and racism. They claim that major corporations that hope to maximize profits have a vested interest in attacking the movement, which, if successful, would significantly raise their costs of production and waste disposal. Activists also complain that the national news media consistently ignore environmental racism, favoring instead sensational stories that do not examine the deeper institutional causes of the environmental disasters featured in the headlines. Finally, proponents also fault the mainstream environmental movement for its fixation on preservation issues.

Despite continuing debates over the meaning of environmental justice and the reality of environmental racism, some politicians have perceived the issues involved as deserving of legislation. In 1992 two members of the U.S. Congress sponsored the Environmental Justice Act, which would have required the Environmental Protection Agency (EPA) to identify and monitor areas with high levels of toxic chemicals. The measure failed, but the legislators' effort brought heightened attention to the issue. Two years later, President Bill Clinton issued Executive Order 12898, which requires federal agencies to pursue environmental justice and acknowledges the existence of environmental racism. The order has had limited impact, but it has drawn more attention to the issue.

One major problem regarding accusations of environmental racism centers on the matter of proof. Courts require evidence that the alleged racism is intentional, and in most cases this is impossible to prove. However, in 1997 a federal judge ruled that suits can be filed on the basis of "disparate impact," which means that the effect of racial discrimination, regardless of the intent, can be used to assess responsibility. This decision was a victory for activists, as it gave them increased opportunities to pursue remedies in the courts. However, some observers noted that the ruling would prompt industrial interests, many of which would not be guilty of polluting, to avoid siting their facilities in minority communities out of fear of expensive lawsuits.

The issue of environmental justice has garnered support because the stated goals of advocates—a clean environment for all and an end to racist industrial and government practices—are attractive to many people. Achieving those goals has proven difficult, however. People tend to desire both justice and the manufactured goods that cause environmental degradation. While most people oppose racism, they understandably have no desire to relocate polluting industries into their own neighborhoods. In addition, as the debate about disparate impact indicates, conflicts over environmental racism may have the unintended consequence of denying low-income people jobs that they desire.

Thomas Clarkin

FURTHER READING

Arnold, Craig Anthony. *Fair and Healthy Land Use: Environmental Justice and Planning.* Chicago: American Planning Association, 2007.

Checker, Melissa. *Polluted Promises: Environmental Racism and the Search for Justice in a Southern Town.* New York: New York University Press, 2005.

Pellow, David Naguib, and Robert J. Brulle, eds. *Power, Justice, and the Environment: A Critical Appraisal of the Environmental Justice Movement.* Cambridge, Mass.: MIT Press, 2005.

Rhodes, Edwardo Lao. *Environmental Justice in America: A New Paradigm.* Bloomington: Indiana University Press, 2003.

Sandler, Ronald, and Phaedra C. Pezzullo, eds. *Environmental Justice and Environmentalism: The Social Justice Challenge to the Environmental Movement.* Cambridge, Mass.: MIT Press, 2007.

Shallcross, Tony, and John Robinson, eds. *Global Citizenship and Environmental Justice.* New York: Rodopi, 2006.

SEE ALSO: Hazardous and toxic substance regulation; Hazardous waste; Incineration of waste products; Landfills; NIMBYism; Right-to-know legislation; Solid waste management policy; Waste management.

Environmental law, international

CATEGORIES: Treaties, laws, and court cases; preservation and wilderness issues; resources and resource management; human health and the environment

DEFINITION: Internationally accepted rules governing the conduct of sovereign nations in their relationships with one another for the international regulation of environmental issues aimed at environmental protection and sustainability

SIGNIFICANCE: International environmental laws regulate the interactions of human beings with the rest of the biophysical or natural environment, with the aim of reducing the negative impacts of human activity on the natural environment.

International environmental laws promote responses to serious threats to the global environment and address some of the most challenging global issues, especially in areas of trade and environment, habitat protection, climate change, oceans and fisheries, water scarcity, endangered species, and access to biological resources. Because pollution does not respect international or political boundaries, environmental law is an important aspect of international relations.

Encompassing agreements among nations, international environmental laws fall into two broad categories: pollution control and remediation, and resource conservation and management. The boundaries of international environmental law are difficult to determine, and therefore it cannot usually be restricted to one specific definition. Rather, it is interdisciplinary and also overlaps and intersects with numerous other areas of research, including ecology, human rights, economics, political science, and navigation.

International environmental law also engages with global efforts to develop commonly accepted environmental and social standards and to govern major investments in developing countries. These efforts involve lawyers and practitioners in the field of environmental law who help to design and enforce global environmental standards.

While the international bodies that proposed, argued, agreed upon, and ultimately adopted existing international agreements vary according to each agreement, certain conferences and commissions have been particularly important, including the United Nations Conference on the Human Environment (1972), the World Commission on Environment and Development (1983), the United Nations Conference on Environment and Development (1992), the World Summit on Sustainable Development (2002), and the United Nations Climate Change Conference, which produced the Copenhagen Accord (2009).

History and Background

Environmental law may be said to have begun in 1306, when England's King Edward I banned the burning of coal in London because of the polluting smoke that it produced. Until the late 1960's, most international agreements that aimed at protecting the environment were only narrowly defined. In 1972, the Stockholm Declaration of the United Nations Conference on the Human Environment paved the way for international agreements that began to promote legislation for the preservation of the environment.

The twentieth century movement known as environmentalism brought about many environmental protection laws at the local, state, national, and international levels. These laws create liabilities for businesses that pollute air, water, or soil or improperly dispose of waste. Legally binding international agreements deal with many types of environmental issues, stretching from terrestrial, freshwater, marine, and atmospheric pollution to the protection of biodiversity and wilderness. Numerous international principles and rules have emerged since the late twentieth century to challenge many of the established rules and principles in the international legal field. Several multilateral environmental agreements have been adopted, and international environmental rules regulate almost every environmental issue imaginable, including marine pollution, hazardous activities, atmospheric pollution, waste management, and access to environmental information.

International environmental law in the twenty-first century has focused in particular on such global environmental problems as the depletion of the ozone layer, transboundary movements of hazardous wastes, the conservation of biological diversity, and the international response to climate change. Principles of ecology, conservation, stewardship, responsibility, and sustainability have all influenced modern international environmental law. Numerous international conventions and other agreements set environmental law concerning many issues, including climate change and global warming (the United Nations Framework Convention on Climate Change, the Kyoto Protocol, and the Copenhagen Accord), sustainable development (the Rio Declaration on Environment and Development), biodiversity (the U.N. Convention on Biological Diversity), transfrontier pollution (the U.N. Convention on Long-Range Transboundary Pollution), marine pollution (the U.N. Convention on the Prevention of Marine Pollution by Dumping of Wastes and Other Matter), endangered species (the U.N. Convention on International Trade in Endangered Species, or CITES), hazardous materials and activities (the Basel Convention on the Control of Transboundary Movements of Hazardous Wastes and Their Disposal), cultural preservation (the Convention Concerning the Protection of the World Cultural and Natural Heritage), desertification (the U.N. Convention to Combat Desertification), and uses of the seas (the U.N. Convention on the Law of the Sea, or UNCLOS).

Global climate change resulting from greenhouse gas emissions has become a particularly important issue in international environmental law. The United Nations Climate Change Conference, held in December, 2009, produced the Copenhagen Accord, which was drafted by the United States, China, India, Brazil, and South Africa. The resulting arguments and unanswered questions caused the U.S. government to agree only tentatively to the proposal, which was not unanimously and officially adopted by all participating countries. The participating countries drafted a document that emphasizes the importance of climate change as one of the most challenging environmental issues of the twenty-first century. Although not yet legally binding, the accord asserts that decisive actions should be taken to limit global temperature increases to less than 2 degrees Celsius (3.6 degrees Fahrenheit). No consensus was reached regarding the reduction of emissions of carbon dioxide—acknowledged by many scientists to be a chief factor in global warming (carbon dioxide is the most important of the "greenhouse gases," gases that contribute to the atmosphere's trapping of radiant heat). Although the contents of the Copenhagen Accord have been opposed by many countries and nongovernmental organizations, and many critics of the accord have asserted that it is only a weak version of what is needed to achieve meaningful action on climate change, by January, 2010, at least 138 countries had signed the agreement.

INTERNATIONAL ENVIRONMENTAL LAW
ORGANIZATIONS

Around the world, intergovernmental and nongovernmental organizations seek to promote the adoption and enforcement of internationally accepted environmental laws. Among these organizations are the following:

- Intergovernmental Panel on Climate Change (IPCC): This panel was established by the World Meteorological Organization and the United Nations Environment Programme. The goal of the IPCC is to engage in total and objective assessment of all scientific, technical, and socioeconomic information relevant to an understanding of the scientific basis of the risks posed by climate change induced by humans, the potential impacts of such risks, options for adaptation to climate change, and ways to reduce the effects of climate change.
- Commission for Environmental Cooperation (CEC): This commission, which was created by the North American Agreement on Environmental Cooperation, a side treaty of North American Free Trade Agreement, addresses regional environmental concerns, helps prevent potential trade and environmental conflicts, and promotes the effective enforcement of environmental laws. Members of the CEC are the United States, Canada, and Mexico.
- United Nations Environment Programme (UNEP): UNEP was established as the environmental conscience of the United Nations system. It provides an integrative and interactive mechanism through which a large number of separate efforts by intergovernmental, nongovernmental, national, and regional bodies in the service of the environment are reinforced and interrelated.
- International Maritime Organization (IMO): This organization, which handles marine pollution regulations, is the specialized agency of the United Nations responsible for improving maritime safety and preventing pollution from ships.

Other intergovernmental organizations with interests in international environmental law include the United Nations Commission on Sustainable Development, the United Nations Educational, Scientific, and Cultural Organization's World Heritage Center, the European Commission's Environmental Directorate General, the European Environment Agency, the World Conservation Union, the International Union for Conservation of Nature, the World Bank, and the World Meteorological Organization.

The Center for International Environmental Law (CIEL) is an example of a nongovernmental organization that works to support the creation and enforcement of environmental laws around the world. This public interest nonprofit law firm, which has offices in the United States and Switzerland, is committed to strengthening existing international environmental laws and using international law and institutions to protect the environment, promote human health, and ensure a just and sustainable society. Other examples of nongovernmental organizations that support international environmental law include the International Environmental Law Committee of the American Bar Association (which considers, informs, and engages its members on public international environmental law, such as global, multilateral, regional and bilateral agreements on the environment), Earthjustice (formerly the Sierra Club Legal Defense Fund), the EnviroLink Network, the Environmental Law Alliance Worldwide, Greenpeace International, the Natural Resources Defense Council, Friends of the Earth International, Resources for the Future, the World Resources Institute, and the World Wide Fund for Nature.

CONTROVERSIES

The definition of "international environmental law" can sometimes generate controversy. Given the broad scope of environmental law, no fully definitive list of environmental laws is possible; the field is made up of a complex of interlocking global, international, national, state, and local treaties, conventions, statutes, regulations, and policies that seek to protect the environment and natural resources affected or endangered by human activities. Together, these instruments attempt to tackle some of the world's most challenging and controversial environment-related issues, including climate change, pollution, resource conservation, trade policies, management of fisheries, and biodiversity preservation.

Controversy arises from the nexus of these goals and economic priorities: From an economic perspective, environmental law may be understood as concerned with the prevention of present and future environmental problems and the preservation of common resources in order to avoid their exhaustion. The limitations and expenses that international environmental laws may impose on commerce are often discussed in the context of the often immeasurable financial benefits of environmental protection. Where policies to promote long-term but difficult-to-measure environmental priorities conflict with policies that promote short-term economic gains (particularly during economic downturns such as the global recession that began in 2008), persons who are suffering economically often find economic goals more compelling and easier to understand than environmental goals.

The concept of environmental justice is also a source of controversy. The development, implementation, and enforcement of environmental laws, regulations, and policies can have disproportionate impacts on particular socioeconomic, ethnic, or national groups of people. Therefore, one of the goals of environmental law must be to clearly define and equitably administer environmental justice for all nations and peoples, by providing each with the same degree of protection from environmental and health hazards and equal access to the decision-making process that seeks to provide healthy environmental living conditions.

Among scientists and politicians, debates are ongoing regarding the existence and causes of climate change—particularly whether it is mainly the result of human activities. If long-term global warming is truly occurring, it will have significant future impacts on agriculture, sea levels, the spread of disease, the deterioration of polar ice (the world's "air conditioner"), species (including human) migration, and biological diversity. If, in addition, global warming can be determined to be primarily the result of human activities (as the IPCC maintains), then the argument can be made that a reduction, or at least an arrest, of human activities contributing to climate change must be implemented. The challenge then becomes how to regulate human-induced global warming with minimal interference in the global economy—and how to do so with minimal disproportionate impacts on particular populations.

That challenge requires creative and innovative law and policy making at all levels of society. Governments around the world must continue to discuss how best to combat climate change and adapt to its effects. Appropriate international environmental laws must be formulated and enforced if goals for the mitigation of negative climate change impacts are to be met. There is a need to gain not only the theoretical but also the proactive support of developing countries for the 2009 Copenhagen Accord, by allowing underprivileged nations their total inputs and helping them to gain an understanding of how to comply with the agreement's complex or misunderstood provisions.

Samuel V. A. Kisseadoo

FURTHER READING

Birnie, Patricia, Alan Boyle, and Catherine Redgwell. *International Law and the Environment.* 3d ed. New York: Oxford University Press, 2009.

Bodansky, Daniel. *The Art and Craft of International Environmental Law.* Cambridge, Mass.: Harvard University Press, 2010.

Guruswamy, Lakshman D., with Kevin L. Doran. *International Environmental Law in a Nutshell.* 3d ed. St. Paul, Minn.: Thomson/West, 2007.

Hunter, David, James Salzman, and Durwood Zaelke. *International Environmental Law and Policy.* 3d ed. New York: Foundation Press, 2006.

Louka, Elli. *International Environmental Law: Fairness, Effectiveness, and World Order.* New York: Cambridge University Press, 2006.

Sands, Philippe. *Principles of International Environmental Law.* 2d ed. New York: Cambridge University Press, 2003.

Weiss, Edith Brown, et al. *International Environmental Law and Policy.* 2d ed. Frederick, Md.: Aspen Law & Business, 2006.

SEE ALSO: Basel Convention on the Control of Transboundary Movements of Hazardous Wastes; Convention on Long-Range Transboundary Air Pollution; Intergovernmental Panel on Climate Change; International Atomic Energy Agency; Johannesburg Declaration on Sustainable Development; Kyoto Protocol; London Convention on the Prevention of Marine Pollution; Rio Declaration on Environment and Development; Stockholm Convention on Persistent Organic Pollutants; United Nations Conference on the Human Environment; United Nations Convention to Combat Desertification; United Nations Environment Programme; United Nations Framework Convention on Climate Change; World Summit on Sustainable Development.

Environmental law, U.S.

CATEGORIES: Treaties, laws, and court cases; preservation and wilderness issues; resources and resource management; human health and the environment

DEFINITION: Federal and state legislation regulating uses of the environment

SIGNIFICANCE: Although a relatively recent aspect of the American legal landscape, environmental law has come to play a dominant role in the use of natural resources. Environmental legislation is often controversial in the political arena, but there is widespread popular support among Americans for protecting the environment through statutes, regulation, and court cases.

At the origin of the American legal system no body of laws existed that directly regulated the environment. Over the course of the twentieth century, however, the nation built up a body of state and federal laws designed to conserve, protect, and restore the environment. Several regulatory agencies are entrusted with the authority to administer these laws. Although there is a large body of environmental law in twenty-first century United States, it is not easy to navigate, as it is divided among federal, state, and even international jurisdictions, is made up of numerous statutes across various legal codes, and is administered in pieces by a host of administrative agencies. For example, the Occupational Safety and Health Act of 1970, which addresses workplace safety issues, has numerous provisions relating to environmental concerns. This law's provisions are administered by both the Department of Labor and the Environmental Protection Agency (EPA).

CONSERVATION LAWS

Sustained environmental legislation in the United States began as efforts to conserve public land. Under early common law, the right to property was considered nearly absolute, even when it included the extravagant consumption of natural resources. The influential eighteenth century British jurist William Blackstone described the rights of a property owner over his or her land as "sole and despotic." The only environmental restraint on use of property was direct damage to the property of another person through the law of nuisance or trespass. Although the emphasis among early American settlers was on bringing natural resources quickly into economic use, there was also a sense, dating from the colonial era, that common lands needed to be conserved. For example, a South Carolina statute from 1671 outlawed the poisoning of waterways with impurities. In 1681 William Penn decreed that for every five trees cut down in his Pennsylvania colony, one had to be conserved. The public trust doctrine obligated states to protect tidal shorelines for the common enjoyment of the public.

With the heavy industrialization of the United States during the late nineteenth century, an awareness began to grow that the nation's natural bounty was becoming endangered. Writers such as Henry David Thoreau and John Muir pointed out the fragile beauty of the land and did much to foster a nascent conservation movement. In 1872 the U.S. Congress established Yellowstone National Park, the first national park, to preserve the land from spoliation. The first federal environmental law is generally considered to be the Rivers and Harbors Act of 1899, which banned pollution of the nation's waterways. The Burton Act of 1905 limited hydroelectric power drawn from Niagara Falls. In 1916 Congress established the National Park Service. Everglades National Park was created in 1947. The Wilderness Act of 1964 allowed for what would eventually be more than 40.5 million hectares (100 million acres) to be set aside as wilderness areas.

By 2010 the United States had created 392 national parks comprising 34 million hectares (84 million acres) of conserved land; 552 national wildlife refuges were also in existence. In addition, various social and

Enacted U.S. Legislation Relating Directly to Climate Change

Year	Act	Effect
1978	National Climate Program Act	Establishes a nationally coordinated program of climate monitoring and prediction
1990	Global Change Research Act	Funds research into global climate change
1997	Byrd-Hagel Resolution	Expresses the sense of the Senate that the United States should join only those climate change treaties that do not harm the domestic economy and that require developing nations, as well as developed nations, to take action against global warming
2005	Energy Policy Act of 2005	Supports voluntary reductions in carbon-intensive activities and the export to developing nations of technologies to reduce carbon intensity

economic legislation passed during the Progressive and New Deal eras had beneficial effects on conservation. For example, the Civilian Conservation Corps was created in 1933 as a New Deal public works program with a focus on the conservation of natural resources.

ENVIRONMENTAL PROTECTION LEGISLATION

Although the conservation laws noted above were important for the preservation of federal lands, they were not concerned with protecting the environment in general. During the 1970's, however, the environmental movement came of age, so much so that this period is sometimes described as the "environmental decade." Writers such as Marjorie Stoneman Douglas and Rachel Carson had earlier publicized the plight of endangered habitats; the nation was shocked in 1969 when the polluted Cuyahoga River in Cleveland caught fire. The first Earth Day was held in 1970. The American people had increasingly become aware that something had to be done to protect a decaying environment.

Congress responded by passing the National Environmental Policy Act of 1969 (actually signed into law on January 1, 1970), which established a framework for comprehensive supervision of the environment. Under this framework Congress passed wide-ranging legislation that constitutes the core of U.S. environmental law. In 1970 Congress enacted amendments to the 1963 Clean Air Act that established regulations on emissions from factories and automobiles. The amendments set minimum standards of air quality and required industries to meet those standards by reducing conventional industrial pollutants. (In 1990 the Clean Air Act was again significantly amended to

include requirements that would cut emission of chlorofluorocarbons, or CFCs, and also reduce acid deposition and acid rain by limiting sulfur dioxide transmissions.) The Clean Water Act (also known as the Federal Water Pollution Control Act Amendments) of 1972 regulates discharge of pollutants and toxic substances into waterways, as well as the filling of wetland areas.

The 1947 Federal Insecticide, Fungicide, and Rodenticide Act was amended in 1972 to regulate the use of pesticides. The Noise Control Act, also signed into law in 1972, addresses excessive noise. Another environment-related law passed in 1972 is the Marine Mammal Protection Act, which protects endangered sea life such as whales, dolphins, sea otters, and seals. Of wider scope was the Endangered Species Act of 1973, which contains provisions aimed at the maintenance of biological diversity and the protection of animal groups in danger of extinction. The Safe Drinking Water Act of 1974 focuses on waters that are the source of drinking water in the United States, mandating basic standards of safety and quality. The Solid Waste Disposal Act and the Resource Conservation and Recovery Act, both enacted in 1976, regulate the treatment and disposal of hazardous wastes. The Toxic Substances Control Act, also passed in 1976, regulates the commercial use and removal of toxic substances such as asbestos, radon, and polychlorinated biphenyls (PCBs).

As important as the laws that were enacted were the administrative agencies entrusted with the laws' implementation and enforcement. Administrative agencies are a modern phenomenon of the American legal system, created in the twentieth century as governmental entities with a mixture of executive, legislative, and judicial functions. Administrative agencies are delegated the authority to enact regulations implementing congressional legislation, to enforce those regulations, and to adjudicate disputed issues. The agency that has the most direct oversight of environmental law is the Environmental Protection Agency, which Congress established in 1970 to consolidate enforcement of various federal laws and duties relating to the environment.

The EPA has functions relating to air, water, and noise pollution and to the handling of toxic and waste substances. It both sets standards and regulations implementing congressional legislation in these areas and enforces these standards with permits, sanctions, lawsuits, and other remedies. The Department of the Interior oversees the national parks and other vital resources. It plays a crucial role in conserving federal lands, forests, and parks; managing irrigation and supplying fresh water; and protecting marine and land wildlife. The Department of Energy runs programs to promote the use of solar energy and the conservation of fossil-fuel resources and to ensure the proper disposal of by-products of energy development, such as radioactive waste. The Department of Agriculture addresses soil issues in forests and on farms. The Council on Environmental Quality, a division of the executive office of the president, is responsible for coordinating governmental responses to environmental issues.

RESTORATION, CLEANUP, AND LITIGATION

The environmental laws enacted in the 1970's focus on protecting the environment from future harm. Once these laws were in place, legislators turned to the question of repairing and restoring damage that had already been done. The Comprehensive Environmental Response, Compensation, and Liability Act of 1980 (CERCLA) created a mechanism (a Superfund) to fund the cleanup of certain environmental disasters—toxic sites that had been caused by waste products. By 1992 more than twelve hundred toxic sites were being cleaned up under Superfund supervision. In addition, CERCLA allows for criminal liability in certain circumstances, in addition to civil penalties.

In 1989 the *Exxon Valdez* oil tanker spilled 11 million gallons of oil onto Prince William Sound off the coast of Alaska. Litigation in the case of *Exxon v. Baker* over the extent of Exxon's liability, including punitive damages, went on for two decades. In response to the spill, Congress enacted the Oil Pollution Act of 1990.

Most U.S. environmental statutes have specific clauses allowing private citizens and entities to sue for environmental harms. Many of the legal debates surrounding these clauses have involved the concept of standing—that is, who has the right to sue. Standing is a traditional legal concept that indicates which persons have suffered sufficient direct harm such that they can become plaintiffs. A major case involving standing was *Massachusetts v. Environmental Protection Agency* (2007), in which the U.S. Supreme Court permitted a coalition of environmental groups and state attorneys general to bring a lawsuit compelling the EPA to determine whether carbon dioxide emissions are an air pollutant under the Clean Air Act.

Perhaps the paradigm environment-related litigation is that concerning asbestos. Hundreds of thousand of Americans are estimated to have contracted asbestos-related cancers. Although there is no question that the nation owes relief to these victims, the long-running asbestos litigation has been criticized. For example, the Manhattan Institute's Center for Legal Policy asserted in a 2003 report that massive asbestos tort litigation was more suited to enriching lawyers than to compensating victims. The Manhattan Institute estimated that of the $70 billion paid out by companies for asbestos claims, $40 billion had gone to lawyers. Companies were driven into bankruptcy by asbestos lawsuits, some of which had little connection to the original asbestos exposure.

Some commentators continue to express fears that widespread environmental lawsuits and mass tort litigation are a boon to lawyers but costly for everybody else. Another lively debate concerns whether governmental supervision of polluters should consist of mandated design and performance standards or market-based incentives.

STATE AND INTERNATIONAL LAWS

The individual U.S. states have environmental laws that run parallel to federal mandates, and the interaction between federal and state laws is complex. Federal environmental laws have priority, but under most federal environmental legislation, state authorities are delegated the power to supervise environmental standards. In other words, even though federal legislation may set the guidelines, state agencies run the programs. For example, the EPA sets recommended standards for water quality under the Clean Water Act, but the states enact specific standards for their localities. In the absence of federal legislation addressing greenhouse gas emissions, numerous states, California in particular, have passed legislation to restrict such emissions. One of the most significant pieces of interstate environmental legislation is the Great Lakes Compact, a legal agreement among eight states that details the use and management of the Great Lakes water supply.

Environmental law in the United States can also be affected by international agreements, especially as environmental problems are of an increasingly global dimension. The best-known examples are agreements regarding greenhouse gases. The 1992 United Nations Framework Convention on Climate Change, a major international effort to reduce greenhouse gas emissions, was followed by a comprehensive extension, the Kyoto Protocol, which took full effect in 2005. The Kyoto Protocol made binding on signatory nations many previously agreed-upon reductions in greenhouse gas emissions; the United States, however, is not a signatory to the protocol. Three international agreements have influenced U.S. legislation regarding depletion of the earth's ozone layer: the Vienna Convention for the Protection of the Ozone Layer (1985), the Montreal Protocol on Substances That Deplete the Ozone Layer (1987), and the London Amendment to the Montreal Protocol (1990).

Howard Bromberg

FURTHER READING

Black, Brian, and Donna Lybecker, eds. *Great Debates in American Environmental History.* 2 vols. Westport, Conn.: Greenwood Press, 2008

Brooks, Karl. *Before Earth Day: The Origins of American Environmental Law, 1945-1970.* Lawrence: University Press of Kansas, 2009.

Buck, Susan. *Understanding Environmental Administration and Law.* 3d ed. Washington, D.C. Island Press, 2006.

Lazarus, Richard. *The Making of Environmental Law.* Chicago: University of Chicago Press, 2004.

Milazzo, Paul. *Unlikely Environmentalists: Congress and Clean Water, 1945-1972.* Lawrence: University Press of Kansas, 2006.

Nagle, John. *Law's Environment: How the Law Shapes the Places We Live.* New Haven, Conn.: Yale University Press, 2010.

Salzman, James, and Barton Thompson. *Environmental Law and Policy.* 2d ed. New York: Foundation Press, 2007.

Sowards, Adam. *The Environmental Justice: William O.*

Douglas and American Conservation. Corvallis: Oregon State Press, 2009.

Zuckerman, Tod I., and Nancy K. Kubasek. *The History of U.S. Environmental Law.* Durham, N.C.: Carolina Academic Press, 2010.

SEE ALSO: Clean Air Act and amendments; Clean Water Act and amendments; Endangered Species Act; Environmental law, international; Environmental torts; Federal Insecticide, Fungicide, and Rodenticide Act; *Massachusetts v. Environmental Protection Agency;* National Environmental Policy Act; Safe Drinking Water Act; Superfund.

Environmental Protection Agency

CATEGORIES: Organizations and agencies; human health and the environment; atmosphere and air pollution; water and water pollution

IDENTIFICATION: U.S. government agency responsible for enforcing many federal environmental laws

DATE: Established on December 2, 1970

SIGNIFICANCE: The U.S. Environmental Protection Agency administers federal laws that protect natural resources such as air, water, and land. Among its many duties, the agency enforces regulations regarding air and water pollutants, oversees programs that promote energy efficiency and conservation, and participates in the cleanup of sites where toxic materials have polluted the natural environment.

The U.S. Environmental Protection Agency (EPA) is responsible for protecting the public health and ensuring a clean environment by safeguarding natural resources, controlling air and water pollution, and regulating the disposal of solid waste in the United States. The agency carries out its mission through its rule-making and enforcement authority granted by the U.S. Congress. The EPA also conducts scientific research, provides environmental education to the public and to private companies, and utilizes the best available scientific information in its quest to reduce environmental risk. States and tribal nations throughout the United States follow the national standards set by the EPA in enforcing their own environmental regulations. The EPA provides grants

(continued on page 486)

U.S. Environmental Protection Agency

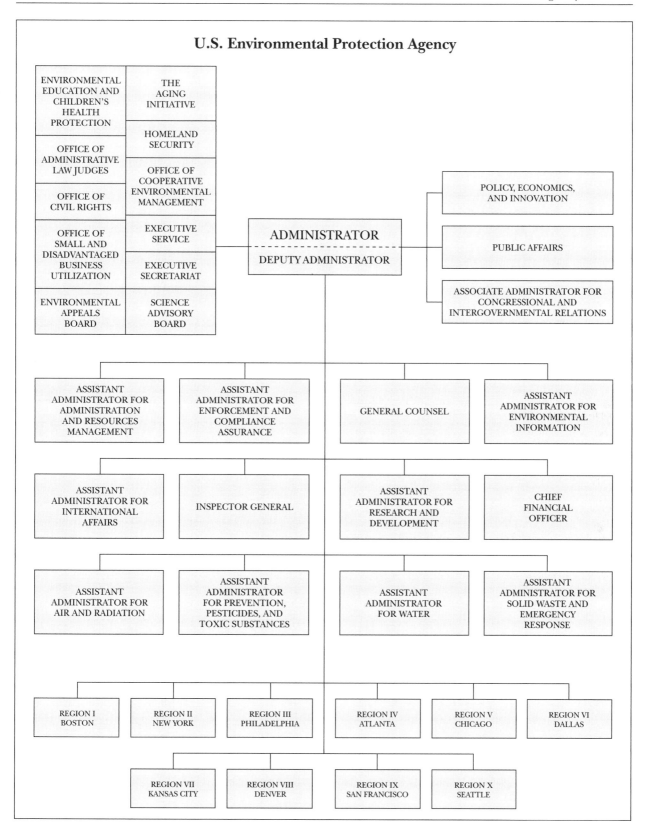

to states, nonprofit entities, and academic institutions to carry out environmental and human public health research, often related to the cleanup of toxic waste sites. The EPA also works with other nations to protect the global environment.

HISTORY

After World War II the growth of industrialization in the United States led to serious air and water pollution and environmental deterioration, which in turn spurred a movement that demanded the adoption of federal laws to protect the public health and clean up the environment. In 1969, President Richard Nixon created a White House committee to consider the existing environmental laws and enforcement agencies in the United States. In addition, Congress passed the National Environmental Policy Act of 1969, which was signed into law on January 1, 1970.

A public policy of achieving harmony between humankind and the environment by assessing the environmental impacts of various federal projects was the impetus behind the National Environmental Policy Act and eventually the establishment of the EPA. The president's committee recommended the creation of an independent environmental agency that would not be influenced by the goals and mandates of other agencies for the purpose of enforcing environmental laws. The Environmental Protection Agency became a reality when Congress consolidated the duties of several federal agencies into one entity in December, 1970. William D. Ruckelshaus became the agency's first administrator.

The EPA was established for the purpose of enforcing many of the environmental laws adopted and amended by the federal government during the 1970's and 1980's as the result of public pressure to clean up the environment. These early laws included the 1970 Clean Air Act amendments, which authorized the EPA to adopt vehicular controls to reduce emissions, set national air-quality standards and attainment goals, and regulate point source air emissions through a permitting and enforcement process. In 1972 the EPA worked in tandem with the U.S. Army Corps of Engineers to set standards for water quality and to regulate discharges into national water resources under the Clean Water Act.

The EPA was named as the lead agency for enforcement of the Federal Insecticide, Fungicide, and Rodenticide Act of 1947, including the act's 1972 and later amendments, which resulted in the ban of the toxic pesticide dichloro-diphenyl-trichloroethane (DDT) in the United States. In addition, the EPA was made responsible for regulating the manufacture, distribution, import, and processing of specific toxic substances through the Toxic Substances Control Act of 1976. The EPA, however, was not given authority over all environmental laws; the U.S. Fish and Wildlife Service remains responsible for enforcement of the Endangered Species Act, and nuclear wastes are regulated by the Department of Energy.

RCRA AND SUPERFUND

Before the adoption of federal environmental laws in the 1970's, the dumping of toxic materials in the United States, mostly illegal, took place with little government control. When the degradation of natural resources became a major concern, the federal government realized that it needed to step in and help to prevent the contamination of the environment with hazardous substances and to clean up old, abandoned toxic waste sites. In 1976 Congress passed the Resource Conservation and Recovery Act (RCRA) to prohibit future dumping of toxic and hazardous substances and granted the EPA authority for legal management of the disposal of such wastes "from cradle to grave." Under the act, the EPA is responsible for regulating the generation, transportation, treatment, storage, and disposal of toxic and hazardous substances. Congress later passed several amendments to RCRA that gave the EPA the power to set standards for nonhazardous solid waste disposal and for the installation and maintenance of underground storage tanks containing hazardous substances such as petroleum products.

By 1980 Congress recognized that contaminated toxic waste sites throughout the United States had become a serious threat to public health and enacted the Comprehensive Environmental Response, Compensation, and Liability Act. The act, known as Superfund, provides for retroactive liability for those deemed to be the parties responsible for the hazardous waste dumping, usually the property owners, even if the dumping took place before passage of the law in 1980. In addition, the law established a trust fund through a tax on the petrochemical industry to cover costs of cleanup not covered by the responsible parties. Superfund provides for short-term toxic waste removal from the most dangerous hazardous waste sites in the United States, which are listed by the EPA on the agency's National Priorities List. In addition, the EPA often employs long-term remedial actions to re-

Federal-Level Environmental Protection

On July 9, 1970, U.S. president Richard M. Nixon delivered the following message to Congress about establishing the Environmental Protection Agency and the National Oceanic and Atmospheric Administration:

To the Congress of the United States:

As concern with the condition of our physical environment has intensified, it has become increasingly clear that we need to know more about the total environment—land, water, and air. It also has become increasingly clear that only by reorganizing our Federal efforts can we develop that knowledge, and effectively ensure the protection, development and enhancement of the total environment itself.

The Government's environmentally-related activities have grown up piecemeal over the years. The time has come to organize them rationally and systematically. As a major step in this direction, I am transmitting today two reorganization plans: one to establish an Environmental Protection Agency, and one to establish, with the Department of Commerce, a National Oceanic and Atmospheric Administration. . . .

The Congress, the Administration and the public all share a profound commitment to the rescue of our natural environment, and the preservation of the Earth as a place both habitable by and hospitable to man. With its acceptance of the reorganization plans, the Congress will help us fulfill that commitment.

rorists in New York City on September 11, 2001; and the 2010 BP *Deepwater Horizon* oil spill in the Gulf of Mexico.

In addition to its enforcement and cleanup functions, the EPA works in many areas to encourage the protection of the environment. Aside from its Energy Star program, which promotes energy efficiency in household appliances and other consumer products, the EPA has undertaken initiatives concerned with the reduction of greenhouse gases and the adoption of air-quality visibility rules. The agency has also conducted risk and peer-review assessments to ensure the safety of high-concern chemicals found in products and in the environment, developed new strategies to protect the quality of public drinking-water supplies, and supported the revitalization and reuse of abandoned and contaminated inner-city brownfields.

duce public health dangers posed by the continuing release of buried hazardous substances at some of the highest-priority sites. In 1986 Congress passed the Superfund Amendments and Reauthorization Act (SARA) to clarify liability issues. Many U.S. states followed suit and adopted their own environmental dumping, enforcement, and cleanup laws.

CHALLENGES AND FUTURE OPPORTUNITIES

Since its establishment, the EPA has demanded compliance with federal environmental laws from private businesses, from states, and from individual cities, such as Cleveland, Ohio; Detroit, Michigan; Los Angeles, California; and Atlanta, Georgia. The EPA has met its enforcement challenges by adopting and imposing strict regulations and often seeking assistance from the courts to obtain compliance. Moreover, the EPA has been in the forefront of cleanup after natural and human-caused environmental disasters, including those related to the hazardous wastes buried under the Love Canal residential development in Niagara Falls, New York; the Three Mile Island nuclear plant core meltdown near Harrisburg, Pennsylvania; the *Exxon Valdez* oil spill in Alaska; the destruction of the World Trade Center towers by ter-

The EPA must carry out its responsibilities of protecting the environment, natural resources, and public health while attempting to avoid negative impacts on economic growth, industrial development, energy production, agriculture, transportation, and international trade—a balance that at times seems impossible to achieve. In its efforts to keep that balance, the EPA has partnered with both public and private companies and institutions to find new and innovative approaches to environmental protection that can satisfy competing goals.

Carol A. Rolf

FURTHER READING

Andrews, Richard N. L. *Managing the Environment, Managing Ourselves: A History of American Environmental Policy.* 2d ed. New Haven, Conn.: Yale University Press, 2006.

Carson, Rachel. *Silent Spring.* 40th anniversary ed. Boston: Houghton Mifflin, 2002.

Collin, Robert W. *The Environmental Protection Agency: Cleaning Up America's Act.* Westport, Conn.: Greenwood Press, 2006.

Landy, Marc K., Marc J. Roberts, and Stephen R. Thomas. *The Environmental Protection Agency: Asking the Wrong Questions—From Nixon to Clinton.* Expanded ed. New York: Oxford University Press, 1994.

McBrewster, John, Frederic P. Miller, and Agnes F. Vandome, eds. *United States Environmental Protection Agency: Federal Government of the United States, National Environment, Environmental Policy of the United States, Energy Star, WaterSense, Safe Drinking Water Act.* Phoenix, Ariz.: Alphascript, 2009.

Scheberle, Denise. *Federalism and Environmental Policy: Trust and the Politics of Implementation.* 2d ed. Washington, D.C.: Georgetown University Press, 2004.

SEE ALSO: Air-pollution policy; Air Quality Index; Automobile emissions; Best available technologies; Clean Air Act and amendments; Clean Water Act and amendments; Hazardous waste; *Massachusetts v. Environmental Protection Agency*; National Environmental Policy Act; Pollution permit trading; Resource Conservation and Recovery Act; Superfund.

Environmental racism. *See* Environmental justice and environmental racism

Environmental refugees

CATEGORIES: Human health and the environment; ecology and ecosystems

DEFINITION: People forced to leave their home sites, permanently or temporarily, because of environmental degradation or environmental hazards resulting from natural or human-induced disasters

SIGNIFICANCE: When environmental conditions force large numbers of people to relocate, the result is often political and social instability. In addition, the ecosystems of the regions to which environmental refugees migrate are in turn stressed by the influx of population.

With accelerating human population growth come increased demands for raw materials, food, water, and energy. In many places, human beings have reduced nature's ability to maintain a healthy carrying capacity for their survival needs. In

addition, environmental degradation from human activities, natural disasters, and changes in global climate have further stressed living conditions for many people. When environmental conditions become so severe that people must move to other areas to seek healthier living conditions to thrive, these people become environmental refugees. Soil erosion, deforestation, pandemic disease, desertification, air pollution, lack of fresh water, and overpopulation are some of the conditions that can push a population out of an area. Natural disasters or human-caused disasters—such as industrial accidents, lack of resources caused by plundering, and war-related environmental degradation—may be the causes of such conditions. Estimates of the numbers of people who are environmental refugees around the world at any one time range from 25 million to 300 million.

Persons who are displaced in this way are often considered nonmilitary threats to security in the regions or countries that receive them. If a region undergoes extreme environmental stresses, its political and social structures also become stressed. Natural disasters, resource depletion, landlessness, and environmental degradation can lead to social unrest and political upheaval. The people who flee these situations as environmental refugees can become a destabilizing factor in international relations if they are perceived as an unacceptable burden on the regions to which they migrate.

HUMAN-INDUCED DISASTERS

Unnatural disasters to the environment are those that are caused directly by human actions; many result in such complete degradation of ecosystems that inhabitants must migrate to survive. A major environmental threat to the well-being of many nations is land degradation. Human actions such as overfarming of erodible soils, depletion of aquifers, overgrazing of lands, deforestation, poor irrigation practices leading to salt deposition, and land capture for urban and industrial growth force many people into landlessness, and landless people become environmental refugees.

One of the most basic unnatural disasters to the environment is war. People who flee fighting in a conflict zone are war refugees, but inhabitants who are forced to leave their home areas because of contamination caused by war are environmental refugees. War and its by-products wreak havoc on the physical, chemical, and biological environment, rendering many ecosystems unfit for habitation. Unexploded ordnance, especially land mines, often make areas unfit to reoccupy or farm. Buildings and industrial sites that have been struck by explosives release toxic substances and airborne pollutants that contaminate water supplies and soils. Military defoliants destroy vast tracts of wood resources and croplands. Battle-generated firestorms combust highly toxic products, decimating downwind wildlife and animal stocks. Residual weapons with radioactive components, such as nuclear bombs and depleted uranium rounds, contaminate environments for years after their usage. Dead bodies contaminate water supplies and promote the spread of disease. These environmental factors all force people to abandon lands contaminated by war.

Pollution resulting from industrial processes, mining operations, or congested living conditions may result in environmental degradation to the point where inhabitants must migrate to healthier environments. Poor air quality, lack of sanitation, and polluted water supplies all result in rapid health decline. Mining wastes are often toxic and can contaminate large swaths of land. Industrial accidents can contaminate large areas, rendering them uninhabitable; examples include the 1984 leak from a pesticide plant in Bhopal, India, and the 1986 explosion of a nuclear power reactor in Chernobyl, Ukraine. Those forced to move from their home sites because of such problems are environmental refugees.

Hurricane Katrina was a large tropical storm, but the massive extent of the damage done to communities struck by Katrina in Louisiana and Mississippi was a direct result of poor engineering, poor land-use management and zoning, negligent preparation, and abysmal postdisaster response. The result was an ecocatastrophe that displaced tens of thousands of inhabitants. Other human-induced disasters resulting in masses of environmental refugees can be correlated to the Indian Ocean tsunami of 2004 and the

Afghan refugees who fled their country because of war and severe drought wait at a Pakistani border post to reenter Afghanistan in 2004. (AP/Wide World Photos)

2010 earthquake in Haiti. In both cases, results of the natural disaster were amplified by human actions prior to the events: building in known hazard zones, poor construction practices, lack of warning systems, and minimal government preparedness.

A consensus exists among scientists that human activities are altering the earth's climate, and it has been suggested that if climate change continues unchecked, billions of people are likely to become increasingly vulnerable to drought, flooding, disease, and famine. Global climate change is expected to alter zones of biodiversity, cause greater extremes in weather patterns, make crops vulnerable to changes in precipitation and temperatures, and raise sea levels, with devastating impacts to nearshore communities and freshwater aquifers. A globally altered climate is likely to have greatest impact in regions that are vulnerable owing to environmental stress: Already stressed marginal lands will respond most severely to changes in precipitation and temperature, and the destitute from these lands will migrate as environmental refugees.

Natural Disasters

Natural disasters are second only to land degradation in creating environmental refugees. Annually, more people are killed and displaced as a result of earthquakes, hurricanes, floods, tornadoes, tsunamis, landslides, volcanic eruptions, avalanches, and wildfires than as a result of warfare. In most cases the effects of these natural disasters are exacerbated by human activities. Most ecosystems in a natural state can handle normal stresses such as severe weather events or earth movements, but as human pressures on the environment increase, ecosystems become less resilient to stress. Human pressures on an ecosystem's soils, forests, landscapes, hydrologic cycle, and carrying capacity often magnify the outcomes of natural events.

Land degradation is the greatest factor in turning a natural event into a catastrophe. Degraded land inhibits an ecosystem's ability to absorb natural stresses, and many areas in which land has been degraded are also prone to natural disasters. Population, cultural, and financial pressures force large numbers of underrepresented and often poor people onto marginal lands, where they try to live off limited resources, and marginal lands exploited by humans trying to thrive at any cost quickly lose their sustainability. Overfarmed and overgrazed soils are easily swept away by high winds. Deforested hillsides become prone to landslides. The destruction of riparian zones and coastal watersheds increases the severity of floods. After a catastrophic natural disaster, diseases often spread rapidly through a population. When these stress conditions become manifest within a region, the population will move, often to land even more marginal than that they left behind.

Scientists have suggested that if the effects of global climate change continue unabated, extensive nearshore communities will be inundated by rises in sea level and coastal flooding, storm tracks will alter monsoons and hurricane strikes, precipitation and drought patterns will shift, island nations in the Pacific and Indian oceans will be submerged, seasonal temperatures will become unpredictable, agricultural zones will alter; and regional diseases will spread to new locations around the planet. As a result, massive numbers of people will need to relocate to thrive, and such large-scale migration will pose a threat to international stability. Many poor and undereducated environmental refugees are likely to want to seek a better life by relocating to developed nations; the developed nations may see these environmental refugees as a security threat and respond accordingly.

Randall L. Milstein

Further Reading

Levitt, Jeremy I., and Matthew C. Whitaker, eds. *Hurricane Katrina: America's Unnatural Disaster.* Lincoln: University of Nebraska Press, 2009.

Myers, Norman. *Ultimate Security: The Environmental Basis of Political Stability.* New York: W. W. Norton, 1993.

Newbold, K. Bruce. *Six Billion Plus: World Population in the Twenty-first Century.* 2d ed. Lanham, Md.: Rowman & Littlefield, 2007.

Renner, Michael. *Fighting for Survival: Environmental Decline, Social Conflict, and the New Age of Insecurity.* New York: W. W. Norton, 1996.

Stott, Philip, and Sian Sullivan, eds. *Political Ecology: Science, Myth, and Power.* New York: Oxford University Press, 2000.

See also: Carrying capacity; Climate change and human health; Deforestation; Desertification; Development gap; Environmental health; Environmental illnesses; Environmental security; Habitat destruction; Overgrazing of livestock; Pandemics; Population growth.

Environmental security

CATEGORIES: Human health and the environment; ecology and ecosystems

DEFINITION: The relationship between a nation's natural resources and livable environments and threats to that nation's social, economic, and political stability

SIGNIFICANCE: When a nation's ecosystems or resources are degraded to a level of unsustainability, the nation becomes destabilized economically and socially. A destabilized nation's need to locate new sources of natural resources or livable environments may trigger military conflicts or initiate the migration of large numbers of people. Either outcome results in the escalation of regional and international tensions.

Environmental security issues result from the need to maintain access to natural resources and livable environments—the foundation of all national economies as well as social and political stability. The basic environmental security of a nation-state involves maintaining the availability of access to sustainable sources of renewable and nonrenewable resources; protecting the soil, water, and air of sovereign territory from becoming unusable; and reducing hazards to the domestic environment as a result of human activities. The security of all these eco-based elements can be compromised by war, terrorism, pollution, unsustainable resource exploitation, overpopulation, ecosystem destruction, and political or philosophical agendas. Because natural resources are a vital component of a nation's economic and social stability and international political potential, environmental security is a vital component of national security.

NATURE OF SECURITY RISKS

Many conflicts between nations arise from local or regional confrontations over access to natural resources, claims of overexploitation of common resources, border-crossing pollution, environmental disturbances perpetrated by one nation that adversely affect another, and environmentally based threats that indirectly extenuate preexisting regional tensions and conflicts. One common source of regional conflict concerning resource access is fresh water: Often one nation may draw down a shared aquifer or, if upstream, draw down a river so that the downstream nation's access is depleted. Shared fishing grounds

constitute another kind of resource that has been at the core of many international conflicts; overexploitation of such grounds has often resulted in opposing navies facing off in attempts to enforce their claims to the harvest.

When a nation disrupts an ecological zone, such as by clear-cutting forests, the ramifications for water quality and microclimatic change directly affect neighboring nations. Industrial pollutants that cross borders and ecological disasters such as nuclear accidents and spills from oil tankers can have outcomes that affect vast areas and escalate regional tensions. The quest for sustainable access to lifestyle-maintaining resources, especially energy resources—natural gas, oil, and coal—has been the basis of numerous conflicts over the past century. Indirect nonconventional environmental threats may include the use of prime agricultural lands to grow nonfood crops such as opium poppies, coca, and cannabis. Similarly, illegal mining operations, often using slave labor, clandestinely harvest emeralds, rubies, and, most notably, diamonds. The profits from these illicit botanicals and "conflict gems" are often used to finance criminal enterprises, civil insurgencies, or terrorist operations—all threats to a nation's security.

Many internal and external conflicts remain ongoing because rebel and terrorist forces finance themselves entirely by plundering natural resources. In some parts of the world, cross-border poaching and theft of exotic animals for their meat and body parts, as well as the illegal taking of large marine mammals on the high seas, have escalated to the point where military forces are used to counter these activities, thus increasing the chance for armed conflicts. Another evolving threat to security is that of environmental terrorism, which can be as limited as the destruction of logging equipment in an attempt to disrupt local commerce or as large in scale as the systematic burning of oil fields to create an environmental catastrophe; both are examples of environmentally based security threats.

ENVIRONMENTAL REFUGEES

One of the more pressing problems of environmental security is that of environmental refugees: people forced to migrate when the environmental conditions in their home areas deteriorate so severely that they can no longer thrive there. Among the environmental triggers that force such migrations are overpopulation, natural disasters, desertification, famine, drought, climate change, human-induced environmental degradation, and resource depletion. While environmental refugees are usually considered nonmilitary security threats, they can become a destabilizing factor in international relations if they place an unacceptable burden on the areas to which they migrate.

Scientists predict that as the expected effects of global climate change manifest, nearshore communities will be inundated by rising sea levels, weather patterns will shift, temperature regimes will alter, growing seasons will become unpredictable, and pests and diseases will invade new regions and hosts. As a result, many people will be displaced, forced to migrate to new environments. Such a mass movement of environmental refugees is likely to stress the environments or resource bases of their end destinations, making the refugees a threat to regional and international stability and security.

Randall L. Milstein

FURTHER READING

Barnett, Jon. *The Meaning of Environmental Security: Ecological Politics and Policy in the New Security Era.* New York: Zed Books, 2001.

Dalby, Simon. *Environmental Security.* Minneapolis: University of Minnesota Press, 2002.

_____. *Security and Environmental Change.* Boston: Polity Press, 2009.

Dodds, Felix, and Tim Pippard, eds. *Human and Environmental Security: An Agenda for Change.* Sterling, Va.: Earthscan, 2005.

Leek, K. Mark, ed. *Cultural Attitudes About the Environment and Ecology, and Their Connection to Regional Political Stability.* Columbus, Ohio: Battelle Press, 1998.

Manwaring, Max G., ed. *Environmental Security and Global Stability: Problems and Responses.* Lanham, Md.: Lexington Books, 2002.

Myers, Norman. *Ultimate Security: The Environmental Basis of Political Stability.* New York: W. W. Norton, 1993.

Renner, Michael. *Fighting for Survival: Environmental Decline, Social Conflict, and the New Age of Insecurity.* New York: W. W. Norton, 1996.

SEE ALSO: Carrying capacity; Ecotage; Environmental economics; Environmental refugees; Fossil fuels; Positive feedback and tipping points; Sea-level changes; Sustainable development.

Environmental torts

CATEGORY: Human health and the environment

DEFINITION: Civil wrongs in which plaintiffs claim monetary awards and injunctive relief for personal injuries and property damages caused by contamination of the environment

SIGNIFICANCE: When contamination of air, land, or water resources result from an environmental tort, damage awards from litigation may involve equitable relief that requires the polluter to cease activities causing the contamination, such as petroleum refining and chemical production, or may result in compensatory damages and in some cases punitive monetary awards to deter future behavior that pollutes the environment.

Environmental torts usually affect multiple persons and properties, often resulting in the filing of class-action lawsuits by hundreds of plaintiffs. Moreover, an action may be filed against multiple defendants, typically industrial entities, that might share in the liability for the claimed injuries. Environmental torts may result in pollution of environmental media, such as in the case of the Love Canal neighborhood of Niagara Falls, New York, where the land and underground water supply were so polluted with toxic substances that home owners had to abandon their properties. Environmental torts may also cause direct personal injuries owing to contact with hazardous substances, many times seen in workplace accidents. Exposure to toxic substances raises a secondary issue in the assessment of personal injury damages, in that some harm caused by toxic substances may not manifest itself at the time of the exposure, and the statute of limitations could preclude the filing of a lawsuit in the future when the injuries are known.

Typical pollutants and toxic substances involved in environmental torts include petroleum products, bacteria, heavy metals such as mercury, and chemicals such as benzene, dioxins, and polychlorinated biphenyls (PCBs). Some of the worst environmental torts have been caused by petrochemical industries, steel mills, and mining companies that have carelessly or deliberately discharged hazardous contaminants into the soil, rivers, and underground water sources. In addition, power and utility companies may be liable for environmental torts if they release dangerous toxic substances into the air. Hospitals and medical facilities may be responsible for tortious bacterial contamination for failing to comply with environmental laws when disposing of medical wastes.

In the United States, the federal government maintains a specialized environmental tort section within the Department of Justice that is responsible for trying environmental tort cases. These cases are usually filed against government contractors, but they may also involve U.S. military facilities that have been used for toxic dumping grounds in the past, thus resulting in serious environmental contamination, especially of underground water supplies. Private litigants filing an environmental tort claim typically engage a specialty law firm. Some private litigants represent environmental advocacy groups that are interested in righting environmental transgressions or in seeking injunctions to stop the continuation of point source environmental contamination. Proving an environmental tort case, however, is complex; to be acceptable in a court of law, the evidence presented must be supported by the scientific community.

CAUSES OF ACTION

Environmental tort actions may be based on both common law and federal and state statutory law. Common-law actions include claims for a nuisance or for trespass, which are mainly used in cases of property damages caused by toxic substances that contaminate the property of another. Both personal and property injuries may be claimed in tortious causes of action for negligence, intentional torts, and strict liability. In a negligence action the plaintiff must show that a breach of duty or carelessness caused the damages claimed. Intentional torts require intent, which means a claimant must allege that the toxic contamination was done purposefully and with knowledge of the damages that could occur. Finally, a strict liability action does not require evidence of causation or intent; it simply requires a plaintiff to show that the injuries suffered were the result of an inherently or abnormally dangerous activity, such as using explosives in an environmentally sensitive area.

Federal environmental laws such as the Toxic Substances Control Act, the Clean Air Act, the Clean Water Act, and the Comprehensive Environmental Response, Compensation, and Liability Act (known as Superfund) provide other avenues for litigating environmental tort cases. Individual states also have comparable laws that provide remedies for environmental wrongs. Many of the cases filed under statutory law involve the dumping of hazardous substances, in-cluding industrial lubricants; the use of toxic substances, such as asbestos, that have resulted in personal injuries; and the discharge of toxins into the environment, including the *Exxon Valdez* oil spill in Alaska.

INTERNATIONAL ENVIRONMENTAL TORTS

Many multinational companies based in the United States conduct business abroad, and some have caused environmental contamination in the locations of their international facilities, especially in countries that do not have many environmental regulations in place, such as developing nations. When a private individual attempts to sue a multinational U.S. company in the United States for an environmental tort that did not occur in the United States, the U.S. courts usually dismiss the claim. Such dismissals are based on a variety of legal grounds, including that the United States has no jurisdiction or right to hear the case, as a plaintiff must seek remedies from the courts in the nation where the tort occurred. Such was the result for many of the lawsuits filed in the United States after the release of a toxic gas at a Union Carbide pesticide plant in Bhopal, India, in 1984.

Industrializing nations such as China, in which environmental contamination is substantially increasing, have begun to adopt stiffer environmental laws. In addition, the international community works to try to curb environmental pollution, and thus prevent potential environmental torts, through environmental treaties and protocols. Some litigants, however, have placed blame on the international community for failure to regulate and mitigate environmental harms and have filed lawsuits against offending nations for such torts as alleged global warming caused by carbon dioxide and other emissions.

Carol A. Rolf

FURTHER READING

Lowry, John, and Rod Edmunds, eds. *Environmental Protection and the Common Law.* Portland, Oreg.: Hart, 2000.

Madden, M. Stuart. *Law of Environmental and Toxic Torts: Cases, Materials, and Problems.* 3d ed. St. Paul, Minn.: Thomson/West, 2005.

Perlman, Cary R. *Environmental Litigation: Law and Strategy.* Chicago: American Bar Association, 2009.

Wilson, Mark W. "Why Private Remedies for Environmental Torts Under the Alien Tort Statute Should Not Be Constrained by the Judicially Created Doctrines of Jus Cogens and Exhaustion." *Environmental Law* 39 (March 22, 2009): 451-480.

SEE ALSO: Asbestos; Bhopal disaster; Birth defects, environmental; Chernobyl nuclear accident; Cuyahoga River fires; *Exxon Valdez* oil spill; Love Canal disaster; Oak Ridge, Tennessee, mercury releases; Oil spills; Three Mile Island nuclear accident.

Environmentalism

CATEGORIES: Philosophy and ethics; activism and advocacy

DEFINITION: Movement devoted to the protection of natural resources from harmful influences

SIGNIFICANCE: The modern movement known as environmentalism has had many successes in a wide variety of areas of concern, including the protection and preservation of untouched natural areas, the conservation of finite natural resources, the development of alternative sources of fuels and other necessities, the promotion of the importance of biodiversity, and the reduction of pollution.

Environmentalism entails advocating for and taking part in activities aimed at preserving and protecting the natural environment. Environmentalists support many different ways of achieving their goals. Among the many concerns of environmentalists are the reduction of the pollution of air, soil, and water; the prevention of the introduction of species of plants and animals into ecosystems to which they are not native; and the prevention of the encroachment of human activities into natural areas.

AMERICAN ENVIRONMENTALISM

When early European settlers arrived in the Western Hemisphere, they exploited every resource they found, including the native populations. Such exploitation continued in the American colonies during English control. After the Revolutionary War, most Americans were committed to environmental exploitation and westward expansion. They swiftly harvested forests and quickly exhausted arable land with nutrient-needy crops. The slaughter of wild animals for food and pelts and of whales for oil, ambergris, and other products was rampant. Discoveries of gold, silver, and other precious minerals were rapidly exploited.

In the nineteenth century some Americans became concerned about this trend, and New England Transcendentalists such as Henry David Thoreau and Ralph Waldo Emerson began to write about the value of the natural environment. Gradually public opinion began to shift, and increasing numbers of Americans began to see the indiscriminate exploitation of natural resources as less than admirable. Grassroots organizations began to form in response to local environmental concerns during the late nineteenth century. John Muir, one of the founders of the Sierra Club and an advocate of preservation of forests in the American West, was instrumental in influencing popular opinion for environmental preservation.

Another important figure during this period was Gifford Pinchot. Pinchot, son of a wealthy land speculator and lumberman, learned about sustainable forestry in Europe. He did not favor wilderness preservation for the sake of scenery or landscape; rather, he was concerned with the conservation of forest resources. Pinchot favored federal ownership and management of public lands. He advocated prudent exploitation of existing forest resources, including the replacement of cut trees with new seedlings. Utilitarian use of the environment became popular with President Theodore Roosevelt, who appointed Pinchot as the first chief of the U.S. Forest Service.

Individual Audubon Society organizations formed in various states late in the nineteenth century, when commercial bird hunting was extensive (many birds were killed so that their feathers could be harvested for the fashion industry). Audubon advocacy was directed toward the preservation of game and wild birds and against the importation of nonnative birds. The various Audubon organizations advocated for passage of the Lacey Act, which would prohibit interstate commerce in illegally captured or protected birds. The act was signed into law on May 25, 1900, by President William McKinley. Many state Audubon groups united to form a national organization in 1905.

In 1933, as part of his New Deal programs during the Great Depression, President Franklin D. Roosevelt created the Civilian Conservation Corps (CCC). This work relief program focused on environmental projects aimed at conserving and developing natural resources. Among their other accomplishments, CCC workers planted millions of trees in the Great Plains and the Midwest.

After the end of World War II in 1945, logging increased dramatically in the United States to meet demands for new construction. The widespread avail-ability of cheap energy also promoted tremendous expansion in industry and mechanized agriculture during this period, and little public concern was expressed about pollution. In 1962, however, fears about the effects of toxic chemicals on human beings and the environment quickly followed publication of Rachel Carson's book *Silent Spring,* which discussed the concentration of toxic and radioactive substances in the food chain, especially the effects of the insecticide dichloro-diphenyl-trichloroethane (DDT) on wildlife and on domesticated animals. The demands for change that followed the publication of Carson's book were met with intense opposition by the U.S. chemical industry, especially by large corporations that manufactured pesticides and herbicides. Corporate publicity unsuccessfully attempted to brand Carson as a fanatic. The publication of *Silent Spring* was an important factor leading to the formation of the Environmental Defense Fund in 1967 and ultimately to the creation of the federal Environmental Protection Agency (EPA) in 1970.

The first Earth Day, on April 22, 1970, was a seminal event. Some twenty million Americans participated in this "teach-in" suggested by Democratic U.S. senator Gaylord Nelson of Wisconsin. Increasing numbers of Americans began to see the pollution of air, water, and soil as a threat to human health, and politicians and industry leaders responded to their concerns.

The mantra "reduce, reuse, and recycle" became popular in the English-speaking world during the late twentieth century as environmentalists sought to encourage the conservation of resources and reduction of the amounts of materials that were being deposited in increasingly scarce landfill space. As the twenty-first century began, many environmentalists emphasized the excessive use of energy and consumption of goods in developed countries. They promoted the reduction of energy consumption, which would require individual lifestyle changes such as reductions in air travel and the use of smaller, more fuel-efficient cars. These suggestions were unpopular with many Americans.

ENVIRONMENTAL ADVOCACY

Prior to the 1960's, various American grassroots groups advocated on behalf of wildlife preservation and opposed such activities as lumbering, dam building, and mining in wilderness areas. After the publication of *Silent Spring,* such efforts expanded to include national campaigns aimed at reducing the pollution caused by toxic chemicals from agricultural and industrial sources. As increasing numbers of environmental groups formed on the national level, their repeated use of legal advocacy led to the establishment of the field of environmental law.

Greenpeace, an international environmental watchdog organization, was formed in 1971 in British Columbia and rapidly became known for its confrontational

A demonstration at the U.S. Capitol in 2009 calls for the plant that provides the Capitol's power to cease its practice of burning coal to generate electricity. A primary concern of the modern environmental movement is the need to find alternatives to polluting fuels. (Roger L. Wollenberg/UPI/LANDOV)

tactics, pitting environmental activists against corporate and government entities. Originally Greenpeace focused primarily on protests against nuclear testing, whaling, and seal hunting, but over time its work evolved to address many other environmental issues as well. In the early years of the twenty-first century, Greenpeace stated that global warming presents the greatest environmental threat to the planet.

By the late twentieth century, Green political parties had become increasingly important carriers of the environmental message. In the United States, the Green Party gained attention with its nomination of Ralph Nader as candidate for U.S. president in 1996; Nader was also the party's candidate in 2000. Twenty-five state Green parties formed the Association of State Green Parties (ASGP) in 1996; the ASGP was replaced by the Green Party of the United States (GPUS), a federation of forty-six state Green parties, in 2001. The GPUS promotes "ten key values" based on "ecological wisdom, social justice, cooperation, and nonviolence." Among the issues the party considers important are "Frankenfoods" (that is, foods created using genetic engineering), corporate farming, inequities between rich and poor, and the problems caused by global warming.

In European countries, increasing numbers of elected local and national offices have been held by Greens, and several different Green parties have been represented in the European Parliament. The European Green parties generally stand for ecological wisdom, social justice, grassroots participatory democracy, and nonviolence, principles similar but not identical to those of the GPUS.

Many environmentalists oppose the use of genetic engineering in plant and animal breeding and have raised objections to the introduction of genetically modified organisms in the production of food products. One controversial example is the use of recombinant bovine somatotropin (rBST), an artificial hormone produced by genetically modified bacteria, to boost milk production in cows. The U.S. Food and Drug Administration (FDA) has stated that the milk of cows treated with rBST is safe for human consumption, but some critics assert that because such milk has a slightly elevated level of insulin-like growth factor (IGF-1), a hormone, it may have detrimental health effects in humans. The European Union, Japan, Australia, New Zealand, and Canada have all deemed rBST to have unacceptable detrimental health effects on cows and have banned its use for humane reasons.

Many environmentalists contend that the claims made for some genetically modified foods are inflated. An example is "golden rice," a genetically modified rice that biosynthesizes beta-carotene (a source of vitamin A) in the grain. This product was engineered to provide vitamin supplementation to the diets of consumers in Africa and Asia, but critics assert that few studies were done on how much vitamin A remains after the rice is cooked. They further argue that a better approach would be to make wider food choices available to people in need, rather than encouraging them to continue their dependence on a rice diet.

Radical environmentalists sometimes take part in criminal activities that are aimed at preventing what they view as harms to the environment; the term often applied to such activities is "ecoterrorism." Some of the tactics used by ecoterrorists, including tree spiking (in which metal spikes are driven into trees to prevent logging with chain saws), are federal crimes in the United States. Among the groups that have been described as ecoterrorist organizations are Earth First!, the Environmental Liberation Front, and the Animal Liberation Front.

ANTIENVIRONMENTALISM

Environmentalism has had a number of well-known critics. Bjørn Lomborg, a Danish academic, was a Greenpeace advocate before he conducted a series of statistical studies of environmentalist claims. His published conclusions were that many claims of impending environmental disaster are grossly overstated.

Within the United States, some individuals and organizations view environmentalists as people opposed to continued technological progress. Author and filmmaker Michael Crichton alleged that environmentalism is a religion. He criticized environmentalists for not using "complexity theory" in environmental management. Crichton was also critical of the idea of global warming, asserting that environmentalists use statistical "tricks" to hide data that contradict the concept that greenhouse gas emissions affect the environment.

Various governments around the world have taken violent actions against environmental activists. An infamous example is the sinking of the Greenpeace ship *Rainbow Warrior* in 1985 in a New Zealand port by operatives of the French General Directorate for External Security. The *Rainbow Warrior* had been shadowing French nuclear vessels to protest nuclear testing in French Polynesia. The French government paid compensation for sinking the ship.

Global Environmentalism and Sustainable Development

Two similar concepts emerged during the 1990's: green development and sustainable development. Green development puts environmental concerns above social and economic concerns. Those who advocate sustainable development call for meeting immediate social, economic, and environmental needs in a way that can be maintained for future generations. Some observers contend that the concept of environmentalism underwent a paradigm shift before the twenty-first century, during which it was replaced by the concept of sustainability.

The promotion of sustainable development is sometimes characterized as an attempt by developed countries to exert "protectionism/paternalism" on less developed regions of the world. Some critics of sustainable development believe that it requires limits to population growth. Other critics argue that development of any kind conflicts with environmentalism.

Instead of funding large (and costly) infrastructure programs, such as building dams, in developing countries, governments and organizations that promote sustainability tend to focus on so-called appropriate technology to provide cheap solutions to everyday problems. Some of these inexpensive efforts, such as the introduction of solar cookers at refugee camps in Sudan's Darfur region, have been relatively successful.

Anita Baker-Blocker

Further Reading

Andrews, Richard N. L. *Managing the Environment, Managing Ourselves: A History of American Environmental Policy.* 2d ed. New Haven, Conn.: Yale University Press, 2006.

Delcourt, Paul, and Hazel Delcourt. *Living Well in the Age of Global Warming: Ten Strategies for Boomers, Bobos, and Cultural Creatives.* White River Junction, Vt.: Chelsea Green, 2001.

Edwards, Andres R. *The Sustainability Revolution: Portrait of a Paradigm Shift.* Gabriola Island, B.C.: New Society, 2005.

Kline, Benjamin. *First Along the River: A Brief History of the United States Environmental Movement.* 3d ed. Lanham, Md.: Rowman & Littlefield, 2007.

Liddick, Don. *Eco-Terrorism: Radical Environmental and Animal Liberation Movements.* Westport, Conn.: Praeger, 2006.

Lomborg, Bjørn. *The Skeptical Environmentalist: Measuring the Real State of the World.* New York: Cambridge University Press, 2001.

Maher, Neil M. *Nature's New Deal: The Civilian Conservation Corps and the Roots of the American Environmental Movement.* New York: Oxford University Press, 2008.

SEE ALSO: Antienvironmentalism; Carson, Rachel; Conservation; Conservation movement; Dichlorodiphenyl-trichloroethane; Environmental education; Genetically modified foods; Genetically modified organisms; Green movement and Green parties; Muir, John; National Environmental Policy Act; Preservation; Sustainable agriculture; Sustainable development.

Erosion and erosion control

CATEGORY: Land and land use

DEFINITION: The loss of topsoil through the actions of wind and water, and the efforts undertaken to mitigate such loss

SIGNIFICANCE: The control of erosion is vital because soil loss from agricultural land is a major contributor to nonpoint source pollution and desertification and represents one of the most serious threats to world food security.

In the United States alone, some two billion tons of soil erode from cropland on an annual basis. About 60 percent, or 1.2 billion tons, is lost through water erosion, and the remainder is lost through wind erosion. This is equivalent to losing 0.3 meter (1 foot) of topsoil from 810,000 hectares (2 million acres) of cropland each year. Although soil is a renewable resource, soil formation occurs at rates of just a few inches per hundred years, which is much too slow to keep up with erosive forces. The loss of soil fertility through erosion is incalculable, as are the secondary effects of pollution of surrounding waters and increase of sedimentation in rivers and streams.

Erosion removes the topsoil, the most productive soil zone for crop production and the plant nutrients it contains. Erosion thins the soil profile, which decreases a plant's rooting zone in shallow soils, and can disturb the topography of cropland sufficiently to impede the operation of farm equipment. Through erosion, nitrates, phosphates, herbicides, pesticides, and other agricultural chemicals are carried into surrounding waters, where they contribute to cultural eutrophication. Erosion also causes the deposit of increased sedimentation in lakes, reservoirs, and streams, which eventually require dredging.

There are several types of wind and water erosion. The common steps in water erosion are detachment, transport, and deposition. Soil particles become detached from soil aggregates, and the particles are carried, or transported, away; in the process, the particles scour new soil particles from aggregates. Finally, the soil particles are deposited when the water flow slows. In splash erosion, raindrops impacting the soil can detach soil particles and hurl them considerable distances. In sheet erosion, a thin layer of soil is removed by tiny streams of water moving down gentle slopes. This is one of the most insidious forms of erosion because the effects of soil loss are imperceptible in the short term. Rill erosion is much more obvious because small channels form on a slope. These small channels can be filled in by tillage. In contrast, ephemeral gullies are larger rills that cannot be filled by tillage. Gully erosion is the most dramatic type of water erosion. It leaves channels so deep that even equipment operation is prevented. Gully erosion typically begins at the bottoms of slopes, where the water flow is fastest, and works its way with time to the top of a slope as more erosion occurs.

Wind erosion generally accounts for less soil loss than does water erosion, but in states such as Arizona, Colorado, Nevada, New Mexico, and Wyoming, it is actually the dominant type of erosion. Wind speeds 0.3 meter (1 foot) above the soil that exceed 16 to 21 kilometers (10 to 13 miles) per hour can detach soil particles. These particles, typically fine to medium-size sand grains fewer than 0.5 millimeters (0.02 inches) in diameter, begin rolling and then bouncing along the soil, progressively detaching more and more soil particles by impact. The process, called sal-

Erosion and sediment in a farm field in Iowa. (NRCS)

tation, is responsible for 50 to 70 percent of all wind erosion. Larger soil particles are too big to become suspended and continue to roll along the soil. Their movement is called surface creep.

The most obvious display of wind erosion is called suspension, when very fine silt and clay particles detached by saltation are knocked into the air and carried for enormous distances. The Dust Bowl of the 1930's was caused by suspended silt and clay in the Great Plains of the United States. It is also possible to see the effects of wind erosion on the downward sides of fences and similar obstacles. Wind passing over these obstacles deposits the soil particles it carries. Other effects of wind erosion are tattering of leaves, filling of road and drainage ditches, wearing of paint, and increasing incidence of respiratory ailments.

CONTROL MEASURES

The four most important factors affecting erosion are soil texture and structure, roughness of the soil surface, slope steepness and length, and soil cover. Several passive and active methods of erosion control involve these four factors. Wind erosion, for example, is controlled through the creation of windbreaks, rows of trees or shrubs that shorten a field and reduce the wind velocity by about 50 percent. Tilling the land perpendicular to the wind direction is also a beneficial practice, as is keeping the soil covered by plant residue as much as possible.

Water erosion is controlled through a number of practices. In the United States, farmers can get help from the federal government-sponsored Conservation Reserve Program to protect highly erosive, steeply sloped land. Tilling land along the contours of slopes aids in preventing erosion, as does the shortening of long slopes by terracing, which also reduces the slope steepness. Permanent grass waterways can be planted in areas of cropland that are prone to water flow. Likewise, grass filter strips can be planted between cropland and adjacent waterways to impede the velocity of surface runoff and cause suspended soil particles to sediment and infiltrate before they can become contaminants.

Conservation tillage practices such as minimal tillage and no-till farming, or zero tillage, have been widely adopted by farmers as a simple means of erosion control. As the names imply, these are tillage practices in which as little disruption of the soil as possible occurs and in which any crop residue remaining after harvest is left on the soil surface to protect the soil from the impacts of rain and wind. The surface residue also effectively impedes water flow, which results in less suspension of soil particles. Because the soil is not disturbed, practices such as no-till farming also promote rapid water infiltration, which reduces surface runoff. Zero tillage is rapidly becoming the predominant practice in southeastern states such as Kentucky and Tennessee, where rainfall levels are high and erodible soils occur.

Mark Coyne

FURTHER READING

Blanco, Humberto, and Rattan Lal. *Principles of Soil Conservation and Management.* New York: Springer, 2008.

Faulkner, Edward. *Plowman's Folly.* Norman: University of Oklahoma Press, 1943.

Field, Harry L., and John B. Solie. "Erosion and Erosion Control." In *Introduction to Agricultural Engineering Technology: A Problem Solving Approach.* 3d ed. New York: Springer, 2007.

Plaster, Edward. *Soil Science and Management.* 5th ed. Clifton Park, N.Y.: Delmar Cengage Learning, 2008.

Schwab, Glen, et al. *Soil and Water Conservation Engineering.* 5th ed. Clifton Park, N.Y.: Delmar Cengage Learning, 2005.

SEE ALSO: Deforestation; Desertification; Dust Bowl; Logging and clear-cutting; Runoff, agricultural; Soil conservation; Strip farming; Sustainable agriculture.

Ethanol

CATEGORY: Energy and energy use

DEFINITION: Fermented distillate of plant life that can be a biofuel when blended with gasoline into gasohol

SIGNIFICANCE: Although ethanol has long been the key ingredient in drinking alcohol and remains an important solvent in the chemical industry, it is of primary interest in the developed world as a fuel source of value in reducing both dependence on imported oil and environmentally destructive carbon emissions emanating from internal combustion engines.

The process for fermenting biomatter into a transportation fuel was developed long ago. As early as 1860 German engineering had produced an engine designed to run on ethanol, or ethyl alcohol, and the first assembly-line automobile produced in the United States (Henry Ford's Model T) was powered by a motor based on that design. Moving a technology from the drawing board to mass deployment, however, does not depend on feasibility alone; other important factors are market conditions and, sometimes, political climate. Both have been decisive in shaping ethanol's history as a transportation fuel.

REVIVING A KNOWN TECHNOLOGY

In the mid-nineteenth century ethanol production and sales constituted a lively business concern in both the United States and Europe, as a supply of non-drinking alcohol was kept in most households for home lighting and cooking purposes. Ethanol still fulfills those roles in parts of the developing world; however, in modern American homes it is most often found in the form of rubbing alcohol. The decline of ethanol began with the high tax on the industrial use of alcohol that the U.S. Congress enacted during the Civil War, but most of the damage was done by the discovery of oil in the United States, the emergence of kerosene as a preferred source of household fuel, and the subsequently discovered advantages of gasoline over ethanol as a motoring fuel.

The ethanol industry's revival in the transportation field began with the Arab embargo on oil shipments to countries that supported Israel in the Yom Kippur War of October, 1973. Panic buying ensued, the Organization of Petroleum Exporting Countries (OPEC) seized the moment to gain control over oil production from the multinational Western oil corporations that had controlled it for half a century, and the embargo raised serious questions as to the reliability of importing oil from Arab producers. When the outbreak of warfare between Iraq and Iran in the fall of 1979 intensified those fears and pushed the price of imported oil to more than $36 per barrel, one response of the U.S. Congress was the Energy Security Act of 1980, in which funds were appropriated to assist the fledgling ethanol industry that had emerged in the United States after 1973. The collapse of oil prices during the early 1980's, however, soon made ethanol production less attractive, as did the early experience of consumers with E90 gasohol (a mixture of 10 percent ethanol and 90 percent gasoline) during this era. A natural solvent, the gasohol cleaned out fuel lines, resulting in the clogging of enough fuel pumps in older cars to earn it a bad reputation.

THE POLITICS OF PROMOTING ETHANOL

From 1990 onward, growing concerns about the need to find green sources of energy, the availability-of-oil issue raised anew by Iraq's 1990 invasion of Kuwait, and the soaring price of oil following the 2003 occupation of Iraq by the United States combined to produce and sustain a renewed interest in ethanol as a transportation fuel throughout the oil-importing world. Politics also played a role. Ethanol does not replace gasoline, and hence its use in low volumes is endorsed by petroleum corporations. Moreover, in low-volume gasohol mixtures such as E90, its burning can be accommodated by existing internal combustion engines. Consequently, the development of the ethanol industry has been supported by an automobile industry under government pressure to become more environmentally friendly. Meanwhile, agriculture-producing U.S. states have lobbied heavily in support of all federal legislation designed to boost ethanol production.

Nevertheless, and despite the subsidies provided by the states and the federal government, and the growing number of ethanol-producing plants in Europe, ethanol has not significantly reduced petroleum imports on either continent. Some critics argue that it may not have made any impact because the processes of planting and harvesting the corn and other grains used as ethanol's feedstock in the developed world are themselves oil-energy-intensive. Environmental costs must also be weighed against the advan-

---

Biofuel Energy Balances

The following table lists several crops that have been considered as viable biofuel sources and several types of ethanol, as well as each substance's energy input/output ratio (that is, the amount of energy released by burning biomass or ethanol, for each equivalent unit of energy expended to create the substance).

BIOMASS/BIOFUEL	ENERGY OUTPUT PER UNIT INPUT
Switchgrass	14.52
Wheat	12.88
Oilseed rape (with straw)	9.21
Cellulosic ethanol	1.98
Corn ethanol	~1.13-1.34

Source: Data from the British Institute of Science in Society.

Gasoline blended with ethanol produces lower carbon emissions than gasoline alone. (AP/Wide World Photos)

tage of reduced carbon emissions that gasohol has over gasoline and diesel fuel. The pesticide "cocktail" that U.S. farmers use to protect the feedstock can harm both wildlife and groundwater, and burning gasohol in internal combustion engines can produce twice as much ground ozone as ordinary gasoline.

Uncertainty over the advantages of ethanol as a transportation fuel has thus dampened the rush to embrace it in the Western world, and Brazil remains the only country that has essentially ended its dependence on imported oil in the transportation sector. Brazil, however, uses sugarcane harvested by low-cost manual laborers and has guaranteed a growing market for ethanol by requiring the manufacture after 2007 of only flexible-fuel vehicles (FFVs) capable of burning mixtures containing as much as 85 percent ethanol. Such legislation is extremely unlikely to be enacted in the developed world, where automotive manufacturers and petroleum corporations can be expected to engage in powerful lobbying efforts

against it. Consequently, ethanol production in the West continues to be measured in millions of gallons produced per year (412 million gallons in all of Europe in 2006, for example), while petroleum imports continue to be recorded in billions of barrels of oil per year.

Joseph R. Rudolph, Jr.

FURTHER READING
Blume, David. *Alcohol Can Be a Gas! Fueling an Ethanol Revolution for the Twenty-first Century.* Santa Cruz, Calif.: International Institute for Ecological Agriculture, 2007.
Boudreaux, Terry. *Ethanol and Biodiesel: What You Need to Know.* McLean, Va.: Hart Energy, 2007.
Freudenberger, Richard. *Alcohol Fuel: A Guide to Making and Using Ethanol as a Renewable Fuel.* Gabriola Island, B.C.: New Society, 2009.
Goettemoeller, Jeffrey. *Sustainable Ethanol: Biofuels, Biorefineries, Cellulosic Biomass, Flex-Fuel Vehicles, and Sustainable Farming for Energy Independence.* Maryville, Mo.: Prairie Oak, 2007.

SEE ALSO: Air-pollution policy; Alternative fuels; Alternatively fueled vehicles; Automobile emissions; Gasoline and gasoline additives; Oil crises and oil embargoes.

Europe

CATEGORIES: Ecology and ecosystems; resources and resource management; atmosphere and air pollution; water and water pollution

SIGNIFICANCE: Because they are heavily populated and highly industrialized, the European nations have significant impacts on the global environment. The decisions made by Europeans regarding the management and use of natural resources have the potential to affect the world's environment in either positive or negative ways.

Under the leadership of the European Union, Europe is intensely involved in efforts to reduce the environmentally harmful effects on the earth that result from human activity in modern industrialized societies. The European countries address these issues not only from a European perspective but from a global perspective as well. European policy makers

People sign a star installed by Friends of the Earth outside the meeting place of a European Union summit in Brussels in 2007, where European leaders pledged to take the lead in the fight against global warming by setting binding targets for cutting greenhouse gas emissions and ensuring that one-fifth of Europe's energy comes from renewable sources such as wind and solar power. (AP/Wide World Photos)

are highly aware of their role in the global community and of the impacts of their decisions and activities on the global environment. For example, as the leaders of industrialized nations, they have accepted a serious commitment to reduce greenhouse gas emissions, which have been determined to be a significant cause of global warming.

Because of the diversity of terrain and the differences in natural resources from one country to another as well as varying country locations, some having major coastal areas and others being totally landlocked, environment-related priorities vary among European nations. All share concerns regarding several major environmental issues, however; these include greenhouse gas emissions and climate change, changes in land use and deforestation, loss of biological diversity, rising sea levels, air pollution, and water pollution (especially eutrophication, or the depletion of oxygen in water).

GREENHOUSE GAS EMISSIONS AND CLIMATE CHANGE

Global warming is a major concern among Europeans. All of the European countries have signed and ratified the Kyoto Protocol, which was set forth by the United Nations Framework Convention on Climate Change. The objective of the agreement is to combat global warming by reducing the amount of greenhouse gases emitted into the air by setting binding target reductions in emissions for each country. European nations have concentrated on the reduction of fossil-fuel use as the primary means of meeting their Kyoto Protocol commitments.

Because coal-fired power plants are the most significant emitters of carbon dioxide (CO_2), the primary greenhouse gas, the reduction and even elimination of the use of coal as a fuel is a major goal in Europe; this trend is supported by environmental groups, including Green political parties. The coal industry, particularly in England, which has had a significant history of coal mining, has attempted to avoid being eliminated completely by developing ways to make coal a cleaner-burning fuel. Two methods that have been suggested are the underground gasification of coal and the trapping and storing of CO_2 emissions underground.

Hydropower has become an important fuel source for power plants in many European nations, especially in countries, such as Norway, that have abundant rivers and lakes. Europe has also investigated the feasibility of using wind and solar power. Another major way in which Europe is attempting to reduce greenhouse gas emissions is by reducing the use of fossil fuels (oil in the form of gasoline and diesel fuel) in the transportation industry. An increase in the number of transport vehicles has resulted in an increase in fuel consumption, which has the potential to nullify previous reductions made in emissions from power plants and other sources. Biofuels have been proposed as alternatives to fossil fuels.

The European Commission (the executive body of the European Union) has made two important proposals regarding the use of renewable resources for the production of energy. The commission has suggested that a target be set requiring that 20 percent of energy used in Europe be produced using renewable resources by 2020. In addition, the commission has proposed that as much as 10 percent of road transport fuels should be biofuels by 2020. Debates continue about the attainability of these goals, and some observers have expressed concerns about the economic effect of increased biofuel production on food-crop production as well as about the environmental impacts of changes in land usage. These concerns do not necessarily make the use of biofuels infeasible, however; producing biofuels from crop and forest residues and wastes would not have adverse effects on the environment.

In regard to the production of biofuels, European nations have considered the effects of such production on ecosystems in other parts of the world. They are aware that the reduction of food-crop production in Europe would necessitate increased production elsewhere to meet the global need, and that this could affect land use, possibly resulting in the destruction of wetlands and rain forests.

Another environmental concern of the nations of Europe is the projected rise in sea levels as a result of global climate change. Rises in sea levels cause coastal deterioration and could result in significant loss of landmass, posing a potential economic threat in the areas of tourism and recreation. The flooding that would result from rising sea levels is also a major concern, particularly to the Netherlands, large portions of which lie near or below sea level. In addition to implementing projects for raising dikes and securing seawalls, the Netherlands has considered dumping millions of tons of sand into the sea to extend its coastline.

CHANGES IN LAND USE AND DEFORESTATION

The relationships of land usage to biodiversity and to air and water pollution are major environmental concerns in Europe. Any activities that alter or destroy ecosystems result in the loss or serious reduction in numbers of the plant and animal species that depend on those ecosystems for habitat. It has been argued that the conversion of increasing areas of arable land to the planting of crops to meet the needs of both food-crop and biofuel-crop production could pose serious threats to biodiversity throughout Europe. Increasing the land used for crop production would have serious impacts on forests and grasslands. Also, the destruction of these types of land areas would remove substantial quantities of the major natural means of removing CO_2 from the air—that is, plants and trees.

The environmental impacts of farming constitute a significant concern among Europeans. Almost one-half of the land area within the European Union is used for agriculture; thus farming methods and animal husbandry have major effects on the European environment. With the increased global demand for food production in response to population growth, farming methods have become more intensive in Europe just as they have worldwide. This has resulted in increased use of fertilizers and pesticides as well as increased land usage for crop production, with serious impacts on both air quality and water quality.

Runoff from fertilized fields has increased the amounts of nutrients in waterways and in surrounding seas, causing harmful effects on both freshwater species and sea life. Residues from pesticides have increased the amounts of toxins in both air and water, destroying aquatic life and causing severe damage to forests. Because of their location, the forests of Switzerland in particular have suffered severe damage from air pollution, with almost 35 percent of the trees seriously affected. Through its system of agricultural programs known as the Common Agricul-

Habitats of Selected Vertebrates of Europe

ture Policy, the European Union has implemented reforms to ameliorate these situations throughout Europe. In addition, the agricultural industry has worked to combat the environmentally adverse effects of farming.

Loss of Biological Diversity

Farming, industrialization, construction of roads and recreational areas, and urban sprawl have all contributed to loss of habitat and its accompanying loss of biodiversity throughout Europe. Although the nations of Europe did not meet a 2010 target for halting loss of biodiversity, significant improvement has been seen. More and more land in Europe has been protected as habitat through both European Union programs and individual country provisions. The Natura 2000 network provides protection from exploitation to 17 percent of the land included in the European Union.

Thirty-nine European countries have programs that ensure protection of habitat, but European biodiversity is still very susceptible to loss, with from 40 percent to 85 percent of habitat and 40 percent to 70 percent of plant and animal species estimated to be at risk. Owing to the ease with which they can be converted to cropland, grasslands and wetlands are at high risk of being lost throughout Europe. Increased use of hydropower also poses a potential threat to biodiversity. Dams and reservoirs destroy both habitats and wildlife populations in the areas they flood; fish populations are threatened by the changing of the flow of rivers caused by dams. Some European countries (including Germany, Norway, and Sweden) have enacted measures to address and alleviate these threats, such as through the installation of fish ladders at dams and the use of mini-hydropower stations.

The loss of marine biodiversity is another environmental concern among the European countries. The three major causes of declines in fish and shellfish populations are global warming, pollution, and overexploitation by the fisheries industry. Global warming has caused certain species of fish to move to cooler waters in more northern regions and has also interfered with spawning owing to unfavorable habitat conditions. Farming and other practices of modern industrialized societies have contributed to declines in populations as well, as agricultural runoff containing pesticides, industrial chemicals, oil spills, and other toxins have caused marked reductions in marine life. Nutrients contained in runoff water from artificially fertilized fields also create unfavorable conditions for fish. In deep-sea waters, these nutrients produce the highly toxic substance hydrogen sulfide.

Overexploitation of the seas in the form of the excessive taking of fish by commercial fisheries is a very significant cause of great reductions in both numbers and species of fish. The European Union addresses this problem through the Common Fisheries Policy, which limits the numbers of fish of particular species that may be taken based on the results of scientific studies. Critics have argued, however, that the impor-

tance of the fisheries industry to the economies of many European nations has led to the setting of these limits at levels that tend to be higher than are actually environmentally sound. This situation, coupled with the numbers of fish taken illegally, continues to present a problem in Europe's reestablishment of aquatic populations.

Shawncey Webb

FURTHER READING

Blennow, Kristina, ed. *Sustainable Forestry in Southern Sweden: The Sufor Research Project.* New York: Haworth Press, 2006.
Deublein, Dieter, and Angelika Steinhauser. *Biogas from Waste and Renewable Resources: An Introduction.* Weinheim, Germany: Wiley-VCH, 2008.
Glover, Leigh. *Postmodern Climate Change.* New York: Routledge, 2006.
Hasenauer, Hubert, ed. *Sustainable Forest Management: Growth Models for Europe.* Berlin: Springer-Verlag, 2006.
Maslin, Mark. *Global Warming: A Very Short Introduction.* 2d ed. New York: Oxford University Press, 2009.
Thorsheim, Peter. *Inventing Pollution: Coal, Smoke, and Culture in Britain Since 1800.* Athens: Ohio University Press, 2006.

SEE ALSO: Carbon dioxide; Coal; Coal-fired power plants; Convention on Biological Diversity; European Environment Agency; Fisheries; Fossil fuels; Global Biodiversity Assessment; Greenhouse gases; Kyoto Protocol; Renewable energy; Sea-level changes; United Nations Framework Convention on Climate Change.

European Diploma of Protected Areas

CATEGORY: Preservation and wilderness issues
IDENTIFICATION: Award given to selected European regions that satisfy certain scientific, aesthetic, or cultural criteria
SIGNIFICANCE: The European Diploma of Protected Areas encourages the protection and preservation of selected regions of Europe and recognizes excellence in environmental success.

The Council of Europe established the European Diploma of Protected Areas in 1965. The diploma is awarded based on a number of criteria, including the following: The area must be of importance for the conservation of Europe's biological diversity (whether because it is home to a large number of species or is the habitat of endangered or threatened species) or for the preservation of the continent's landscape diversity, or it must be a site of remarkable natural or geographic phenomena (such as spectacular geological sites, noteworthy paleontological sites, or other historical areas).

Areas that are awarded the diploma must be protected by the laws of the countries in which they are found, must be clearly designated and maintained by specific plans, and must have sufficient staff and resources for their protection. Managers of the areas supervise the protected zones to ensure their maintenance and meet regularly together to discuss common concerns. By 2010 the Directorate of Culture of the Council of Europe had awarded the diploma to almost seventy areas in more than two dozen countries, including Austria, Belarus, Belgium, the Czech Republic, Finland, France, Germany, Greece, Hungary, Italy, Luxembourg, the Netherlands, Romania, Russia, Spain, Sweden, Switzerland, Ukraine, and the United Kingdom.

The awarded areas include the Karlstejn National Nature Reserve in the Czech Republic. The reserve is an important site both biologically and historically, not only for the country but also for the continent as a whole. It contains a number of unique species, especially insects and mollusks, some of which are endangered and others of which are at the northern limit of their habitat. The area has been inhabited for some three thousand years and is the site of the famous Karlstejn castle, built in the fourteenth century. Other protected areas are the Fair Isle National Scenic Area in Scotland, an almost treeless island between Shetland and Orkney; the Minsmere Nature Reserve in eastern England; Cretan White Mountains National Park in Greece; and Boschplaat Nature Reserve in the Netherlands.

One of Germany's awarded protected areas is the notorious Berchtesgaden National Park, where Adolf Hitler had his palatial mountain retreat. The directorate awarded the diploma to this region because of "the exceptional quality of its landscapes, the richness of its flora and fauna and the diversity of its natural sites."

Another area, the only one shared by two countries, is the Germano-Luxembourg Nature Park, which straddles the Our and Sauer rivers. The area is undisturbed by industrialization and contains many cultural and natural sites of interest. Its landscape of high plateaus and ravines in the north and hills and streams in the south captures the features of the Ardennes and Lorraine areas.

Frederick B. Chary

FURTHER READING

Hunkeler, Pierre. *European Diploma for Protected Areas.* Strasbourg, France: Council of Europe, 2000.

Keulartz, Josef, and Gilbert Leistra, eds. *Legitimacy in European Nature Conservation Policy: Case Studies in Multilevel Governance.* New York: Springer, 2007.

SEE ALSO: Biosphere reserves; Conservation policy; Ecotourism; Europe; European Environment Agency; International Union for Conservation of Nature; National forests; National parks; Nature preservation policy; Nature reserves; Preservation; World Heritage Convention.

European Environment Agency

CATEGORIES: Organizations and agencies; ecology and ecosystems; resources and resource management

IDENTIFICATION: Agency of the European Union that collects and disseminates information about the environment

DATE: Established on December 10, 1990

SIGNIFICANCE: The European Environment Agency plays a key role in the European Union's programs to protect and improve the environment in Europe. Through its sharing of the data it collects the agency also significantly contributes to the development and implementation of policies to benefit the environment on a global basis.

The European Environment Agency (EEA) was created in 1990 by a European Union (EU) regulation. In 1993 Copenhagen, Denmark, was chosen as the location for the agency, which began operations in 1994. EEA operates under an executive director and a management board assisted by an advisory committee of scientists. EEA's main function is to provide objective, dependable information about the state of the environment, environmental trends, various policies that have been implemented, and the effectiveness of those policies. EEA also attempts to anticipate environmental problems that may arise in the future. The agency addresses the major environmental issues of global importance, including climate change, air pollution, biodiversity, health concerns in relation to the environment, and the necessity for sustainable production and consumption.

As an organization created by the European Union, EEA assists the EU and its member countries and various institutions operating within the EU, such as the European Parliament and the Committee of the Regions, in the making of decisions and policies that affect the environment. In addition, EEA makes information about the environment available to EU candidate countries, EEA cooperating countries, and the EU's neighboring countries under the EU Neighbourhood Policy. It works cooperatively with the U.S. Environmental Protection Agency, Environment Canada, and the State Environmental Protection Agency of China. EEA also works with international organizations such as the World Health Organization, the World Bank, and the United Nations to promote environmentally sound policies. EEA makes its information available to all segments of the general public and is of particular assistance to individuals and groups in the business and academic communities.

The regulation that created EEA in 1990 also established the European Environment Information and Observation Network, known as Eionet, which is coordinated by EEA. Eionet collects and records data, which it then collates and assesses. The network is composed primarily of environmental agencies and ministries and other environmental organizations of member countries of EEA. In addition to Eionet, EEA has established the European Topic Centre, which collects and analyzes data on specific topics such as climate change and land use. EEA also works in cooperation with the EU Statistical Office and the EU Joint Research Centre. In partnership with the European Commission and its member countries, EEA participates in a program to interconnect their databases through a shared environmental information system.

In 2004 EEA developed the first environmental management program introduced by an EU entity, a system designed to help EEA employees understand their roles in the agency and to encourage them to participate in projects to improve the environment. EEA went on to help other agencies develop management systems.

Shawncey Webb

FURTHER READING

Dilling, Rasmus. "Improving Implementation by Networking: The Role of the European Environment Agency." In *Implementing EU Environmental Policy: New Directions and Old Problems,* edited by Christoph Knill and Andrea Lenschow. New York: Manchester University Press, 2000.

European Environment Agency. *The European Environment: State and Outlook 2005.* Luxembourg: Office for Official Publications of the European Communities, 2005.

Hunter, Janet R., and Zachary A. Smith. *Protecting Our Environment: Lessons from the European Union.* Albany: State University of New York Press, 2005.

SEE ALSO: Biodiversity; Ecological economics; Environment Canada; Environmental Protection Agency; Europe; European Diploma of Protected Areas; Global Biodiversity Assessment; United Nations Environment Programme; World Health Organization.

European Green parties

CATEGORIES: Activism and advocacy; philosophy and ethics

DEFINITION: European political parties that seek to influence society toward greater consciousness of environmental issues

SIGNIFICANCE: The Green movement has been particularly strong in Europe, and the efforts of European Green parties have had important impacts on governments' approaches to environmental issues as well as on policy making in the related areas of economics, social justice, and foreign relations.

Ecological parties, or Green parties, can be found across the globe. In Europe many of these parties promote sustainable development, environmental justice, improvement of the quality of life for all individuals, foreign policy centered on peaceful means, and the reorientation of the European Union to emphasize social and environmental issues rather than focusing solely on economic issues. Green parties are fairly unique among political parties inasmuch as they do not always align themselves easily at one end or another of the left-right political spectrum; they often target issues that fall on both sides. Without actually upsetting the cleavage structures of established par-

Germany's Green Party announces its candidates for forthcoming European Parliament elections in January, 2009. The candidates on the stage hold a sign that translates as "We make Europe clear!" (AP/Wide World Photos)

ties, Greens add new dimensions with their focus on the environment and social justice.

Most Green parties can trace their origins to the environmental movement of the 1960's, the protest movements of the 1970's, and the peace movement throughout the 1980's. Concerns regarding environmental degradation and the testing of nuclear weapons led to increased support of these movements, but it was not until the 1970's that true political entities were formed around these issues. The first recognized ecology party in the world was established in Great Britain in 1973. Over time the party gained political power, and by 1989 the United Kingdom Greens were able to secure 15 percent of the votes yet were still unable to attain any seats in Parliament. Throughout the 1980's and 1990's more Green parties emerged, and many gained legislative seats. By 2010 more than thirty-five countries throughout Europe had Green political parties; more than sixty-five such parties were in existence worldwide. The most successful Green parties, in terms of seats in legislatures, have been those in Western Europe.

THE GERMAN GREENS AND THE EUROPEAN PARLIAMENT

Although not the first to form or the first to gain political representation, the German Green Party is widely regarded as the mother of all Green parties. The German Greens (Die Grünen) were the first ecological party to gain large-scale representation in the federal parliament. In 1983 they garnered the support of almost one million voters throughout Germany and were awarded 28 seats out of 497, or 5.6 percent. In 2009 the German Green Party received roughly 10.7 percent of the total vote.

The European Parliament operates as a legislative body for Europe. Members of the parliament tend to align on the basis of ideological interests instead of by national identity. Recognizing the benefits of forming political alliances within the European Parliament, Greens from various nations began to pair with other individuals and parties. In 1984 European Greens formed a coalition with regionalists who favored devolution (the return of powers to the subnational units of government), creating the Rainbow Group. This coalition dissolved in 1989, and the Greens became the Green Group. These two groups later reunited and became the Group of the Greens—European Free Alliance, with a total representation of fifty-five members in the seventh European Parliament (2009-2014).

Other Green Parties

Operating under a majoritarian electoral system, the Green Party in France (Les Verts) received as much as 10.6 percent of the popular vote in the parliamentary elections of 1989. The Irish Green Party (Comhaontas Glas) has been in existence since 1981. Promoting public transportation, an environmentally friendly economy, clean politics, and an honest tax package, the party has managed to maintain roughly 4 percent of the national vote.

The Scandinavian countries tend to favor proportional representation systems, and that has helped their Green parties succeed in gaining representation. In Finland the Green League (Vihreä liitto) has continually increased its number of seats in the parliament. The Danish Greens ran their first campaign in 1985. Their platform focuses on a variety of issues, including the creation of a society free of violence, grassroots democracy, and a desire to end poverty. When the Swedish Green Party (Miljöpartiet de Gröna) gained seats in the parliament in 1988, it was a landmark event, as the Green Party was the first new party to enter the Swedish parliament in seventy years. The scope of the issues addressed by the Swedish party is much broader than that of many other ecological parties; the Swedish Greens emphasize decentralization, direct democracy, social justice, gender equality, and placing the environment before short-term economic interests.

Kathryn A. Cochran

Further Reading

Bomberg, Elizabeth E. *Green Parties and Politics in the European Union.* New York: Routledge, 1998.

Burchell, Jon. *The Evolution of Green Politics: Development and Change within European Green Parties.* Sterling, Va.: Earthscan, 2002.

Carter, Neil. *The Politics of the Environment: Ideas, Activism, Policy.* 2d ed. New York: Cambridge University Press, 2007.

Cassola, Arnold, and Per Gahrton, eds. *Twenty Years of European Greens, 1984-2004.* Brussels: European Federation of Green Parties, 2003.

Dobson, Andrew. *Green Political Thought.* New York: Routledge, 1995.

Hanley, David L. *Beyond the Nation State: Parties in the Era of European Integration.* New York: Palgrave Macmillan, 2008.

Müller-Rommel, Ferdinand, and Thomas Pogunkte, eds. *Green Parties in National Governments.* Portland, Oreg.: Frank Cass, 2002.

See also: Antinuclear movement; Environmental ethics; Environmental justice and environmental racism; European Environment Agency; Green movement and Green parties.

Eutrophication

Category: Water and water pollution
Definition: Overenrichment of a water body with nutrients
Significance: The process of eutrophication, in which a body of water receives an excessive amount of nutrients that accelerates aquatic plant growth, leads to a reduction in the oxygen in the water that fish need to survive. Continued enrichment can lead to excessive algal growth, which in turn impairs fisheries, limits recreational uses of the water, and can create major problems for water-supply purveyors.

From a geologic perspective, lakes are relatively temporary features of a landscape. For example, the Great Lakes were formed over the last two million years by four or more major advances of huge ice sheets. Only relatively recently (about twelve thousand years ago) did these lakes develop into their current shapes.

Most lakes go through a series of trophic (nutrition-related) states that can take thousands of years before the lake basins eventually fill in with sediment. Oligotrophy is the first trophic stage; it is characterized by clear water, low plant productivity because of limited nutrient inputs, and high levels of dissolved oxygen throughout the water column. Crater Lake in Oregon is a good example of an oligotrophic lake. The next stage is mesotrophy, which has moderately clear water and moderate plant productivity and lower levels of oxygen in the hypolimnion, which is the lowest level in a lake. In the third stage, eutrophy, excess nutrients are present, transparency is reduced, algal scums start to appear, and oxygen (which fish need) is not present in the summer in the hypolimnion. Hypereutrophy is the final stage; in this stage algal scums dominate in the summer, few macrophytes (plants that can be seen without magnification) exist, and the hypolimnion is devoid of oxygen.

Anthropogenic (human-caused) changes in lakes result from excessive inputs of sediment, fertilizers (such as nitrates and phosphates), pesticides from agricultural practices, runoff from urban sewage, and industrial effluents such as lead and mercury. The inevitable result is accelerated or cultural eutrophication, which can occur in decades as compared to the thousands of years seen under natural conditions.

Reservoirs that are used for water-supply purposes are particularly sensitive to algal blooms. If sufficient light and warmth are present, which is common in the high sun season, thermal stratification occurs, and the lower zone of the reservoir (where the hypolimnion is located) becomes anaerobic. This lack of oxygen enables algal nutrients, such as silica, phosphorus, and ammonia, to come from bottom sediments, which in turn increases more algal growth, and the cycle continues. The reservoir surface becomes covered with algae, which then requires either physical removal or large applications of copper sulfate, which in itself can cause problems.

Eutrophication can occur in streams and marine environments as well as in lakes and reservoirs. Coastal waters, bays or estuaries that are partially enclosed, and shallow seas provide excellent opportunities for unbridled nutrient enrichment. For example, excess nutrients emanating from the extensive agricultural area in the midwestern United States flow down the Mississippi and Missouri rivers to where the rivers empty into the Gulf of Mexico. As a result, a hypoxic (depleted of oxygen) area known as a dead zone appears on a fairly regular basis off the mouth of

Dead fish float in Greenwich Bay, Rhode Island, in 2003. The fish kill was caused by eutrophication of the bay. (Tom Ardito/IAN Image Library)

the Mississippi River in the gulf. Although it varies in size, this dead zone has been known to cover an area of approximately 21,000 square kilometers (8,000 square miles), which is larger than the entire state of Massachusetts.

Robert M. Hordon

Further Reading

Cech, Thomas V. *Principles of Water Resources: History, Development, Management, and Policy.* 3d ed. New York: John Wiley & Sons, 2010.

Cunningham, William P., and Mary Ann Cunningham. *Principles of Environmental Science: Inquiry and Applications.* 4th ed. New York: McGraw-Hill, 2008.

Gray, N. F. *Drinking Water Quality: Problems and Solutions.* 2d ed. New York: Cambridge University Press, 2008.

Laws, Edward A. "Cultural Eutrophication: Case Studies." In *Aquatic Pollution: An Introductory Text.* 3d ed. New York: John Wiley & Sons, 2000.

North American Lake Management Society and Terrene Institute. *Managing Lakes and Reservoirs.* Madison, Wisc.: Author, 2001.

See also: Agricultural chemicals; Cultural eutrophication; Dams and reservoirs; Dead zones; Pesticides and herbicides; Runoff, agricultural; Runoff, urban; Sedimentation; Septic systems; Sewage treatment and disposal; Water treatment; Watershed management.

Everglades

Categories: Places; ecology and ecosystems; preservation and wilderness issues

Identification: Sparsely inhabited subtropical wetlands located in southern Florida

Significance: The Everglades are home to many different species of animal and plant life, some of which are found nowhere else in the world. Since the late twentieth century, efforts have been under way to repair the damage done to the region's fragile wetlands ecosystem by human activities.

The subtropical wetlands region known as the Everglades encompasses an area of approximately 607,000 hectares (1.5 million acres), about one-fifth of which—some 121,400 hectares (300,000 acres)—was dedicated as the Everglades National Park in 1947. This ecologically fragile area is actually a river more than fifty miles wide and just a few inches deep.

The northern portion of the Everglades resembles a prairie covered with shallow water. Most of it is filled with saw grass, an invasive plant with sharp, serrated edges that can reach heights exceeding 3 meters (10 feet). On areas of higher ground, called tree islands, plants such as royal palm, bustic, gumbo limbo, and live oak flourish. Near the southern section of the Everglades the landscape is covered with salt marshes that are home to stands of mangrove. The mangrove swamps are the natural habitat of deer, fish, pelicans, heron, alligators, snakes, and the Florida panther.

In 1915 construction was undertaken on a two-lane highway that would cross the Everglades, linking the Gulf of Mexico in the west to the Atlantic Ocean in the east. Although there was considerable commercial interest in creating such a thoroughfare, progress on it was slow. It was not until April 25, 1928, that this road, named the Tamiami Trail, was completed. Part of U.S. Highway 41, it runs for 266 kilometers (165 miles) across the northern edge of what later became Everglades National Park. Subsequently Interstate 75 was built as part of the national highway system; the section of this interstate that runs through the Everglades is nicknamed Alligator Alley.

Agricultural operations arrived in the Everglades shortly after the end of World War I in 1918 with the establishment of sugar plantations and farms. Canals were built southeast from Lake Okeechobee to provide water for the plantations and farms and for the communities that grew up around them.

In the late 1940's, after World War II, the U.S. government began to develop an interest in preserving the ecology of this sensitive area. In 1947 Marjory Stoneman Douglas published *The Everglades: River of Grass*, a book that awakened many Americans to the uniqueness of the region and aroused the concerns of naturalists and conservationists, many of whom had grown increasingly aware of how people with business and commercial interests in southern Florida had begun to despoil the area's extraordinary natural wonder. Some of these people wanted to drain the Everglades and use the land for farming and real estate development. The sugar plantations, truck farms, and housing tracts they envisioned would destroy valuable habitat for dozens of species of plants and wildlife and would require enormous quantities of water for their efficient operation.

Despite public protests against the destruction of

this valuable wetland area, by the 1960's those who wished to drain the swamp had made significant inroads. The U.S. Army Corps of Engineers had already built a canal that diverted fresh water from the Kissimmee River, which had followed its natural course into Lake Okeechobee, to towns and cities bordering the Everglades. Fresh water was becoming a scarce commodity in the region, and as agriculture and human habitation increased, the diminishing water supply was being degraded by harmful chemicals used in farming.

By the last decade of the twentieth century, following considerable efforts on the part of those wishing to protect the area and preserve its ecology, the Army Corps of Engineers agreed to work on the restoration of the Kissimmee River to its former, natural course. The Corps also began working under the Comprehensive Everglades Restoration Plan to return to their previous condition those areas in which the Corps had erected artificial barriers and constructed canals to drain the wetlands.

R. Baird Shuman

FURTHER READING

Douglas, Marjory Stoneman. *The Everglades: River of Grass.* 60th anniversary ed. Sarasota, Fla.: Pineapple Press, 2007.

Grunwald, Michael. *The Swamp: The Everglades, Florida, and the Politics of Paradise.* New York: Simon & Schuster, 2006.

Levin, Ted. *Liquid Land: A Journey Through the Florida Everglades.* Athens: University of Georgia Press, 2003.

Lisagor, Kimberly, and Heather Hansen. "The Everglades, Florida." In *Disappearing Destinations: Thirty-seven Places in Peril and What Can Be Done to Help Save Them.* New York: Vintage Departures, 2008.

Solecki, William D., et al. "Human-Environment Interactions in South Florida's Everglades Region: Systems of Ecological Degradation and Restoration." *Urban Ecosystems* 3 (October, 1999): 305-343.

Bush Takes Action to Help Restore the Everglades

The principal aim of the Comprehensive Everglades Restoration Plan is to capture and clean much of the water that flows unused to the sea and deliver it wherever it is needed most. The following is President George W. Bush's statement on his administration's commitment to the project:

On June 4, 2001, I joined the Governor of Florida in visiting the Everglades. The Everglades and the entire south Florida ecosystem are a unique national treasure. The restoration of this ecosystem is a priority for my Administration, as well as for Governor Bush. Today we are very pleased to solidify our commitment and full partnership in this unprecedented endeavor by signing a joint agreement to ensure that adequate water supplies will be available to benefit State and federally owned natural resources.

The Water Resources Development Act of 2000 authorized the Comprehensive Everglades Restoration Plan. The Plan has a projected cost of $7.8 billion over 30 years, the largest ecosystem restoration project ever undertaken. The Plan establishes a unique 50/50 cost-sharing partnership between the State of Florida and the Federal Government.

A critical component of the Plan relates to the supply and management of water for multiple uses in South Florida—restoration, municipal, agricultural, and flood control. The Congress determined that the overarching objective of the Plan is the restoration, preservation, and protection of the South Florida ecosystem, while providing for other water-related needs of the region, including water supply and flood protection.

Because the Federal Government's primary interest is in restoration and protection of the federally owned natural resources in the State, the Congress called for the President and the Governor to agree formally that the State would reserve under State law for each restoration project water sufficient to meet the needs of the South Florida ecosystem, including Everglades National Park, the Big Cypress National Preserve, and other natural areas owned by the State and Federal Government. The reservation of water under State law will be included in the Project Implementation Report for each project and will be consistent with the Plan.

My Administration is deeply committed to the Federal/State Everglades partnership, and the Department of the Interior and the Army Corps of Engineers will have important roles in this effort.

Taylor, Caroline. *Restoring the River of Grass.* Miami: National Audubon Society, 1996.

Toops, Connie. *The Florida Everglades.* Stillwater, Minn.: Voyageur Press, 1998.

SEE ALSO: Endangered Species Act; Endangered species and species protection policy; Fish and Wildlife Act; Great Swamp National Wildlife Refuge; Habitat destruction; National Wildlife Refuge System Administration Act; Wetlands; Wilderness areas.

Exotic species. *See* **Introduced and exotic species**

Experimental Lakes Area

CATEGORY: Water and water pollution

IDENTIFICATION: Research facility that studies environmental problems, especially acidification, in lakes in eastern Canada

DATE: Established in 1968

SIGNIFICANCE: The research conducted at the Experimental Lakes Area sheds light on the causes of water pollution and suggests ways of maintaining the health of freshwater lakes.

Increasing acidification of the water in lakes in Ontario and Nova Scotia led to significant declines of fish populations in those lakes during the 1960's through the 1980's. As a result, Canadian environmentalist David Schindler founded the Experimental Lakes Area (ELA) research facility in 1968 in order to investigate acidification and other environmental problems that were affecting the lakes of eastern Canada. The goals of the ELA are fourfold: to develop better understanding of global threats to the environment through knowledge gained from ecosystem, experimental, and scientific research; to monitor and demonstrate the impacts of human activities on watersheds and lakes; to develop appropriate responsibility for the preservation, restoration, and enhancement of ecosystems; and to promote environmental protection and conservation for ecosystems through education. The ELA includes fifty-eight small lakes and their drainage basins, plus three additional stream segments.

The ELA is operated by the Central and Arctic Region of the Canadian Department of Fisheries and Oceans from its Freshwater Institute in Winnipeg. Because the ELA research facility is located in a sparsely inhabited region of southern Ontario, it is relatively unaffected by external human influences and industrial activities. As such, it serves as a natural laboratory for the study of physical, chemical, and biological processes and interactions operating on an ecosystem over a large area and a multiyear time scale.

With renewed operating support in the late 1990's, the ELA took on several new experimental studies in addition to whole-lake acidification experiments and eutrophication recovery studies. For example, one ELA ecosystem study investigated additions of trace amounts of mercury to one of the ELA lakes. It is generally believed that high concentrations of methylmercury, the most toxic form of mercury, in fish in remote lakes is caused by elevated inputs of atmospheric inorganic mercury deposited directly into lakes and indirectly through their watersheds. The ELA has provided researchers with the opportunity to investigate this hypothesis as well as an alternative that suggests that geologic mercury is the most important source of mercury to remote lakes; such mercury originates from the weathering of mineral deposits in lake basins.

ELA scientists have also examined the effects of climate change, dissolved organic carbon, and ultraviolet radiation on lakes. One study investigated the effects of experimentally deepening the mixed layer of a lake, thereby exposing more of the lake water and

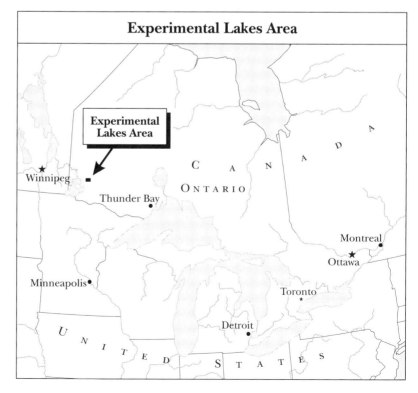

Experimental Lakes Area

organisms to surface radiation and simulating the effects observed in natural ELA lakes during two decades of warming between 1970 and 1990. Another area of research has involved reducing the natural color of an ELA lake to investigate the effects of increased ultraviolet radiation on lake life-forms. Other projects have included studies of the effects of persistent toxic substances—such as cadmium and hydrocarbons—in lakes, contributions of forest materials to lake nutrient inputs, and the alteration of food-chain processes caused by human intervention.

Alvin K. Benson

FURTHER READING

Laws, Edward A. *Aquatic Pollution: An Introductory Text.* 3d ed. New York: John Wiley & Sons, 2000.

O'Sullivan, P. E., and Colin S. Reynolds, eds. *The Lakes Handbook.* 2 vols. Malden, Mass.: Blackwell, 2004-2005.

Resetarits, William J., Jr., and Joseph Bernardo, eds. *Experimental Ecology: Issues and Perspectives.* New York: Oxford University Press, 1998.

SEE ALSO: Acid deposition and acid rain; Ecosystems; Environment Canada; Lake Erie; Mercury; Water pollution; Water quality.

Externalities

CATEGORIES: Atmosphere and air pollution; water and water pollution

DEFINITION: Effects that are imposed on third parties who were not voluntarily involved in the activity or transaction creating them

SIGNIFICANCE: Parties who voluntarily engage in market transactions expect to benefit from them, but if a transaction imposes costs on unwilling third parties, the transaction might be unfair to those parties and might create a net loss for society.

When one resident of an apartment building has a party, his neighbors may be bothered by the noise. If an office building is built next to a resort hotel, shading the latter's swimming pool, the hotel may lose much of its business. If a pig farm is close to a housing development, the noise, smells, and flies that go along with the farming operation may reduce the property value of the houses. These are examples of externalities.

Consider a manufacturer of automobiles, who must pay for the land, equipment, materials, labor, utilities, advertising, and myriad other costs that are necessary for the production process. Pollution is also a cost that results from the manufacturing process. Suppose the manufacturer is able to avoid paying the cost of disposing of the pollutants by dumping them into the air or water near the plant, so that the costs are instead borne by the surrounding community or downstream landowners. The manufacturer would be externalizing the pollution cost—that is, making someone else pay for it.

A rational manufacturer will maximize profit by increasing the level of production so long as each unit can be sold for more than its cost. Suppose an automaker calculates that producing one additional car will add $20,000 to costs, but the car can be sold for $20,200. It makes sense to make that additional car. The manufacturer should, in fact, increase production until it is no longer possible to make a profit by making additional units. However, what if the external cost caused by pollution amounts to $500 of harm to the environment for each unit produced? If the manufacturer can externalize that cost by allowing others to absorb it, the calculation of the profit-maximizing number of units is not affected. The social cost (manufacturing cost plus harm caused by pollution) for the marginal units actually exceeds the value of the automobile, as measured by its purchase price. The manufacturer thus has incentive to produce more than the socially optimal amount of pollution and more than the optimal number of automobiles.

If forced to "internalize" the pollution cost—that is, to pay for it—the manufacturer would reduce the amount of pollution and the number of units produced to the socially optimal level. One way in which governments force manufacturers to internalize pollution costs is by compelling specific pollution controls. For example, governments may mandate cleaner production processes or the safe disposal of liquid effluents. Another approach governments take is to impose taxes on effluents equal to the amount of harm the pollution causes. If manufacturers can find ways to reduce the pollution at costs lower than the taxes, they will do so.

In many externality cases it is not obvious who should absorb the costs. In the case of the pig farm, should the farmer have to compensate the home owners or move the farming operation to another location? Should home owners who knew they were moving in next to a pig farm have the right to complain about the negative environmental impacts of the farm? Courts are commonly asked to resolve such conflicts. They often decide the issue by requiring the party who can solve the problem at lowest cost to do so.

Howard C. Ellis

Further Reading

Friedman, David D. *Law's Order: What Economics Has to Do with Law and Why It Matters.* Princeton, N.J.: Princeton University Press, 2001.

Winston, Clifford. *Government Failure Versus Market Failure: Microeconomic Policy Research and Government Performance.* Washington, D.C.: Brookings Institution Press, 2006.

See also: Air pollution; Airborne particulates; Environmental economics; Environmental law, U.S.; Environmental torts; Land-use planning; Noise pollution; Pollution permit trading; Property rights movement; Water pollution.

Extinctions and species loss

Category: Animals and endangered species

Definitions: Extinction is the complete local, regional, or global die-off of a species; species loss is the reduction in biodiversity owing to extinction

Significance: Although extinctions occur naturally, human activities have accelerated the extinctions of some species. Early humans have been implicated in the complete disappearance of some prehistoric species, and in recent centuries the increasing influences of human population growth and technology on the biological ecosystems of the world have caused extinction rates to skyrocket to thousands of times the background rate.

In general, the term "extinction" refers to global extinction, the elimination of a species from the earth, but species can also be said to be extinct locally or regionally. The term "extirpation" is sometimes used to describe local or regional extinction. By some estimates, before humans arrived on the scene animal and plant species disappeared from the planet at an average rate of about one per year. According to the International Union for Conservation of Nature and Natural Resources (IUCN), the modern extinction rate is anywhere from one thousand to ten thousand or more times higher than the expected natural rate.

The fossil record indicates that throughout the history of life on the earth extinctions have resulted primarily from the planet's changing surface and climate. The rates of speciation (the creation of new species) and extinction have remained fairly constant over millions of years. However, five mass extinctions have occurred over the course of the earth's history. The most famous, if not the largest, extinction event occurred 65 million years ago at the end of the Mesozoic era, when a large number of land and marine animals, including the dinosaurs, suddenly disappeared from the fossil record. Some scientists credit this extinction to a large meteor that hit the earth and subsequently caused climate changes that devastated some life-forms. Other scientists credit the demise of dinosaurs to the evolutionary superiority of mammals. The current accelerated loss of species is sometimes called the "sixth extinction crisis." It is the only mass extinction attributable to a single species: human beings.

Causes of Recent Extinctions

As human populations grow, people clear forests and land at increasing rates to build homes, businesses, farms, and infrastructure. This process disrupts the habitats that are home to microbes, plants, and animals, which all have unique ecological relationships to one another. Habitat destruction is one of the most prominent causes of extinctions in recent history. With the destruction of one species, there is the potential for the secondary extinction of other species that are dependent on the first species for reproduction or food. For instance, a bird species may disperse seeds by eating the fruit of a plant and expelling the seeds in its droppings as it travels through its home range. If the bird dies off, the plant loses an important means of propagation. If a prey species becomes extinct, predators may not be able to find enough food to survive. If a major predator becomes extinct, there may be a population explosion among prey species, followed by a catastrophic depletion of food sources for those species and a subsequent die-off.

Highway construction and other development can cause fragmentation of a habitat, preventing migration between the two fragments. This can cause a spe-

Recoveries from Near Extinctions

In the wild, herds of American bison (*Bison bison*) were decimated to extinction east of the Mississippi River by 1830, and in the west by 1897. When the American Bison Society formed in 1905, only about 25 protected bison remained in Yellowstone National Park. A census of captive herds taken in 1908 indicated that 1,116 bison were held in 15 zoos and private collections. Starting in 1907 with fifteen animals from the New York Zoological Park, captive bison were relocated to newly established reserves. By 1933, the bison herds numbered 4,404 animals. Today, bison are once again common in their western range.

Likewise, the European bison, or wisent (*Bison bonasus*), became extinct in the wild in the 1920's. The International Society for the Protection of Wisent was founded in Europe in 1923. The Frankfurt Zoo staff, and later the Warsaw Zoo staff, established a studbook for wisent in 1932. Cooperative breeding programs among European zoos increased wisent herds in captivity, which were then used to restock the wild forests. Another animal saved from extinction by European zoos was the Przewalski's horse (*Equus caballus przewalskii*). When the Polish general for whom the horse is named found the last herd in western Mongolia in the 1870's, he brought some of the horses back to Poland. From this initial captive stock, zoos bred enough horses to release them back into their natural habitats in the wild.

The Catholic missionary Armand David, who also collected natural history specimens for the natural history museum in Paris, found a herd of Père David's deer (*Elaphurus davidianus*) at a Chinese royal game park in 1865. Not sure of what they were, he obtained some living deer and brought them back to the London and Berlin Zoos. While the captive herds were increasing at European zoos, the wild herd in China had died out by 1920, due to flooding of the park by the nearby Yellow River in 1894, and decimation of the herd for food during the Boxer Rebellion of 1900. To further increase the deer population, the eleventh Duke of Bedford gathered the European animals onto his Woburn Abbey estate between 1893 and 1895. This propagation effort increased the herd to 88 animals in 1914 and 255 by 1948. By 1986, there were some 1,500 deer in England, and some were sent back to China to repopulate their natural habitat, including the park where the last herd was originally found.

Arabian oryx (*Oryx leucoryx*) were hunted to extinction on or about October 18, 1972. However, this possibility was anticipated by the Fauna Preservation Society, which captured four animals before the last remaining herd was gone. A captive propagation program began at the Phoenix Zoo, in Arizona, and in 1975 the herd had increased to forty-five. Some of these were sent to other zoos to form additional herds. By 1986, there were 172 oryx in thirteen American collections, six European collections had another 30 animals, and ten collections in the Middle East had 528 oryx. Reintroduction of the oryx to its native habitat began in 1978 and has achieved good success; however, hunting has again become a problem. Nevertheless, despite a need for improved conservation education and protection, the Arabian oryx remains in the wild.

Vernon N. Kisling, Jr.

cies to die out if neither area is capable of supporting a viable population of the species. This is also true of habitats that are decreasing in size. In the United States at the close of the twentieth century, 99.8 percent of the nation's tallgrass prairies had been destroyed, 50 percent of the wetlands were gone, and 98 percent of virgin and old-growth forests had been cut. At least five hundred native species had become extinct, with tremendous losses in the populations of animals such as wolves, black bears, bison, and cougars.

Tropical rain forests are home to anywhere from one-half to two-thirds of the world's plants and animals. By some calculations, tropical rain forests lose approximately four to six thousand or more species per year because of deforestation. Many of the lost species are unique to the tropical forests. This destruction has also contributed to a serious decline in the numbers of many migratory songbirds that winter in Central and South America. The songbirds' northern summer habitats are becoming fragmented as well, causing more losses in their populations.

Even when a species is not completely obliterated by development, extinction can occur if the population size becomes too small to recover. Such a population is said to fall into an extinction vortex. This problem may occur if there are too few females left in the population to breed, or, if the habitat is too fragmented, if individuals are not able to locate partners with which to mate. If enough mates cannot be found, genetic inbreeding can destroy the viability of the species. For instance, a 2009 study of moss carder bumblebees living on the Hebridean islands off the west coast of Scotland found that many generations of inbreeding had left the insects more susceptible to parasitic disease than their mainland counterparts. Inbreeding among these bumblebees had also resulted in an increase in infertile males.

Small populations that are vulnerable to environmental fluctuations may also fall into an extinction vortex. For example, extremely harsh winter weather can wipe out an entire species that has already been reduced to a small population. Larger groups, by contrast, are better able to survive adversity. The chance of being destroyed by environmental fluctuations increases exponentially with decreasing population size.

The smallest population of a species that is able to stay above the extinction vortex is often called the minimum viable population (MVP). If a population declines below this size, it is usually only a matter of time before breeding problems and climatic fluctuations will destroy the whole population. Likewise, if a habitat is reduced to a point where it is unable to support the MVP in an adverse year, the population will vanish.

POLLUTION, OVERUSE, AND OVERHUNTING

Pollution can kill many plants and animals and at the same time alter and destroy habitats. Acid rain and air pollution are detrimental to forests and forest animals. Sediment and excessive nutrients that run off into lakes, rivers, and bays often have adverse effects on aquatic life. Pesticides that persist in the environment, such as dichloro-diphenyl-trichloroethane (DDT), have caused large losses and near extinction among some birds, notably meat- and fish-eating species such as bald eagles, ospreys, peregrine falcons, and brown pelicans. These species were particularly susceptible to DDT because the effects of chemicals are amplified as they become increasingly concentrated in the fatty tissue of organisms at higher trophic levels in the food chain.

Exotic species are animals or plants that have been introduced into an area to which they are not native. It has been estimated that the introduction of exotic species has contributed to approximately 40 percent of all animal extinctions worldwide since 1600. Because the new exotic species may not have any natural predators or competitors in its new habitat, it can dominate the new ecosystem and reduce the populations of many native species. Islands are particularly vulnerable to invasive exotic species. The brown tree snake, a mildly venomous reptile, was introduced to the Pacific island of Guam during World War II, probably as an accidental stowaway on a military cargo ship from Papua New Guinea. By the end of the century, the snakes had decimated most of Guam's native species of birds and small mammals and reptiles. Among the island's native fauna, nine of eighteen bird species, two of three bat species, and four of ten lizard species had been extirpated, and more had become uncommon or rare. Because the snake has no natural predator on Guam, and overall lizard populations have remained high enough to provide the snakes with an ample food supply, in some areas the snake has maintained population densities of nearly thirteen thousand individuals per square mile.

During the nineteenth century whales were harvested at a rate of fewer than one hundred every three years. With the development of faster boats and more efficient weapons, however, by 1933 the whaling industry was killing thirty thousand whales per year; by 1967 that number had more than doubled. It is interesting to note that 2.5 million barrels of whale oil were harvested in 1933, as opposed to only 1.5 million barrels in 1967. This is because the larger whales, the blues and fins, had been hunted to the brink of extinction by 1967. An international whaling moratorium observed by many nations since 1986 has allowed the partial recovery of some species. However, many may lack the genetic diversity to thrive in their pollution-stressed environment.

In the United States, overhunting has led to the extinction or near extinction of many species, such as the American bison. Early in the nineteenth century the passenger pigeon was one of the most abundant birds on earth. Alexander Wilson, a renowned ornithologist, is said to have observed a flock of passenger pigeons that took several hours to fly by him. He estimated that the flock, which appeared to be approximately 1.6 kilometers (1 mile) wide and 386 kilometers (240 miles) long, was composed of two billion birds. The passenger pigeon became extinct in 1914, largely because of overhunting by market hunters. The hunters used nets, guns, and even dynamite to trap the birds, which were viewed as a culinary delicacy.

Many species are under government and international protection but are still hunted because the economic incentive of selling skins, horns, or bone outweighs the risk of a small fine or short prison sentence. In the early twenty-first century, a Bengal tiger skin can fetch $35,000 on the black market, and rhinoceros horns used in traditional medicines and aphrodisiacs can sell for as much as $1,200 per kilogram. A poacher can earn millions of dollars annually.

PEST CONTROL, MONOCULTURES, AND WILD PLANTS

Species can become endangered or even extinct if they must compete with the human population for food. Farmers in Africa have killed elephants to prevent the elephants from eating or trampling food crops. In the United States the Carolina parakeet was exterminated by farmers in 1914 because it fed on fruit crops, and 98 percent of the nation's prairie dog population has been exterminated with poisons so that horses and cattle do not break their legs stepping into the rodents' burrows. In turn, the prairie dog's primary predator, the black-footed ferret, has come close to extinction with the massive reduction of the population of its food source.

Extinctions and species losses are critical problems for the human population because human life is dependent on the biodiversity of species. One area in which this is apparent is agriculture. Of the world's estimated seventy-five hundred species of edible plants, farmers have developed high-yielding monocultures of only about thirty species, each with a minimum of genetic variation. These monocultures lack the vigor of wild plants, which are constantly developing new ways to adapt to adverse conditions and fend off the animals and microorganisms that attack them. When a monoculture crop fails because of disease or other problems, plant breeders must go back to wild species to find the traits that their crop needs to thrive and breed these characteristics into the crop species. If wild species are not protected, their gene pools will dwindle and the species will eventually disappear.

Wild plants, especially in the tropical rain forests, have important uses as medicine. In 1960 the rosy periwinkle, a shrub that grows in Madagascar, was found to contain two chemicals that revolutionized the treatment of childhood leukemia and Hodgkin's disease. The use of these drugs brought about a 95 percent remission rate in children who previously had little chance of surviving leukemia. Paclitaxel (originally known as taxol), a drug extracted from the bark of the Pacific yew tree, has been valuable in the treatment of Kaposi's sarcoma and breast, ovarian, and small-cell cancers. A plant related to the periwinkle, rauwolfia, provides an alkaloid that has been widely used in the control of high blood pressure. Digitalis, which comes from the foxglove plant, is a highly effective drug used for centuries in the treatment of various heart conditions. Of the world's estimated 400,000 plant species, only a few thousand have been studied for medicinal use. It is believed that the world's tropical forests may be home to thousands of plants that have cancer-fighting properties. Some 25 percent of the active ingredients in the anticancer drugs currently in use come from organisms native to rain forests.

Toby Stewart and Dion Stewart
Updated by Karen N. Kähler

FURTHER READING

Barrow, Mark V., Jr. *Nature's Ghosts: Confronting Extinction from the Age of Jefferson to the Age of Ecology.* Chicago: University of Chicago Press, 2009.

Ehrlich, Paul R., and Anne H. Ehrlich. *Extinction: The Causes and Consequences of the Disappearance of Species.* New York: Ballantine Books, 1981.

Goudie, Andrew. *The Human Impact on the Natural Environment: Past, Present, and Future.* 6th ed. Malden, Mass.: Blackwell, 2006.

Kaufman, Les, and Kenneth Mallory, eds. *The Last Extinction.* 2d ed. Cambridge, Mass.: MIT Press, 1993.

McGavin, George. *Endangered: Wildlife on the Brink of Extinction.* Buffalo, N.Y.: Firefly Books, 2006.

Novacek, Michael J. *Terra: Our 100-Million-Year-Old Ecosystem—and the Threats That Now Put It at Risk.* New York: Farrar, Straus and Giroux, 2007.

Raup, David M. *Extinction: Bad Genes or Bad Luck?* New York: W. W. Norton, 1991.

Ward, Peter D. *Under a Green Sky: Global Warming, the Mass Extinctions of the Past, and What They Mean for Our Future.* New York: Smithsonian Books/Collins, 2007.

Wilson, Edward O. *The Diversity of Life.* New ed. New York: W. W. Norton, 1999.

SEE ALSO: American bison; Balance of nature; Biodiversity; Condors; Dodo bird extinction; Endangered species and species protection policy; Habitat destruction; Introduced and exotic species; Passenger pigeon extinction; Whaling.

Exxon Valdez oil spill

CATEGORIES: Disasters; water and water pollution

THE EVENT: Grounding of the supertanker *Exxon Valdez* in a shallow stretch of Prince William Sound off the coast of Alaska, resulting in the largest oil spill in U.S. history

DATE: March 24, 1989

SIGNIFICANCE: The *Exxon Valdez* spill caused enormous environmental damage to coastal areas in southeastern Alaska and to marine life in Prince William Sound. Populations of both large and small marine mammals, birds, fish, and mollusks were devastated, as was plant life.

Oil spills are common occurrences in U.S. waters. According to a U.S. Coast Guard report, about ten thousand oil spills occur in and around U.S. waters each year, totaling 15 million to 25 million gallons of oil. These oil spills may result from drilling accidents, as was the case with the major spill that took place in Santa Barbara, California, in 1969, or they may be related to problems associated with supertankers such as the *Exxon Valdez*.

The *Exxon Valdez* was a single-hulled supertanker operated by the Exxon Corporation. The ship was 300 meters (987 feet) long and cost about $125 million to build. It was equipped with state-of-the-art instruments for depth sounding, guidance, and navigation. On March 23, 1989, the vessel was loaded with more than ten million barrels, or approximately 420 million gallons, of North Slope crude. The oil had been transported about 1,300 kilometers (800 miles) from Prudhoe Bay near the Arctic Circle to the Port of Valdez in southern Alaska. Shortly after midnight on Friday, March 24, the ship left port and traveled west and southwest down the Valdez fjord to the vicinity of Bligh Island in Prince William Sound, where it ran aground on Bligh Reef.

At the time the ship left port, conditions for sailing were ideal: light winds, calm seas, and good visibility. However, because of miscalculations by the offi-cers in charge, the ship hit a chain of rocks about 4 kilometers (2.5 miles) west of Bligh Island. The rocks tore a gash in the tanker hull and allowed an estimated 10 million to 11 million gallons of crude oil to escape before noon. The oil spread across 4,600 square kilometers (1,776 square miles) of water in Prince William Sound and the Gulf of Alaska. Approximately 5,100 kilometers (3,169 miles) of shoreline received some oiling. The oiled areas included a number of fishing villages, a national forest, state and national parks, national wildlife refuges, critical habitat areas, and a state game sanctuary.

The damage to coastal areas in southeastern Alaska and to marine life in Prince William Sound was enormous. The spill affected the livelihoods of villagers along the west side of the sound, the Alaska and Kenai peninsulas, the Kodiak Archipelago, and part of Cook Inlet. Populations of both large and small marine

On March 26, 1989, two days after the Exxon Valdez *ran aground, the* Exxon Baton Rouge *(smaller ship) pulled alongside in an attempt to offload crude oil.* (AP/Wide World Photos)

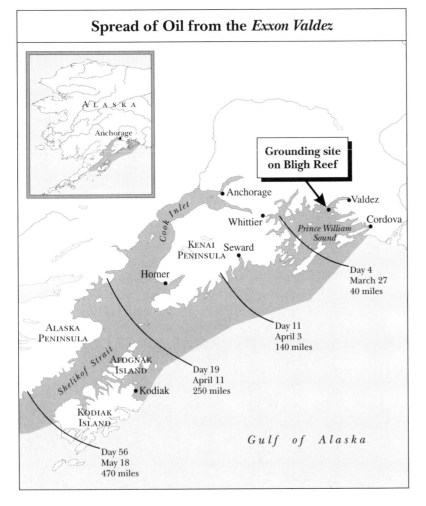

Spread of Oil from the *Exxon Valdez*

the marine life in Prince William Sound, this proved to be generally true; an exception was the area's population of harbor seals, which did not recover and stabilize until the mid-1990's.

RECOVERY EFFORTS

Immediately after the spill, various techniques were used to consume or disperse the oil slick and the oil coating surrounding beaches. Attempts to ignite the oil met with limited success because they occurred after most of the lighter, more volatile components of the oil had evaporated, leaving behind emulsified, pancakelike layers of crude oil. Boat skimmers, which are designed to confine and collect oil with floating booms, recovered less than 5 percent of the oil. Such booms are most effective when wave height is below 1 meter (3.28 feet). The application of chemical dispersants to the oil slick also was not effective. In cold water, the mousselike layers of oil residue are almost impossible to break up chemically.

The techniques used to wash beach rocks ranged from hand-applied cold water to steam cleaning. The latter was effective in some areas. Exxon workers substantially reduced the oil coating on beach rocks at some islands in the northern part of Prince William Sound. This method may have done more harm than good, however, because most of the small, beneficial organisms along the beach and in the tidal zone were killed by the hot waters. Meanwhile, tons of oiled gravel underlying mussel beds were removed from Prince William Sound and the Kenai Peninsula by recovery team workers and local residents. The oiled gravel was replaced with clean sediment.

Perhaps the most effective technique used to clean the oil was bioremediation. Workers sprayed a fertilizer solution along miles of the cobbled beaches in Prince William Sound in an effort to promote the growth of naturally occurring, oil-eating microbes. Such microorganisms consist primarily of spiral-type bacteria that can rapidly develop under ideal condi-

mammals, birds, fish, and mollusks were devastated, as was plant life. Thousands of dolphins, sea lions, sea otters, and harbor seals died or were otherwise adversely affected by the fouling of their environment. An estimated 250,000 to 300,000 seabirds were also killed by the oil spill. These included harlequin ducks, pigeon quillemotes, common murres, and marbled murrelets. A significant number of bald eagles (more than one hundred), Canadian geese, and cormorants also perished. Economically important fish, such as pink and sockeye salmon and Pacific herring, as well as a large number of small forage fish (capelin, pollock, sandlance, and smelt), died as a result of the spill. Shellfish, such as clams, crabs, oysters, and shrimp, were also killed or threatened.

According to G. Tyler Miller, Jr., and Scott Spoolman, writing in *Environmental Science* (2010), "Many forms of marine life recover from exposure to large amounts of crude oil within about 3 years." For tions. Most of these encapsulated cells reproduce asexually by splitting in half, a process known as fission. These oil-metabolizing bacteria can double their number in less than twenty-four hours.

As a result of the *Exxon Valdez* spill, both criminal charges and civil damage claims were filed against Exxon by the state of Alaska and the United States. In the civil settlement, Exxon was required to pay $900 million over a ten-year period. A state-federal trustee council consisting of six members was designated to administer the settlement and coordinate studies of the spill's effects on wildlife and the environment. In an agreement concerning the criminal charges, Exxon was originally required to pay a fine of $250 million. However, because of Exxon's cooperation with governmental agencies during the cleanup and its quick payment of most private claims, $125 million of the fine was forgiven.

Donald F. Reaser

FURTHER READING

Fingas, Merv. *The Basics of Oil Spill Cleanup.* 2d ed. Boca Raton, Fla.: CRC Press, 2001.

Keeble, John. *Out of the Channel: The Exxon Valdez Oil Spill in Prince William Sound.* 2d ed. Cheney: Eastern Washington University Press, 1999.

Kvenvolden, Keith, et al. "Ubiquitous Tar Balls with a California-Source Signature on the Shorelines of Prince William Sound, Alaska." *Environmental Science and Technology* 29 (October, 1995): 2684-2694.

Loughlin, Thomas R., ed. *Marine Mammals and the Exxon Valdez.* San Diego, Calif.: Academic Press, 1994.

Miller, G. Tyler, Jr., and Scott Spoolman. *Environmental Science: Problems, Concepts, and Solutions.* 13th ed. Belmont, Calif.: Brooks/Cole, 2010.

Ornitz, Barbara E., and Michael A. Champ. *Oil Spills First Principles: Prevention and Best Response.* New York: Elsevier, 2002.

Ott, Riki. *Not One Drop: Betrayal and Courage in the Wake of the Exxon Valdez Oil Spill.* White River Junction, Vt.: Chelsea Green, 2008.

SEE ALSO: *Amoco Cadiz* oil spill; *Argo Merchant* oil spill; Bioremediation; Oil spills; *Sea Empress* oil spill; Tobago oil spill; *Torrey Canyon* oil spill.

F

Federal Insecticide, Fungicide, and Rodenticide Act

CATEGORIES: Treaties, laws, and court cases; human
health and the environment
THE LAW: U.S. legislation that established the role
of the federal government in regulation of
pesticides
DATE: Enacted on June 25, 1947
SIGNIFICANCE: The Federal Insecticide, Fungicide,
and Rodenticide Act established firm guidelines for
the registration and use of pesticide products in the
United States, resulting in a significant reduction in
harm to humans and other animals from exposure to
pesticides.

The Federal Insecticide, Fungicide, and Rodenticide Act of 1947 (FIFRA) represented the culmi-
nation of events in which the food and chemical industries in the United States began a merger of
practices. In the late nineteenth century, American
consumers began to be alarmed—with some justification—by farmers' practice of applying insecticides
and other toxic chemicals to crops haphazardly. The
Insecticide Act of 1910 and various so-called pure
food and drug laws were early attempts to regulate
such chemicals.

During the 1940's, a large number of toxic chemicals were developed as a result of the war effort. With
the end of World War II, industries attempted to find
other uses for chemicals that were no longer in short
supply. Some of the chemicals had potential applications as insecticides or pesticides, and it was logical for
the manufacturers to look for uses for them in the
food industry. At the time, it was estimated that nearly
200 million bushels of corn and small grain could be
added to the market if a population of some 140 million rats could be brought under control.

With FIFRA's passage by Congress in 1947, the federal government began to oversee the labeling and
marketing of all pesticides and fungicides entering
the U.S. marketplace. The act required a scientific review by members of the Department of Agriculture of
any chemicals introduced into the interstate market
and also required such chemicals to be packaged with

proper labeling of claims and precautions. Revisions
to FIFRA during the 1970's strengthened federal
oversight of pesticides and fungicides, including requiring the registration of new products with the Environmental Protection Agency.

Richard Adler

SEE ALSO: Agricultural chemicals; Biofertilizers; Biopesticides; Dichloro-diphenyl-trichloroethane; Environmental law, U.S.; Hazardous and toxic substance
regulation; Pesticides and herbicides.

Federal Land Policy and Management Act

CATEGORIES: Treaties, laws, and court cases; land
and land use
THE LAW: U.S. federal law governing how the Bureau
of Land Management manages public lands
DATE: Enacted on October 21, 1976
SIGNIFICANCE: The Federal Land Policy and Management Act of 1976 is the guiding law for the development, enhancement, and protection of designated
public lands in the United States. The act mandates
that the Bureau of Land Management administer
these lands in a sustainable way to ensure their use
for generations to come.

In 1976, as a result of years of neglect of public lands
that had allowed the proliferation of a number of
problems, such as vandalism and destruction of natural resources, lack of sanitation facilities, littering, and
overuse, the U.S. Congress enacted the Federal Land
Policy and Management Act (FLPMA). For the first
time a law provided jurisdiction for the management
of public lands under one federal government agency,
the Bureau of Land Management (BLM), which operates within the Department of the Interior. Prior to the
passage of FLPMA in 1976, the BLM was managing
public lands under a number of different laws; the new
legislation gave the BLM a unified way of managing
public lands, which are defined as lands that are owned
by the federal government, excluding lands that are

controlled by Native American tribes or set aside for national forests, national parks, and military uses.

FLPMA repealed many obsolete laws related to the management of public lands and gave the BLM new tools for administering such lands. One important purpose of the law is to enable the federal government to retain ownership of public lands while allowing some exchanges of lands and even sales in specific cases. The act's policy declarations specify, among other things, that public lands and their resources must be inventoried periodically and systematically; that all lands that have not previously been designated for any specific uses must be reviewed; that the lands shall be managed on a multiple-use basis as guided by public land-use planning; that they shall be managed in such a way as to protect the quality of their scientific, scenic, historical, ecological, environmental, and archaeological values; and that they shall be managed in a manner that recognizes the nation's need for domestic sources of food, fiber, timbers, and minerals.

FLPMA also states that the federal government is to receive fair market value for the use of public lands and their resources, and that it shall provide payments to compensate state and local governments for any burdens created as a result of the immunity of federal lands from state and local taxation. The act includes provisions covering the disposal of public lands, the acquisition of nonfederal lands, and exchanges of public and private lands; it also addresses regulations concerning the protection of public land areas that have critical environmental concerns.

Public lands in the United States total approximately 105.2 million hectares (260 million acres), the majority of these being in the West. These lands represent approximately 40 percent of the federally owned land, 12 percent of the U.S. land area, and 20 percent of the land situated between the Rocky Mountains and the Pacific Ocean. Most of these lands are located in the states of Nevada, Utah, Wyoming, Idaho, and Oregon. Among other uses, public lands support grazing for livestock on more than 55.4 million hectares (137 million acres) in eleven western states.

Lakhdar Boukerrou

FURTHER READING

Allen, Leslie. *Wildlands of the West: The Story of the Bureau of Land Management.* Washington, D.C.: National Geographic Society, 2002.

Loomis, John B. *Integrated Public Lands Management: Principles and Applications to National Forests, Parks,*

Wildlife Refuges, and BLM Lands. 2d ed. New York: Columbia University Press, 2002.

Skillen, James. *The Nation's Largest Landlord: The Bureau of Land Management in the American West.* Lawrence: University Press of Kansas, 2009.

SEE ALSO: Bureau of Land Management, U.S.; Department of the Interior, U.S.; Forest and range policy; Forest management; Grazing and grasslands; Land-use policy; Multiple-use management; Range management; Wildlife management.

Fish and Wildlife Act

CATEGORIES: Treaties, laws, and court cases; resources and resource management; animals and endangered species

THE LAW: U.S. federal legislation regulating fish and wildlife resources

DATE: Enacted on August 8, 1956

SIGNIFICANCE: Although the Fish and Wildlife Act focused on the commercial fishing industry in the United States, it also laid out protections for sportfishing and expanded general public opportunities for access to fish and wildlife resources.

The Fish and Wildlife Act of 1956 established the U.S. Fish and Wildlife Service within the Department of the Interior; the act also created the positions of director and assistant secretary of the Fish and Wildlife Service, with both positions to be appointed by the president of the United States with Senate approval. The act focused on the commercial fishing industry and provided administration to make sure that American citizens would maintain the right to fish for recreational purposes. The act created the Bureau of Commercial Fisheries, which would oversee the $10 million Fisheries Loan Fund. This fund would be used to invest in operations relating to fisheries; to provide for preservation, restoration, and equipment for fishing vessels; and to study concerns in the fisheries themselves.

Annual research money of up to $5 million was another provision of the act. The reports generated by the research would cover topics such as the domestic production of fish and fish goods, the foreign production of such goods as they affect American industries, the biological information needed to understand these industries, and the creation of more fish and wildlife refuges. Also to be funded were investigations into the impacts of pesticides, fungicides, insecticides, and herbicides on fish and wildlife, including determination of the levels of such chemicals that could be dangerous.

The act was implemented to reserve and plan for the proper management of the valuable renewable resources of fish, shellfish, and wildlife, which contribute to the U.S. economy and food supply. Other stated justifications for the act were that it would generate employment opportunities, would fortify the national defense with trained sailors and available ships, and would improve the general health and physical fitness of sportsmen who could serve in times of military necessity. Amendments to the act in the years since its passage have added requirements for the secretary of the interior to develop wildlife refuges and to provide education programs for the public.

Theresa L. Stowell

FURTHER READING

Hathaway, Jessica. "Fifty Years Ago (Fishing Back When)." *National Fisherman*, September, 2006, 8.

McKay, David. "Environmental Politics." In *American Politics and Society*. 7th ed. Malden, Mass.: Wiley-Blackwell, 2009.

SEE ALSO: Fish and Wildlife Service, U.S.; Fish kills; Fisheries; National Wildlife Refuge System Administration Act; Wildlife refuges.

Fish and Wildlife Service, U.S.

CATEGORIES: Organizations and agencies; animals and endangered species

IDENTIFICATION: Federal government bureau responsible for protecting fish and animal habitats

DATE: Established in 1940

SIGNIFICANCE: The U.S. Fish and Wildlife Service traces its roots to some of the world's oldest natural resource conservation programs. The agency is responsible for conserving, protecting, and enhancing fish, wildlife, and plant species and their habitats, and its environmental stewardship includes administration of the Endangered Species Act.

During the late nineteenth century, resource use in the United States developed at such a rapid pace that it led to abuse of the environment. Among the resources most depleted were the national fisheries. Major declines in fish populations and commercial fish catches led the U.S. Congress to establish the Commission on Fish and Fisheries in 1871. Its initial responsibilities were to study and reverse the decline in food fishes, restore streams, improve fisheries, and stock lakes and streams.

Concern for wildlife in relation to agriculture also led to the establishment in 1885 of the Division of Economic Ornithology and Mammalogy within the U.S. Department of Agriculture. The division's early focus was the study of the role of birds in controlling insect pests and the geographical distribution of animal and plant species throughout the nation. Before the end of the nineteenth century, the expanding division was renamed the Bureau of Biological Survey.

Through merger and reorganization, the Commission on Fish and Fisheries and the Bureau of Biological Survey eventually became the Fish and Wildlife Service (FWS) of the Department of the Interior in 1940. During this evolution, a variety of federal laws were passed that strongly influenced the direction and authority of the FWS over environmental resources. The establishment of wildlife refuges was begun under President Benjamin Harrison. In 1900 the Lacey Act was passed, under which authority the federal government began to enforce laws regarding interstate and foreign commerce in wildlife. During Theodore Roosevelt's presidency, many more wildlife refuges were created from public domain lands. The Migratory Bird Treaty Act of 1918 significantly expanded the service's authority by involving two foreign governments, Canada and Mexico, in wildlife protection. The Migratory Bird Conservation Act of 1929, which established refuges for migratory waterfowl management, provided authority from which the National Wildlife Refuge System developed over the following decades.

The 1930's saw the passage of two important acts that ultimately would fall under the oversight of the FWS. The Migratory Bird Hunting and Conservation Stamp Act of 1934, often called the Duck Stamp Act, requires waterfowl hunters to purchase stamps, revenues from which are used to acquire and protect important wetlands. Over its history, the duck stamp program has generated enough revenue to protect approximately 1.8 million hectares (4.5 million acres) of waterfowl habitat. The Federal Aid in Wildlife Restoration Act of 1937 (commonly known as the Pittman-Robertson Act), which has been identified by some writers as the most significant wildlife act in U.S. history, established an excise tax on firearms, archery equipment, and ammunition used in hunting, as well as a manufacturers' tax on handguns. These tax dollars, which are specified for wildlife and fisheries, are apportioned to state wildlife agencies through the FWS.

One of the FWS's best-known responsibilities is coadministration of the Endangered Species Act (ESA), a duty it shares with the National Oceanic and Atmospheric Administration. Congress passed the ESA in 1973 to protect endangered plants and animals domestically and internationally. The FWS determines whether populations of freshwater and land species are threatened or endangered and designates critical habitats.

Other FWS responsibilities include operating the National Wildlife Refuge System (which encompasses some 38 million hectares, or 93 million acres), cooperating with state fish and game agencies and Native American tribal officials, assisting foreign governments with international conservation efforts, establishing university extension programs regarding fish and wildlife, and collecting, maintaining, and distributing statistical information on fisheries, wetlands, and estuaries. To carry out all of these duties, the FWS has field and regional offices across the country. The scope of the agency's activities reflects the wide range of environmental issues that affect fish and wildlife resources.

Jerry E. Green
Updated by Karen N. Kähler

FURTHER READING

Clarke, Jeanne Nienaber, and Daniel McCool. *Staking Out the Terrain: Power and Performance Among Natural Resource Agencies.* 2d ed. Albany: State University of New York Press, 1996.

Hays, Samuel P. *A History of Environmental Politics Since 1945.* Pittsburgh: University of Pittsburgh Press, 2000.

U.S. Fish and Wildlife Service. *Shared Commitments to Conservation: 2007 Annual Financial Report of the U.S. Fish and Wildlife Service.* Washington, D.C.: U.S. Department of the Interior, 2007.

SEE ALSO: Darling, Jay; Endangered Species Act; Fish and Wildlife Act; Migratory Bird Act; National Wildlife Refuge System Administration Act; Wetlands; Wildlife management; Wildlife refuges.

Fish kills

CATEGORY: Animals and endangered species

DEFINITION: Significant, sudden mass die-offs of fishes

SIGNIFICANCE: The cause of a fish kill is rarely obvious, and its determination often requires complex analyses of water quality, fish tissues, and environmental and human events that have occurred in the watershed.

Notification of a fish kill usually occurs after, rather than during, the event, in which case causative agents may have dissipated or become diluted, and the tissues of the affected fish may have decomposed. Consequently, formal field investigation involves trained personnel, fishery biologists, fish pathologists, and conservation officers. Relevant field investigation includes the collection of factual and precise data. These are obtained through careful observa-

The Greenwich Bay, Rhode Island, fish kill of 2003 was caused by a lack of oxygen in the water owing to eutrophication. More than one million fish died during this event. (Tom Ardito/IAN Image Library)

tion, interview of witnesses, and meticulous sampling of water and dead fish. Because fish kills caused by human activity may result in litigation, samples and other evidence must be collected according to prescribed methods and quickly transported to an analytical laboratory.

About one-half of reported fish kills are the results of such natural causes as oxygen depletion; blooms of toxic algae; infectious diseases caused by bacteria, viruses, protozoans, or parasites; stranding of fish schools; overabundant runoff of silt or ash from forest fires; toxic runoff water; and life-cycle events. In most cases, combinations of causative factors are present. For example, during the summer, fish can become physiologically stressed when water temperatures exceed the fish's optimal limits. If all other environmental factors remain optimal, the fish may tolerate the increase in temperature. Oxygen solubility is lowered as temperature increases, however; thus the concentration of dissolved oxygen is lower at a time when stress requires the fish to increase their oxygen uptake. Also, the increased metabolism of all other biological organisms puts additional demands on the limited supply of oxygen. The killer in this case is not temperature but oxygen depletion, the primary culprit in fish kills.

Oxygen depletion can also be caused by input of high organic loads, algae blooms and their degradation, low water flow, and ice cover on a lake or pond during the winter. Snow cover reduces light penetration, which inhibits photosynthesis, the only source of dissolved oxygen in ice-covered lakes. This results in winter fish kills. Natural fish kills in coastal waters are often caused by blooms of microorganisms that generate red tides and other epidemiological effects.

A variety of human activities can cause significant fish kills. The dumping of treated and untreated sewage waste into natural waters artificially increases the nutrient load of the system. This practice in itself does not cause a fish kill, but it stimulates increased production and biodegradation processes, which reduce dissolved oxygen below critical levels. A more obvious cause is the introduction into a water system of toxic pollutants such as pesticides, heavy metals, and industrial wastes. Alterations of the landscape of a watershed can increase water temperature and siltation; silt reduces the efficiency of oxygen uptake by fishes' gills. Dams impede fish migration and, during low water levels, create pockets of deoxygenated water. The turbines of hydroelectric power plants and other

water-intake facilities are another detriment to fish. The disposal of undesirable fish by commercial fishing operations often generates fish kill reports.

Richard F. Modlin

FURTHER READING

Helfman, Gene S. *Fish Conservation: A Guide to Understanding and Restoring Global Aquatic Biodiversity and Fishery Resources.* Washington, D.C.: Island Press, 2007.

Walters, Carl J., and Steven J. D. Martell. *Fisheries Ecology and Management.* Princeton, N.J.: Princeton University Press, 2004.

SEE ALSO: Acid deposition and acid rain; Agricultural chemicals; Cultural eutrophication; Dams and reservoirs; Dead zones; Eutrophication; Thermal pollution; Water pollution.

Fisheries

CATEGORIES: Animals and endangered species; resources and resource management; agriculture and food

DEFINITION: Locations where one or more species of fish are caught by human fishers

SIGNIFICANCE: The status of the world's fisheries is a chief concern of many environmental nonprofit organizations, governmental agencies, and international entities such as the United Nations and multinational trade organizations because of fisheries' crucial importance in protecting marine environments. The collapse of fisheries poses serious threats to biodiversity, local and global economies, food security, and the ecological health of the oceans and of small fresh- and saltwater bodies.

Fisheries, in the most basic way, signify the relationship between humans and fish. Since the dawn of civilization, humans have taken fish for consumption from all accessible bodies of water, traditionally without much concern about depleting resident populations. In modern times, however, overfishing has led to the collapse of major consumed fish species from stocks that once seemed ever-replenishing. Perhaps the most poignant example is the Northern Atlantic cod fishery collapse during the early 1990's. Although measures were taken to limit catch size some thirty

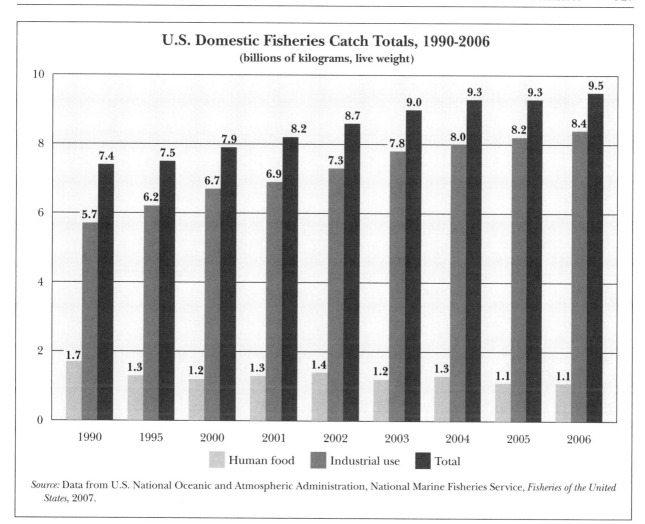

U.S. Domestic Fisheries Catch Totals, 1990-2006
(billions of kilograms, live weight)

Human food   Industrial use   Total

Source: Data from U.S. National Oceanic and Atmospheric Administration, National Marine Fisheries Service, *Fisheries of the United States,* 2007.

years previously, and most of the fishers in the region obeyed the new rules, the fishery still collapsed. This event was seen as a warning signal to the world; it caused policy makers to take greater notice of the epidemic of overfishing and sparked worldwide efforts to take more effective measures to protect fisheries as important resources.

Overfishing is a serious threat to biodiversity, local and global economies, food security, and the ecological health of the oceans and of small fresh- and saltwater bodies. According to the World Wide Fund for Nature, 52 percent of the world's fisheries had been completely exploited by 2010, and 24 percent were overexploited, depleted, or on the road to recovery from depletion. It has been predicted that stocks of all fished species will collapse by 2048 if early twenty-first century trends continue. The proper management of

fisheries is essential to the maintenance of sustainable populations of commercial species of fish for both ecosystem integrity and the use of future generations.

Sustaining edible fish populations has benefits to both the environment and to those who rely upon fish-derived nutrients for sustenance. According to the United Nations Food and Agriculture Organization, fish and shellfish provide essential nutrition for 3 billion people and at least 50 percent of total animal protein and minerals for 400 million people in the world's poorest countries. Aside from the economic returns of the fishing industry, many rely on fish as a major food source, and further fisheries collapses could result in food crises within reliant communities. This means that responsible management of fisheries is also an environmental justice issue, as mismanagement caused by thirst for higher monetary returns in the short term could deprive the world's hungry of an important source of nutrition.

In addition, many of the most overfished species are high on the food chain, meaning they prey upon smaller species of fish and play crucial roles in maintaining the equilibria of other fish populations and the larger marine ecosystem. The collapse of further fisheries would therefore not only spell disaster for communities that depend on a continuous supply of fish but also cause the degradation of marine ecosystems through a trickle-down effect in the food web.

ECOLOGICAL IMPORTANCE

The relationship between humans and the fish they catch and consume is basically similar to any predator-prey relationship. It is dissimilar to the relationship between humans and domesticated food animals, which are bred in captivity, kept in managed numbers, and are in no threat of extinction. The latter relationship most closely resembles a form of mutualism, in which both populations are maintained, although one partner in this case is exploited by the other. Regardless of any ethical issues raised by the practice, domesticated animals live in populations sustained by humans. Fish, in contrast, are usually taken from nature, except in the case of farmed fish, which account for about 40 percent of seafood eaten. Since the other 60 percent is hunted, and many species are located in remote marine areas without national ties, excessive fishing occurs frequently and is mostly undocumented. Species under the threat of extinction can be taken without repercussion, and many are sold under the monikers of more common fish. Also, unlike the case of domesticated animals, the harvesting of fish compromises many different species. Once one species has collapsed, commercial fishing can shift to another species, usually without major recognition from global consumers. Commercial fishing therefore poses a threat to multiple fish species at once.

A good example of the multilevel ecosystem effects of irresponsible fishing is the tuna, a carnivorous fish; a number of related species of tuna are found around the world. Tuna have a propensity to swim close to dolphin pods, presumably because the dolphins afford protection from sharks and other predators. Fishers have traditionally relied on dolphin pods to lead them to schools of tuna, and dolphins have often been caught and strangled in the fishers' nets. When the public at large became aware of this practice, the resulting outcry led to the adoption of technology de-signed to reduce dolphin catch. Also important ecologically, the widespread decline of tuna species could potentially cause unnatural explosions in tuna prey populations. Certainly as well, if dolphin deaths were to approach a large number, dolphin populations would also be in danger, and species that are connected with them in the food web would likewise be subject to change. More often than not, the decline of many commercial fish species could result in similar top-down trophic (nutrition-related) effects.

SOCIOECONOMIC IMPORTANCE

The negative effects of overfishing are felt most severely in developing countries where the populations depend on fishing for sustenance and nutrition. The populations of many traditionally fished species have declined because of unrestricted extraction, and aquatic ecosystems on which communities rely for good catches year after year have become increasingly degraded. More than 500 million people in developing countries depend on fish for sustenance, and this number is only expected to increase as populations increase.

Features of coastal lands such as estuaries, coral reefs, mangrove forests, and sea grass beds are essential for many fish to spawn and provide the fish with shelter and habitat as well. The health of such areas is an important element of productive fisheries, and maintaining healthy fisheries in turn ensures the perpetuity and value of these areas. Mangrove forests are natural barriers to destructive waves caused by storms, and they reduce coastal erosion by holding sediments within their roots. Coral reefs and sea grass beds harbor high levels of marine biodiversity and host a variety of ecosystem functions. Estuaries help with nutrient cycling and provide spawning grounds for fish and habitat for migratory birds. Further, all these kinds of coastal features are threatened not only by degradation caused by human activities but also by climate change.

SOLUTIONS TO OVERFISHING

Fish farming, or aquaculture, is one option for alleviating the damage done by commercial fishing. In contrast to the capture of wild fish, in aquaculture commercial aquatic animals are raised under controlled conditions for maximum yield. Aquaculture is akin in many ways to agriculture in theory and practice. As some commercial species can be bred in captivity, aquaculture offers the possibility of relieving wild populations from overfishing, potentially allowing their numbers to recover.

Aquaculture operations are generally designed with yield in mind, however, and they can produce many of the same kinds of negative environmental impacts that industrial agriculture operations can cause and for which they are criticized. The use of antibiotics and poor waste handling by aquaculture operations lead to the pollution of waterways and, in the case of marine operations, the oceans as well. Pharmaceutical pollution is known to cause a plethora of negative effects in marine organisms, and high waste loads lead to algal blooms and other devastating downstream problems.

The feed for carnivorous farmed fish must come from other fish, and although some breeds are being developed that tolerate grain feed, a large percentage will still be fish-based, thus making the whole process arguably energy-inefficient and failing to seal a more sustainable circle of inputs and outputs. Further, fish that escape from aquaculture operations pose threats to the native aquatic ecosystems in which they are raised, as they are more often than not invasive. Regardless of the potential hazards to the environment, aquaculture was one of the world's fastest-growing food-production systems in the early years of the twenty-first century.

Responsible consumerism in seafood is a growing movement, spearheaded in the United States by the Monterey Bay Aquarium's Seafood Watch Program, which offers consumers information—through print pocket guides and the aquarium's Web site—on the various kinds of fish available, noting which are sustainable choices and which are better avoided. The reasoning is that if consumers purchase mostly seafood that originates from well-managed fisheries, seafood from overfished areas will be subject to negative market pressures.

Some observers have argued that the most important change that can be made to protect the world's fisheries is a shift to sustainable development in poor countries and increased food security for people who must rely on overfished resources to survive. Only when people are guaranteed adequate nutrition can they make choices about their diets. Even if continuing a destructive practice will ultimately mean an end to resource production altogether, communities that depend on seafood protein will fish where they must to survive.

Jamie Michael Kass

FURTHER READING

Black, Kenneth D., ed. *Environmental Impacts of Aquaculture*. Boca Raton, Fla.: CRC Press, 2000.

Field, John G., Gotthilf Hempel, and Colin P. Summerhayes, eds. *Oceans 2020: Science, Trends, and the Challenge of Sustainability*. Washington, D.C.: Island Press, 2002.

Hart, Paul J. B., and John D. Reynolds, eds. *Handbook of Fish Biology and Fisheries*. Malden, Mass.: Wiley-Blackwell, 2002.

Miller, G. Tyler, Jr., and Scott Spoolman. "Sustaining Aquatic Biodiversity." In *Living in the Environment: Principles, Connections, and Solutions*. 16th ed. Belmont, Calif.: Brooks/Cole, 2009.

Reddy, M. P. M. *Ocean Environment and Fisheries*. Enfield, N.H.: Science Publishers, 2007.

Walters, Carl J., and Steven J. D. Martell. *Fisheries Ecology and Management*. Princeton, N.J.: Princeton University Press, 2004.

SEE ALSO: Aquaculture; Commercial fishing; Coral reefs; Dolphin-safe tuna; Fish and Wildlife Act; Fish and Wildlife Service, U.S.; Fish kills.

Flavr Savr tomato

CATEGORIES: Biotechnology and genetic engineering; agriculture and food

DEFINITION: Genetically engineered tomato developed by Calgene, a California company

SIGNIFICANCE: As the first genetically engineered food product on the market, the Flavr Savr tomato was closely monitored by food-safety advocates and environmentalists.

Tomatoes are an important agricultural crop, sold both as fresh produce and as ingredients for the food-processing industry. Tomatoes are typically harvested while they are still hard and green, so that they can survive shipping and storage without bruising and crushing, damage that makes them susceptible to rotting. They are then treated with ethylene gas—the natural ripening agent—to induce softening and a red color change, but such tomatoes do not taste like vine-ripened fruit.

To alter the tomato's ripening process and enhance flavor, scientists at Calgene in Davis, California, used genetic material known as antisense deoxyribonucleic acid (DNA). When inserted into tomato plants, this DNA interferes with the natural plant gene responsible for tomato softening. Since the normal genetic message is blocked, production of a critical protein that breaks down pectin is diminished, and the ripening process is inhibited. Calgene named the tomato it developed in this way the Flavr Savr; it could be kept on the vine longer to turn red and develop better flavor, but the fruit remained firm. The fruit was later softened by ethylene treatment.

In 1991 Calgene asked the U.S. Food and Drug Administration (FDA) to examine data to determine the Flavr Savr tomato's safety. In May, 1994, the FDA approved the tomato, and almost immediately Calgene began marketing it under the brand name of MacGregor. Although Calgene was not required to label the tomato as genetically engineered, the company did so; it also offered information to consumers regarding the genetic alterations.

In 1992 the FDA ruled that genetically engineered foods do not require premarket approval or special labeling. This ignited fears among some critics that future genetically modified foods would not be as thoroughly tested as was the Flavr Savr tomato. Some were especially concerned about the possible presence of bacterial antibiotic resistance genes in genetically engineered foods. In approving the Flavr Savr tomato, the FDA concluded that these genes, which serve as markers to determine if an organism has been successfully modified, did not significantly differentiate the tomato from others. However, such markers might be allergenic and could potentially be transferred to intestinal bacteria in people who eat the tomato. The development of new antibiotic-resistant strains of bacteria was already recognized as a serious medical problem.

The long-term environmental impacts of genetically engineered crops are as yet unknown, but two major concerns have been raised. First is the possibility that engineered plants could become "superweeds" that would damage ecosystems as well as agricultural lands; this, in turn, might necessitate additional herbicide use. Second is the fear that the foreign genes present in engineered plants could be transferred to wild relatives growing in the vicinity, with unpredictable consequences.

Concerns raised by consumer groups and environmental activists, a barrage of press coverage using terms such as "Frankentomato," and general public distrust of technology all contributed to the poor mar-

ket performance of the Flavr Savr tomato. Additionally, when grown commercially, the plants did not have acceptable yields or disease resistance, and the tomatoes did not withstand the shipping process as well as expected. All of these factors contributed to Calgene's discontinuing production of the Flavr Savr tomato in 1997.

Diane White Husic

FURTHER READING

Avise, John C. *The Hope, Hype, and Reality of Genetic Engineering: Remarkable Stories from Agriculture, Industry, Medicine, and the Environment.* New York: Oxford University Press, 2004.

Martineau, Belinda. *First Fruit: The Creation of the Flavr Savr Tomato and the Birth of Biotech Food.* New York: McGraw-Hill, 2001.

SEE ALSO: Biotechnology and genetic engineering; Genetically altered bacteria; Genetically engineered pharmaceuticals; Genetically modified foods; Genetically modified organisms; Horizontal gene transfer.

Flood Control Act

CATEGORIES: Treaties, laws, and court cases; water and water pollution; resources and resource management

THE LAW: U.S. federal law designed to direct and coordinate significant water development projects in the Missouri River basin

DATE: Enacted on December 22, 1944

SIGNIFICANCE: To help control chronic flooding along the Missouri River and to help irrigate the Great Plains, the Flood Control Act of 1944 provided a sweeping vision for reconceiving the river basin and a practical apparatus for the federal construction of dams and levees to achieve that reconstruction.

Although the Dust Bowl of the southern plains received far more national attention during the economic catastrophe of the 1930's, chronic flooding along the Missouri River was responsible for disastrous property losses in the same period (the river drains more than one-sixth of the continental United States). Navigation on the river was unreliable, and the unpredictable depth of the river (at different points it was a meandering stream, at others a broad and swift river) meant that it generated little hydropower. Further, despite its considerable length (more than 4,000 kilometers, or 2,500 miles), the river was largely neglected as an irrigation source for the arid northern plains states.

The Flood Control Act of 1944, passed in the second session of the Seventy-eighth Congress, was breathtaking in its scope even among the numerous large-scale public works projects authorized by Franklin D. Roosevelt's presidential administration. Never had river management on such a scale been undertaken by the federal government. The act envisioned nothing less than the redesign of the Missouri River; it aimed to develop the river's considerable commercial potential for hydroelectric energy and for the agriculture industry while at the same time protecting both the river itself and its wildlife and fish resources. It was known as the Pick-Sloan Act, named for two men largely responsible for proposing the enormous reach of the legislation: General Lewis Pick, who directed the U.S. Army Corps of Engineers, and William Glenn Sloan, who served in the Department of the Interior.

The Flood Control Act authorized the Army Corps of Engineers, in consultation with specific cabinet officers and the governors of affected states, to direct water development projects along the broad reach of the Missouri: irrigation projects through the Department of the Interior, flood-control projects with direct impacts on navigation through the Department of the Army, and projects designed to protect against soil erosion and river sedimentation through the Department of Agriculture. In addition, the Corps of Engineers would develop and maintain public parks and recreational facilities along the river.

The most immediate impacts of the act were realized in dozens of dams (and modifications of existing dams), as well as in miles of levees erected along the main stem of the Missouri. Those dams, in turn, created more than fifty lakes that continue to be used as recreational facilities (for boating, fishing, and swimming) as well as reservoirs for generating hydroelectric power. As public works legislation, the Flood Control Act was responsible for creating thousands of jobs, particularly among veterans returning from World War II. The act in turn authorized the Pick-Sloan Missouri River Basin Program to coordinate river projects.

Although the water-control projects authorized by the act provided long-term control of flooding in the Missouri basin as well as improved navigation on the river, the legislation also raised significant questions concerning the reach of the federal government into local development. Thousands of Native Americans were displaced from their homes by the projects, and hundreds of thousands of hectares of tribal lands were flooded, most notably those of the Lakota and Dakota tribes.

Joseph Dewey

FURTHER READING

Andrews, Richard N. L. *Managing the Environment, Managing Ourselves: A History of American Environmental Policy.* 2d ed. New Haven, Conn.: Yale University Press, 2006.

O'Neill, Karen M. *Rivers by Design: State Power and the Origins of U.S. Flood Control.* Durham, N.C.: Duke University Press, 2006.

Schneiders, Robert Kelley. *Unruly River: Two Centuries of Change Along the Missouri.* Lawrence: University Press of Kansas, 1999.

Thorson, John E. *River of Promise, River of Peril: The Politics of Managing the Missouri River.* Lawrence: University Press of Kansas, 1994.

SEE ALSO: Dams and reservoirs; Department of the Interior, U.S.; Environmental engineering; Environmental law, U.S.; Floodplains; Floods; Hydroelectricity; Mississippi River; Sedimentation; Water use.

Floodplains

CATEGORIES: Land and land use; water and water pollution

DEFINITION: Low-lying areas adjacent to river channels that become partially or completely covered with water when the rivers overflow their banks

SIGNIFICANCE: Floodplains occupy an important part of landforms around rivers covering large areas, especially within humid and tropical climatic settings. They also house riparian wetlands, acting as buffers to flooding. Floodplains provide habitat for many land and aquatic life-forms.

Floodplains filter water and provide silt and nutrients that make them fertile places. Perhaps the most famous examples are the fertile floodplains of

Rivers

Roughly 70 percent of Earth's surface is covered with water. About 97 percent of this water is salt water in oceans, and 2 percent is fresh water in glaciers and groundwater. Only 1 percent is fresh water in rivers, lakes, and water vapor. Fresh water is recycled again and again by evaporation and condensation in a hydrologic cycle that creates and maintains rivers and lakes.

When water vapor condenses and falls as rain, it seeps into the ground, filling pores in soil and rocks. After all pores fill, excess water pools on the surface, flowing downhill due to gravity. Then it becomes streams that coalesce into rivers. Each river is a large body of water, re-generated over and over by the hydrologic cycle, flowing through low ground areas that become channels (river courses) and valleys. Because Earth's oceans are lower than the land, rivers flow down into them.

A river and the composition of its banks change as movement toward the ocean both creates and destroys land. A river's course has upper, middle, and lower parts. The upper course, near the river's source, flows downhill and carves deep valleys, as well as waterfalls wherever tough rock resists erosion. Next, the river flows onto plains in its middle course, and the gently sloping land slows it, leading to sideways erosion and snakelike bends. The last part of a river, its lower course, near entry to an ocean, is often an estuary. Here, the slow flow of the river across virtually flat land causes settling of sediment, forming mud flats and sandbanks.

Egypt's Nile Valley region, which have supported civilizations for several millennia. Floodplains also provide fresh water and backwaters to wetlands, and they also dilute salts, thereby improving the health of the habitat for fish, bird, and plant populations that inhabit the floodplains.

Floodplains are good for food production such as rice cultivation. Farmers graze their livestock on the grasslands in floodplains, and fresh fruits and cash crops are grown in floodplains, which are often very fertile and easy to cultivate. In tropical settings, timber is harvested on floodplains, and nontimber forest resources such as animals and plants are used for foods and medicines as well as construction materials.

TYPES AND ECOLOGY

Riverine flooding can cover vast areas, many of which are among the most diverse biologically productive ecosystems on earth. Three types of floodplains are identified based on temperature: temperate stochastic, temperate seasonal, and tropical seasonal. Within floodplains, algae appear to provide the most important source of primary production within the grazer web. The flow regime is very important in determining the physical habitat for biotic composition. The shape, size, and the formation of

features such as deltas, riffles, runs, pools, and backwaters that tend to shift are linked to the flow regimes of rivers. Certain aquatic life-forms have their early life stages in floodplains, and the types of fauna and flora within floodplains can be as diverse as in any other ecosystems. Owing to the highly dynamic nature of floodplain terrains, varied species may be seen on the same floodplains over the course of years.

Floodplains contain several kinds of geomorphologic features, including oxbow lakes, point bars, areas of dead water, and braided channels. Swamps, among other types of riparian wetlands, can also be found in floodplains. Floodplains can be classified into different types depending on their morphology. Several methods of classification are used, but the simplest and most common is based on the fluvial styles: gravel-dominated, sand-dominated with high sinuosity, and sand-dominated with low sinuosity.

ENVIRONMENTAL THREATS

As global warming increases, some floodplains may see more flooding, which will greatly affect local populations because of land subsidence and increases in water level. For example, Bangkok, the capital of Thailand, is sinking at a rate of 2 to 5 centimeters (0.79 to 1.97 inches) per year because of sediment compression and compaction owing to increased human activities. The elevation of New Orleans, Louisiana, is also dropping at a rate of about 5 centimeters per year.

Urbanization affects the hydrology of floodplains, either by reducing water through withdrawal or by adding to it through importation. It also alters the water chemistry by introducing chemicals, sediments, and other form of pollutants, including increases in temperature, all of which affect the biotic richness of floodplains. The nutrients brought into play through flood activities can also be altered through changes in land-use patterns. Changes within a drainage basin (watershed) affect the production and supply of organic materials in floodplains. High levels of biodiversity can provide stability to floodplain ecosystems and help protect them from human-caused impairments.

The expansion of urban areas into floodplains and wetlands alters the onset, duration, distribution, speed, quantity, and quality of floodwaters. Among the human activities that lead to increases in flooding are deforestation and the removal of stabilizing vegetation along riverbanks. Human-built structures along or near rivers affect the flow direction, resulting in deflection of the water, or reduce storage. Storm drains, housing developments, and pavements increase the rate of rainfall runoff to rivers, thereby increasing the rate of flooding. The straightening of river channels increases the rate at which water is transported. Another human activity that affects floodplains is the dumping of sediment loads from farms or construction sites into rivers, which decreases channel depth and increases the area covered by floodwaters. With increasing changes in land use, a watershed approach to floodplain management becomes imperative.

Impairment of floodplain waters can have adverse effects on coastal ecosystems, as these waters end up in lakes or oceans. The quality of the water in rivers has a great impact on the quality of the water in nearby coastal areas; the waters of the Amazon, for example, can be traced several miles into the Atlantic Ocean. Contaminants carried in such waters ultimately affect large biological populations.

Solomon A. Isiorho

Further Reading

Bridge, John S. *Rivers and Floodplains: Forms, Processes, and Sedimentary Record.* Malden, Mass.: Blackwell, 2003.

Millius, Susan. "Losing Life's Variety." *Science News,* March 13, 2010, 20-25.

Richards, Keith, James Brasington, and Francine Hughes. "Geomorphic Dynamics of Floodplains: Ecological Implications and a Potential Modelling Strategy." *Freshwater Biology* 47, no. 4 (2002): 559-579.

Tockner, Klement, and Jack A. Stanford. "Riverine Flood Plains: Present State and Future Trends." *Environmental Conservation* 29, no. 3 (2002): 308-330.

See also: Amazon River basin; Aswan High Dam; Biodiversity; Floods; Great Swamp National Wildlife Refuge; Habitat destruction; Marsh, George Perkins; Nile River.

Floods

CATEGORIES: Weather and climate; water and water pollution

DEFINITION: High water flows that usually emanate from rivers, streams, or drainage ditches whose waters overtop their normal confines and spill out over lands that are normally dry

SIGNIFICANCE: Floods are among the most common and widespread natural disasters in terms of human hardship and economic loss. Almost all countries of the world are subject to annual flooding of different magnitudes.

Floods are primarily caused by heavy rainfall. Several hydraulic factors exacerbate the problem of discharging the enormous volume of water that collects in catchment areas. These factors include very low gradient of a river, the loss of channel capacity because of siltation, inadequate dredging of riverbeds, and the disruption of existing drainage systems by road construction without adequate culverts and by the building of unplanned houses. Unplanned urbanization, rapid population growth, and conversion of agricultural lands to other uses also contribute to flooding. Other causes of floods include melting of snow, the greenhouse effect and global warming, tides, elevated sea levels, storm surges, and tsunamis.

Floods account for one-third of all global disasters involving geophysical hazards. Occurrences of severe floods in human-occupied regions often involve loss of human life and property along with disruption of ongoing activities in affected rural and urban communities. In remote, unpopulated regions, natural floodplains can be significantly changed though not damaged by floods.

In the United States, floods represent the most costly extreme natural events. Since 1900 floods have been responsible for the deaths of more than ten thousand persons in the United States, and the economic consequences of floods are estimated to be in the billions of dollars per year. Despite the implementation of numerous flood mitigation measures to reduce the adverse impacts of these events, U.S. flood fatalities and losses have consistently increased over time.

Types of Floods

Floods are classified into three major types based on their origins: river floods, flash floods, and coastal

A father and son paddle through flooded streets after heavy rains in Poquoson, Virginia, in 2009. (AP/Wide World Photos)

floods. River floods occur when runoff from heavy or prolonged rainfall or melting of mountain snow causes rivers to rise high enough to overflow their banks or levees and spread water to floodplain zones. The extent of influence of such floods varies with distance from the river and depends on the type and amount of sediment load the river carries. In cold regions where there is a lot of accumulation of snow, river flooding often occurs as a result of snow melting or when ice or floating debris causes a jam. Such debris flows tend to occur primarily in the summer. Floods also sometimes occur when the water table rises above ground level; such flooding often leads to overspill flooding of rivers and streams.

Flash floods, which are localized extreme events, occur suddenly under a broad range of climatological and geographical conditions. They can occur within a few minutes or hours and may involve excessive rapid rainfall, a dam failure, a sudden release of water that had been held by an ice jam, slow-moving thunderstorms, or heavy rains such as those from hurricanes or tropical storms. Flash floods normally occur when the ground has become completely saturated with water, making further absorption impossible. The resultant overflow causes flooding. Flash floods can be very violent; the heavy, rapid flows of water can move

boulders, tear out trees, destroy buildings and bridges, and trigger catastrophic mudslides. Flash floods are frequently associated with other events, such as riverine floods on larger streams and mudslides. Areas prone to flash flooding include steep canyons, urbanized areas, and arid and semiarid regions.

Coastal floods are divided into two subtypes. Tidal floods occur twice a day in low-lying areas adjacent to estuaries and tidal waters; the areas affected by tidal flooding vary seasonally. Storm-surge floods are associated with tropical cyclones or hurricanes; flooding occurs when strong offshore winds push water from an ocean onto the land. Such floods are season-specific.

All types of flood events lead to inundations of water that carries solids, either in suspension or in solution. The depth and velocity of the water in such inundations vary. The impacts that floods have, through erosion or deposition of soil and through social and economic losses, depend on a combination of factors, including the quality, depth, and velocity of the water; seasonality and frequency of the flooding; and the type of flooding.

Low-lying plains in close proximity to rivers and coastal areas are commonly considered flood-prone zones. Generally, coastal areas and river valleys have large human populations, and these residents may be particularly vulnerable to damage from flooding. The level of a population's vulnerability to flood risk is often dependent on the level of development of the economy in the country affected. In case of developed countries, the risk of flooding depends on the amount of urban development in flood-prone zones rather than population density. In less developed regions, persons who live in flood-prone areas often have a high risk of exposure to floods.

POSITIVE ASPECTS

Outside the Western world, floods are not always viewed negatively and regarded as natural disasters. The people of Bangladesh, for example, perceive flooding as both a resource and a hazard. In that country, a normal flood (*barsha*) resulting from typical monsoon rainfall is considered a resource. It is beneficial in the sense that it makes the land productive by providing necessary moisture and fresh fertile silt to the soil. Moreover, fish caught during flood season constitute the main source of protein for many. Millions of Bangladeshis depend on fisheries in the floodplains for their livelihood, and the life cycles of some key fish species depend on the species' ability to migrate between the rivers and the seasonal floodplains.

Abnormal flooding (*bonna*), which occurs once every few years and results from excessive rainfall, is regarded by Bangladeshis as an undesirable and damaging phenomenon. It causes widespread damage to standing crops and property and costs animal and human lives. An analysis of crop production in Bangladesh found that despite considerable crop damage from flooding, there can be compensatory increases in rice production in areas not affected by the floods. Moreover, affected areas can experience increases in rice production in the following dry season; these increases can be attributed to the extra moisture resulting from the year of high floods as well as deposits of fertile silt on the floodplains.

STRATEGIES FOR DAMAGE REDUCTION

To reduce the loss of lives and property, people who live in flood-prone areas often undertake a host of measures, known as flood adjustments, at the household level. For example, in Bangladesh farmers resow or replant crops immediately after receding floods. They also sell land or other assets at distress prices to buy food or other necessary items. In the United States, people in flood-prone zones place sandbags along riverbanks and around homes to protect their property from flooding. They also take steps to protect their possessions from flood damage, such as moving important materials from the basement to the first or second floor and relocating electrical circuits and outlets from floor level to waist or ceiling level. Structures that are generally impermeable to water may be modified with watertight barriers or closures at openings to keep out floodwaters. Waterproofing sealants may be applied to walls or floors to reduce permeability.

Other effective measures that are taken to prevent structures from being damaged by floods are known as property relocation, property elevation, and walling. In property relocation, an existing structure is physically moved to land that is unlikely to flood. Property elevation involves the retrofitting of existing buildings in flood-prone areas so that they are raised above the likely flood level. Constructing walls high enough to keep floodwaters out around individual buildings or small clusters of buildings can also be an effective way of keeping flood damage to a minimum.

The measures that communities and governments undertake to modify the hazards associated with flooding include the control of flood flows through reservoir storage, levees, and channel improvement. Dikes, levees, and flood walls are all structures that hold the water off the land and confine it within a main channel of flow. Dikes are the most commonly used method of flood protection around the world. Extensive levee systems tend to raise flood levels because they block off flood channels, decreasing the channels' conveyance of water. Dams are widely used in the United States to minimize the damage caused by floods. Although dams are a good mitigation measure against flooding, they do not necessarily prevent flooding from occurring.

Often local governments adopt regulatory measures aimed at reducing flood damage, such as land-use zoning and building codes. Other flood preparedness and mitigation measures taken by governments include providing flood warnings to potential victims, evacuating people to safer places, and encouraging people to buy flood insurance. Such measures, however, have not been widely and effectively introduced in developing countries that are prone to severe flooding.

Sohini Dutt

FURTHER READING

Few, Roger, and Franziska Matthies, eds. *Flood Hazards and Health: Responding to Present and Future Risks.* Sterling, Va.: Earthscan, 2006.

Miller, G. Tyler, Jr., and Scott Spoolman. "Water Resources." In *Living in the Environment: Principles, Connections, and Solutions.* 16th ed. Belmont, Calif.: Brooks/Cole, 2009.

Smith, Keith, and David N. Petley. "Hydrological Hazards: Floods." In *Environmental Hazards: Assessing Risk and Reducing Disaster.* 5th ed. New York: Routledge, 2009.

Smith, Keith, and Roy Ward. *Floods: Physical Processes and Human Impacts.* New York: John Wiley & Sons, 1998.

Wescoat, James L., Jr., and Gilbert F. White. *Water for Life: Water Management and Environmental Policy.* New York: Cambridge University Press, 2003.

SEE ALSO: Dams and reservoirs; Flood Control Act; Floodplains; Rain gardens; Stormwater management; Weather modification.

Fluoridation

CATEGORY: Water and water pollution

DEFINITION: The treatment of community water-supply systems with fluoride as a public health measure

SIGNIFICANCE: The issue of whether it is a safe practice for communities to add fluoride to water supplies to prevent tooth decay remains controversial among some environmentalists.

Fluoridation of water supplies was first introduced in the United States in the 1940's as a preventive measure to reduce tooth decay, which was a serious and widespread problem in the early twentieth century. Since that time, many cities have taken the step of adding fluoride to their public water-supply systems, but the merits and drawbacks of fluoridation have long been subjects of debate. Proponents of fluoridation claim that it has dramatically reduced tooth decay in Americans, but opponents of the practice have not been entirely convinced of its effectiveness, and some are concerned about possible health risks that may be associated with fluoridation. The decision to fluoridate drinking water generally rests with local governments and communities.

Fluoride is the water-soluble, ionic form of the element fluorine. It is present naturally in most water supplies at low levels, generally less than 0.2 part per million (ppm), and nearly all food contains traces of fluoride. Tea contains more fluoride than most foods, and fish and vegetables also have relatively high levels. The findings of many scientific studies suggest that water containing a concentration of about 1 ppm fluoride, in contrast with water containing less fluoride, dramatically reduces the incidence of tooth decay.

Tooth decay occurs when acids in the mouth dissolve the protective enamel outer coating of a tooth, creating a hole, or cavity. These acids are present in food and can also be formed by bacteria that convert sugars into acids. The American diet has long included large quantities of sugar, which is a significant factor in the high incidence of tooth decay. By contrast, studies reveal that tooth decay is less common among people in primitive cultures; these findings have been attributed to these cultures' more natural diets.

Early fluoridation studies conducted between 1930 and 1950 demonstrated that fluoridation of public water systems produced a 50 to 60 percent reduction in tooth decay and that no immediate health risks were associated with increased fluoride consumption. Consequently, many communities quickly moved to fluoridate their water, and fluoridation was endorsed by most major health organizations in the United States.

OPPOSITION TO FLUORIDATION

Strong opposition to fluoridation began to emerge in the 1950's, as some people asserted that the possible side effects of consuming fluoride had not been adequately investigated. This concern was not unreasonable, given that high levels of ingested fluoride can be lethal. It is not unusual, however, for a substance that is lethal at high concentration to be safe at low levels, as is the case with most vitamins and trace elements. Opponents of fluoridation were also concerned on moral grounds; they argued that fluoridation represents compulsory mass medication.

Since the 1960's, heated debates have arisen over the issue of fluoridation across the United States. Critics have pointed to the harmful effects of large doses of fluoride, including bone damage, and the special risks fluoride may pose for some people, such as those with kidney disease and others who are particularly sensitive to toxic substances. Between the 1950's and the 1980's, some scientists suggested that fluoride may have a mutagenic effect—that is, it may be associated with human birth defects, including Down syndrome.

Controversial claims that fluoride can cause cancer were also raised in the 1970's, most notably by biochemist John Yiamouyiannis, who asserted that U.S. cities with fluoridated water had higher rates of death from cancers than did cities with unfluoridated water. Fluoridation proponents were quick to discredit his work by pointing out that he had failed to take other factors into consideration, such as the levels of known environmental carcinogens. Most scientific opinion suggests that the link between cancer and fluoride is a tenuous one. Nevertheless, it is a link that cannot be ignored completely, and a number of respected scientists continue to argue that the benefits of fluoridation are not without potential health risks.

A 1988 article in the American Chemical Society publication *Chemical and Engineering News* gained considerable attention with its suggestion that scientists opposing fluoridation were more credible than had

A protester makes his opinion of water-supply fluoridation known during a city council meeting in Worcester, Massachusetts, in 2001. (AP/Wide World Photos)

previously been acknowledged. By the 1990's even some fluoridation proponents began suggesting that observed reductions in tooth decay as a result of water fluoridation may have been at levels of only around 25 percent. Other factors—such as education, improvements in dental hygiene, and the addition of fluoride to some foods, salt, toothpastes, and mouthwashes—may also contribute to the overall reduction in levels of tooth decay.

While there is little doubt that fluoride does reduce tooth decay, the exact degree to which fluoridated water contributes to the reduction remains unanswered. It also remains unclear what, if any, side effects are associated with the ingestion of fluoride in water at the level of 1 ppm over many years. Whether or not any risks associated with fluoridation are small, any risk level may not be acceptable to everyone.

Since the 1960's and 1970's, Americans' concerns about environmental and health issues have been growing, and it has often been difficult, if not impossible, for scientists to measure—and to explain to the complete satisfaction of the public—the potential hazards posed by small amounts of chemical substances in the environment. The fact that, as of 2006, only about 69 percent of U.S. residents were living in communities with fluoridated water supplies is indicative of the continuing caution regarding this issue. In 1993 the National Research Council published a report on the health effects of ingested fluoride that included information on an attempt to determine whether the Environmental Protection Agency's maximum recommended level of 4 ppm for fluoride in drinking water should be modified. The report concluded that this level is appropriate but stated that further research may indicate a need for revision. The report also noted inconsistencies in the scientific studies of fluoride toxicity and recommended further research in this area.

The development of the fluoridation issue in the United States has been closely observed by other countries. Dental and medical authorities in Australia, Canada, New Zealand, and Ireland have endorsed fluoridation of water supplies, although not without considerable opposition from various groups. In Western Europe fluoridation has been greeted less enthusiastically, and scientific opinion in some countries, such as France, Germany, and Denmark, has concluded that it is unsafe. As a result, few Europeans drink fluoridated water.

Nicholas C. Thomas

FURTHER READING

De Zuane, John. *Handbook of Drinking Water Quality*. 2d ed. New York: John Wiley & Sons, 1997.

Martin, Brian. *Scientific Knowledge in Controversy: The Social Dynamic of the Fluoridation Debate*. Albany: State University of New York Press, 1991.

National Research Council. *Health Effects of Ingested Fluoride*. Washington, D.C.: National Academy Press, 1993.

Reilly, Gretchen Ann. "The Task Is a Political One: The Promotion of Fluoridation." In *Silent Victories: The History and Practice of Public Health in Twentieth-Century America*, edited by John W. Ward and Christian Warren. New York: Oxford University Press, 2007.

Stewart, John Cary. *Drinking Water Hazards*. Hiram, Ohio: Envirographics, 1989.

Weinstein, L. H., and A. W. Davison. *Fluorides in the Environment*. Cambridge, Mass.: CABI, 2004.

SEE ALSO: Drinking water; Water quality; Water treatment.

Food and Agriculture Organization

CATEGORIES: Organizations and agencies; agriculture and food

IDENTIFICATION: Agency of the United Nations concerned with eliminating hunger and improving human nutrition by enhancing efficiency in the production and distribution of food and agricultural products around the world

DATE: Established on October 24, 1945

SIGNIFICANCE: The Food and Agriculture Organization increasingly works to achieve its objectives by promoting sustainable agricultural practices that minimize deleterious environmental consequences.

The United Nations Food and Agriculture Organization (FAO) is the oldest and largest permanent specialized agency of the United Nations. Established at the end of World War II, FAO was initially concerned with feeding people whose countries had been devastated by war. It also sought to minimize the boom-and-bust cycles that long characterized agricultural markets. FAO has operated in the context of the rapid rise in human population that distinguished the second half of the twentieth century and has concentrated on working in the developing countries of the world.

In conjunction with the World Health Organization (WHO), FAO establishes international food standards, providing an objective basis for the evaluation of food consumption patterns. It also convenes conferences to address problems of hunger, malnutrition, and food security, seeking viable solutions and stimulating their adoption. In 1960, for example, FAO launched the Freedom from Hunger Campaign to mobilize nongovernmental support for the issue. In 1981 it initiated the first World Food Day, observed by 150 countries, to promote awareness of the continuing problem.

FAO coordinates efforts by governments and technical agencies to develop programs in agriculture, nutrition, forestry, and fisheries, as well as related economic and social policy, including rural development. In so doing, the agency attempts to provide impartial recommendations for such programs. To that end, it conducts research into areas requiring further knowledge and documentation. It also provides direct technical assistance and develops materials for education in food production, processing, transportation, and consumption. In addition, FAO maintains worldwide statistics on the production, trade, and consumption of agricultural commodities. Specifically, in 1986 it started the Agrostat database, a highly comprehensive source of agricultural information and statistics for all countries. It also publishes numerous periodicals and yearbooks.

Long-term solutions to food security must be sustainable; that is, they must be able to be maintained indefinitely. Examples of unsustainable practices that degrade the environment and make it less likely that future generations will be able to feed themselves include overfishing, overgrazing, overexploitation of water supplies, overdependence on exhaustible fossil fuels, and pollution of air and water. Sustainable development, which has long undergirded the activities of FAO, was formally recognized as one of the agency's eight departments when FAO was reorganized in 1994. The organization plays a crucial role in the intertwined triumvirate of population growth, food security, and the environment.

James L. Robinson

FURTHER READING

Staples, Amy L. S. *The Birth of Development: How the World Bank, Food and Agriculture Organization, and*

World Health Organization Changed the World, 1945-1965. Kent, Ohio: Kent State University Press, 2006.

Vapnek, Jessica, and Melvin Spreij. "United Nations Specialized Agencies." In *Perspectives and Guidelines on Food Legislation, with a New Model Food Law.* Rome: Food and Agriculture Organization of the United Nations, 2005.

SEE ALSO: Population growth; Sustainable agriculture; Sustainable development; Sustainable forestry; World Health Organization.

Food chains

CATEGORY: Ecology and ecosystems
DEFINITION: A linear depiction of food items and the organisms that consume them
SIGNIFICANCE: The concepts of food chains and food webs illustrate the interdependence of various species and how environmental contaminants can have consequences at many levels.

A food chain comprises food items and the organisms that, in turn, consume them. In a lake, for example, the food chain may begin with algae, which are eaten by microscopic zooplankton, which are, in turn, eaten by zooplanktivorous fishes, which are eaten by piscivorous fishes, which may ultimately be consumed by humans.

The original source of energy for almost every food chain is the sun. At the base of a food chain is a primary producer, such as a terrestrial plant or aquatic algae. Organisms that consume plants are the primary consumers (or herbivores), which are themselves consumed by secondary consumers (predators), which are consumed by tertiary consumers, and so on. Organisms that are the same number of feeding levels away from the original source of energy are said to be at the same trophic level. Also important to the concept of a food chain is the energy transfer between adjacent trophic levels, which is not very efficient: Only about 10 percent of the energy at one trophic level is actually transferred to the next one. This makes much less energy available for animals than for the primary producers in a system.

A number of factors make natural systems more complex than is suggested by the basic concept of a food chain. First, most organisms consume more than one type of food and are consumed by more than one type of predator. Such complexity leads to the concept of a food web, in which a series of food chains are interrelated in a weblike arrangement. Second, some organisms can feed on more than one trophic level at the same time. For example, bluegill sunfish can feed on zooplankton (primary consumers), predatory invertebrates (secondary consumers), and even young fishes (secondary or tertiary consumers). Third, changes occur in the foods that an organism consumes and the predators that can consume an organism as it grows, a concept termed the "ontogenetic niche." For example, a frog begins life by hatching from an aquatic egg and living for a period of time as an aquatic larvae. After a metamorphosis (change in form), it becomes a terrestrial frog, clearly consuming different food types and facing consumption from an array of predators different from those that threatened it in the aquatic larval stage. All of these factors combine to make the concepts of food chains and food webs much more complex than is apparent from initial simple definitions.

BIOACCUMULATION AND BIOMAGNIFICATION

Two important issues related to the concepts of food chains and food webs from an environmental perspective are bioaccumulation and biomagnification, specifically as related to some pesticides. The group of pesticides known as organochlorides includes dichloro-diphenyl-trichloroethane (DDT), which was widely used during the twentieth century to control pests such as mosquitoes. While DDT helped eliminate small pest organisms at low concentrations, these concentrations were not harmful to larger organisms. Because DDT is not biodegradable, however, it is not broken down or eliminated by organisms; rather, it is stored in fat within those organisms to which it is not lethal. As animals continued to eat food containing DDT, they accumulated more and more of it, increasing the concentration of DDT in their bodies above that in the surrounding environment. This process is known as bioaccumulation.

Those organisms in which DDT bioaccumulated were eventually consumed by their predators, which further increased the concentration of the pesticide as it proceeded up the food chain. This process is called biomagnification. For some organisms at lower or intermediate trophic levels, the pesticide was not present at concentrations that were harmful. How-

Dodos and Bats: Cases of Disrupted Ecosystems

Within complex ecosystems, animals and plants interact with each other in very intricate and dynamic ways. Plants provide animal species with nutritious fruits. In return, animals help to disperse their seeds. These mutually beneficial relationships sustain plant and animal populations and ultimately the ecosystem itself.

A pair of organisms, the dodo bird and the tambalacoque tree, coexisted on the island of Mauritius in the Indian Ocean for a long time. When humans arrived at the island, the slow dodos were easy prey. By 1681, humans had hunted the dodo to extinction. The damage, however, reached far beyond just dodos. It turned out that dodos helped disperse the seeds and promote the germination of the tambalacoque trees of Mauritius. Dodos ate the tambalacoque fruit before it had a chance to rot, thoroughly cleaning the seeds and thereby protecting them from infections by destructive fungi. Dodos also dispersed the seeds all over the island, ensuring that some of them reached fertile soil for germination. With the dodos gone, the fruit rapidly rots, the seeds are destroyed by fungi, and the tambalacoque trees are seriously threatened.

In tropical forests, bats serve as the most important agents for seed dispersal. Bats may eat up to twice their weight in fruit in a day. They also fly more than twenty miles per night, defecating the seeds in flight. After passing through bats' digestive tract, seeds have more than a 95 percent germination rate, as compared to 10 percent in seeds planted directly from fruit. As bat populations (and those of other fruit-eating animals) decline dramatically, seed dispersal has stopped or been significantly reduced. Tropical fruits are rotting on the forest floor or sending up doomed sprouts under the shade of their parents and other canopies. Some types of tropical forests may not survive.

BIODIVERSITY

Food chains are also important in the area of biodiversity. Two aspects of biodiversity are important from the perspective of food chains and food webs—species extinction and introduction of exotic species. All living organisms occupy particular niches in what they eat, where they live, what eats them, and so on. Because organisms interact with other organisms, as implied by the concepts of food chains and food webs, the loss of a species through extinction opens a gap in the food web that might eliminate food for a predator or eliminate prey animals' control over their own food resources. Similarly, the introduction of an exotic species that did not previously occur in that food chain or food web would add competitive or predatory interactions that did not occur previously. Both of these situations alter the food web such that new interactions must be developed and a new system equilibrium reached. An example is the introduction of the sea lamprey to Lake Michigan, which led to the decline and eventual disappearance of many native fish species because of the lamprey's predatory influence. Based on an understanding of the food web that occurred in Lake Michigan, biologists were able to reintroduce a suite of predators, many of which were nonnative, that established a food web that was similar to the one that had existed prior to lamprey introduction.

Dennis R. DeVries

ever, as its concentration increased up the food chain, harmful levels were eventually reached. For mammals, this did not always lead to direct harm to adults but sometimes caused the death of their young. In addition, many predatory birds that fed on fish in which DDT had become concentrated (for example, ospreys, bald eagles, peregrine falcons, and brown pelicans) produced thin-shelled eggs, resulting in the deaths of unhatched chicks during incubation.

Another related consequence for the food web was that the target pest was not the only organism affected by the pesticide. Nontarget organisms could also be affected, and predator-prey balances that existed in the food web could be altered, releasing prey from predatory control (if a predator was reduced), potentially leading to another pest problem. As a consequence, DDT was banned in many developed nations, including the United States, in the early 1970's. In the decades since this ban, scientists have documented that populations of affected predatory birds have displayed dramatic recovery.

FURTHER READING

Enger, Eldon D., and Bradley F. Smith. *Environmental Science: A Study of Interrelationships.* 12th ed. Boston: McGraw-Hill Higher Education, 2010.

Kitchell, James F., and Larry B. Crowder. "Predator-Prey Interactions in Lake Michigan: Model Predictions and Recent Dynamics." *Environmental Biology of Fishes* 16, nos. 1-3 (1986): 205-211.

Morgan, Kevin. "Greening the Realm: Sustainable Food Chains and the Public Plate." *Regional Studies* 42, no. 9 (2008): 1237-1250.

Myers, Richard. "Environmental Chemistry." In *The Basics of Chemistry*. Westport, Conn.: Greenwood Press, 2003.

See also: Biodiversity; Biomagnification; Dichloro-diphenyl-trichloroethane; Extinctions and species loss; Introduced and exotic species.

Food irradiation

Category: Agriculture and food

Definition: A process in which nuclear radiation is used to sterilize foods

Significance: Although many studies since the 1950's have concluded that the irradiation of food poses no danger to those who consume the food, the process is still the subject of some debate.

Food irradiation is used to reduce spoilage of food and to decrease the incidence of illness from contaminated food. The U.S. Food and Drug Administration (FDA) has certified that irradiation is safe for many foods, including spices, fresh fruit, fish, poultry, and hamburger meat. Opponents of food irradiation focus on the inherent hazards of nuclear technology—particularly the production, transportation, and disposal of radioactive materials—and criticize the possible creation of harmful radiation products in foods.

The most common radioactive source used for food processing is cobalt 60, which is produced through the irradiation of ordinary cobalt metal in a nuclear reactor. The shape of the source is typically a bundle of many thin tubes mounted in a rack. The source is kept in a building with thick walls and under 4.6 meters (15 feet) of water for shielding. The food to be irradiated is packaged and put on a conveyor belt. The operator raises the cobalt-60 source out of the water by remote control as the food packages slowly travel past the source on the moving belt. Typical exposure time ranges from three to thirty minutes, depending on the type of food, the required dose, and the source intensity. Radiation dose is measured in a unit called the kilogray (kGy), where one kGy equals 1,000 Grays, and one Gray equals 100 radiation absorbed doses (rads). (The Gray is named for

British radiation biologist Louis Harold Gray. The rad is an older unit still commonly used in the medical profession.) When radiation passes through the food, it interacts with atomic electrons and breaks chemical bonds. Microorganisms that cause food spoilage or illnesses are inactivated, so they cannot reproduce.

Fresh strawberries, sweet cherries, and tomatoes have a normal shelf life of only seven to ten days. Research has shown that a radiation dose of 2 kGy can double the shelf life of these fruits without affecting the flavor. Trichinosis parasites in fresh pork can be controlled with a dose of 1 kGy. Doses up to 3 kGy are used to destroy 99.9 percent of salmonella in chicken meat and *Escherichia coli* in ground hamburger. A dose of less than 0.1 kGy is sufficient to interrupt cell growth in onions and potatoes to prevent undesirable sprouting in the spring. Larger doses, up to 30 kGy,

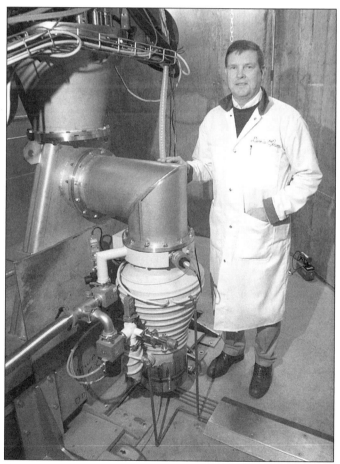

Dennis Olsen, director of Iowa State University's irradiation research program, stands next to a linear accelerator. The device is used to apply doses of radiation to food products for the purpose of eliminating harmful microorganisms. (AP/Wide World Photos)

are used to eliminate insects, mites, and other pests in spices, herbs, and tea. The American Dietetic Association supports food irradiation as an effective technique to reduce outbreaks of food-borne illnesses, which cause several million cases of sickness and more than nine thousand fatalities annually in the United States.

STUDIES OF IRRADIATION EFFECTS

Irradiated food does not become radioactive. It does not "glow in the dark," as some opponents of irradiation have claimed. Hundreds of animal feeding studies with irradiated foods have been done since the late 1950's. Irradiated chicken, wheat, oranges, and other foods have been fed to four generations of mice, three generations of beagles, and thousands of rats and monkeys. No increases in cancer or other inherited diseases have been detected in comparison to control groups eating nonirradiated food. In one experiment, several thousand mice were fed nothing but irradiated food. After sixty generations—about ten years—the cancer rate for the experimental group was no greater than for the control group.

In studies looking for potentially harmful by-products from radiation, chemical analyses of irradiated foods have found small quantities of benzene and formaldehyde. However, canning, cooking, and baking have been shown to create these same by-products even more abundantly. Based on the accumulated evidence, food irradiation has been endorsed by an impressive list of organizations: the American Medical Association, the U.S. Department of Agriculture, the American Diabetic Association, the World Health Organization, the United Nations Food and Agriculture Organization, and the FDA. Astronauts on space missions have eaten irradiated foods since 1972, and many hospitals use irradiated foods for patients with impaired immune systems.

OPPOSITION TO IRRADIATION

The antinuclear movement was born after World War II when the United States and the Soviet Union were conducting nuclear weapons tests that released large amounts of radioactive fallout into the atmosphere. Public pressure eventually led to the Limited Test Ban Treaty in 1963. Two serious accidents at nuclear power plants, one at Three Mile Island (1979) in the United States and the other at Chernobyl (1986) in the Soviet Union, dramatized the hazards of nuclear technology for the general public. In its litera-

ture, a consumer activist organization called Food and Water, based in Walden, Vermont, connects fear of the atomic bomb with food irradiation by showing a picture of a mushroom cloud hovering over a plate filled with food. The caption says, "The Department of Energy has a solution to the problem of radioactive waste. You're going to eat it." The most serious safety issues relating to food irradiation lie in the production of radioactive cobalt; in the protection of workers who transport, install, and use the source; and in the eventual disposal of radioactive waste. Such issues, however, do not have the same impact on consumers as the implied claim that foods may become tainted by irradiation. Commercial food processors have been deterred from installing irradiation facilities by fear of negative publicity and a potential consumer backlash.

The FDA requires that irradiated foods packaged for sale in the United States be labeled with a special symbol (the Radura) and the statement "treated with radiation." It is up to consumers to decide if the benefits of a safer food supply outweigh the potential hazards of an expanded nuclear industry.

Hans G. Graetzer

FURTHER READING

Gibbs, Gary. *The Food That Would Last Forever: Understanding the Dangers of Food Irradiation.* Garden City Park, N.Y.: Avery, 1993.

Hilgenkamp, Kathryn. "Food Safety Concerns." In *Environmental Health: Ecological Perspectives.* Sudbury, Mass.: Jones and Bartlett, 2006.

Louria, Donald B., and George G. Giddings. "Point-Counterpoint: Controversy over Food Irradiation." In *Human Biology,* by Daniel D. Chiras. 5th ed. Sudbury, Mass.: Jones and Bartlett, 2005.

Satin, Morton. *Food Irradiation: A Guidebook.* Lancaster, Pa.: Technomic, 1996.

Wagner, Henry, and Linda Ketchum. *Living with Radiation: The Risk, the Promise.* Baltimore: The Johns Hopkins University Press, 1989.

Wood, Olivia Bennett, and Christine M. Bruhn. "Position of the American Dietetic Association: Food Irradiation." *Journal of the American Dietetic Association* 100, no. 2 (2000): 246-253.

SEE ALSO: Antinuclear movement; Environmental illnesses; Food and Agriculture Organization; Food quality regulation; Genetically modified foods; Nuclear and radioactive waste.

Food quality regulation

CATEGORIES: Agriculture and food; human health
and the environment

DEFINITION: Legally enforceable standards and
procedures aimed at controlling the safety of
food products

SIGNIFICANCE: Laws regulating food quality for public
health purposes often promote agricultural methods that are beneficial to the environment in that
they protect soil, water, and the atmosphere from
toxins and support conservation and sustainability.

Since the mid-nineteenth century, legislators have
passed laws regulating food quality to address specific safety concerns and to remove from the marketplace foods identified as hazardous to public health.
In 1906 the U.S. Congress passed the Pure Food and
Drug Act and created the U.S. Food and Drug Administration (FDA) to regulate food quality by establishing standards and procedures for food producers to
implement in fields and factories. The U.S. Department of Agriculture's Food Safety Inspection Service
evaluates meats and poultry products. The National
Marine Fisheries Service, Department of Homeland
Security, U.S. Centers for Disease Control and Prevention, and public health departments also regulate aspects of U.S. food supplies. Food manufacturers
worldwide have adopted the Hazard Analysis and Critical Control Points (HACCP) system to aid them in
detecting contamination hazards and to enhance the
quality of the foods they produce.

LEGISLATION AND CONCERNS

Significant U.S. legislation of the twentieth century
concerning food quality includes the Federal Food,
Drug, and Cosmetic Act (FFDCA), passed in 1938,
and the Federal Insecticide, Fungicide, and Rodenticide Act (FIFRA), approved in 1947. In 1996 the U.S.
Congress passed the Food Quality Protection Act
(FQPA) after legislators concerned about the dangers
of pesticides in foods evaluated existing food regulations and determined that FFDCA and FIFRA did not
provide adequate protection for the public. Worried
that food products that had been exposed to chemical
pesticides in fields and other agricultural sites could
contain enough toxic residues to harm children's development and exacerbate existing health conditions,
including allergies, the legislators created FQPA. The
new law implemented more rigorous limitations and

controls on the use of certain chemicals in food-
producing agriculture than had FIFRA and FFDCA
and delegated enforcement of FQPA provisions to the
Environmental Protection Agency (EPA).

Some U.S. regulations address the risks to food
quality presented by chemicals, hormones, and antibiotics in the food consumed by livestock raised for
meat or dairy products. Aware that chemical threats
could decrease supplies of safe food, government officials have urged farmers to explore ways in which they
can combat insects and weeds without using pesticides and herbicides. They have also suggested that
farmers use fertilizers with minimal chemical content. In response to such suggestions and to their own
concerns about the environment, many farmers have
adopted natural methods of fighting insects and
weeds, often combining several methods in what is
known as integrated pest management.

Some regulations concerning food quality in the
United States are based in the standards for water
quality established by the Safe Drinking Water Act.
Food quality regulations include procedures that
food producers must follow when food supplies and
agricultural lands come into contact with contaminated water, such as that produced by flooding.

The FDA's Center for Food Safety and Applied Nutrition (CFSAN) issues compliance documentation,
revised every three years, to guide the inspectors who
evaluate various foods, ranging from dairy products
to canned items, for conformity to regulations regarding the presence in food products of pesticides and
other chemicals and toxins harmful to the environment. CFSAN also distributes food quality guidelines
and guidelines regarding food labeling requirements, another element of food quality regulation.
The Nutrition Labeling and Education Act of 1990
sets requirements for the proper, truthful labeling of
food products, including the nutrition labels that
must appear on packaged foods; this act also limits the
health claims that manufacturers can make for products based on the products' ingredients. Other regulations address the use of dyes, flavorings, preserva-
tives, and other additives in foods.

Developments in the field of genetic engineering
have complicated food quality regulation efforts and
have offered both positive and negative possibilities
in food production involving both plants and animals.
Bioengineering methods can alter environments; for
example, some pests (insects, weeds, or pathogens)
develop resistance to traits bred into plants or live-

An inspector for the U.S. Department of Agriculture checks the quality of eggs at a facility in Maine. (AP/Wide World Photos)

stock to repel those pests, and this results in the use of ever-increasing amounts of chemicals to kill the pests. The creation of the Flavr Savr tomato through genetic engineering in the 1990's resulted in a ruling by the FDA that bioengineered fruits and vegetables are not hazardous; the FDA simultaneously announced a food additive regulation permitting the insertion of the kanamycin resistance gene in plants.

Concerns with food quality and regulations to ensure food safety vary among the world's nations. The European Union has enacted food quality laws that address the use of pesticides, the use of hormones, food irradiation, food additives, and the presence in foods of such substances as polychlorinated biphenyls (PCBs) and heavy metals. In 2010 China's Ministry of Health established stricter regulations concerning food additives after several incidents involving Chinese food products led to deaths. U.S. laws require the monitoring of all imported food products and the rejection of items determined to be unsafe or otherwise unacceptable. Among the types of foods that inspectors must reject are any made from endangered animal or plant resources and any made using manufacturing practices that are damaging to the environment.

SECURITY AND ENVIRONMENTAL PROTECTION MEASURES

The Public Health Security and Bioterrorism Preparedness and Response Act of 2002 (also known as the Bioterrorism Act) requires food product manufacturers in the United States to document where they acquire the agricultural resources they process into foods and the locations where they ship those foods, so that officials can efficiently trace any contaminants that may be found in those products. In 2007 the FDA created its Food Protection Plan, which enacted measures to ensure the security of the environmental resources used to produce food, particularly soil and water, against possible harm by terrorists.

Many food manufacturers attempt to use production methods that are compatible with environmental protection and preservation, such as by conserving energy and water and recycling waste products. Guidelines have been developed to encourage such practices while also protecting food quality. The National Pollutant Discharge Elimination System provides poultry and meat producers with guidelines for treating effluent to purge toxins without damaging the environment, and the EPA distributes the Multimedia Environmental Compliance Guide for Processors. Some manufacturers use biosensors to evaluate foods for undesirable substances or pathogens, enabling the identification of contaminated agricultural sources so that those environments can be cleansed of toxins and restored.

ENVIRONMENTAL CHANGES AND FOOD QUALITY

Climate changes can have impacts on food quality. Scientists have noted that global warming affects the distribution of microorganisms, including pathogens, to new ecosystems, and indigenous plants and animals utilized for food products can become contaminated because they have no immunities to the invasive microorganisms. Scientists have suggested that warmer temperatures in oceans may have contributed to oysters becoming contaminated with the bacteria *Vibrio*, and heat extremes are associated with the expansion of the bacteria *Listeria*. Insects and microbes evolve as a result of climatic changes, resulting in the need for different control methods for agricultural processes and regulations to ensure that those techniques are not detrimental to the quality of the foods produced.

Direct environmental contamination caused by human activities can also result in special needs for food quality regulation. Oil spills, for instance, can contaminate marine ecosystems to such an extent that any food animals harvested from these ecosystems can be unsuitable for human consumption. The BP *Deepwater Horizon* oil spill in the Gulf of Mexico in 2010 provides an example. The FDA released a statement in response to the spill, noting possible damage to the quality of shellfish living in the Gulf Coast owing to their exposure to polycyclic aromatic hydrocarbons (PAHs), which are compounds found in oil. The FDA stressed that food processors accepting shellfish from gulf waters should be certain that the shellfish were harvested by licensed operators who had recorded procurement details from the areas that had been approved to remain open for fishing. The FDA also emphasized that food processors must follow established fishery products regulations in utilizing HACCP to detect contaminants detrimental to food quality. Inspectors were tasked with evaluating all varieties of Gulf Coast seafood specimens to determine whether they were of suitable quality for consumption before they were approved for distribution to food services.

Elizabeth D. Schafer

FURTHER READING

Clute, Mark. *Food Industry Quality Control Systems*. Boca Raton, Fla.: CRC Press, 2009.

Fortin, Neal D. *Food Regulation: Law, Science, Policy, and Practice*. Hoboken, N.J.: John Wiley & Sons, 2009.

Hawthorne, Fran. *Inside the FDA: The Business and Politics Behind the Drugs We Take and the Food We Eat*. Hoboken, N.J.: John Wiley & Sons, 2005.

Hoffmann, Sandra A., and Michael R. Taylor, eds. *Toward Safer Food: Perspectives on Risk and Priority Setting*. Washington, D.C.: Resources for the Future, 2005.

Jha, Veena, ed. *Environmental Regulation and Food Safety: Studies of Protection and Protectionism*. Ottawa, Ont.: International Development Research Centre, 2005.

Josling, Tim, Donna Roberts, and David Orden. *Food Regulation and Trade: Toward a Safe and Open Global System*. Washington, D.C.: Institute for International Economics, 2004.

Roberts, Cynthia A. *The Food Safety Information Handbook*. Westport, Conn.: Oryx Press, 2001.

SEE ALSO: Agricultural chemicals; Biofertilizers; Biopesticides; Biotechnology and genetic engineering; Bovine growth hormone; Centers for Disease Control and Prevention; Flavr Savr tomato; Food irradiation; Genetically modified foods; Integrated pest management; Organic gardening and farming; Pesticides and herbicides.

Foreman, Dave

CATEGORIES: Activism and advocacy; preservation and wilderness issues

IDENTIFICATION: American environmental activist and author

BORN: October 18, 1946; Albuquerque, New Mexico

SIGNIFICANCE: As one of the cofounders of the radical environmental group Earth First! and through his continued leadership in less radical organizations, Foreman has had a great deal of influence on the environmental movement.

Dave Foreman, the son of Benjamin and Lorane Foreman, was born in 1946. His father was a pilot in the U.S. Air Force, and the family moved often. They were living in Blythe, California, in 1964 when Foreman graduated from high school. He attended junior college for one year and enrolled the following fall at the University of New Mexico in Albuquerque. His involvement in politics began in high school when he did volunteer work for Republican senator Barry Goldwater. In 1966 Foreman was the state chairman of the conservative organization Young Americans for Freedom and a vocal supporter of U.S. involvement in the Vietnam War. In 1967 Foreman enlisted in the Marines Corps, but he found life in the military unsatisfactory, and he fled into the mountains of New Mexico. Eventually he turned himself in and served time in prison; he was dishonorably discharged from the service.

In 1970 Foreman took to the woods again, supporting himself with odd jobs while spending time backpacking and rafting. He joined an environmental organization called the Black Mesa Defense Fund, and his knack for waging effective campaigns against developers was apparent. The Wilderness Society employed Foreman in 1973 as its principal consultant in New Mexico, and he continued to lead campaigns against polluters and land developers. By 1980 he was one of the Wilderness Society's most visible members.

Foreman, however, felt the need to become more confrontational. In 1980 he resigned from the Wilderness Society and cofounded, with several others, Earth First!, a radical environmental group that used direct action to combat environmental destruction. The founders were inspired in part by Edward Abbey's novel *The Monkey Wrench Gang* (1975), which details the exploits of a small group of people who destroy construction equipment to stop environmental destruction in the southwestern United States. One of the early actions conducted by Earth First! was the unfurling of a 300-foot piece of black plastic down the front of Glen Canyon Dam to represent a crack in the structure. In 1985 Foreman published *Ecodefense: A Field Guide to Monkeywrenching*, a volume of detailed instructions on how to destroy heavy machinery, deface billboards, and spike trees (drive long metal spikes into trees to prevent logging with chain saws).

Earth First!, however, developed problems in keeping its aims directed to environmental protection, and in 1989 Foreman left the group. Around the same time he was arrested on charges of conspiring to topple power lines in Arizona. The case was heard in 1991, and Foreman pled guilty to a reduced conspiracy charge with delayed sentencing and an agreement that the charge would be reduced after five years of good behavior. In 1995 Foreman accepted a three-year term as a director of the Sierra Club, but he resigned in 1997 to devote more time to the Wildlands Project, which he had initiated in the early 1990's. The Wildlands Project founded a think tank in 2003, the Rewilding Institute, which has as its stated mission "to develop and promote the ideas and strategies to advance continental-scale conservation in North America."

Kenneth H. Brown

FURTHER READING

Foreman, Dave, and Bill Haywood, eds. *Ecodefense: A Field Guide to Monkeywrenching.* 3d ed. Chico, Calif.: Abbzug Press, 2002.

Scarce, Rik. *Eco-Warriors: Understanding the Radical Environmental Movement.* Updated ed. Walnut Creek, Calif.: Left Coast Press, 2006.

SEE ALSO: Earth First!; Ecotage; Ecoterrorism; Glen Canyon Dam; Monkeywrenching; Wilderness Society.

Forest and range policy

CATEGORIES: Forests and plants; resources and resource management

DEFINITION: High-level governmental plans of action for managing the use of forests and grazing lands

SIGNIFICANCE: Many national and regional governments have established legal policies regarding the use of forests and rangelands. Forest and range policy typically seeks to maintain and protect biodiversity while setting guidelines for the sustainable use of natural resources. The lack of sound policy can lead to major problems such as deforestation, overgrazing, and desertification.

Rangeland supplies forage for grazing and browsing animals. According to 2009 statistics from the United Nations Food and Agriculture Organization, 3.4 billion hectares (8.4 billion acres) are used for grazing worldwide. Pastureland occupies 26 percent of the ice-free land on earth. Roughly one billion cattle, buffalo, sheep, and goats graze on rangelands. These animals are important in converting forages into milk and meat, which provide nourishment for people around world. Forests cover approximately 31 percent of the earth and provide humans with lumber, fuelwood, spices, chocolate, tropical fruits, nuts, latex rubber, and valuable chemicals that constitute the active ingredients of prescription and nonprescription pharmaceuticals, including powerful anticancer drugs. Rangelands and forests also function as important ecosystems that play a vital role in providing food and shelter for wildlife, serving as watersheds, controlling erosion, and purifying the atmosphere. Forests and rangelands have been facing alarming rates of destruction and degradation at the hands of humans.

PROTECTING FORESTS AND RANGELANDS

The 6 billion hectares (14.8 billion acres) of forest estimated to have existed eight thousand years ago have been reduced to 4 billion hectares (9.9 billion acres) by human conversion of forestland to cropland, pastureland, cities, and nonproductive land. Forests, if properly maintained or left alone, are the most productive and self-sustaining ecosystems that land can support. Tropical rain forests are the natural habitat for at least 50 percent of the species on earth. Harvard biologist Edward O. Wilson has stated that, with tropical deforestation condemning an estimated fifty thousand species to extinction annually, 25 percent of the earth's species could become extinct within a fifty-year period.

Many national governments have established legal policies for protecting forest habitats and the important biological diversity found within them. National parks and reserves provide protection for both forests and rangelands. Some countries have local, regional, and national laws that protect particular forests or prohibit the clearing, burning, or logging of forests. China, which suffered from erosion and terrible floods as a result of centuries of deforestation, began an impressive reforestation campaign during the 1960's. By 1999, the nation's tree plantation area had reached 46.7 million hectares (115.4 million acres). As an additional environmental protection measure, in 1998 the Chinese government banned logging in large forested areas in Sichuan province. Planting efforts were increased in 2002 but were suspended in 2009 in marginal arable lands that could be used for food production.

South Korea had lost almost all of its forested land by the end of the Korean War. Thanks to concentrated reforestation efforts during the 1970's and 1980's, by the beginning of the twenty-first century some 65 percent of the country's land area was forested. In post-World War II Japan, strict environmental laws enabled the nation to reforest roughly two-thirds of its land area. However, Japan's reliance on imported timber to achieve its goals of domestic forest conservation has contributed to deforestation in the Tropics and the American northwest. Even with the world's many success stories in reforestation, efforts to protect and sustainably manage forests and rangelands still need to be increased if forests are to be saved.

MULTIPLE-USE POLICIES

Protecting forestland involves an interdisciplinary approach. In the United States, 78 million hectares (193 million acres) of the country's forests are publicly owned and managed by the U.S. Forest Service. Passage of the Multiple Use-Sustained Yield Act of 1960 marked the beginning of the nation's official policy of managing public forests to protect wildlife and fish habitats and watersheds while at the same time providing outdoor recreation, range, and timber for current and future generations. The Forest and Rangeland Renewable Resources Planning Act (RPA) of 1974, amended by the National Forest Management Act (NFMA) of 1976, directs that management plans must be developed for forests and rangelands to ensure that resources will be available on a sustained basis. Management policies must aim to sustain and protect biodiversity; old-growth forests; riparian areas; threatened, endangered, and sensitive species; rangeland; water and air quality; access to forests; and wildlife and fisheries habitat.

The Forest Service provides inexpensive grazing lands for more than two million cattle, sheep, goats, horses, and burros every year, supports multimillion-dollar mining operations, maintains a network of roads eight times longer than the U.S. interstate highway system, and allows access to nonreserved national forestlands for commercial logging. The Forest Service is responsible for producing plans for multiple uses of national lands. Assessments of forest and rangeland natural resources are conducted every ten years, and interim updates are issued between the assessments.

Sustainability policies require that the net productive capacity of a forest or rangeland does not decrease with multiple-use practices. Soil productivity must thus be maintained, which involves keeping erosion, compaction, and displacement by mining or logging equipment and other motorized vehicles within tolerable limits. The maintenance of productivity further requires that a large percentage of the forest remain undeveloped so that soils and habitats, as well as tree cover, remain undisturbed and in their natural state.

The RPA and NFMA, along with the Endangered Species Act (ESA) of 1973, mandate policies that encourage the proliferation of species native to and currently living in forests. Even though forests and rangelands are required to be multiple-use areas, policy maintains that no threatened, endangered, or sensitive species should suffer adverse impacts as a result of any human uses. Species habitats within forests are to remain well distributed and free of barriers that can cause fragmentation of animal populations and ultimately species loss. If human activities result in fragmented areas in a forest, corridors are constructed to connect the forest patches so that members of given species are not isolated from one another and viable populations can exist.

In the case of natural disasters, the Forest Service creates artificial habitats to encourage the survival of species. When Hurricane Hugo devastated the Francis Marion National Forest in South Carolina in 1989, winds snapped 90 percent of the trees with active woodpecker cavities in some areas of the forest. This habitat destruction caused 70 percent of the endan-

gered red-cockaded woodpecker population to disappear. The Forest Service and university researchers created nesting and roosting cavities to save the woodpeckers, and within a four-year period the population had made a dramatic recovery.

TIMBER, OIL, AND MINERAL LEASING

In U.S. forests, logging activities are covered by the NFMA and its amendments. Forested land must be evaluated for its ability to produce commercially usable timber without negative environmental impact. There must be reasonable assurance that stands managed for timber production can be adequately restocked within five years of the final harvest. Furthermore, irreversible resource damage must not be allowed to occur. Policy also requires that the silvi-

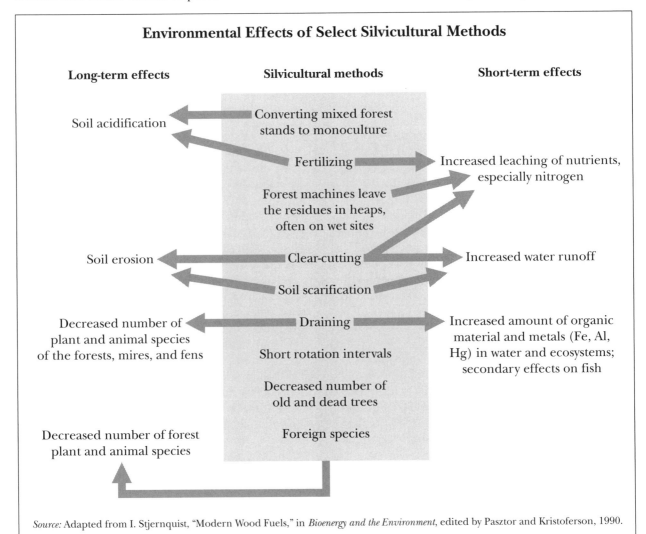

Environmental Effects of Select Silvicultural Methods

Long-term effects	Silvicultural methods	Short-term effects

Soil acidification ← Converting mixed forest stands to monoculture

Fertilizing → Increased leaching of nutrients, especially nitrogen

Forest machines leave the residues in heaps, often on wet sites

Soil erosion ← Clear-cutting → Increased water runoff

Soil scarification

Decreased number of plant and animal species of the forests, mires, and fens ← Draining → Increased amount of organic material and metals (Fe, Al, Hg) in water and ecosystems; secondary effects on fish

Short rotation intervals

Decreased number of old and dead trees

Decreased number of forest plant and animal species ← Foreign species

Source: Adapted from I. Stjernquist, "Modern Wood Fuels," in *Bioenergy and the Environment*, edited by Pasztor and Kristoferson, 1990.

culture practices that are best suited to the land-management objectives of the area be used. Cutting practices are then monitored.

Multiple-use practice under the NFMA also allows forests to be available for oil, gas, and mineral leasing. Certain lands have been exempted from mineral exploration by acts of Congress or executive authority, but the search for and production of mineral and energy sources remain under the jurisdiction of the Forest Service, which must provide access to national forests for mineral resource activities. The Federal Onshore Oil and Gas Leasing Reform Act of 1987 gave the Forest Service increased authority in making lease decisions.

The 1990's were characterized by a trend toward restricting logging in the United States in order to protect habitats and preserve older stands of trees. In the 1993 Renewable Resource Assessment update, the Forest Service found that timber mortality was 24.3 percent and was still interfering with biological diversity. Some forested areas were withdrawn from timber production because of their fragility. The early years of the twenty-first century, by contrast, were marked by a series of logging-friendly administrative changes to forestry rules. Environmental organizations quickly countered these shifts in policy with lawsuits that blocked their implementation.

Pest and Weed Control

Insecticides are sometimes used in attempts to ensure the health of forestland. Policy in the United States requires the use of safe pesticides and encourages the development of integrated pest management (IPM) plans. Any decision to use a particular pesticide must be based on an analysis of its effectiveness, specificity, environmental impact, economic efficiency, and effects on humans. The application and use of pesticides must be coordinated with federal and state fish and wildlife management agencies to ensure that no harm occurs to either fish or wildlife. Pesticides can be applied to areas that are designated as wilderness only when their use is necessary to protect or restore resources in the area. Other methods of controlling disease include removing diseased trees and vegetation from the forest, cutting infected areas from plants and removing the debris, treating trees with antibiotics, and developing disease-resistant plant varieties.

Forest Service policy on IPM was revised in 1995 to emphasize the importance of integrating noxious weed management into the forest plan for ecosystem analysis and assessment. Noxious weed management must be coordinated in cooperation with state and local government agencies, as well as private landowners. Noxious weeds include invasive, aggressive, or harmful nonindigenous or exotic plant species. They are generally poisonous, toxic, or parasitic, or may carry insects or disease. The Forest Service is responsible for the prevention, control, and eradication of noxious weeds in national forests and grasslands.

One strategy for promoting weed-free forests that has been implemented in several western U.S. states requires pack animals on national forestland to eat state-certified weed-free forage. In North Dakota, goats have been used to help control a noxious weed, leafy spurge, that crowds out native plants. Leafy spurge is unpalatable and toxic to cattle, but goats can safely include it in their diet. In a five-year weed-control study, goats grazed on designated spurge patches during the day and were kept in portable corrals during the night. The study found that the goats effectively reduced stem densities of spurge patches to the extent that livestock forage plants were able to reestablish themselves. Another weed-eradication strategy involves the use of certified weed-free straw and gravel in construction and rehabilitation efforts within national forests. Biocontrols, herbicides, and controlled burning are also commonly used during IPM operations in forests.

Other Protection Issues

The plant communities that grow in riparian areas (that is, natural watercourses and their banks) often serve as important habitats for a large variety of animals and birds and also provide shade, bank stability, and filtration of pollution sources. It is therefore important that these areas remain in good ecological condition. Riparian areas and streams are managed according to legal policies for wetlands, floodplains, water quality, endangered species, and wild and scenic rivers.

Access to open roads in forests may sometimes be curtailed to preserve habitat and return land to a more natural state. Dirt roads in national forests are often closed when road sediment pollutes riparian areas and harms fish populations. Forest and rangeland roads are also closed to prevent disruption of breeding or nesting colonies of various species of animals. Seemingly harmless human endeavors—such as seeking mushrooms, picking berries, or hiking in the forest—can cause problems for calving elk and nesting eagles, for example.

Studies have shown that the vast numbers of roads, and accompanying vegetation loss, on Bureau of Land Management (BLM) and Forest Service lands have made it too easy for humans to kill elk. The animals are in danger not only of being targeted by hunters but also of being struck by motor vehicles. In Oregon, researchers found few mature bulls in the elk population. Lack of mature bulls can cause many problems, including disruption of breeding seasons and conception dates and decreases in calf survival rates. Younger bulls generally breed later and over a longer period of time than mature bulls, a tendency that results in a calving season that lasts longer. Calves that are born late in the spring do not have enough time to feed on high-quality forage before winter. Long calving seasons also make the calves more vulnerable to predators, such as coyotes, bears, and mountain lions.

Fire management is important to healthy forests. In many cases fires are prevented or suppressed, but prescribed fires are used to protect and maintain ecosystem characteristics. Some conifers, such as the giant sequoia and the jack pine, release their seeds for germination only after being exposed to intense heat. Lodgepole pines do not release their seeds until they have been scorched by fire. Ecosystems that depend on the recurrence of fire for regeneration and balance are called fire climax ecosystems. Prescribed fires are used as a management tool in these areas, which include some grasslands and pine habitats. Where overzealous fire suppression has interrupted natural cycles of burn and recovery, the buildup of forest undergrowth causes the wildfires that do occur to be hotter, more intense, and more difficult to control.

In 1964 the U.S. Congress passed the Wilderness Act, which mandates that certain federal lands be designated as wilderness areas. These lands must remain in their natural condition, provide solitude or primitive types of recreation, and (with a handful of exceptions) be at least 2,023 hectares (5,000 acres) in area. They usually contain ecological or geological systems of scenic, scientific, or historical value. No roads, motorized vehicles, or structures are allowed in these areas. Furthermore, no commercial activities are allowed in designated wilderness areas except livestock grazing and limited mining endeavors that began before the areas received their official designation.

GRAZING PRACTICES AND PROBLEMS

About half of the world's land area is rangeland—some 6.5 billion hectares (16.1 billion acres)—although roughly half of that is too dry, cold, or remote for livestock grazing. The grasses in rangelands are known for their deep, complex root systems, which make the grasses hard to uproot. When the tip of the leaf is eaten by a grazing animal, the plant quickly regrows. Each leaf of grass on the rangeland grows from its base, and the lower half of the plant must remain for the plant to thrive and survive. As long as only the top half of the grass is eaten, grasses serve as renewable resources that can provide many years of grazing. Each type of grassland is evaluated based on grass species, soil type, growing season, range condition, past use, and climatic conditions. These conditions determine the herbivore carrying capacity—that is, the maximum number of grazing animals a rangeland can sustain and remain renewable.

Overgrazing occurs when too many animals are allowed to graze in one area for too long or herbivore numbers exceed the carrying capacity. Grazing animals tend to eat their favorite grasses first and leave the tougher, less palatable plants. If animals are allowed to do this, the vegetation begins to grow in patches, allowing cacti and woody bushes to move into vacant areas. As native plants disappear from the range, weeds also begin to grow. As the nutritional level of the forage declines, hungry animals pull the grasses out by their roots, leaving the ground bare and susceptible to damage from hooves. This process initiates the desertification cycle. With no vegetation present, the soil become vulnerable to erosion, as rain quickly drains off the land and does not replenish the groundwater. Overgrazing and livestock-related soil compaction and erosion have degraded an estimated 70 percent of all grazing land in the world's dry areas. Other common factors contributing to rangeland loss include conversion into croplands or urban developments. In tropical areas where rangeland is increasing, gains are more often than not the result of conversion of rain-forest lands.

The United States entered the twenty-first century with roughly one-third of its land area represented by rangeland. Some 27 percent of this was owned and managed by the BLM, about 7 percent by the Forest Service. Rangeland held by private owners or state or local governments accounted for the remainder. Efforts to preserve rangelands in the United States include close monitoring of herbivore carrying capacity and removal of substandard ranges from the grazing cycle until they recover. Grazing practices such as cattle and sheep rotation help to preserve the renewable quality of rangelands. Grazing is also managed with consideration to season, moisture, and plant growth conditions. Noxious weed encroachment is controlled, and native forages and grasses are allowed to grow.

Most rangelands in the United States are short-grass prairies located in the western part of the nation. These lands are further characterized by thin soils and low annual precipitation. They experience numerous environmental stresses. Woody shrubs, such as mesquite and prickly cactus, often invade and take over these rangelands as overgrazing or other degradation occurs. Such areas are especially susceptible to desertification. Off-road recreational vehicles such as motorcycles, dune buggies, and four-wheel-drive trucks can also damage the vegetation on ranges. By the late twentieth century, overgrazing and other poor practices had left much of the nation's rangeland in unsatisfactory condition.

Steps toward improving the health of rangelands include restoring and maintaining riparian areas and priority watersheds. These areas are monitored on a regular basis, and adjustments are made if their health is jeopardized by sediment from road use or degradation of important habitats caused by human activity. In the United States, the Natural Resources Conservation Service teaches private landowners how to burn unwanted woody plants on rangelands, reseed with perennial grasses that help hold water in the soil, and rotate grazing of cattle and sheep on rangelands so that the land is able to recover and thrive. Such methods have proven to be successful.

Toby Stewart and Dion Stewart
Updated by Karen N. Kähler

FURTHER READING

Bryner, Gary C. *U.S. Land and Natural Resources Policy: A Public Issues Handbook.* Westport, Conn.: Greenwood Press, 1998.

Clarke, Jeanne Nienaber, and Daniel McCool. *Staking Out the Terrain: Power and Performance Among Natural Resource Agencies.* 2d ed. Albany: State University of New York Press, 1996.

Culhane, Paul J. *Public Lands Politics: Interest Group Influence on the Forest Service and the Bureau of Land Management.* Baltimore: The Johns Hopkins University Press, 1981.

Grice, A. C., and K. C. Hodgkinson, eds. *Global Rangelands: Progress and Prospects.* New York: CABI, 2002.

Hirt, Paul W. *A Conspiracy of Optimism: Management of the National Forests Since World War Two.* Lincoln: University of Nebraska Press, 1994.

Platt, Rutherford H. *Land Use and Society: Geography, Law, and Public Policy.* Rev. ed. Washington, D.C.: Island Press, 2004.

Robbins, William G., and James C. Foster, eds. *Land in the American West: Private Claims and the Common Good.* Seattle: University of Washington Press, 2000.

Rowley, William D. *U.S. Forest Service Grazing and Rangelands: A History.* College Station: Texas A&M University Press, 1985.

United Nations Food and Agriculture Organization. *State of the World's Forests 2009.* Rome: Author, 2009.

SEE ALSO: Deforestation; Desertification; Forest management; Forest Service, U.S.; Grazing and grasslands; Logging and clear-cutting; National forests; Rain forests; Range management; Sustainable forestry.

Forest management

CATEGORIES: Forests and plants; resources and resource management

DEFINITION: Policy making and supervision related to the ways in which various resources contained in forestlands are used and protected

SIGNIFICANCE: The world's forests provide lumber for homes, fuelwood for cooking and heating, and raw materials for making such products as paper, latex rubber, dyes, and essential oils. Forests are also home to millions of plants and animal species and are vital in regulating climate, purifying the air, and controlling water runoff. Issues surrounding the management of these important resources are the subject of ongoing discussion.

Thousands of years ago, before humans began clearing the forests for croplands and settlements, forests and woodlands covered almost 6.1 billion hectares (15 billion acres) of the earth. By the end of the twentieth century, approximately 16 percent of the forests had been cleared and converted to pasture, agricultural land, cities, and nonproductive land. The remaining 4.6 billion hectares (11.4 billion acres) of forests covered approximately 30 percent of the earth's land surface.

Clearing forests has severe environmental consequences. It reduces the overall productivity of the land, and nutrients and biomass stored in trees and leaf litter are lost. Soil once covered with plants, leaves, and snags becomes prone to erosion and drying. When forests are cleared, habitats are destroyed and biodiversity is greatly diminished. Destruction of forests causes water to drain off the land instead of being released into the atmosphere by transpiration or percolation into groundwater. This can cause major changes in the hydraulic cycle and ultimately in the earth's climate. Forests also remove a large amount of carbon dioxide from the air; thus the clearing of forests causes more carbon dioxide to remain in the air, upsetting the delicate balance of atmospheric gases.

RAIN FOREST DESTRUCTION

The destruction of tropical rain forests is of great concern. These forests provide habitats for at least 50 percent (some estimates are as high as 90 percent) of the total stock of animal, plant, and insect species on earth. They supply one-half of the world's annual harvest of hardwood and hundreds of food products, such as chocolate, spices, nuts, coffee, and tropical fruits. Tropical rain forests also provide the main ingredients in 25 percent of the world's prescription and nonprescription drugs, as well as 75 percent of the three thousand plants identified as containing chemicals that fight cancer. Many industrial materials, such as natural latex rubber, resins, dyes, and essential oils, are also harvested from tropical forests.

Tropical forests are often cleared by individuals, groups, or companies with the intent of producing pastureland for large cattle ranches, establishing logging operations, constructing large plantations, growing drug crops such as marijuana or coca plants, developing mining operations, or building dams to provide power for mining and smelting operations. In 1985 the United Nations Food and Agriculture Organization's Committee on Forest Development in the Tropics developed the Tropical Forestry Action Plan to combat these practices. Fifty nations in Asia, Africa, and Latin America adopted the plan, which sought to develop sustainable forest methods and protect precious ecosystems. The Tropical Forestry Action Plan was later replaced by the Tropical Forestry Action Programme.

Several management techniques have been suc-cessfully applied to tropical forests. Sustainable logging practices and reforestation programs have been established on lands that allow timber cutting, with complete bans of logging on virgin lands. Certain regions have set up extractive reserves to protect land for the native peoples who live in the forests and gather latex rubber and nuts from mature trees. Sections of some tropical forests have been set aside as national reserves, which attract tourists while preserving trees and biodiversity. Developing countries have also been encouraged to protect their tropical forests by using a combination of debt-for-nature swaps and conservation easements. In debt-for-nature swaps, nations act as custodians of their tropical forests in exchange for foreign aid or relief from debt. Conservation easements involve tropical countries protecting specific habitats in exchange for compensation from other countries or from private organizations.

Another management technique involves putting large areas of forestlands under the control of indigenous peoples who use swidden or milpa agriculture. These traditional, productive forms of slash-and-burn agriculture follow multiple-year cycles. Each year a forest plot approximately 1 hectare (2.5 acres) in size is cleared to allow the sun to penetrate to the ground. Leaf litter, branches, and fallen trunks are burned and leave a rich layer of ashes. Fast-growing crops such as bananas and papayas are planted and provide shade for root crops, which are planted to anchor the soil. Finally, crops such as corn and rice are planted. Crops mature in a staggered sequence, thus providing a continuous supply of food. The natives' use of mixed perennial polyculture helps prevent insect infestations, which can destroy monoculture crops. After one or two years the forest begins to take over the agricultural plot. The native farmers continue to pick the perennial crops but essentially allow the forest to reclaim the plot for the next ten to fifteen years before clearing and planting the area again.

U.S. FORESTS AND MANAGEMENT

Forests cover approximately one-third of the land area of the continental United States; American forests constitute 10 percent of the forests in the world. Only about 22 percent of the commercial forest area in the United States lies within national forests. The rest is managed primarily by private companies that grow trees for commercial logging. The land managed by the U.S. Forest Service provides inexpensive grazing lands for more than three million cattle and

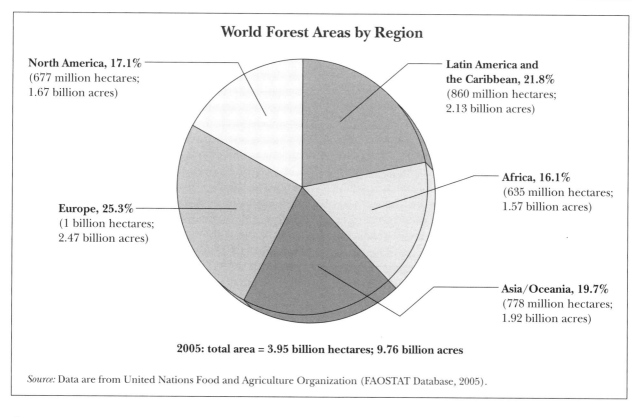

World Forest Areas by Region

North America, 17.1%
(677 million hectares;
1.67 billion acres)

Latin America and
the Caribbean, 21.8%
(860 million hectares;
2.13 billion acres)

Africa, 16.1%
(635 million hectares;
1.57 billion acres)

Europe, 25.3%
(1 billion hectares;
2.47 billion acres)

Asia/Oceania, 19.7%
(778 million hectares;
1.92 billion acres)

2005: total area = 3.95 billion hectares; 9.76 billion acres

Source: Data are from United Nations Food and Agriculture Organization (FAOSTAT Database, 2005).

sheep every year, supports multimillion-dollar mining operations, and consists of a network of roads eight times longer than the U.S. interstate highway system. Almost 50 percent of American national forestland is open for commercial logging, and nearly 14 percent of the timber harvested in the United States each year comes from national forestlands. Total wood production in the United States has caused the loss of more than 95 percent of the old-growth forests in the lower forty-eight states. This loss includes not only high-quality wood but also a rich diversity of species not found in early-growth forests.

National forests in the United States are required by law to be managed in accordance with principles of sustainable yield. The U.S. Congress has mandated that forests be managed for a combination of uses, including grazing, logging, mining, recreation, and protection of watersheds and wildlife. Healthy forests also require protection from pathogens and insects. Sustainable forestry, which emphasizes biological diversity, provides the best management. Other management techniques include removing only infected trees and vegetation, cutting infected areas and removing debris, treating trees with antibiotics, developing disease-resistant species of trees, using insecti-

cides and fungicides, and developing integrated pest management plans.

Two basic systems are used to manage trees: even-aged and uneven-aged. Even-aged management involves maintaining trees in a given stand that are about the same age and size. Trees are harvested at the same time, then seeds are planted to provide for a new even-aged stand. This method, which tends toward the cultivation of a single species or monoculture of trees, emphasizes the mass production of fast-growing, low-quality wood (such as pine) to give a faster economic return on investment. Even-aged management requires close supervision and the application of both fertilizers and pesticides to protect the monoculture species from disease and insects.

Uneven-aged management maintains trees at many ages and sizes to permit a natural regeneration process. This method helps sustain biological diversity, provides for long-term production of high-quality timber, allows for an adequate economic return, and promotes a multiple-use approach to forest management. Uneven-aged management also relies on selective cutting of mature trees and reserves clear-cutting for small patches of tree species that respond favorably to such logging methods.

HARVESTING METHODS

The use of a particular tree-harvesting method depends on the tree species involved, the site, and whether even-aged or uneven-aged management is being applied. Selective cutting is used on intermediate-aged or mature trees in uneven-aged forests. Carefully selected trees are cut in a prescribed stand to provide for a continuous and attractive forest cover that preserves the forest ecology.

Shelterwood cutting involves removing all the mature trees in an area over a period of ten years. The first harvest removes dying, defective, or diseased trees. This allows more sunlight to reach the healthiest trees in the forest, which will then cast seeds and shelter new seedlings. When the seedlings have turned into young trees, a second cutting removes many of the mature trees, leaving enough mature trees to provide protection for the younger trees. When the young trees become well established, a third cutting harvests the remaining mature trees, leaving an even-aged stand of young trees from the best seed trees to mature. When done correctly, this method leaves a natural-looking forest and helps both to reduce soil erosion and to preserve wildlife habitat.

Seed-tree cutting harvests almost every tree at one site with the exception of a few high-quality, seed-producing, and wind-resistant trees, which function as seed sources to generate new crops. This method allows a variety of species to grow at one time and aids in erosion control and wildlife conservation.

Clear-cutting removes all the trees in a single cutting. The clear-cut may involve a strip, an entire stand, or patches of trees. The area is then replanted with seeds to grow even-aged or tree-farm varieties. More than two-thirds of the timber produced in the United States, and almost one-third of the timber in national forests, is harvested by clear-cutting. A clear-cut reduces biological diversity by destroying habitat, can make trees in bordering areas more vulnerable to winds, and creates an area that may take decades to regenerate.

FOREST FIRES

Forest fires can be divided into three types: surface, crown, and ground fires. Surface fires tend to burn only the undergrowth and leaf litter on the forest floor. Most mature trees easily survive these kinds of fires, as does wildlife. These fires occur every five years or so in forests with an abundance of ground litter and help prevent more destructive crown and ground fires. Such fires can even release and recycle valuable mineral nutrients, stimulate certain tree seeds, and help eliminate insects and pathogens.

Crown fires are very hot fires that burn both ground cover and tree tops. They normally occur in forests that have not experienced fires for several decades. Strong winds allow these fires to spread from deadwood and ground litter to treetops. They are capable of killing all vegetation and wildlife, leaving the land prone to erosion. Ground fires are most common in northern bogs. They can begin as surface fires but burn peat or partially decayed leaves below the ground surface. They can smolder for days or weeks before anyone notices them, and they are difficult to douse.

Natural forest fires can be beneficial to some species of trees, such as the giant sequoia and the jack pine, which release seeds for germination only after being exposed to intense heat. Grassland and pine forest ecosystems that depend on fires to regenerate are called fire climax ecosystems. They are managed for optimum productivity with prescribed fires.

The Society of American Foresters has begun advocating a concept called new forestry, in which ecological health and biodiversity, rather than timber production, are the main objectives of forestry. Advocates of new forestry propose that any given site should be logged only every 350 years, wide buffer zones should be left beside streams to reduce erosion and protect habitat, and logs and snags should be left in forests to help replenish soil fertility. Proponents of such forestry also endorse the involvement of private landowners in the cooperative management of lands.

Toby Stewart and Dion Stewart

FURTHER READING

Bettinger, Pete, et al. *Forest Management and Planning.* Burlington, Mass.: Academic Press, 2009.

Colfer, Carol J. Pierce. *The Equitable Forest: Diversity, Community, and Resource Management.* Washington, D.C.: Resources for the Future, 2005.

Crow, Thomas R. "Landscape Ecology and Forest Management." In *Issues and Perspectives in Landscape Ecology,* edited by John Wiens and Michael Moss. New York: Cambridge University Press, 2005.

Davis, Lawrence S., et al. *Forest Management: To Sustain Ecological, Economic, and Social Values.* 4th ed. Boston: McGraw-Hill, 2001.

McNeely, Jeffrey A., et al. *Conserving the World's Biological Diversity.* Washington, D.C.: Island Press, 1990.

Robbins, William G. *American Forestry*. Lincoln: University of Nebraska Press, 1985.

SEE ALSO: Biodiversity; Conservation easements; Deforestation; Forest and range policy; Logging and clear-cutting; National forests; Rain forests; Sustainable forestry; Watershed management.

Forest Service, U.S.

CATEGORIES: Organizations and agencies; forests and plants; resources and resource management

IDENTIFICATION: Federal agency responsible for managing national forests

DATE: Established in 1905

SIGNIFICANCE: The U.S. Forest Service was established primarily to manage the nation's timber resources, but its scope broadened as the nation's environmental awareness expanded. The agency's activities in the twenty-first century range from local flood control to reforestation efforts.

The mission of the U.S. Forest Service is to conserve national forests for multiple uses. The agency has the difficult task of balancing the needs of the nation's commercial forest products industries with the desire of the general public to pursue a variety of recreational activities in forests. In all cases the Forest Service attempts to follow what professional foresters and research scientists believe are the most environmentally sustainable management policies. Housed within the U.S. Department of Agriculture, the Forest Service directly manages 78 million hectares (193 million acres) of national forests and grasslands, as well as sharing oversight with state agencies for 202 million hectares (500 million acres) of nonfederal rural and urban forests.

Under the administration of President Theodore Roosevelt, Congress created the U.S. Forest Service in 1905 in response to a perceived impending timber famine. Nineteenth century lumbermen left a swath of barren, eroding land in their wake as they clear-cut their way across the North American continent from New England to the Pacific Northwest. Forest resources that had appeared inexhaustible at the start of the nineteenth century seemed about to vanish forever at the opening of the twentieth century. Navigable streams became choked with silt and debris when rain poured across deforested hillsides, and flash floods devastated riverside towns.

In a comparatively short time, the Forest Service successfully reversed the loss of forest habitat within the United States as it met the goals of its original mandate to ensure a sustainable forest reserve for future generations. As public awareness of other environmental issues—such as the importance of maintaining watersheds and preserving old-growth forests for the continued viability of certain wildlife species—increased during the 1970's and 1980's, the Forest Service found itself facing growing criticism for policies that apparently favored logging interests over ecosystem preservation. For example, environmental activists argued that because the commercial forest industry had access to millions of hectares of privately owned plantation forests, logging should cease in the national forests. Frustrations among both timber and environmental interests mounted during the early twenty-first century, when the presidential administration of George W. Bush made forestry policy revisions that eased logging restrictions, only to be challenged and defeated in court. Criticisms of the Forest Service persist, often overshadowing significant gains the agency has made in maintaining forest health and promoting global environmental stability.

In addition to managing national forests and undertaking research projects, the Forest Service provides advice and training to private landowners and state and local governments in areas such as reforestation, community forestry, and agroforestry. Through its research stations, the Forest Service develops environmentally sound harvesting methods, known as best management practices (BMPs), for loggers to follow when removing timber from forests. BMPs include directions for building haul roads in ways that discourage erosion, maintaining buffer zones between logged areas and nearby rivers and streams, and adhering to an optimum size and shape when clear-cutting to minimize loss of wildlife habitat. The Forest Service's technical experts also work with the U.S. Agency for International Development on forest management projects in other countries and have been active in trying to prevent the loss of forest habitat in environmentally sensitive regions around the world.

Increases in the number, extent, and intensity of wildfires, a trend that began during the late twentieth century, have forced the Forest Service to expend more and more of its resources on suppressing and preventing fires. Longer and warmer growing seasons, combined with more home construction near natural areas, have made fire management an increasingly complex, risky, and expensive undertaking. By the early years of the twenty-first century more than 40 percent of the Forest Service's annual budget was going toward fire control, leaving the agency with fewer resources to devote to other programs such as research, recreation, forest restoration, and control of pests, diseases, and invasive species.

Nancy Farm Männikkö
Updated by Karen N. Kähler

FURTHER READING

Andrews, Richard N. L. *Managing the Environment, Managing Ourselves: A History of American Environmental Policy.* 2d ed. New Haven, Conn.: Yale University Press, 2006.

Hays, Samuel P. *The American People and the National Forests: The First Century of the U.S. Forest Service.* Pittsburgh: University of Pittsburgh Press, 2009.

Hirt, Paul W. *A Conspiracy of Optimism: Management of the National Forests Since World War Two.* Lincoln: University of Nebraska Press, 1994.

U.S. Forest Service. *The U.S. Forest Service: An Overview.* Washington, D.C.: Author, 2009.

SEE ALSO: Conservation policy; Forest and range policy; Forest management; National Forest Management Act; National forests; Old-growth forests; Pinchot, Gifford; Sustainable forestry.

Forest Summit

CATEGORIES: Forests and plants; resources and resource management

THE EVENT: Meeting of government officials, representatives of the timber industry, and environmental groups to resolve issues related to the management of federally owned forests in the Pacific Northwest

DATE: April 3, 1993

SIGNIFICANCE: The plan of forest management that was adopted as an outcome of the Forest Summit failed to please either environmentalists or loggers.

In June, 1990, the northern spotted owl was declared an endangered species. In May, 1991, U.S. District Judge William Dwyer ruled that the presiden-

tial administration of George H. W. Bush was deliberately violating the Endangered Species Act by failing to develop a plan that would adequately protect the owl from extinction. The judge placed an injunction against logging on approximately 1.2 million hectares (3 million acres) of federally owned old-growth forestland in Northern California, Oregon, and Washington until an acceptable plan could be developed. The Forest Summit, held in Portland, Oregon, and officially known as the Portland Timber Summit, brought together disputing parties in the conflict between loggers and supporters of protection of the habitat of the northern spotted owl. The aim was to allow participants to present their perspectives and to encourage them to identify mutually acceptable new directions in forest management.

President Bill Clinton held the daylong summit with Vice President Al Gore, cabinet members, Environmental Protection Agency (EPA) administrators, and other federal officials. Prior to the summit, environmental activists and representatives of the timber industry and logger groups made a number of efforts to "sell" their perspectives on the logging situation to administration officials. On the day before the conference, government officials received helicopter, airplane, and four-wheel-drive vehicle tours of the forests. Employees of timber mills were given the day of the summit off, and many were bused to Portland to picket the conference hall with signs proclaiming the need to save jobs and protect families.

VIEWS PRESENTED

The conference was expressly designed as a forum for participants to present their perspectives on the conflict. Loggers tended to focus on the negative effects of the logging ban on logging communities. In these presentations, loggers were often depicted as environmentalists with a vested interest in a healthy forest ecology and as modern-day Paul Bunyans unable to pass their culture on to the next generation. Logging company officials often focused on the economic difficulties they faced and on the country's need for housing lumber and other wood products. Representatives of small logging companies described how they had successfully adapted to changes in timber supplies and markets. Native Americans described the traditional and continuing significance of the forest in their lives.

Environmentalists suggested that the economic problems of logging communities were not caused by

President Bill Clinton, right, and Vice President Al Gore confer during a session of the Forest Summit. (AP/Wide World Photos)

the restriction of cutting on federal land but rather were a consequence of general economic conditions, such as the low number of housing starts and a general decline in the logging industry. Environmentalists claimed that the loggers' desire for federal old-growth timber was a result of mismanagement of private lands. They also asserted that logging-related jobs in the United States could be saved if log exports to foreign countries were reduced.

Option 9

Following the summit, the task of developing a management plan was turned over to thirty-seven physical scientists and economists. The teams developed ten options, ranging from option 1, the "save it all" option, which would allow 190 million board feet of lumber to be cut and would save all the federally owned old-growth forests, to option 10, which would permit the cutting of 1.84 billion board feet. In June, 1994, a committee of senior federal officials began considering the options and prepared decision memoranda for the president. Clinton selected option 9, which set annual harvests at 1.2 billion board feet, a level about one-fourth of 1980's harvests but more

than twice the level permitted since the northern spotted owl had been declared an endangered species. This option was known as the "efficiency option" because it focused on watersheds as the basic building blocks of the ecosystem and included measures designed to protect dwindling salmon stocks.

The management plan also called for the provision of $1.2 billion over five years to offset economic losses from logging. The plan eliminated 6,000 jobs in 1994 but created more than 8,000, retaining an additional 5,400 jobs. The option also established ten adaptive management areas in which local community and government groups would work together to allow logging and protect wildlife. Salvage logging, the cutting of fire- and insect-damaged trees, and thinning would be permitted in some sections of old-growth forest if an interagency team determined that these practices would not be detrimental to the northern spotted owl's habitat.

Reactions to Option 9

Clinton introduced his timber management plan with the prophetic words, "Not everyone is going to like this plan. . . . Maybe no one will." Although many observers judged the plan to be a fair solution, none of the parties directly involved publicly expressed satisfaction with it. Loggers did not believe the promise of economic aid, especially at a time when the federal government was attempting to balance the budget through spending cuts. Loggers further concluded that the plan was unfair because it contained few provisions for meeting their main concern of preserving logging jobs and the logging-based economies of their small towns. Retraining funds were offered for the purpose of equipping loggers with the skills to be employed in other jobs. Timber industry leaders believed that the plan was based on faulty assumptions about forest productivity and that it placed undue restrictions on logging.

Environmentalists believed that the plan was unfair because it provided for more cutting than had been permitted under the court injunction and it had loopholes that would be used to permit even more logging. They believed that timber companies would use the provision to allow salvage logging to cut green trees, as definitions of salvage logging had been stretched in the past. Whistle-blowers within the U.S. Forest Service had leaked a memorandum that explicitly directed employees to allow green cutting to happen. Workers in the Agriculture Department Inspector General's Office had also found documents indicating that Forest Service officials may have in the past made questionable agreements with logging company officials prior to timber sales. Environmentalists further believed that expected cuts in the number of Forest Service staff would make policing of the plan more difficult.

Representatives of all sides in the dispute stated that they were considering bringing lawsuits against the plan. One basis for a lawsuit was the lack of public participation in and review of the plan's development. Although the Portland Timber Summit was a putative effort to bring the sides together to discover mutually agreeable solutions, once the summit had concluded, the alternative management options were developed by experts working in seclusion. The summit gave participants voice by allowing them to articulate their ideas, but the participants did not have the power to determine the specifics of the selected management plan.

On June 6, 1994, Judge Dwyer ruled that the option 9 management plan satisfactorily addressed concerns about protection of the northern spotted owl that had prompted the original injunction. Despite the continued threat of lawsuits, U.S. Forest Service officials proceeded with plans to sell timber in the disputed areas.

George Cvetkovich and Timothy C. Earle

FURTHER READING

Dietrich, William. *The Final Forest: The Battle for the Last Great Trees of the Pacific Northwest.* New York: Simon & Schuster, 1992.

Ervin, Keith. *Fragile Majesty: The Battle for North America's Last Great Forest.* Seattle: Mountaineers Books, 1989.

Spies, Thomas A., and Sally L. Duncan, eds. *Old Growth in a New World: A Pacific Northwest Icon Reexamined.* Washington, D.C.: Island Press, 2009.

Stout, Benjamin B. *The Northern Spotted Owl: An Oregon View, 1975-2002.* Victoria, B.C.: Trafford, 2006.

U.S. Forest Service. Forest Ecosystem Management Assessment Team. *Forest Ecosystem Management: An Ecological, Economic, and Social Assessment.* Washington, D.C.: Author, 1993.

SEE ALSO: Endangered Species Act; Endangered species and species protection policy; Forest management; Logging and clear-cutting; National forests; Northern spotted owl; Old-growth forests.

Fossey, Dian

CATEGORIES: Activism and advocacy; animals and endangered species
IDENTIFICATION: American zoologist and author
BORN: January 16, 1932; San Francisco, California
DIED: December 26, 1985; Virunga Mountains, Rwanda
SIGNIFICANCE: Fossey influenced views of animal behavior and the need for animal protection through her writings about the mountain gorillas of Central Africa and her passionate attempts to save the gorillas from poachers.

Dian Fossey was born in 1932, the daughter of George Fossey III, an insurance agent, and Kitty Fossey, a homemaker. When Fossey was six years old her parents divorced, and she grew up with her mother and her stepfather, Richard Price, a building contractor. After high school, Fossey enrolled in a veterinary medicine program at the University of California, Davis, while supporting herself with low-paying jobs. Her academic difficulties with chemistry and physics courses led her to transfer to San Jose State College, where she earned her bachelor's degree in occupational therapy in 1954. After her postcollege clinical training, she became the director of the occupational therapy department at the Kosair Crippled Children's Hospital in Louisville, Kentucky.

Fossey's love for Africa was inspired by a book on mountain gorillas written by American zoologist George Schaller. In 1963 Fossey took a bank loan to finance a seven-week safari trip to Africa. At Olduvai Gorge in Tanzania, she met with anthropologists Mary Leakey and Louis S. B. Leakey, who were involved with a search for hominid fossils. Fossey's first encounter with a mountain gorilla had a tremendous impact on her. After the end of her African trip, she returned to Kentucky and resumed her work with disabled children. Three years later Leakey arrived in Louisville and convinced Fossey that she should study the gorillas in the wild as part of a long-term expedition. She agreed and, after paying a short visit to Jane Goodall—the British ethologist was studying chimpanzees in Tanzania—during which she learned methods of data collection, Fossey set up her first campsite and work station at Kabara, Zaire (now the Democratic Republic of the Congo).

Fossey managed to approach and study the gorillas in the remote mountain areas for about seven months, until political unrest took over Zaire. On July 10, 1967, she was arrested by armed guards; she was kept in custody for two weeks, during which time she was raped repeatedly. She eventually managed to escape and found refuge in Uganda. In 1970 she enrolled at Cambridge University in England, where she earned her doctorate in zoology six years later. Immediately thereafter Fossey traveled to Rwanda, where she stayed until 1980, when she accepted a visiting as-

---

Fossey's First Gorillas

In her memoir Gorillas in the Mist *(1983), Dian Fossey vividly describes a pivotal moment in her life:*

I shall never forget my first encounter with gorillas. Sound preceded sight. Odor preceded sound in the form of an overwhelming musky-barnyard, humanlike scent. The air was suddenly rent by a high-pitched series of screams followed by the rhythmic rondo of sharp *pok-pok* chestbeats from a great silverback male obscured behind what seemed an impenetrable wall of vegetation. Joan and Alan Root, some ten yards ahead on the forest trail, motioned me to remain still. The three of us froze until the echoes of the screams and chestbeats faded. Only then did we slowly creep forward under the cover of dense shrubbery to about fifty feet from the group. Peeking through the vegetation, we could distinguish an equally curious phalanx of black, leather-countenanced, furry-headed primates peering back at us. Their bright eyes darted nervously from under heavy brows as though trying to identify us as familiar friends or possible foes. Immediately I was struck by the physical magnificence of the huge jet-black bodies blended against the green palette wash of the thick forest foliage.

Most of the females had fled with their infants to the rear of the group, leaving the silverback leader and some younger males in the foreground, standing tense with compressed lips. Occasionally the dominant male would rise to chestbeat in an attempt to intimidate us. The sound reverberated throughout the forest and evoked similar displays, though of lesser magnitude, from gorillas clustered around him. Slowly, Alan set up his movie camera and proceeded to film. The openness of his motions and the sound of the camera piqued curiosity from other group members, who then treed to see us more clearly. As if competing for attention, some animals went through a series of actions that included yawning, symbolic-feeding, branch-breaking, or chestbeating. After each display, the gorillas would look at us quizzically as if trying to determine the effect of their show. It was their individuality combined with the shyness of their behavior that remained the most captivating impression of this first encounter with the greatest of the great apes.

---

sociate professorship at Cornell University. She continued to act as the project coordinator at the Karisoke Research Center, which she had founded in Rwanda in September, 1967.

While at Cornell, Fossey became aware of increased poaching of gorillas in Rwanda and the rapid deterioration of the research center. She returned to Karisoke in June, 1983, with the aim of improving the situation. She actively took over management of the center but failed to keep the support of the National Geographic Society. Her incessant, passionate anti-poaching activities created many enemies. On December 26, 1985, Fossey was found dead from machete wounds in her camp in the Virunga Mountains.

She was buried in the gorilla cemetery she had built near the camp. Wayne Richard McGuire, an American wildlife researcher, was prosecuted as the primary suspect in her death but fled Rwanda.

Fossey's life was driven by the fighting of poachers and continuation of mountain gorilla conservation efforts that covered a period of almost twenty years in the mountains of Zaire, Uganda, and Rwanda. Poachers, slaughters, and revolutions, as well as loneliness, were part of her daily routine. At the Karisoke Research Center she studied more than fifty gorillas that she described as rather peaceful, charging at humans only when threatened. Fossey was acknowledged as the world's leading authority on the mountain gorillas after the 1983 publication of her book *Gorillas in the Mist*, which dramatically enlarged the contemporary knowledge of gorilla habits, communication, and social structure. Fossey believed that gorillas are altruistic and regal animals whose family structure is unbelievably strong.

Soraya Ghayourmanesh

FURTHER READING

De la Bédoyère, Camilla. *No One Loved Gorillas More: Dian Fossey—Letters from the Mist*. Washington, D.C.: National Geographic Society, 2005.

Fossey, Dian. *Gorillas in the Mist*. 1983. Reprint. New York: Mariner Books, 2000.

SEE ALSO: Animal rights; Mountain gorillas; Poaching.

Fossil fuels

CATEGORIES: Energy and energy use; resources and resource management

DEFINITION: Fuels formed over long spans of time from buried dead organisms

SIGNIFICANCE: Fossil fuels are the most widely used sources of energy production throughout the world and are essential to human activity in modern society. The use of such fuels, however, is the cause of significant environmental degradation through air and water pollution and habitat destruction. The carbon dioxide emitted by the burning of fossil fuels has been linked to global warming.

Fossil fuels—coal, oil, and natural gas—are found in the earth's crust. They are the result of the decomposition of the remains of dead plants and animals under heat and pressure as they were covered with sediment, becoming part of the earth's crust, as either landmass or seabed. These fuels are nonrenewable resources; they have required millions of years to form.

Fossil fuels are composed of high percentages of carbon and hydrocarbons, and the ratio of carbon to hydrogen varies considerably from one type of fossil fuel to another. A gas such as methane has a low ratio and burns quickly, whereas a substance such as anthracite coal, composed almost entirely of carbon, has a lower ratio and burns more slowly. When burned, all fossil fuels produce large amounts of energy, and this characteristic led them to play a significant role in the industrialization and modernization of the world. According to the Energy Information Association, in 2007 fossil fuels accounted for the production of 86.4 percent of the energy consumed worldwide. Although fossil fuels are capable of meeting the energy production needs of the world, they are a resource that is diminishing, and their extraction and use both cause considerable environmental problems.

The major concern in regard to the use of fossil fuels is their emission of carbon dioxide (CO_2), one of the greenhouse gases that has been linked to global warming. According to the U.S. Department of Energy, the burning of fossil fuels produces almost twice as much CO_2 as natural processes can absorb each year.

COAL AND EFFECTS ON THE ENVIRONMENT

The three types of coal—anthracite, bituminous, and lignite—are retrieved either by deep-shaft underground mining or by opencast (surface) mining. Both types of mining cause considerable damage to the area mined, as they destroy land and pollute air and rivers. The pollution from lignite mining is particularly harmful to forests.

When coal is burned it emits a number of harmful substances, including sulfur dioxide, nitrogen oxide, mercury, particulates, and carbon dioxide. Sulfur dioxide, nitrogen oxide, and particulates contribute to the formation of acid rain, which can cause respiratory illness. Mercury that enters rivers, streams, or lakes combines into the chemical methylmercury. It is highly toxic to water plants, to fish, and to animals and people who consume the fish. It is, however, the CO_2 produced by the burning of coal that is of the greatest environmental concern. CO_2 is the major pollutant that causes global warming, and coal-fired power plants are responsible for the greatest amount of CO_2 released into the air. The transport of coal further contributes to the release of pollutants into the air. Although some coal is transported as slurry through pipelines, the majority of coal is transported by train using diesel-fueled locomotives, which in turn emit more CO_2 and other pollutants.

ENVIRONMENTAL IMPACTS OF OIL

Crude oil or petroleum is composed of hydrogen and carbon compounds. It is a liquid form of fossilized biomass derived from the decomposition of dead plants and animals found in underground reservoirs in sedimentary basins on land areas and in seabeds. Oil is extracted from the earth by pumping, using wells and oil rigs, and it is transported long distances through pipelines or in ships. The major danger to

the environment from these activities is the occurrence of oil spills, which have serious impacts on wildlife, especially marine life, seabirds, and sea mammals.

Gasoline, diesel fuel, jet fuel, and liquid propane gas are all derived from oil. It is also the feedstock or raw material from which plastics, polyurethane, and many other products are made. Oil plays a vital role in the everyday life of the modern world. Gasoline, diesel, and jet fuel are the most commonly used sources of energy in transportation. Gasoline is the primary fuel used in private cars; diesel fuel is used in freight trucks, in train engines, and in such heavy equipment as construction and farm vehicles. Diesel is also the fuel of choice for other kinds of machinery used in agriculture and construction. The generators used to

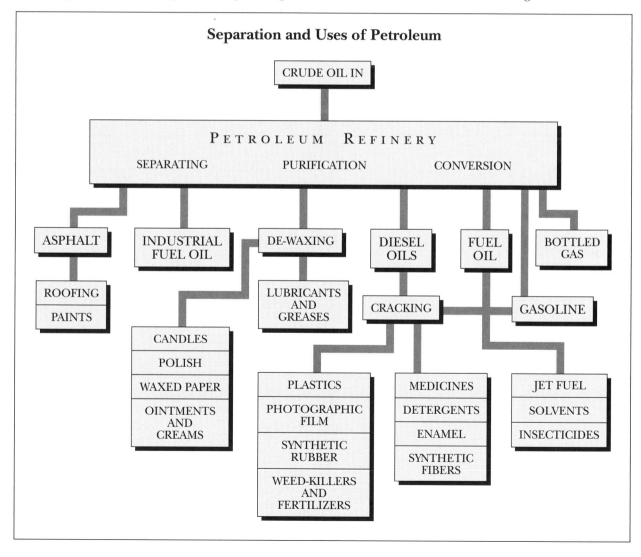

provide electricity to such facilities as hospitals and nursing homes in times of emergency are usually powered by diesel fuel. Propane, which is a cleaner-burning fuel than either gasoline or diesel, is used in indoor equipment.

Oil's greatest impacts on the environment come from its use as fuel. In 2008, consumption of oil worldwide amounted to 85.4 million barrels per day. When oil is burned in any of its forms, it emits a number of harmful substances into the atmosphere. These include CO_2, carbon monoxide (CO), sulfur, and lead. Lead in particulate matter is classified as a carcinogen. Sulfur dioxide contributes to the formation of acid rain, which is harmful to animals, plants, and human beings. The nitrogen oxide and volatile organic compounds found in emissions from the burning of oil are among the causes of ground-level ozone. Many of these pollutants contribute to lung irritation, asthma, bronchitis, and lung disease.

World Proved Fossil Fuel Reserves, 2003			
REGION	OIL	COAL	NATURAL GAS
Far East	8,041	140,362	16,317
Middle East and North Africa	105,662	1,322	71,385
Europe	11,845	135,783	48,433
Sub-Saharan Africa	6,500	33,348	4,497
North America	6,493	125,311	6,209
Central America and West Indies	2,550	690	1,037
South America	14,120	8,601	5,608
Oceania	611	41,748	2,679
Total	156,700	501,172	158,198

Source: International Energy Agency.
Note: In millions of metric tons of oil equivalent.

ENVIRONMENTAL IMPACTS OF NATURAL GAS

Natural gas is composed of hydrogen and carbon; it is primarily methane. Like coal and oil, natural gas is formed over millions of years from the decomposing remains of plants and animals covered by sand and silt and subjected to heat and pressure. Deposits of natural gas are found in both landmasses and seabeds. Extraction of the gas, achieved through drilling and the establishment of wells, has impacts on the wildlife in the area of the drilling through the disruption of habitat. Natural gas is transported by pipelines to refineries. At both drilling sites and along the pipelines, leaks can occur that can result in serious explosions.

Natural gas has many uses. It is used to generate electricity, as a fuel in industry, and for heating homes and powering home appliances. Natural gas also serves as a raw material for producing a wide variety of products ranging from fertilizers to medicines. When burned, natural gas emits fewer pollutants, especially sulfur and nitrogen, than either oil or coal. It does, however, produce CO_2.

EFFORTS TO REDUCE NEGATIVE IMPACTS

Fossil fuels continue to play an important role in all major areas of the world's economy. The generation of electricity, transportation, construction, and food production are all highly dependent on the use of fossil fuels. Governments around the world and the fossil-fuel industries are involved in ongoing efforts to combat the adverse effects of fossil fuels on the environment. Both by passing legislation within their own borders, such as the Clean Air Act in the United States, and by entering into international agreements such as the Kyoto Protocol, governments have set limits on the amounts of pollutants that may be emitted from fossil fuels and have set targets for reducing greenhouse gas emissions. Through research and new technology, as well as habitat reclamation programs, the fossil-fuel industries have worked to reduce the negative impacts of fossil fuels on the environment.

Because coal-fired power plants are the greatest emitters of CO_2, the primary greenhouse gas, citizens' groups in several countries, especially the United Kingdom, have argued for the elimination of coal mining and of the use of coal as a fuel. Methane has replaced coal as the fuel of choice in a number of power plants. The coal industry has responded with strong efforts to develop technologies aimed at reducing CO_2 emissions from the burning of coal and improving mining techniques to reduce adverse effects on land and communities in the areas where coal is mined. The reclamation of land at strip mines is one of these efforts.

In the United Kingdom, the technique of extracting coal through underground coal gasification has been investigated. The procedure involves injecting a mixture of steam and oxygen down a borehole through which gas is extracted from the coal and brought to the surface. This technique, however, may cause contamination of underground water supplies in onshore locations and the collapse of the burned-out coal seams both on- and offshore.

Coal-fired power plants using a technology known as integrated gasification combined cycle (IGCC) have also been introduced; IGCC plants convert coal into a synthetic gas before it is used to generate electricity. These plants have the capability of reducing pollutant emissions significantly, including the capture and separation of up to 95 percent of CO_2, which is then stored underground.

Other technologies and procedures implemented by the oil and gas industries to reduce the negative effects of their products on the environment include horizontal drilling techniques that can increase the area from which oil or gas can be extracted from one well. Reducing the number of wells drilled reduces the impacts on habitat and wildlife. In addition, the oil and gas industries use double-hulled tankers and double-lined pipelines to help reduce oil spills. In response to governmental mandates, oil companies have funded research that has resulted in the reformulation of gasoline and diesel fuels to reduce the emissions they produce.

Shawncey Webb

FURTHER READING

Archer, David. *Global Warming: Understanding the Forecast.* Malden, Mass.: Wiley-Blackwell, 2006.

Higman, Christopher, and Maarten van der Burgt. *Gasification.* 2d ed. Boston: Elsevier, 2008.

Kelley, Ingrid. *Energy in America: A Tour of Our Fossil Fuel Culture.* Lebanon, N.H.: University Press of New England, 2008.

Martin, Raymond S., and William L. Leffler. *Oil and Gas Production in Nontechnical Language.* Tulsa, Okla.: Pennwell, 2005.

Shogren, Jason F. *The Benefits and Costs of the Kyoto Protocol.* Washington, D.C.: AEI Press, 1999.

Williams, A., et al. *Combustion and Gasification of Coal.* New York: Taylor & Francis, 1998.

SEE ALSO: Automobile emissions; Carbon dioxide; Carbon dioxide air capture; Coal; Coal-fired power plants; Global warming; Greenhouse gases; Habitat destruction; Oil drilling; Oil spills.

Franklin Dam opposition

CATEGORIES: Activism and advocacy; preservation and wilderness issues

IDENTIFICATION: Resistance to the building of a hydroelectric dam proposed for the Franklin and Gordon river system in southwestern Tasmania, Australia

DATES: 1978-1983

SIGNIFICANCE: In stopping the construction of a hydroelectric dam that promised an industrial boon but threatened to destroy thousands of square miles of Tasmania's temperate rain forest, a grassroots coalition of environmental activists achieved through the Australian court system the first significant legal victory in international efforts to protect natural reserves against development.

In October, 1978, Tasmania's powerful energy conglomerate, the Hydro-Electric Commission, announced ambitious plans to build four dams along the Franklin and Gordon rivers to boost Tasmania's entire energy output by more than 20 percent, a bold move that supporters claimed would guarantee thousands of jobs and catapult the tiny Australian island state of Tasmania into a major industrial economy. The reaction from local environmentalists, however, was immediate. Just five years earlier, the same coalition, known as the Tasmanian Wilderness Society, had failed to stop the construction of a similar dam that had, in turn, destroyed Lake Pedder. The dam proposed for the Gordon River would flood the Franklin River, a wild and barely charted river, as well as thousands of hectares of ancient unspoiled wilderness.

By mid-1980, the antidam coalition was sufficiently organized to stage a massive rally that paralyzed Hobart, Tasmania's capital. The sitting Labor government, which generally supported the project, attempted a compromise, proposing that the dam be moved downriver. The effect, the environmentalists argued, would be the same: the obliteration of the Franklin River basin. Their cause was made more urgent when, in early 1981, caves were discovered in the basin that dated back eight thousand years and contained artifacts and cave paintings made by ancient Aborigines.

A national referendum in late 1981 did little to quash the controversy—voters were asked only to choose between sites rather than to decide on the dam idea itself. When Tasmanians subsequently elected a pro-dam Liberal government, it appeared the dam would be constructed. Conservationists, however, took the fight to the Australian mainland, where they staged a savvy media campaign, citing the Franklin River wilderness area's recent designation as a World Heritage Site by the United Nations Educational, Scientific, and Cultural Organization (UNESCO); World Heritage Sites are cultural and natural sites deemed invaluable to the global community. With bulldozers moving into position, the conservationists organized a blockade of the planned site of the dam that ran from November, 1982, to June, 1983. The blockade created an international outcry as more than one thousand protesters were arrested (to flood Tasmania's limited prison system, the arrested protesters refused bail).

The turning point came when the charismatic Bob Hawke was elected Australia's prime minister in March, 1983. The new Labor government, citing the World Heritage Site designation as an overriding mandate from the international community, ordered work on the dam to stop. This action raised thorny constitutional questions about the reach of the federal government, and the Tasmanian government challenged the federal government's authority in Australia's High Court.

On July 1, 1983, the High Court handed down a narrow (four-to-three) victory to the commonwealth government (and by extension to the environmentalists), finding that the federal government had the power to act to enforce any international treaty. The decision ended efforts to dam the Franklin and Gordon river system, and the campaign to stop the dam came to be regarded as the first successful campaign conducted by what over the next decade would become the Green movement.

Joseph Dewey

FURTHER READING

Buckman, Gary. *Tasmania's Wilderness Battles: A History.* Sydney: Allen & Unwin, 2008.
Harding, Dennis, and Michelle Dale. *Wilderness in Tasmania: The Untouched Land.* Talbot, Tasmania: Tasmanian Book, 2000.
Lines, William J. *Patriots: Defending Australia's Natural Heritage.* St. Lucia: University of Queensland Press, 2006.
McCully, Patrick. *Silenced Rivers: The Ecology and Politics of Large Dams.* Enlarged ed. London: Zed Books, 2001.

SEE ALSO: Australia and New Zealand; Dams and reservoirs; Ecosystems; Environmental economics; Environmental law, international; Green movement and Green parties; Hydroelectricity; Rain forests; United Nations Educational, Scientific, and Cultural Organization; World Heritage Convention.

Free market environmentalism

CATEGORY: Philosophy and ethics
DEFINITION: Approach to environmental protection that relies on property rights and market forces rather than on government regulation
SIGNIFICANCE: Free market environmentalism presents an incentive-based approach to environmental protection that can lower the costs of pollution control and habitat protection. By providing a new way of looking at environmental issues, free market environmentalism has helped regulatory agencies develop alternative strategies beyond command and control.

The Industrial Revolution brought a new level of material wealth to many people, but it also brought a new set of problems. One problem was the pollution created as a by-product of manufacturing. By the end of the nineteenth century, citizens began to look to government for protection from pollution. The first U.S. regulation aimed at water pollution, the Rivers and Harbors Act, was enacted in 1899.

While Americans were becoming increasingly aware of the problems caused by pollution, the dominant force in the first half of the twentieth century was economic development. In 1962 Rachel Carson's book *Silent Spring* brought the dangers of pollution to public attention, and in 1968 ecologist Garrett Hardin introduced the concept of the "tragedy of the commons" to describe how individuals pursuing their own self-interests tend to overuse open-access resources, such as common grazing areas or ocean fisheries. Hardin's idea and Carson's book helped spark a surge in environmental regulation in the United States. The Environmental Protection Agency (EPA) was created in 1970, followed by significant environmental legislation including the 1970 amendments to the Clean Air Act of 1963, the Clean Water Act (1972), and the Endangered Species Act (1973).

Armed with new legislation and administrative resources, the federal government began to address pollution problems using a general approach known as command and control. Standards were set for acceptable levels of various pollutants, and government told industries what antipollution technologies and methods they must use. While these efforts were supported by many citizens, critics argued that such regulation by government is inefficient and too costly and that it places unacceptable constraints on individual liberty. Early pollution regulations tended to include strict control standards that were costly for industries to meet, and as a result some factories were forced to shut down. During the 1980's, American opinion began to shift back toward a position favoring less government regulation, as evidenced by the landslide victory of Ronald Reagan in the 1980 presidential election. In addition, the limitations of any government attempts to protect natural resources were exposed as the environmental degradation in the Soviet Union and the Eastern Bloc nations came to light. Pollution levels were higher in the government-regulated areas of Poland, Hungary, and the Soviet Union than they were in the market-oriented United States and Western Europe.

FOUNDATIONAL IDEA

With government's ability to protect the environment in question, critics of regulation began to develop the notion that market processes might be more effective than regulation. The foundational idea in free market environmentalism is that defined property rights for natural resources will end overuse and environmental abuse. Individual property owners have a strong incentive to maintain the value of their property, so they will tend to protect their resources.

Free market environmentalism views humans through the classic economics perspective—that is, humans are guided by self-interest, not good intentions. If a forest is owned by an individual, family, or corporation, it is in the resource owner's self-interest to protect the forest. When government (the public) owns the forest, some individuals will use the forest in ways that benefit them in the short run, even though the actions harm the forest in the long run. With no direct ownership, some people are willing to trade individual short-run benefits for collective long-run harm. Further, there is no guarantee that public ownership will actually protect the forest. The government agency created to protect the resource may

make poor choices, may be captured by special interest groups, or may become overly bureaucratic.

THE COASE THEOREM

The intellectual roots of free market environmentalism can be traced back to a 1960 article written by Ronald H. Coase titled "The Problem of Social Cost." The basic idea, known as the Coase theorem, is that under certain conditions defined property rights and the existing legal system will produce efficient levels of environmental protection without government regulation. For example, suppose that the residents of a small community located on a pristine lake have defined property rights for clean lake water. If a firm pollutes the lake, it is responsible for compensating the community, just as a driver responsible for an automobile accident must compensate the innocent driver of the other car. Through negotiation, the community and firm would agree on an acceptable level of pollution and an acceptable compensation. If the firm produces more than the negotiated level of pollution, the community can sue for damages.

While the Coase theorem is valuable for presenting a new way of thinking about pollution, its actual application is limited. The theorem holds that efficient outcomes are possible when property rights are defined, when informed negotiation is possible, and when transaction costs are low. These conditions, present in the lake example, are never present in the real world. Suppose a property right for clean drinking water were given to the citizens of New York City. Could nine million people engage in informed negotiations with the thousands of firms discharging byproducts anywhere in the entire watershed? The transaction costs—proving what damage was created by which individual act of pollution—would be an impossible barrier. The Coase theorem is also limited in dealing with pollution that crosses political borders. How would the millions of citizens living in the Southwest region of the United States negotiate with thousands of Mexican factories creating air pollution that drifts across the border?

Even though the usefulness of the Coase theorem is limited, defined property rights do effectively protect natural resources. In Europe, public access is limited for areas supporting recreational activities such as hunting and fishing. Anyone who wants to catch salmon in Scotland must pay a significant daily fishing fee to a landowner. On high-quality streams, such a fee may reach $500 per day during the peak season.

People will pay the fee only if the likelihood of their actually catching a trophy salmon is high, so the landowner has a strong incentive to protect the river from habitat degradation or overuse. In the United States, the public has access to at least part of any river with salmon runs. The tragedy of the commons then unfolds, with too many anglers overfishing the resource.

SHIFTS IN THINKING

Environmental groups such as the Nature Conservancy and the National Audubon Society have recognized the problem of common access and have begun to buy or lease environmentally sensitive areas so that they can directly control how the land is used. With recognition of the importance of defined property rights to environmental protection, increasing numbers of books and articles favoring free market environmentalism have been published. Conservative think tanks, including the Political Economy Research Center, the American Enterprise Institute, and the CATO Institute, support conferences, workshops, and publications on the topic.

As new ideas emerged, the EPA rethought its pollution-control tactics and began to experiment with incentive-based strategies that relied on markets to achieve environmental improvements. For example, the Clean Air Act amendments of 1990 created a marketable permit program to reduce the sulfur dioxide emissions that were contributing to the acid-rain problem in the northeast United States. Coal-fired power plants in the Midwest and Northeast were given allocations of permits, with each permit allowing a plant to emit one ton of sulfur dioxide. Unlike under the old command-and-control approach, individual plants were given considerable latitude in deciding how best to reduce their pollution. Plants that reduced their emissions below their permit limits were allowed to sell their unused permits to other plants; thus the plants had a powerful economic incentive to reduce pollution as much as possible. The emissions permit trading approach was successful, with plants decreasing sulfur dioxide levels by 50 percent at an annual cost savings estimated at $1 billion relative to compliance under the former command-and-control strategy.

Although the sulfur dioxide trading program is viewed as a success, most mainstream environmentalists remain skeptical of free market environmentalism as the central approach to environmental protection. The chief criticism is that many environmental prob-lems involve circumstances that do not allow defined property rights. How would one create individual property rights for the deep ocean, the total atmosphere, or rivers that cross national boundaries? In the absence of defined property rights, the legal remedy is not available to those hurt by pollution or overuse. Critics also question whether the free market approach would adequately protect endangered species or ecosystems that have limited economic value. A clear incentive exists for a landowner to protect salmon fishing habitat, but what is the market incentive to protect the habitat of the American burying beetle or the Key Largo woodrat?

By providing a new way of looking at environmental issues, free market environmentalism helped regulatory agencies develop alternative strategies beyond command and control. Academics and policy makers are interested in developing more incentive-based strategies that use the power of markets to protect the environment in a cost-effective manner. Free market environmentalism will continue to shape public policy as governments around the world seek solutions to environmental problems.

Allan Jenkins

FURTHER READING

Anderson, Terry L., and Donald R. Leal. *Free Market Environmentalism.* Rev. ed. New York: Palgrave, 2001.

Bliese, John R. E. "'Free Market Environmentalism': Environmentalism for Conservatives?" In *The Greening of Conservative America.* Boulder, Colo.: Westview Press, 2002.

Coase, Ronald H. "The Problem of Social Cost." *Journal of Law and Economics* 3 (October, 1960): 1-44.

Dryzek, John S. *The Politics of the Earth: Environmental Discourses.* 2d ed. New York: Oxford University Press, 2005.

Vig, Norman J., and Michael E. Kraft, eds. *Environmental Policy: New Directions for the Twenty-first Century.* 7th ed. Washington, D.C.: CQ Press, 2010.

SEE ALSO: Carson, Rachel; Clean Air Act and amendments; Coal-fired power plants; Conservation easements; Environmental economics; Environmental torts; Hardin, Garrett; Nature Conservancy; Pollution permit trading; Privatization movements; Property rights movement; *Silent Spring*; Tragedy of the commons.

Freon

CATEGORY: Atmosphere and air pollution
DEFINITION: A nontoxic, nonflammable refrigerant gas
SIGNIFICANCE: Freon and other chlorofluorocarbons served a number of purposes in many industries, but it was found that they were harmful to the earth's ozone layer, and nations around the world united in banning their use.

Freon is the Du Pont Corporation's trade name for a compound used as a refrigerant. Freon, which was introduced in 1930, is an example of a class of gases known as chlorofluorocarbons (CFCs), carbon compounds that contain fluorine and chlorine. They are derivatives of simple alkanes—such as methane, ethane, propane, and butane—through direct or selective ultraviolet halogenation using chlorine or fluorine gas.

Freon found extensive uses in industry. CFCs have served as dispersing gases in aerosol cans, in the preparation of foamed plastics, and, primarily, as refrigerants. Their manufacture, together with that of the closely related halons, rose in the mid-1970's and peaked in 1986 with the production of almost 1.25 million tons. At that time these compounds were universally used in aerosol products ranging from insecticides to shaving foams and hair sprays, as well as in the insulation of buildings and as cleaning solutions for circuit boards and other electronic parts. One of them, bromotrifluoromethane (Freon 13B1), was used as a fire extinguisher in situations where the use of water had to be avoided, such as electrical fires.

Common members of this family of chemicals include trichlorofluoromethane (Freon or CFC 11), dichlorodifluoromethane (CFC 12), and 1,2-dichloro 1,1,2,2-tetrafluoroethane (CFC 114). They are all either gases or low-boiling liquids at room temperature and are virtually insoluble in water. They are generally dense, easily liquefied, not flammable, thermally stable, virtually odorless, and inexpensive to manufacture. They do not undergo decomposition via the ordinary chemical reactions that take place in the troposphere. As a result, they were seen as ideal for use as propellants in aerosol cans of deodorants, hair sprays, and various commercially available food products. Their relative inertness toward other chemicals allows them also to persist in the atmosphere, causing environmental problems.

Because they are water-insoluble, rain cannot dissolve them and wash them down to the ground. As a result, they drift upward into the stratosphere and the ozone layer, which they reach after approximately seven to ten days. They may stay in the stratosphere for several decades, absorb the sun's ultraviolet light, and yield free radicals, which appear to undergo chemical reactions that lead to the depletion of the ozone layer. Although ozone is toxic to human lungs, its presence in the stratosphere is critical in protecting the earth from the harmful ultraviolet part of the electromagnetic radiation associated with sunlight. If the ozone layer gets thin, exposure to ultraviolet radiation will exponentially increase the cases of skin cancer and other diseases while at the same time destroying crops and other plants.

CFCs have been found to escape into the atmosphere from old refrigerators and air-conditioning units. Most industrialized countries have banned their use and have replaced them with methylene chloride or nonhalogenated hydrocarbons, such as isobutane. The flammability of those hydrocarbons and the suspected carcinogenicity of methylene chloride have created an incentive for the development of CFC substitutes. In 1970 the U.S. Congress passed legislation aimed at curbing the sources of air pollution by setting standards for air quality. In 1987 more than twenty nations signed the Montreal Protocol, an agreement to downscale production of CFCs, with the intent of eventually eradicating their use. By 2009, all members of the United Nations had ratified the protocol, and it was expected that if all nations comply with their obligations under ongoing revisions of the protocol, the ozone layer will recover by 2050.

Soraya Ghayourmanesh

FURTHER READING

Dauvergne, Peter. "Refrigerating the Ozone Layer." In *The Shadows of Consumption: Consequences for the Global Environment*. Cambridge, Mass.: MIT Press, 2008.
Joesten, Melvin D., John L. Hogg, and Mary E. Castellion. "Chlorofluorocarbons and the Ozone Layer." In *The World of Chemistry: Essentials*. 4th ed. Belmont, Calif.: Thomson Brooks/Cole, 2007.
Parson, Edward A. *Protecting the Ozone Layer: Science and Strategy*. New York: Oxford University Press, 2003.

SEE ALSO: Aerosols; Air pollution; Chlorofluorocarbons; Clean Air Act and amendments; Ozone layer.

Friends of the Earth International

CATEGORIES: Organizations and agencies; activism and advocacy

IDENTIFICATION: International network of environmental organizations

DATE: Established in 1969

SIGNIFICANCE: Friends of the Earth International, the world's largest grassroots environmental network, focuses on the economic and development aspects of sustainability. With its membership weighted toward groups in the developing world, the organization brings global attention to issues such as economic justice, climate justice, and food sovereignty.

Friends of the Earth was founded in the United States in 1969 by David Brower after his departure from the Sierra Club because of that organization's reluctance to challenge the construction of nuclear power plants. Friends of the Earth became Friends of the Earth International (FOEI), an international network, in 1971 following a meeting of like-minded environmental activists from the United States, the United Kingdom, Sweden, and France. By 2010 it had grown into an international network with affiliates in seventy-seven countries and more than five thousand local activist groups.

Unlike many other global nongovernmental organizations dealing with environmental problems and issues, FOEI is structured from the bottom up as a confederation of national organizations or groups. The national organizations are themselves multitiered networks comprising grassroots activists as well as national pressure groups engaged in campaigns for various environmentally progressive and sustainable policies. Activists and groups at all levels are involved in educational activities and research projects.

National and local groups and organizations affiliated with FOEI are required to act independently of established political parties, religious organizations, and any other associations. They are expected to be open and democratic and to uphold nondiscriminatory policies in their practices and internal structures. They are encouraged to cooper-

ate and work with other like-minded organizations pursuing the same goals. National affiliates of FOEI typically work on issues affecting their own countries; they may also choose to participate in any of the international campaigns of FOEI that they deem relevant or important to them. Similarly, local, grassroots activists may choose to work on local, national, or international issues, as they see fit.

National affiliates of FOEI may choose to name themselves Friends of the Earth followed by country name (in parentheses), such as Friends of the Earth (U.S.) and Friends of the Earth (Canada), or they may alternatively use an equivalent translation in the na-

A demonstrator with Friends of the Earth in England takes part in a protest outside the Houses of Parliament against genetically modified crops. (AP/Wide World Photos)

tional language, such as Les Amis de la Terre (France) and Amigos de la Tierra (Spain and Argentina). About half of the member groups work under their own names, reflecting thereby an independent foundation and subsequent joining of the FOEI network, such as WALHI (FOE Indonesia), ERA (FOE Nigeria), and NSCN (FOE Norway). Some groups that use the name Friends of the Earth are not members of FOEI; for example, Friends of the Earth (Hong Kong) is not affiliated with FOEI.

FOEI is supported by a secretariat based in Amsterdam and an executive committee that is elected by all member groups at a general assembly held every two years. All overall policy and priority decisions are made at the general assembly.

FOEI views environmental issues in their social, political, and human rights contexts. Its campaigns and activities are typically situated beyond the traditional area of the conservation movement and seek to address the economic and development aspects of sustainability. Although FOEI originated in the United States and Europe, its membership has become more heavily weighted toward groups in the developing world. Some of the organization's priorities in the early twenty-first century include economic justice, climate justice and alternative energy, food sovereignty, and biodiversity.

All of FOEI's international campaigns incorporate elements of three core themes: protection of human and environmental rights, protection of biodiversity, and repayment of the ecological debt owed by rich, industrialized countries to those they have exploited. FOEI is particularly engaged in work on the issues of desertification, protection of Antarctica, water quality and distribution, maritime mining, and nuclear power.

Nader N. Chokr

FURTHER READING

Carter, Neil. *The Politics of the Environment: Ideas, Activism, Policy.* 2d ed. New York: Cambridge University Press, 2007.

Friends of the Earth. *Climate Change: Voices from Communities Affected by Climate Change.* Washington, D.C.: Author, 2007.

_____. *Climate Debt: Making Historical Responsibility Part of the Solution.* Washington, D.C.: Author, 2005.

Porritt, Jonathon, ed. *Friends of the Earth Handbook.* New ed. London: Macdonald Optima, 1990.

SEE ALSO: Antarctic and Southern Ocean Coalition; Brower, David; Earth First!; Environmental education; Environmental ethics; Greenpeace; Group of Ten; League of Conservation Voters.

G

Gaia hypothesis

CATEGORIES: Philosophy and ethics; ecology and ecosystems

DEFINITION: Theory that the earth functions as a single living, self-regulating system made up of the planet's living organisms and their material environment

SIGNIFICANCE: When first introduced during the early 1970's, the Gaia hypothesis attracted only slight attention in the scientific world, and much of that was critical. Since that time, however, the hypothesis has been given more serious consideration in the fields of geoscience, ecological science, and environmental ethics.

Natural scientist James Lovelock formulated the Gaia hypothesis in the 1960's in reaction to his work for the National Aeronautics and Space Administration (NASA) concerning the detection of life on Mars; he began to publish articles on the hypothesis in the early 1970's. The hypothesis is named for the primal Greek goddess personifying the earth. According to the Gaia hypothesis, the earth's physical and biological processes are linked to a complex entity in a feedback system that optimizes the conditions for life. Well-established examples of homeostatic mechanisms in this system are the regulation of the global temperature, the atmospheric composition, and the salinity of the oceans. In all of these mechanisms, life seems to play an important role in completing feedback loops.

When it was first presented, the Gaia hypothesis was strongly criticized in the scientific world for ascribing to the earth some goals and purposes, and thus interpreting all processes on earth in a teleological way. In answer to the critics, Lovelock developed Daisyworld, a simple computer model of a hypothetical planet that shows that the earth does not require a predetermined goal but only feedback loops for homeostasis. Other aspects of the Gaia hypothesis remain highly controversial; these include Lovelock's idea of coevolution between biological and nonbiological systems. Since the early 1990's, however, the hypothesis has been considered seriously within the fields of geoscience, ecological science, and environmental ethics.

Claudia Reitinger

SEE ALSO: Biodiversity; Deep ecology; Ecosystems; Environmental ethics; *Limits to Growth, The*; Lovelock, James.

Ganges River

CATEGORIES: Places; water and water pollution

IDENTIFICATION: River that flows from west to east across the plains of north India from headwaters at Gangotri in the Himalayas to the Bay of Bengal

SIGNIFICANCE: Bringing life to a fertile plain that has hosted human habitation since the dawn of humanity, the Ganges is prominent in Indian lore. With Himalayan snowmelt and rains from both the southwest and northeast monsoons, the Ganges remains a viable waterway and irrigation source year-round, sustaining more than 400 million people, despite the fact that it is highly polluted.

The Ganges, also known as the Ganga, originates from an ice cave at the foot of the Gangotri glacier in the western Himalayas and drains an area of more than 1 million square kilometers (386,000 square miles) into the Bay of Bengal, more than 2,500 kilometers (1,550 miles) away. The Ganges and its Himalayan tributaries—Bhagirathi, Alaknanda, Mandakini, Gandak, Ghāghara, and Kosi—host heavy tourist and pilgrim traffic. The Ganges Canal cuts across the densely populated Doab (two rivers) region from Haridwar to the Yamuna River. The cities of Delhi and Agra, and Mathura and Vrindavan of Mahabharata lore, are on the banks of the Yamuna, which joins the Ganges at Allahabad. This is called Triveni Sangam (three-river confluence) in memory of the ancient Sarasvati River, which disappeared into the earth, probably as the result of seismic events. The Ganges

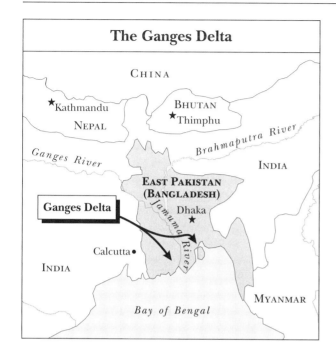

The Ganges Delta

species to survive far upstream. Further, soil erosion and diversion of the river's water for irrigation have reduced water levels and inhibited navigation.

The Indian government undertook the Ganga Action Plan (GAP), an effort to reduce pollution in the river, in 1985; among other steps, sewage treatment plants were built, but little progress was made. A second GAP phase was later introduced; it included tough laws requiring industries to build effluent treatment plants, as well as provisions for proactive community education, reforestation, and passive gravity-fed filtering systems.

Narayanan M. Komerath and Padma P. Komerath

FURTHER READING

Hollick, Julian Crandall. *Ganga: A Journey down the Ganges River.* Washington, D.C.: Island Press, 2007.

Mukherjee, D., M. Chattopadhyay, and S. C. Lahiri. "Water Quality of the River Ganga (the Ganges) and Some of Its Physico-chemical Properties." *Environmentalist* 13, no. 3 (September, 1993): 199-210.

Rao, R. J. "Biological Resources of the Ganga River, India." *Hydrobiologia* 458 (2001): 159-168.

Ray, P. *Ecological Imbalance of the Ganga River System: Its Impact on Aquaculture.* Delhi: Daya, 1998.

SEE ALSO: Agricultural chemicals; Drinking water; Ecotourism; Floods; Irrigation; Population growth; Runoff, agricultural; Runoff, urban; Sedimentation; Sewage treatment and disposal; Urban sprawl; Water pollution; Watershed management.

flows past the cities of Kanpur, Ramnagar, and Varanasi. In the northeast Indian state of Bihar, the tributaries Gomati, Ghāghara, Gandak, Bhagmati, Kosi, and Phalgu (on the banks of which is found the Buddhist holy city of Gaya) join the Ganges, and it flows past the state capital city of Patna.

The city of Kolkata (formerly known as Calcutta) is on the navigable Hugli River, an important channel of the Ganges that branches southward at the Bangladesh border. North of the Faridpur district of Bangladesh, the Ganges joins the lower Brahmaputra River and they become the Padma River, which joins the Meghna River and branches off through the vast Sundarbans (beautiful forest) delta to the Bay of Bengal.

The ancients revered the purity of the Ganges, and the river continues to be worshiped as a provider of life and annual cleanser by flood; it has been estimated that two million people engage in ritual bathing in the Ganges every day. The river's ecosystem—which is home to more than 140 unique fish species and 90 amphibian species and is also the site of five bird sanctuaries—has become threatened over time by high levels of pollution. Among the causes of the river's decline have been invasions, colonial poverty, industrialization, urbanization, agricultural runoff containing fertilizers, and efficient mass transport that brings millions of tourists. Pollution threatens many freshwater species native to the Ganges, such as the Ganges River dolphin, and allows invasive marine

Garbology

CATEGORY: Waste and waste management
DEFINITION: The study of materials discarded by humans
SIGNIFICANCE: By studying the items that people discard, anthropologists and other scientists can learn about human consumption habits. Environmentalists are particularly interested in the reduction of waste materials through lowered consumption and recycling.

The informal term "Garbology" is most commonly used in reference to the archaeological study of refuse, but it can include any activity in which refuse is closely studied. No university has an official garbology program, but many people call themselves garbologists. Despite the formal definition of garbage as moist refuse, in practice garbology includes the study of all types of household and commercial refuse. William Rathje is recognized as the foremost expert in garbology. Combining his anthropological and archaeological expertise, Rathje created and directed the Tucson Garbage Project at the University of Arizona. From 1973 onward, he and his team of researchers analyzed more than fifteen thousand samples of household refuse from around the world. Rathje and Cullen Murphy coauthored *Rubbish: The Archaeology of Garbage* (1992), which describes the more interesting findings from these Garbage Project investigations.

A typical Garbage Project study uses recently collected refuse from a particular neighborhood, keeping the names of refuse generators anonymous. Garbage Project researchers then comb through this garbage, sorting it into 150 categories and recording information about individual items' sizes, weights, and other characteristics. These garbage sorts provide a wide range of consumer information, ranging from the quantity of recyclable materials thrown away to trends in liquor consumption over time.

Other people have popularized garbology by examining the discards of famous people. In the 1970's A. J. Weberman sifted through musician Bob Dylan's garbage for many months and wrote magazine articles about his findings. He and his associates at the National Institute of Garbology promote their garbological investigations of luminaries such as Jacqueline Kennedy Onassis, Spiro Agnew, and Dustin Hoffman. A 1975 *National Enquirer* article featuring journalist Jay Gourley's analysis of Secretary of State Henry Kissinger's garbage sparked widespread controversy about journalistic standards and invasion of privacy.

Likewise, law-enforcement and intelligence agents analyze the discards of suspected criminals and political figures to build evidence about the activities of their subjects. Elevating garbology to international intrigue, the Central Intelligence Agency (CIA) reportedly paid one thousand dollars in the 1950's for the refuse from a single Soviet airliner. Although the U.S. Constitution protects against illegal search and seizure of private property, the Supreme Court ruled in 1988 that garbage left for collection can be searched without a warrant. Some states and municipalities have laws to prevent refuse scavenging, although such statutes may not be rigidly enforced.

The opportunity to look for useful items in refuse has led to the growing popularity of "Dumpster diving," the practitioners of which may or may not consider themselves garbologists. Some of these individuals are desperate for food and necessities, but others are environmentalists who dislike seeing usable items go to waste.

The term "garbology" is sometimes used to designate other refuse-related activities as well. Persons who work in the fields of refuse collection and recycling may refer to themselves as garbologists. Garbology research has been conducted on the types of litter strewn on highways and in other public places so that antilittering programs can then target key sources of litter. Even artists who use items found in trash heaps or garbage dumps to make "trash art" may refer to themselves as garbologists.

Andrew P. Duncan

FURTHER READING

Melosi, Martin V. *Garbage in the Cities: Refuse, Reform, and the Environment.* Rev. ed. Pittsburgh: University of Pittsburgh Press, 2005.

Rathje, William L., and Cullen Murphy. *Rubbish! The Archaeology of Garbage.* 1992. Reprint. Tucson: University of Arizona Press, 2001.

Rogers, Heather. *Gone Tomorrow: The Hidden Life of Garbage.* New York: New Press, 2005.

SEE ALSO: Landfills; Recycling; Resource recovery; Waste management.

Gasoline and gasoline additives

CATEGORIES: Energy and energy use; pollutants and toxins

DEFINITION: Liquid fuel produced from petroleum, and chemicals that may be added to such fuel to change the way it performs, cleans, lubricates, or appears

SIGNIFICANCE: Gasoline plays a significant role in the transportation industry throughout the world, and gasoline additives play an important role in increasing both octane and performance of gasoline-fueled vehicles. The burning of gasoline, however, has many negative impacts on the environment and on human health.

Gasoline is a liquid fuel that contains a mixture of liquid hydrocarbons. Refined from crude oil, gasoline is shipped to storage terminals from refineries in pipelines that are also used to transport other liquid products. It is then trucked from the storage terminals to retail outlets.

Gasoline varies widely in its composition owing to refineries' mixing of crude oil from various locations, the mixing that may occur between gasoline and other products shipped in shared pipelines, and the various additives mixed in by individual companies. The invention of the automobile in the late nineteenth century transformed gasoline from a useless by-product of oil refining to a valuable transportation fuel. By the 1920's automobiles, gasoline, and gasoline stations were major elements of everyday life in the United States. Gasoline companies began putting additives in gasoline for marketing purposes, to distinguish their products from those of competitors.

Among the many different types of gasoline additives are hybrid compound blends, alcohols that act as oxygenators, ethers, antioxidants, and antiknock agents. Many gasoline additives produce emissions that are harmful to the environment and can cause health problems; others are beneficial to the environment in that they reduce air pollution. The first gasoline additive used was tetraethyl lead, which increased the octane of the gasoline and also acted as a lubricant. Benzene and toluene have been commonly used as additives. Methyl tertiary-butyl ether (MTBE) is an additive that increases octane and reduces the emission of pollutants by increasing the amount of oxygen in gasoline.

Gasoline and the Environment

When gasoline is burned it releases a number of harmful pollutants into the air, including carbon monoxide, nitrogen oxide, particulate matter, and carbon dioxide, a major greenhouse gas that has been linked to global warming. In addition, the leakage of gasoline from storage tanks and from pipelines has the potential to contaminate water supplies.

Both gasoline and the additives it contains can pose health hazards when the gasoline is burned or vaporized. Exposure to lead, which was until the 1970's the additive of choice, is known to cause cancer and other serious health problems. Benzene vapor from gasoline can cause headaches and dizziness, and high levels can cause rapid heart rate and tremors; benzene is also a carcinogen. Exposure to toluene can cause headaches, respiratory irritation, and nausea.

MTBE, which replaced lead as an octane enhancer in 1979, is an oxygenate that has had positive effects on the environment. It reduces the amount of toxic emissions as it causes more of the gasoline to burn. Inhalation of MTBE is not believed to pose a hazard to health; however, some concerns have been expressed regarding the additive's potential to contaminate drinking water, and it has been banned in many U.S. states.

Environmental Regulations

As early as 1924, exposure to lead was determined to be a serious health hazard. Nevertheless, the levels of lead in gasoline increased through the 1950's as motorists wanted larger, faster cars, and increasing the amount of lead in the gasoline increased both octane and engine performance. As a result of further research and greater awareness of the damage to both the environment and human health caused by the use of leaded gasoline, the 1970 amendments to the Clean Air Act set air-quality standards and included a plan for eliminating lead in gasoline. This legislation also set limits of acceptability for the amount of carbon monoxide, hydrocarbons, and nitrogen oxides in vehicle emissions. In 1976, all new vehicles were required to be equipped with catalytic converters to reduce pollutants in emissions. Only unleaded gas, which had been introduced in 1975, could be used in the vehicles with catalytic converters. The manufacture of these cars did much to phase out the use of lead in gasoline. On January 1, 1996, the use of leaded gasoline in on-road vehicles was banned in the United States.

Many other laws have been enacted that include provisions aimed at making gasoline less harmful to the environment and less of a health hazard. The 1990 amendments to the Clean Air Act set a target date of 1995 for all gasoline sold in certain metropolitan areas to be reformulated so that it would burn cleaner. This measure was taken to reduce ground-level ozone, a serious problem in many large cities, such as Dallas, Chicago, and New York. In 1990, the U.S. Environmental Protection Agency (EPA) also required the replacement of all buried single-lined gasoline storage tanks with double-lined tanks to decrease the possibility of leakage. In 2006, the reduction of sulfur levels in gasoline became a priority. Refineries were required to reduce sulfur levels in gasoline, and new vehicles were mandated to have pollution-control devices that need low sulfur content to operate properly.

The negative effects of gasoline and gasoline additives on the environment and on health are issues of major concern throughout the world. Although leaded gasoline is still used in Asia, Africa, and some Latin American countries, many countries, such as Brazil, have converted to the use of pure ethanol as a motor vehicle fuel and have reduced or eliminated lead usage. As of January 1, 2000, leaded gasoline was no longer available in Europe except in a few eastern and southeastern countries. In 2009, the European Union passed legislation requiring gasoline stations to install systems for capturing and recycling toxic emissions.

Shawncey Webb

FURTHER READING

Dinan, Terry, and David Austin. *Reducing Gasoline Consumption: Three Policy Options.* New York: Novinka Books, 2003.

Reynolds, John G., and M. Rashid Khan. *Designing Transportation Fuels for a Cleaner Environment.* Philadelphia: Taylor & Francis, 1999.

United Nations Environment Programme. *Phasing Lead Out of Gasoline: An Examination of Policy Approaches in Different Countries.* New York: United Nations Publications, 2000.

SEE ALSO: Alternative fuels; Alternatively fueled vehicles; Automobile emissions; Carbon dioxide; Catalytic converters; Clean Air Act and amendments; Nitrogen oxides; Smog; Water pollution.

Gene patents

CATEGORY: Biotechnology and genetic engineering

DEFINITION: Patents granted by governments on specific gene sequences or on methods for obtaining or using such sequences

SIGNIFICANCE: Proponents of gene patents argue that they stimulate research and innovation by rewarding scientific effort and allowing companies to protect their investments in research. Critics, in contrast, assert that patenting genes prevents the free flow of scientific information and may inhibit scientific and medical research, including the development of diagnostic tests and disease treatments.

To be patentable, an idea (or invention or product) must meet four basic criteria: It must be useful (that is, it must have some practical use), it must be novel (not known before), it must be nonobvious (not

an idea easily developed by someone trained in the given field), and it must be described in the patent application in enough detail to enable someone trained in that area to use the idea for the stated purpose. In general, products of nature are not considered patentable. Deoxyribonucleic acid (DNA) products are patentable because they have been purified or modified in some way not found in nature. It is estimated that about 4,000 (or about 20 percent) of human genes have been patented. Some 47,000 patents involving genetic material have been granted.

In 1977 the University of California applied for patents for insulin and growth hormone genes. In the 1980 case *Diamond v. Chakrabarty* the U.S. Supreme Court ruled that life-forms modified through genetic engineering are patentable because they have been changed in some way through human intervention and are not likely to be found in nature. The insulin patent was granted in 1982, and in 1987 the growth hormone patent was granted. These patents were awarded following the general patent rule that vaccines and natural products, including hormones, made into useful forms are patentable. Insulin was the first recombinant product approved in 1982 for marketing. Genentech holds the patent for cloned human insulin, which was licensed to the pharmaceutical company Eli Lilly.

In 2009 the American Civil Liberties Union (ACLU) and the Public Patent Foundation (which is affiliated with the Benjamin N. Cardozo School of Law of Yeshiva University) filed a lawsuit in the U.S. District Court of Southern New York against the U.S. Patent and Trademark Office, Myriad Genetics, and the University of Utah Research Foundation, holders of the patents for *BRCA1* and *BRCA2*—genes associated with increased risk of inherited breast and ovarian cancers. Plaintiffs represented by the ACLU included patients, doctors, universities, and medical associations such as the American College of Medical Genetics. The claim stated that gene patents violate the First Amendment to the U.S. Constitution and patent law because as products of nature, genes should not be patentable. The plaintiffs maintained that gene patents limit scientific research, the free flow of information, and the development of diagnostic tests. The ACLU also noted that the high price charged by Myriad for diagnostic testing to determine a woman's *BRCA1* and *BRCA2* gene sequences could prevent women from obtaining genetic testing to determine their risk of cancer. In March, 2010, the court

German Greenpeace activists, dressed as businessmen with cloned "Dolly" sheep masks, gather in front of the European Patent Office in Munich in December, 1999, to protest the possibility of pharmaceutical companies obtaining patents on human organs, hormones, and genetically modified animals for commercial use. (AP/Wide World Photos)

found in favor of the plaintiffs and struck down the patents; an appeal of the decision was expected.

Susan J. Karcher

FURTHER READING

Caulfield, Timothy, Tania Bubela, and C. J. Murdoch. "Myriad and the Mass Media: The Covering of a Gene Patent Controversy." *Genetics in Medicine* 9, no. 12 (December, 2007): 850-855.

Gaisser, Sibylle, et al. "The Phantom Menace of Gene Patents." *Nature* 458 (March 26, 2009): 407-408.

Holman, Christopher M. "Patent Border Wars: Defining the Boundary Between Scientific Discoveries and Patentable Inventions." *Trends in Biotechnology* 25, no. 12 (December, 2007): 539-543.

SEE ALSO: Biotechnology and genetic engineering; Cloning; *Diamond v. Chakrabarty*; Flavr Savr tomato; Genetically altered bacteria; Genetically engineered pharmaceuticals; Genetically modified foods; Genetically modified organisms.

General Mining Act

CATEGORIES: Treaties, laws, and court cases; land and land use

THE LAW: U.S. federal legislation granting persons and companies the right to stake mining claims on federal land without having to pay royalties to the government

DATE: Enacted on May 10, 1872

SIGNIFICANCE: The General Mining Act of 1872 served to protect the interests of prospectors who had filed mining claims on federal lands during the preceding two decades. The law also expanded mining rights that had previously been restricted to minerals such as gold and silver.

The precedent for the General Mining Act of 1872 dated to the late 1840's and principally affected prospectors during the gold rush of 1849 in California. Following the Treaty of Guadalupe Hidalgo, which officially ended the Mexican-American War

(1846-1848) and allowed the United States to acquire Alta California, laws pertaining to prospecting and mining in the region were largely in limbo. Mexican laws no longer applied in what was now American territory, and American laws remained vague. The importance of the issue became apparent with the discovery of gold in California in January, 1848. The following month, the military governor of California, Richard Barnes Mason, formally abrogated any preexisting laws pertaining to mining but failed to establish any legal precedents for American law; the result was largely chaos. The gold rush of 1849 brought thousands of prospectors to the region, and within five years more than one billion dollars' worth of gold had been discovered on public land.

The riches being drawn from the mines caught the attention of members of the U.S. Congress. Following the Civil War, some members of Congress had proposed laws to seize mines in western U.S. states and territories, arguing that the miners were squatters on public land, with no rights of ownership. In June, 1865, bills were introduced to seize the mines, with the threat of army troops being sent to ensure there would be no resistance. In response, representatives from the West argued that the miners were performing a public service by encouraging migration into new regions and providing significant economic benefits to the nation. The westerners prevailed, and miners were given official ownership. This legislation, proposed by Jerome Chaffee, representative from the territory of Colorado, passed in July, 1866.

The so-called Chaffee laws, which applied largely to the mining of gold, silver, and copper, became the basis for the General Mining Act of 1872, which also included "other valuable deposits" such as platinum and, years later, was amended to include oil, gas, and uranium deposits. As well as protecting individual rights, the law was intended to encourage the movement of Americans into what were largely unpopulated regions. Both individuals and companies could establish mining claims without being forced to pay royalties to the federal government. The act also granted to the miners extralateral rights—that is, miners could file claims not only on the mines themselves but also on the right of ownership for any subsurface extensions of the veins in their mines. Given that such extensions frequently extended beneath mines owned by other claimants, extensive litigation often resulted.

Later changes in the law, as well as new acts that su-perseded the 1872 provisions, established regulations on the removal and use of resources such as oil and natural gas and also provided protection for national parks and other historic sites. In 1994 the U.S. Congress established budget restrictions on mining claims, which resulted in the Bureau of Land Management's being prevented from accepting new applications.

Richard Adler

FURTHER READING

Bakken, Gordon Morris. *The Mining Law of 1872: Past, Politics, and Prospects.* Albuquerque: University of New Mexico Press, 2008.

Burke, Barlow, and Robert Beck. *The Law and Regulation of Mining: Minerals to Energy.* Durham, N.C.: Carolina Academic Press, 2009.

SEE ALSO: Acid mine drainage; Department of the Interior, U.S.; Environmental law, U.S.; Mine reclamation; Reclamation Act; Strip and surface mining; Surface Mining Control and Reclamation Act.

Genetic engineering. *See* Biotechnology and genetic engineering

Genetically altered bacteria

CATEGORY: Biotechnology and genetic engineering

DEFINITION: Bacteria that humans have manipulated at the genetic level to possess specific properties or carry out certain functions

SIGNIFICANCE: The genetic modification of bacteria has made possible the manufacture of medically important human proteins such as insulin and growth hormone. Genetically altered bacteria have also been used as a means to introduce into plants genetic material that increases the plants' resistance to disease, pests, or freezing.

The ability to alter bacteria genetically is the outcome of several independent discoveries. In 1944 Oswald Avery and his coworkers demonstrated gene transfer among bacteria using purified deoxyribonucleic acid (DNA), a process called known as transformation. In the 1960's the discovery of restriction enzymes permitted the creation of hybrid molecules of DNA. Such enzymes cut DNA molecules at specific sites, allowing fragments from different sources to be joined within the same piece of genetic machinery. Restriction enzymes are not species-specific in choosing their targets. Therefore, DNA from any source, when treated with the same restriction enzyme, will generate identical cuts. The treated DNA molecules are allowed to bind with each other, while a second set of enzymes called ligases are used to fuse the hybrids. The recombinant molecules may then be introduced into bacteria cells through transformation. In this manner, the cell acquires whatever genetic information is found in the DNA. Descendants of the transformed cells will be genetically identical, forming clones of the original.

The most common forms of genetically altered DNA are bacterial plasmids, small circular molecules separate from the cell chromosome. Plasmids may be altered to serve as appropriate vectors for genetic engineering, usually containing an antibiotic resistance gene for selection of only those cells that have incorporated the DNA. Once the cell has incorporated the plasmid, it acquires the ability to produce any gene product encoded on the molecule. The resulting artificially produced DNA is called recombinant DNA. The first bacterium to be genetically altered through recombinant DNA technology for medical purposes, *Escherichia coli*, contained the gene for the production of human insulin. Prior to creation of the insulin-producing bacterium in the 1970's, diabetics were dependent on insulin purified from animals. In addition to being relatively expensive, insulin obtained from animals produced allergic reactions among some diabetes patients. Insulin obtained from genetically altered bacteria, by contrast, is identical to human insulin. Subsequent recombinant DNA research has led to the manufacture of a variety of human proteins, including human growth hormone, parathyroid hormone, several kinds of interferons, many monoclonal antibodies, hepatitis B surface antigen, clotting factors, and granulocyte colony-stimulating factor.

Genetically altered bacteria may also serve as vectors for the introduction of genes into plants. The bacterium *Agrobacterium tumefaciens*, the etiological agent for a plant disease called crown gall, contains a plasmid known as Ti. Following infection of the plant cell by the bacterium, the plasmid is integrated into the host chromosome, becoming part of the plant's genetic material. Any genes that were part of the plasmid are integrated as well. Desired genes can be introduced into the plasmid, promoting pest or disease resistance within plants infected by the bacterium.

In April, 1987, scientists in California sprayed strawberry plants with genetically altered bacteria to improve the plants' freeze resistance; this event marked the first deliberate release of genetically altered organisms in the United States to be sanctioned by the Environmental Protection Agency (EPA). The release of the bacteria represented the climax of more than a decade of public debate over what would happen when the first products of biotechnology became commercially available. Fears centered on the creation of bacteria that might radically alter the environment through elaboration of gene products not normally found in such cells. Other concerns included the creation of super bacteria with unusual resistance to conventional medical treatment.

Despite these fears, approval for further releases of genetically altered bacteria soon followed, and the restrictions on release were greatly relaxed. By 1991 permits for field tests of more than 180 genetically altered plants and microorganisms had been granted. Between 1987 and 2004 more than 10,000 trials were conducted at more than 39,000 sites, and more than sixty biotechnology products entered the market. Among future planned tests are clinical trials of the use of a modified *Streptococcus mutans* in fighting dental cavities. Scientists have modified this bacterium, responsible for tooth decay in its unaltered form, so that it does not produce the lactic acid that ordinarily erodes tooth enamel. In animal tests, it has been found that the modified bacterium eventually replaces the *S. mutans* naturally occurring in the mouth.

In general, "red biotechnology" (the application of biotechnology in medicine) tends to generate less controversy than "green biotechnology" (use of biotechnology in food production). In the United States, anyone intending to produce or import genetically altered microorganisms for commercial purposes must submit a notice to the EPA, which assesses whether the organism constitutes an unreasonable risk to human health or the environment.

Richard Adler
Updated by Karen N. Kähler

FURTHER READING

Drlica, Karl. *Understanding DNA and Gene Cloning: A Guide for the Curious.* 4th ed. Hoboken, N.J.: John Wiley & Sons, 2004.

Food and Agriculture Organization and World Health Organization. *Safety Assessment of Foods Derived from Genetically Modified Microorganisms.* Geneva: World Health Organization, 2001.

Han, Lei. "Genetically Modified Microorganisms: Development and Applications." In *The GMO Handbook: Genetically Modified Animals, Microbes, and Plants in Biotechnology,* edited by Sarad R. Parekh. Totowa, N.J.: Humana Press, 2004.

Stemke, Douglas J. "Genetically Modified Organisms: Biosafety and Ethical Issues." In *The GMO Handbook: Genetically Modified Animals, Microbes, and Plants in Biotechnology,* edited by Sarad R. Parekh. Totowa, N.J.: Humana Press, 2004.

Watson, James D., et al. *Recombinant DNA: Genes and Genomes—A Short Course.* 3d ed. New York: W. H. Freeman, 2007.

SEE ALSO: Biopesticides; Biotechnology and genetic engineering; *Diamond v. Chakrabarty*; Gene patents; Genetically engineered pharmaceuticals; Genetically modified foods; Genetically modified organisms.

Genetically engineered pharmaceuticals

CATEGORIES: Biotechnology and genetic engineering; human health and the environment

DEFINITION: Drugs created through the manipulation of genetic materials

SIGNIFICANCE: Genetic technology is being used to develop state-of-the-art drugs to treat human diseases and health conditions. Genetic engineering can create carefully targeted drugs that offer greater effectiveness and potency than conventional pharmaceuticals while causing fewer side effects.

Genetics is the scientific study of heredity—the biological factors that determine the characteristics of all living things. All reproductive life-forms develop under the laws of genetics. The basis of genetics is the gene, a tiny unit of matter that determines some identifiable characteristic of an individual. Genes are located in fixed positions on chromosomes (molecular chains) that reside in the center of cells. A major part of the chromosome is deoxyribonucleic acid (DNA), which is responsible for transmitting genetic information, in the form of genes, when new life is created. This transfer of traits applies to organisms of all sizes, from microscopic to larger and more complex systems, such as humans. Inherited traits include color of hair, eyes, and skin, as well as susceptibility to various ailments.

As the complex chemical-biological process of reproduction takes place at the gene-chromosome level, it is possible for a random processing error called a mutation to be introduced. Mutant cells may cause many human defects and diseases. Genetic abnormalities, also known as birth defects, include such ailments as hemophilia (resistance to blood clotting), color blindness, anatomical defects, speech disorders, hormonal disorders, brain disorders, and psychiatric illness. Aside from birth defects, genetic cellular mutations can occur anytime during a lifetime. Normally the body contains certain controlling genes that destroy mutant genes that spontaneously appear. If these controlling genes become defective, the mutant genes can take over the body, as occurs with cancerous tumors.

GENETIC TECHNOLOGY IN MEDICINE

Pharmaceuticals are drugs used to treat human diseases and conditions. The term "engineered drug" implies that scientific principles and manufacturing processes are applied in creating the drug. A genetically engineered pharmaceutical is a specialized drug made by the application of specific genetic principles. Gene-based technology is used to investigate, test, and apply state-of-the-art pharmaceuticals to invasive and widespread diseases. Its potential for fighting illnesses such as cancer and acquired immunodeficiency syndrome (AIDS) is of particular interest to medical science.

For example, scientists have developed ways to control cancer cells using genetic medicines instead of killing the cancer with radiation, conventional drugs, or surgical removal. Herceptin, a genetically engineered drug approved by the U.S. Food and Drug Administration (FDA), is used in treating certain breast cancers. Herceptin is an antibody engineered to attack specific cancer cells, helping to reduce the cancer tumor by keeping a particular protein from reproducing. Some tumor cells are inherently resistant to the drug or become resistant over time.

In the past the production of natural body chemicals required the harvesting of the needed chemicals from human or animal materials. Supplies of such sources are sometimes minimal, and concentrating chemicals from human and animal tissue can also multiply the chances of carrying diseases from those sources. For instance, during the 1960's through the mid-1980's some children suffering from growth failure were treated with human growth hormone (HGH) extracted from the pituitary glands of human cadavers. In 1985 three adults in the United States who had been treated with this HGH during childhood died from Creutzfeldt-Jakob disease (CJD), a rare, incurable, and fatal brain disease with a long incubation period. Other recipients of HGH, both in the United States and abroad, also contracted CJD, apparently from HGH that was contaminated with the infectious agent that causes CJD.

Genetic engineering of substances such as growth hormone circumvents many traditional problems. The genes for producing the desired chemicals can be implanted in the genetic code of plants or microorganisms, especially the benign bacterium *Escherichia coli*. These sources can enable high-volume production at high levels of concentration. Because plants and microorganisms are very different from people, the chance of spreading disease through this production method is minimal.

Genetically Engineered Vaccines

Genetically engineered pharmaceuticals can increase the body's production of naturally occurring chemical substances and supply toxins to attack targeted pathogens (disease-causing viruses, bacteria, fungi, or parasites). Vaccines work by triggering the body's immune system, which then defends itself. Compared with traditional methods of creating vaccines, genetic engineering enables faster development of safer vaccines.

A live attenuated vaccine is one that contains the living pathogen, but in a weakened form that cannot induce illness. Recombinant DNA technology can be used to remove key genes from microbes to render them harmless. This method has been used to engineer a vaccine against *Vibrio cholerae*, the bacterium that causes cholera. Another live attenuated vaccine, one used against the rotavirus pathogen responsible for serious diarrheal illness in infants and young children, was created through a technique that combines genes from different strains of the pathogen in a way that makes a harmless simulator virus. At the cellular or molecular level, the simulator virus appears to the

immune system to be a pathogenic invader, causing antibodies to develop that attack the active pathogen.

A subunit vaccine does not employ the entire pathogenic organism; rather, it relies on its antigens, substances that trigger the body's immune system. For this technique, select antigens or portions thereof are used to provoke an immune response. While the antigens could be harvested from laboratory-grown microbes, recombinant DNA technology makes it possible to manufacture the antigen molecules. These parts of the pathogen's genetic code are inserted into common baker's yeast, a harmless microbe. Hepatitis B subunit vaccine uses a portion of the protein coat surrounding the virus's DNA. Because the rest of the microbe is not included, the possibility of an adverse reaction is greatly reduced.

In the creation of conjugate vaccines, the bacterial antigens selected for the immune response they provoke are not easily recognized by the body. These are joined with easily recognizable antigens located on a harmless bacterial shell and injected into the body to trigger the immune system. Conjugate vaccines for children have been developed against middle-ear infections and other diseases caused by pneumococci, a group of common bacteria. In 2000 the FDA licensed Prevnar, a conjugate vaccine that targets the seven most common types of pneumococci causing invasive disease in infants and toddlers. Ten years later, with other types of pneumococci becoming increasingly common in young children, the FDA licensed a conjugate vaccine to replace Prevnar, one that targets thirteen pneumococcal strains.

Naked DNA vaccines have been tested on diseases such as AIDS and some cancers. This experimental method involves injecting a person with some of a pathogen's DNA—specifically, with the genes that code for antigens. Some of the body's cells accept this added DNA as instructions to produce antigens, thereby triggering immune response. A similar experimental technique, recombinant vector vaccination, employs attenuated microorganisms that act as carriers for the DNA while further stimulating immune response.

The use of genetically engineered pharmaceuticals has raised concerns about their possible effects on the environment. In 1989 the Virginia Department of Health approved a field test of baits spiked with genetically engineered oral vaccine to control the spread of rabies in raccoons. Health officials were worried, however, about the possible danger to humans posed by the vaccinia virus used as the vaccine; thus an island location was chosen to prevent the possible spread of vaccinated animals to larger, mainland populations. Several researchers expressed concerns about the long-term effects of releasing a nonnative virus into the environment; although vaccinia had been used for many years to prevent smallpox, little was known about its host range or its ability to cause disease. The U.S. Department of Agriculture concluded in a 1991 report, however, that laboratory and field tests had shown the genetically engineered rabies vaccine to have had no adverse effects on any species. In the same report, the department approved further field tests on the grounds that such tests were safe and posed no significant environmental risk. This rabies-control method has since become common practice in North America and Europe.

Robert J. Wells
Updated by Karen N. Kähler

FURTHER READING

Aldridge, Susan. *The Thread of Life: The Story of Genes and Genetic Engineering.* 1996. Reprint. New York: Cambridge University Press, 2000.

Castilho, Leda R., et al., eds. *Animal Cell Technology: From Biopharmaceuticals to Gene Therapy.* New York: Taylor & Francis, 2008.

Crommelin, Daan J. A., Robert D. Sindelar, and Bernd Meibohm, eds. *Pharmaceutical Biotechnology: Fundamentals and Applications.* 3d ed. New York: Informa Healthcare, 2008.

Gad, Shayne Cox, ed. *Handbook of Pharmaceutical Biotechnology.* Hoboken, N.J.: John Wiley & Sons, 2007.

Groves, M. J., ed. *Pharmaceutical Biotechnology.* 2d ed. Boca Raton, Fla.: Taylor & Francis, 2006.

Guzman, Carlos Alberto, and Giora Z. Feuerstein, eds. *Pharmaceutical Biotechnology.* New York: Springer, 2009.

Rehbinder, E., et al. *Pharming: Promises and Risks of Biopharmaceuticals Derived from Genetically Modified Plants and Animals.* Berlin: Springer, 2009.

Walsh, Gary. *Biopharmaceuticals: Biochemistry and Biotechnology.* 2d ed. New York: John Wiley & Sons, 2003.

SEE ALSO: Biotechnology and genetic engineering; Cloning; Gene patents; Genetically altered bacteria; Genetically modified organisms.

Genetically modified foods

CATEGORIES: Biotechnology and genetic engineering; agriculture and food

DEFINITION: Foods derived from living organisms that have been modified by gene-transfer technology

SIGNIFICANCE: Applications of genetic engineering in agriculture and the food industry have the potential to increase world food supplies, reduce environmental problems associated with food production, and enhance the nutritional value of certain foods. However, these benefits are countered by food safety concerns, the potential for ecosystem disruption, and fears of unforeseen consequences resulting from the alteration of natural selection.

Humans rely on plants and animals as food sources and have long used microbes to produce foods such as cheese, bread, and fermented beverages. Conventional techniques such as cross-hybridization, production of mutants, and selective breeding have resulted in new varieties of crop plants and improved livestock with altered genetics. However, these methods are relatively slow and labor-intensive, are generally limited to intraspecies crosses, and involve a great deal of trial and error.

Recombinant deoxyribonucleic acid (DNA) techniques developed in the 1970's enable researchers to make rapid, specific, and predetermined genetic changes. Because the technology also allows for the transfer of genes across species and kingdom barriers, an infinite number of novel genetic combinations are possible. The first transgenics (animals and plants containing genetic material from other organisms) were developed in the early 1980's. In 1986 the United States and France conducted the first field trials of transgenic plants, which involved tobacco engineered to contain an herbicide-resistance gene. In 1990 the U.S. Food and Drug Administration (FDA) approved chymosin, an enzyme used in the production of dairy products such as cheese, as the first substance produced by engineered organisms to be used in the food industry. That same year a transgenic bull was developed that had been engineered to pass certain human genes along to his offspring; those genes would enable his female descendants to produce human milk proteins for infant formula. The well-publicized Flavr Savr tomato obtained FDA approval in 1994. By 1996 field trials had been conducted for more than thirty-five hundred genetically engineered crop plants.

BENEFITS OF GENETIC MODIFICATION

The goals for altering food-crop plants by genetic engineering fall into three main categories: to create plants that can adapt to specific environmental conditions to make better use of agricultural land, increase yields, or reduce losses; to increase quality, nutritional value, or flavor; and to alter transport, storage, or processing properties for the food industry. Many genetically engineered crops are also sources of ingredients for processed foods and animal feed.

Herbicide-resistant plants such as the Roundup Ready soybean can be grown in the presence of glyphosphate, an herbicide that normally destroys all plants with which it comes in contact. Beans from these plants have been approved for food-industry use in several countries. Herbicide-resistant rice and sugarbeets, corn and potatoes made insect-resistant thanks to a bacterial gene that encodes for a pesticidal protein (*Bacillus thuringiensis*, or *B.t.*), and a viral disease-resistant squash are but a few other examples of genetically engineered food crops on the market. Scientists have also created plants that produce healthier unsaturated fats and oils rather than saturated ones, coffee plants that produce beans that are caffeine-free without processing, zucchinis equipped with insect proteins that make their smell and taste unappealing to insect pests, and cold-tolerant tomatoes that owe their frost resistance to a fish gene.

Animals can also be genetically engineered food sources, although transgenic research involving plants presents fewer technical challenges. One of the best-known applications of genetic engineering in animals involves recombinant bovine growth hormone (rBGH; also known as recombinant bovine somatotropin, or rBST) synthesized by bacteria containing the BGH gene found in cattle. When rBGH is given to cows as a supplement, their milk production can increase 11 to 16 percent. The FDA approved the use of rBGH in 1993, and a 2007 survey by the U.S. Department of Agriculture found that roughly 17 percent of the dairy cows in the United States were being treated with rBGH.

Cloning is not genetic engineering per se, as it does not involve adding, removing, or modifying genes. Rather, it is a method for copying an organism's genetic traits to create one or more living replicas of the organism. While cloning is too expensive to be used to produce food animals, it can be employed to replicate breeding animals that can pass desirable traits to their offspring. The technology used to clone Dolly the sheep in 1996 represented a significant advancement. In 2001, by which time other mammal clones had been created, the FDA asked livestock producers and researchers to keep animal clones and their offspring out of the food supply until a food consumption risk assessment could be conducted. In 2008 an FDA report concluded that meat and milk from cow, pig, and goat clones and their offspring are as safe as any other animal products.

Genetically Modified Crop Plants Unregulated by the U.S. Department of Agriculture

CROP	PATENT HOLDER	GENETICALLY ENGINEERED TRAIT(S)
Canola	AgrEvo	herbicide tolerance
Corn	AgrEvo	herbicide tolerance
	Ciba-Geigy	insect resistance
	DeKalb	herbicide tolerance; insect resistance
	Monsanto	herbicide tolerance; insect resistance
	Northrup King	insect resistance
Cotton	Calgene	herbicide tolerance; insect resistance
	DuPont	herbicide tolerance
	Monsanto	herbicide tolerance; insect resistance
Papaya	Cornell	virus resistance
Potato	Monsanto	insect resistance
Soybean	AgrEvo	herbicide tolerance
	DuPont	altered oil profile
	Monsanto	herbicide tolerance
Squash	Asgrow	virus resistance
	Upjohn	virus resistance
Tomato	Agritope	altered fruit ripening
	Calgene	altered fruit ripening
	Monsanto	altered fruit ripening
	Zeneca	altered chemical content in fruit

Source: U.S. Department of Agriculture Animal and Plant Health Inspection Service (APHIS).

In a greenhouse at the U.S. Agricultural Research Service in Beltsville, Maryland, plant physiologist Autar Mattoo examines tomato plants that have been genetically engineered to enhance the phytonutrient content and longevity of the fruit. (©Scott Bauer/USDA)

Genetically engineered microbes are used for the production of food additives such as amino acid supplements, sweeteners, flavors, colorings, vitamins, and thickening agents. In some cases, these substances previously had to be obtained from slaughtered animals. Altered organisms are also used for improving fermentation processes in the food industry.

DISADVANTAGES AND CONTROVERSIES

Food safety and quality are at the center of the controversies related to genetically engineered foods. Concerns include the possible introduction of new toxins or allergens into human and animal diets and changes in the nutrient composition of foods. Proponents of using genetic engineering to modify food products argue that it would enable the enhancement of some foods' nutritional value.

Critics opposed to allowing genetically modified products to enter the environment cite the danger of outcrossing, or "genetic pollution"—that is, the transfer of genetic material to wild relatives—leading to the development of new plant diseases. The emergence of "superweeds," either the engineered plants themselves or new plant varieties formed by the transfer of recombinant genes conferring various types of resistance to wild species, is another concern. These weeds, in turn, would compete with valuable plants and have the potential to destroy ecosystems and farmland unless stronger poisons were used for eradication.

Opponents also note that crops engineered to be herbicide-resistant encourage excessive use of herbicides. Where glyphosphate herbicides are liberally applied, the local weed population has come to be dominated by glyphosphate-tolerant species. *B.t.*-resistant insect pests have also been found attacking engineered crops.

Use of rBGH has become controversial because of concerns regarding the health of treated cows and the safety of the milk. Cows receiving rBGH are more susceptible to udder infections than are other cows, and thus are treated with more antibiotics and other drugs; those who consume the milk ingest residues of those drugs as well as residues of the hormones themselves. Canada, all members of the European Union, Japan, New Zealand, and Australia have banned the use of rBGH in cows producing milk intended for human consumption.

Cloning technologies have been similarly controversial. Animal rights issues, vegetarian and religious objections to animal-based components in plant foods, worries about infectious agents that could be transferred to humans, and concerns regarding the application of animal cloning technologies to produce human beings have all hindered developments in this field.

Environmental problems such as deforestation, erosion, pollution, and loss of biodiversity have all resulted, in part, from conventional agricultural practices. Proponents of genetically engineered crops point out that such crops could allow better use of existing farmland and lead to decreased reliance on pesticides and fertilizers. A large percentage of crops worldwide are lost each year to drought, temperature extremes, and pests. Plants have already been engineered to exhibit frost, insect, disease, and drought resistance. Such alterations can increase yields, allow food to be grown in areas that are too dry or infertile to support nonengineered crops, and have positive impacts on the world food supply, particularly in the face of global climate change.

Also, it is argued, genetic engineering might be necessary to develop food sources that can survive rapidly changing environmental conditions. Pollution, global climate change, and increased ultraviolet irradiation because of decreased stratospheric ozone result in stress conditions for living organisms, and all have impacts on agriculture. The processes of natural selection and adaptation may be too slow to keep up with the food needs of the earth's ever-increasing human population.

Opponents counter that genetic modification does not automatically lead to higher production, nor does higher production automatically reduce hunger. It is poverty, they maintain, that determines who goes hungry, not the availability of food. The production costs of genetically modified crops are beyond the means of many small farmers in the developed world and most farmers in the developing world. One of the most widely voiced criticisms of genetically modified foods thus centers on the ethics of allowing consolidated control by agribusiness of the world's food production.

Diane White Husic
Updated by Karen N. Kähler

FURTHER READING

Falkner, Robert, ed. *The International Politics of Genetically Modified Food: Diplomacy, Trade, and Law.* New York: Palgrave Macmillan, 2007.

Fedoroff, Nina V., and Nancy Marie Brown. *Mendel in the Kitchen: A Scientist's View of Genetically Modified Foods.* Washington, D.C.: Joseph Henry Press, 2004.

Lambrecht, Bill. *Dinner at the New Gene Café: How Genetic Engineering Is Changing What We Eat, How We Live, and the Global Politics of Food.* New York: St. Martin's Griffin, 2001.

Lurquin, Paul. *High Tech Harvest: Understanding Genetically Modified Food Plants.* Boulder, Colo.: Westview Press, 2002.

McHughen, Alan. *Pandora's Picnic Basket: The Potential and Hazards of Genetically Modified Foods.* New York: Oxford University Press, 2000.

Martineau, Belinda. *First Fruit: The Creation of the Flavr Savr Tomato and the Birth of Biotech Food.* New York: McGraw-Hill, 2001.

Nestle, Marion. *Safe Food: Bacteria, Biotechnology, and Bioterrorism.* Berkeley: University of California Press, 2004.

Rissler, Jane, and Margaret Mellon. *The Ecological Risks of Engineered Crops.* Cambridge, Mass.: MIT Press, 1996.

Weasel, Lisa H. *Food Fray: Inside the Controversy over Genetically Modified Food.* New York: AMACOM, 2009.

SEE ALSO: Agricultural chemicals; *Bacillus thuringiensis*; Biotechnology and genetic engineering; Cloning; Flavr Savr tomato; Genetically altered bacteria; Genetically modified organisms; Green Revolution; Intensive farming; Pesticides and herbicides.

Genetically modified organisms

CATEGORY: Biotechnology and genetic engineering

DEFINITION: Living organisms whose genetic compositions have been altered through technology

SIGNIFICANCE: Organisms can be engineered for use in scientific research, human and veterinary medicine, industry, agriculture, and environmental remediation. Despite the beneficial applications, potential risks and ethical issues associated with the technology have led to controversy and restricted its use in some countries.

With the advent of recombinant deoxyribonucleic acid (DNA) technology in the 1970's came the ability to modify and create genes and to transfer genetic material between unrelated species in a rapid and specific manner. The development of new traits is no longer limited to mutation or natural selection from a limited pool of genes; to alter an organism, scientists can introduce genetic traits from any species. Scientists first developed transgenic animals and plants in the early 1980's using genetic engineering in combination with techniques such as cell fusion, tissue culture, in vitro fertilization, and embryo transplantation. Clonal propagation, possible in cell and tissue culture for years, was successfully applied to mammals with the birth of Dolly the sheep in 1996.

APPLICATIONS

Industry has made widespread use of genetically modified organisms (GMOs). Modified microbes are used in fermentation processes and to produce food ingredients. In the chemical industry, GMOs are engineered to produce reagents and novel catalysts and to convert hazardous waste into harmless or useful substances. The result has been increased efficiency of certain industrial processes and decreased waste. In 1982 the U.S. Food and Drug Administration (FDA) approved the use of human insulin derived from genetically engineered bacteria. Subsequently, the pharmaceutical industry has made significant use of GMOs to develop new drugs, vaccines, and diagnostic tests.

Genetic engineering has also been used extensively in agriculture. Products of engineered organisms are used to protect plants from frost and insects, manipulate lactation and growth processes in livestock, and improve animal health. In 1986 regulators approved the release of the first genetically engineered crop plant, tobacco, in the United States. By 2006 seed producers had applied to the U.S. Department of Agriculture for permission to field-test almost 11,600 genetically engineered varieties. Plants have been designed to resist disease, drought, frost, insects, and herbicides; other uses of the technology have fo-

cused on improving the nutritional value and flavor of foods. Plants have even been engineered to produce synthetic rubber, plastics, vaccines, and renewable fuels.

One of the first examples of an organism genetically engineered to address environmental problems was a bacterium created to degrade oil. This oil-eating bacterium was at the center of an important U.S. Supreme Court case (*Diamond v. Chakrabarty*, 1980) in which it was ruled that the bacterium was a living invention and thus patentable. Such bacteria have proven effective in cleaning up oil spills both in the oceans and on land.

The use of living organisms to remove toxic chemicals from the environment is known as bioremediation. Genetically engineered microorganisms (GEMs) are essential to this process because many pollutants are human-made chemicals that cannot normally be degraded by living organisms. GEMs contain new or modified enzymes that enable them to digest the pollutants and convert them to nontoxic compounds such as carbon dioxide and water. GEMs are also advantageous in that they can be used for in situ (in

Members of a South Korean civic group in Seoul participate in a 2008 rally against the importation of corn that has been genetically modified. (AP/Wide World Photos)

place) treatments including on-site soil decontamination, detoxification of wetlands and streams, and groundwater purification, eliminating some of the technical problems associated with other remediation methods. GEMs are also used to recover minerals from mining and industrial wastes.

Genetically modified plants are also used to remove environmental pollutants. Researchers have genetically altered poplar trees to absorb trichloroethylene, a toxic solvent and common groundwater contaminant, from a liquid solution in quantities roughly thirty times greater and at rates one hundred times faster than naturally occurring poplars. The engineered trees break the solvent down into nontoxic components. In the future, genetically enhanced phytoremediation—the use of trees, grasses, and other plants to remove contaminants—may be employed to clean up sites contaminated by hydrocarbons, heavy metals, and radioactivity.

Concerns and Regulation

The modification of natural selection and the disruption of ecosystems are the major concerns associated with the introduction of genetically engineered organisms into the environment. In addition, not all of the consequences of this technology are predictable because of a lack of data on the stability of artificial genetic changes, the tendency of DNA manipulation to induce mutations in organisms, and the natural complexity of organisms. Unforeseen environmental problems posed by GMOs are intractable, because once introduced into the environment, they may be impossible to remove and isolate. Critics of the technology make reference to "genetic wastes" that are able to propagate, mutate, and migrate.

The possibility of gene flow or escape—the transfer of genes from the modified organisms to related species in the wild—is of concern. Antibiotic resistance genes used as markers in the development of transgenic organisms might be transferred to bacteria, leading to new treatment-resistant strains. (While scientists regard this as unlikely, the remote possibility has led to the use of alternative types of marker gene, such as one that causes the plant to fluoresce under ultraviolet light.) Modified viruses used in many recombinant DNA techniques might escape to create new disease-causing agents. Transgenic organisms designed to be more vigorous or any new species created by the gene transfer may have selective advantages over native species, leading to the disruption of natu-

ral balance in ecosystems and possibly exacerbating biodiversity restriction. However, even without human interference with genetic material, the exchange of genes can occur in nature.

Underlying some criticisms of the genetic engineering of organisms are ethical concerns related to the fair treatment of animals and the possiblitiy that the technology could be used to modify humans selectively. Furthermore, critics note the potential for the misuse of the technology in human experimentation, the development of biological weapons, and acts of terrorism.

In the United States three federal agencies evaluate new genetically engineered crops. Under the Plant Protection Act of 2000, the Department of Agriculture is responsible for agricultural and environmental safety associated with genetically modified crops. The FDA has oversight of safety aspects where human food and animal feed are involved. The Environmental Protection Agency evaluates food safety and environmental quality where genetically modified plants have insect resistance or other pesticidal properties. Within the European Union (EU), the European Food Safety Agency provides scientific advice regarding food and animal feed safety. Between 1999 and 2003 the EU imposed a moratorium on genetically modified imports. Six EU member states have banned the cultivation of GMOs within their borders: Austria, France, Greece, Hungary, Germany, and Luxembourg.

In 2000 more than 130 countries adopted the Cartagena Protocol on Biosafety, which entered into force in 2003. The objective of this supplementary agreement to the Convention on Biological Diversity (CBD) is to contribute to the safe transfer, handling, and use of living modified organisms (LMOs) such as genetically engineered plants, animals, and microbes that cross international borders. The protocol is an effort to protect biodiversity and human health on a global scale without causing unnecessary disruption to the world food trade. Under the protocol, LMOs intended for introduction into the environment may not be imported into a country without that country's informed consent. The United States is not a party to the CBD and thus cannot become a party to the Cartagena Protocol; however, the United States did participate in negotiations for this international treaty.

Diane White Husic and H. David Husic
Updated by Karen N. Kähler

FURTHER READING

Bodiguel, Luc, and Michael Cardwell, eds. *The Regulation of Genetically Modified Organisms: Comparative Approaches.* New York: Oxford University Press, 2010.

Glick, Bernard J., Cheryl L. Patten, and Jack J. Pasternak. *Molecular Biotechnology: Principles and Applications of Recombinant DNA.* 4th ed. Washington, D.C.: ASM Press, 2009.

International Union for Conservation of Nature. *Genetically Modified Organisms and Biosafety: A Background Paper for Decision-Makers and Others to Assist in Consideration of GMO Issues.* Gland, Switzerland: Author, 2004.

Nelson, Gerald C., ed. *Genetically Modified Organisms in Agriculture: Economics and Politics.* San Diego, Calif.: Academic Press, 2001.

Pardey, Philip G., ed. *The Future of Food: Biotechnology Markets and Policies in an International Setting.* Washington, D.C.: International Food Policy Research Institute, 2002.

Rifkin, Jeremy. *The Biotech Century: Harnessing the Gene and Remaking the World.* New York: Jeremy P. Tarcher/Putnam, 1998.

Stewart, C. Neal. *Genetically Modified Planet: Environmental Impacts of Genetically Engineered Plants.* New York: Oxford University Press, 2004.

SEE ALSO: Bioremediation; Biotechnology and genetic engineering; Cloning; *Diamond v. Chakrabarty*; Flavr Savr tomato; Gene patents; Genetically altered bacteria; Genetically engineered pharmaceuticals; Genetically modified foods.

Geoengineering

CATEGORY: Weather and climate

DEFINITION: Large-scale efforts to modify the earth's climate as a way of mitigating global warming

SIGNIFICANCE: The prospect of geoengineering is remote because of the great expense and impracticality of implementing many of the projects that have been proposed, but the fact that these ideas are being discussed is an indication of the urgency of the problem of global warming.

Several ideas involving geoengineering on a grand scale have been proposed as ways to ameliorate anthropogenic (human-caused) global warming. In 2006 Paul Crutzen of the Max Planck Institute for Chemistry in Germany, who won a Nobel Prize in 1995 for showing how industrial gases damage the earth's ozone shield, gave a fresh airing to ideas described during the 1980's by Wallace S. Broecker, a geoengineering pioneer at Columbia University. Broecker had theorized that humans could cool the earth by spreading tons of sulfur dioxide into the stratosphere, as erupting volcanoes occasionally do. The injections, he calculated during the 1980's, would require a fleet of hundreds of jumbo jets and, as a by-product, could increase acid rain. Other scientists argued that sulfur conveyed to the stratosphere, which is dry, would not return to the earth's surface as acid rain. Crutzen estimated that the annual cost of spreading the sulfur could be as high as fifty billion dollars, or about 5 percent of the world's annual military spending. The sulfur dioxide would have to be refreshed at least twice per month, as the previous load dispersed.

PROPOSED METHODS OF ATMOSPHERIC MODIFICATION

The so-called sulfur solution had also been raised in a 1991 National Academy of Sciences report on climate change mitigation strategies that concluded with consideration of atmospheric modification. Other proposed strategies included the placing of mirrored platforms in orbit to reflect sunlight, the use of guns or balloons to add dust to the stratosphere to reflect sunlight, and the placing of billions of aluminized, hydrogen-filled balloons in the stratosphere, also to reflect incoming solar radiation.

Additional strategies that have been proposed include the use of aircraft with engines modified to be less efficient (making them intentionally polluting) to maintain a dust cloud between the earth and the sun. In 2006 Roger P. Angel, an astronomer at the University of Arizona, suggested a plan that would involve putting into orbit trillions of small, very thin lenses that would bend sunlight away from the earth.

Such ideas for modifying the earth's atmosphere probably will remain limited to an intellectual parlor game, because the costs of carrying any of them out would be prohibitively high. In addition to monetary costs, the environmental costs of filling the stratosphere with sulfur dioxide or other pollutants would far outweigh the benefits, even in an increasingly warm and humid world. In the case of the use of sulfur dioxide, scientists have observed that sulfuric acid also tends to attract chlorine atoms, creating a chemical combination that could assist chlorofluorocarbons (CFCs) in devouring stratospheric ozone.

Another proposed solution to global warming that would use sulfur involves the burning of sulfur in ships and power plants to form sulfate aerosols. Mikhail Budyko, a Russian climatologist, proposed a massive atmospheric infusion of sulfur that would form enough sulfur dioxide to wrap the earth in radiation-deflecting thin white clouds within a few months. Budyko posited that the net effect would be to cool the earth in a fashion similar to the massive eruption of the volcano Mount Tambora in 1815, which ejected enough sulfur into the air to produce, in 1816, "the year without a summer." Crops across New England and upstate New York were devastated by frosts that continued into the summer months, and in the far northern part of New York State frosts were reported into June.

The Atmospheric Chemistry of Sulfur

Suspended particulates caused by emissions of sulfur dioxide and some other urban air pollutants (aerosols) increase the reflectivity of the earth, and thus usually exert a cooling influence on planetary temperatures. When Mount Pinatubo erupted in 1991, about 20 million tons of sulfur were injected into the atmosphere, enough to cool the earth's near-surface atmosphere about 0.5 degree Celsius (0.9 degree Fahrenheit) for a year or two, or roughly the increase attributable to global warming during the previous century. The physical challenge of lifting that much sulfur into the atmosphere year after year would be very great, especially as increases in the levels of greenhouse gases over time would require more of it.

Further, an increase in atmospheric particulate matter does not always exert a cooling influence. Researchers working with the National Oceanic and Atmospheric Administration (NOAA) have assembled data indicating that periodic increases in atmospheric dust concentrations during the glacial periods of the past 100,000 years may have resulted in significant regional warming, and that this warming may have triggered some of the abrupt climatic changes observed in paleoclimate records.

Criticisms of Geoengineering Proposals

Aside from causing atmospheric conditions that might contribute to the creation of acid rain, filling the upper atmosphere with sulfur could also deplete stratospheric ozone and reduce overall precipitation, most notably during the African and Indian monsoons, which are crucial for hundreds of millions of subsistence farmers. Implementation of the "sulfur solution" also would reduce pressure on human populations to reduce their greenhouse emissions, making it likely that such emissions would increase. As a result, if the process of injecting sulfur into the atmosphere ever stopped, global warming would accelerate rapidly.

Some scientists object to geoengineering on the grounds that it would remove the pressure on the world's governments to deal with the problem of global warming at its source—that is, to use energy sources other than fossil fuels. Biochemist Meinrat Andreae of the Max Planck Institute for Chemistry has compared putting sulfur into the stratosphere as a solution to global warming to giving a junkie a fix. Kenneth Caldeira of the Carnegie Institution Department of Global Ecology at Stanford University has stated that the biggest risk of geoengineering is that it eliminates the pressure to decrease greenhouse gases. Others argue that geoengineering could make the earth dependent on a continuing human-provided sulfur "fix," and if the supply of sulfur should falter, the earth could heat up quickly within a few years.

Scientists have also noted that intentionally polluting the atmosphere to counter surface warming quickly gives rise to other problems. As T. M. L. Wigley of the National Center for Atmospheric Research has observed, global warming and increases in carbon dioxide concentration pose two threats: one from climate change directly and another from increasing acidity of the oceans. While geoengineering might provide a temporary fix for temperature change, it would do nothing about ocean acidity.

Bruce E. Johansen

Further Reading

Ackerman, A. S., et al. "Reduction of Tropical Cloudiness by Soot." *Science* 288 (May 12, 2000): 1042-1047.

Archer, David. *The Long Thaw: How Humans Are Changing the Next 100,000 Years of Earth's Climate.* Princeton, N.J.: Princeton University Press, 2009.

Kerr, Richard A. "Pollute the Planet for Climate's Sake?" *Science* 314 (October 20, 2006): 401-403.

Kolbert, Elizabeth. "Hosed: Is There a Quick Fix for the Climate?" *The New Yorker,* November 16, 2009, 75-77.

National Research Council. *Geological and Geotechnical Engineering in the New Millennium: Opportunities for Research and Technological Innovation*. Washington, D.C.: National Academies Press, 2006.

Wigley, T. M. L. "A Combined Mitigation/Geoengineering Approach to Climate Stabilization." *Science* 314 (October 20, 2006): 452-454.

SEE ALSO: Acid deposition and acid rain; Airborne particulates; Carbon dioxide; Global warming; Greenhouse effect; Greenhouse gases; Iron fertilization; Volcanoes and weather; Weather modification.

Geographic information systems

CATEGORY: Resources and resource management
DEFINITION: Software-based mapping systems capable of linking qualitative and quantitative geographic information from databases with interactive computerized maps
SIGNIFICANCE: Geographic information systems, which enable the computerized spatial analysis of various phenomena, have many environment-related applications. For example, they are useful for tracking nonpoint pollution, mapping out the positions of various biomes, monitoring the spread or dispersal of river sediments, and observing the migratory and nesting patterns of various animal species.

The use of geographic information system (GIS) technology in spatial analysis is a relatively new area of the field of applied geographic studies. Rudimentary GIS technology became available in the 1970's and was employed mostly for corporate and university use. It was not until the arrival of object-oriented operating systems and the spread of the personal computer, however, that GIS became a user-friendly software application. Today both online and hard-drive-resident GIS software programs are being used for a variety of applications.

If a question being investigated has a spatial dimension to it, a GIS can no doubt be used to express the data in a mapped format. The power of the GIS for environmental investigation comes from its ability to display data easily and quickly in a mapped form. In addition, the GIS can manipulate the data by applying various forms of statistical analysis and showing the changes and predictions of the phenomenon on the computerized map.

GIS software is often coupled with Global Positioning System (GPS) technology to identify absolute locations of points or areas of various phenomena or features in the field. The coordinates (latitude and longitude) of these observations are then brought into the GIS and automatically displayed in the form of a map.

GIS COMPONENTS

Like the application of a spreadsheet with a database, GIS uses data and then attaches the data to a form of geographic coordinate system to allow for expression of the data in a spatial or mapped form. The basic GIS is composed of a computer software program that can open existing digital map files and manipulate the active features within these files. The software also can build digital maps when it is supplied with spatial digital data. Although the utility of a digital map has many advantages over a hard copy, GIS software also makes possible the production of large- and small-scale hard-copy maps for display purposes.

The power of GIS software is derived from its unique handling of spatial information. GIS software takes data and expresses them in the form of mapping layers that can be laid over one another to compose a map showing various features in the contexts of time and space. Furthermore, these features can be linked as hot spots to provide users of the resulting map with additional information about the features. Conventional mapping protocols such as color, shading, and line also can be utilized interactively by the user of a GIS. Additionally, zooming in and out of features and layers is easily accomplished. By using advanced applications and algorithms, a GIS user can quickly and easily model spatial information to show impacts of the variables in mapped form.

One of the major challenges of early GIS use was that many maps existed only on paper, and they needed to be converted to digital format. Map digitizer hardware was used to scan paper maps into a digital format, in a tedious process that required the digitizer operator to trace over each detail on each map using a device similar to a conventional computer mouse, called a puck. Digitizers are still used in some cases, but large-area map scanners have taken over much of this work. Additionally, with the advancements in GPS technology that took place in the 1990's, the process of getting features from the real world into digital format was revolutionized. Modern

GPS systems allow operators to collect points and polygons in the field in the form of geographic coordinates and bring these into GIS databases.

ENVIRONMENTAL APPLICATIONS

The processes of environmental research, environmental assessment, and the aggregation and collection of ecological assets have all been greatly enhanced by the use of GIS mapping software. Examples of applications of GIS software in environmental research are diverse. Energy managers have utilized GIS to locate optimal power-generating sites for wind turbines. Wildlife managers use the analysis capabilities of GIS technology to help them manage wildlife within conservation areas by mapping data on such things as nesting sites and game runs. Resource and ecological agencies have used GIS in the mapping of nonpoint pollution sources.

GIS technology also has a considerable presence within urban planning agencies. The software is used to manage transportation systems, to locate potential sites for urban parks, and to maintain database records for zoning map purposes. Fire districts use GIS applications in managing their dispatch and communications systems. GIS has been a crucial tool for on-site analysis in some emergency management applications, and it has been used to chart the flow of hazardous chemical spills as well as to alert community residents who might be in danger from these environmental hazards.

M. Marian Mustoe

FURTHER READING

Amdahl, Gary. *Disaster Response: GIS for Public Safety.* Redlands, Calif.: ESRI Press, 2001.

Clarke, Keith C., Bradley O. Parks, and Michael P. Crane, eds. *Geographic Information Systems and Environmental Modeling.* Upper Saddle River, N.J.: Prentice Hall, 2002.

Davis, David E. *GIS for Everyone: Exploring Your Neighborhood and Your World with a Geographic Information System.* 3d ed. Redlands, Calif.: ESRI Press, 2003.

Easa, Said, and Yupo Chan, eds. *Urban Planning and Development Applications of GIS.* Reston, Va.: American Society of Civil Engineers, 2000.

Greene, R. W. *Confronting Catastrophe: A GIS Handbook.* Redlands, Calif.: ESRI Press, 2002.

MacArthur, R. "Geographic Information Systems and Their Use for Environmental Monitoring." In *Environmental Monitoring and Characterization,* edited by Janick F. Artiola, Ian L. Pepper, and Mark Brusseau. Burlington, Mass.: Elsevier Academic, 2004.

SEE ALSO: Earth resources satellites; Urban planning; U.S. Geological Survey; Wildlife management.

Geothermal energy

CATEGORY: Energy and energy use

DEFINITION: Renewable source of heat and power that draws on the earth's internal heat

SIGNIFICANCE: Although the extraction of geothermal energy is not without some negative environmental impacts, geothermal resources represent a renewable, low-carbon energy source that can be used for a variety of energy services.

Geothermal energy is predominantly the product of three sources: decay of radioactive isotopes in the earth's crust, heat remaining from the time of the earth's formation, and incoming solar radiation. The creation of an extractable geothermal resource requires a heat source, a reservoir in which to collect heat, and an insulating barrier to preserve accumulated heat. Though the most highly valued reservoirs feature temperatures of several hundred degrees Celsius, lower-temperature resources, those under 150 degrees Celsius (302 degrees Fahrenheit), can still be economical to utilize. In addition to temperature, the depth and the presence or absence of water also help determine the use and usefulness of geothermal resources.

The presence of water in geothermal reservoirs creates hydrothermal, or hot water, resources. Human beings have used hot springs for therapeutic bathing throughout history, and for several millennia, communities located near warm water sources have used those resources directly for space heating. In the modern world, networks of piped hot groundwater are used for a variety of purposes, including space and district heating, greenhousing and aquaculture, and industrial drying and processing. The appropriateness of using a hydrothermal resource for a given direct-heating purpose is dependent on several factors, including water temperature and mineral content, and proximity to the site of use. If used efficiently, a single flow of hot water can provide multiple

Geothermal Power Plants

Geothermal power plants use three different technologies to convert hydrothermal fluids to electricity: dry steam, flash steam, and binary cycle. The first geothermal power generation plants were of the dry steam type. As of the early twenty-first century, flash steam plants were the most common type in operation. Most geothermal plants in the future will be binary-cycle plants.

- **Dry steam power plants:** Steam goes directly to a turbine that drives a generator to produce electricity. Steam technology is used at the Geysers field in Northern California, the world's largest single source of geothermal power.

- **Flash steam power plants:** Hydrothermal fluids above 182 degrees Celsius are used to make electricity. Fluid is sprayed into a tank that is held at a pressure much lower than the fluid, so that the fluid vaporizes (or flashes) rapidly. The vapor drives a turbine that in turn drives a generator to produce electricity. Liquid that remains in the tank is flashed again in another tank to produce more energy.

- **Binary-cycle power plants:** Energy is extracted from the moderate-temperature water (204 degrees Celsius) that exists in most geothermal areas through a process in which hot geothermal fluid and a second fluid with a boiling point much lower than that of water are passed through a heat exchanger. The secondary fluid flashes to vapor from the heat of the geothermal fluid, and the vapor drives a turbine, which drives a generator.

Source: U.S. Department of Energy, Office of Energy Efficiency and Renewable Energy, Geothermal Technologies Program, "Geothermal Power Plants" (2004).

heating services. Worldwide, geothermal energy provides almost 30,000 megawatts of heating services.

Geothermal heating and cooling of buildings is feasible in the absence of hydrothermal resources. In many places in the world, the top 3 meters (10 feet) of the ground remain a near-constant 10 to 15 degrees Celsius (50 to 59 degrees Fahrenheit) throughout the year. In summer or winter, when the ambient air temperature is substantially higher or lower than that of the subsurface, a heat pump can be used to circulate air through a system of underground pipes to cool or warm that air before its reintroduction into the building.

POWER PLANT TYPES

The first power plant that used geothermal energy to create electricity was built in 1904, and by 2009 plants producing more than 10,000 megawatts of generation capacity had been installed around the world. Different power plant types have been designed to create electricity from different hydrothermal resource types. When hydrothermal resources exist as steam (that is, hot water vapor), they can be used directly in conventional turbine generators. When liquid water is present at high temperatures (above 150 degrees Celsius) and under high pressure, a flash steam power plant pumps the water into low-pressure tanks, where the water "flashes" (depressurizes and vaporizes), producing steam that drives a turbine. When hydrothermal resources are at relatively low temperatures (below 150 degrees Celsius), a binary-cycle power plant that employs a secondary working fluid, such as *n*-pentane or ammonia, can be employed. In such a plant, the heat from the hydrothermal resource is used to boil the secondary fluid, thereby providing the vapor required to operate the turbine.

Because most unexploited hydrothermal resources occur at lower temperatures, future geothermal power plants built around natural hydrothermal resources are likely to employ the binary-cycle design. Nearly all major geothermal power plants built before 2010 are located at tectonic plate junctions, sites that are prone to high terrestrial heat flow. Accessing this heat flow often requires drilling boreholes up to 3 kilometers (1.9 miles) deep. Next-generation enhanced geothermal systems (EGS) are being designed that can access hot, dry rock up to 10 kilometers (6 miles) below the surface and, through the injection of water, artificially create a hydrothermal resource. EGS technology could allow large geothermal power plants to be built in areas far from plate junctions.

ADDITIONAL CONSIDERATIONS

A feature of geothermal energy that distinguishes it from other renewable energy sources, such as sunshine or wind, is that it is almost continuously available. For this reason, geothermal power plants present a viable option for base-load electricity generation, a power plant class long dominated by coal and nuclear fuels.

Although geothermal energy is a highly available resource, care must be taken not to mine heat at a faster rate than it is replenished by natural processes, as this practice degrades the geothermal resource and diminishes the efficiency of electricity production. Similarly, many geothermal power plants must be designed to reinject any water extracted by pumping so as to maintain the stock of on-site hydrothermal resource.

Geothermal energy is often touted by environmentalists as offering a way to generate electricity without releasing large volumes of greenhouse gases, unlike fossil-fuel-powered plants. Geothermal energy extraction does, however, have other environmental consequences. Depending on the system, utilizing geothermal energy can require drilling holes ranging in depth from a couple of meters to several kilometers. This process can change surface morphology, disrupt plants and wildlife, and pose a threat to groundwater sources. The operation of geothermal power plants can promote the release of toxic gases (such as hydrogen sulfide and ammonia) and minerals (such as arsenic and mercury). The discharge of spent hydrothermal resources into waterways can lead to thermal pollution, and the drilling of deep wells, such as those required for enhanced geothermal systems, may lead to increases in seismic activity.

Although the total amount of thermal energy stored in the earth is immense, much of that energy is either too deep or too diffuse to be extracted readily. However, several estimates of the power available in economically accessible geothermal reservoirs support the claim that geothermal energy can contribute a substantial portion of the energy needed to meet worldwide demand, either through direct use for heating and cooling or through indirect use for electricity production. Using existing technologies with high-quality geothermal resources, geothermal plants can produce power and heat at prices competitive with such widely used energy sources as coal and natural gas.

Joseph Kantenbacher

Further Reading

Dickson, Mary H., and Mario Fanelli, eds. *Geothermal Energy: Utilization and Technology.* Bangalore, India: United Nations Educational, Scientific, and Cultural Organization, 2003.

DiPippo, Ronald. *Geothermal Power Plants: Principles, Applications, Case Studies, and Environmental Impact.* 2d ed. Burlington, Mass.: Butterworth-Heinemann, 2008.

Gupta, Harsh, and Sukanta Roy. *Geothermal Energy: An Alternative Resource for the Twenty-first Century.* Boston: Elsevier, 2007.

MacKay, David J. C. *Sustainable Energy—Without the Hot Air.* Cambridge, England: UIT Cambridge, 2009.

SEE ALSO: Alternative energy sources; Heat pumps; Power plants; Renewable energy; Thermal pollution.

Gibbons, Euell

CATEGORIES: Activism and advocacy; forests and plants; agriculture and food

IDENTIFICATION: American ethnobotanist and nature writer

BORN: September 8, 1911; Clarksville, Texas

DIED: December 29, 1975; Sunbury, Pennsylvania

SIGNIFICANCE: Gibbons improved the public image of wild food foraging and thus of environmentalism in general, as his staid, avuncular image made environmentalism acceptable to Americans who had tended to perceive environmental activism as subversive.

By becoming a best seller, Euell Gibbons's first book, *Stalking the Wild Asparagus* (1962), moved the art of collecting and preparing wild foods out of the realm of the eccentric and into the mainstream of popular culture. Gibbons's writing, as well as his lifetime of research both in the field and at the Pendle Hill Quaker Study Center in Wallingford, Pennsylvania, offered original arguments for the necessity of maintaining the balance of nature. Whereas many environmentalists blamed a fundamentalist reading of the biblical mandate to "subdue the world" for many of the ecological problems of the twentieth century, Gibbons placed equal blame on a misunderstanding of Darwinism. Much of the notion of conquering nature that Gibbons, like most ecologists, despised stemmed, he thought, from misunderstanding of the concept of natural selection. "Survival of the fittest," Gibbons taught, does not always mean "survival of the strongest" but rather "survival of the most cooperative." A forager who uproots a plant is not a destroyer from a larger ecological point of view, since such an action scatters seeds and invariably creates more plants than it destroys. By depicting the human forager as a vital part of the plant's reproductive cycle, Gibbons offered an ideological framework for understanding natural selection without the insistence on predator and prey.

Gibbons's impact on ecological thought went beyond the influence of his books, however. Both the Boy Scouts of America and the U.S. Navy hired him to teach foraging as a survival skill, though Gibbons disliked that context for his discipline. His advocacy of wild foods was culinary rather than survivalist, and he resented the romanticized notion of going into the woods with no provisions and "living off the land." Besides, as a Quaker and pacifist, he resented the military uses to which his skills might be put by the Navy. Nevertheless, he welcomed the opportunity to spread the gospel of cooperation with nature. Another forum for his ideas was a series of seminars he ran at Ithaca College and Bucknell University in the 1970's.

Euell Gibbons teaching a survival course at a boys' camp. (Time & Life Pictures/Getty Images)

With these seminars he assured a wide distribution of his methods by training college students in the first half of the seminar and then guiding them in training high school students in the second half.

Though not formally educated in botany—his college studies were in anthropology and creative writing—Gibbons read virtually every article and book, scholarly and popular, that dealt with the gathering, preparation, and nutritive value of wild plants. He coined the term "ethnobotany" because he saw his field as recovering the botanical and culinary knowledge of Native American culture.

John R. Holmes

FURTHER READING

Cunningham, Anthony B. *Applied Ethnobotany: People, Wild Plant Use, and Conservation.* Sterling, Va.: Earthscan, 2001.

Gibbons, Euell. *Stalking the Wild Asparagus.* 25th anniversary ed. Putney, Vt.: A. C. Hood, 1987.

SEE ALSO: Balance of nature; Environmental ethics.

Gibbs, Lois

CATEGORIES: Activism and advocacy; human health and the environment

IDENTIFICATION: American environmental activist

BORN: June 25, 1951; Buffalo, New York

SIGNIFICANCE: Gibbs united her community by forming the Love Canal Homeowners Association and leading efforts to compel state and federal officials to relocate residents whose homes were compromised by exposure to toxic waste.

In 1977 Lois Gibbs discovered that the elementary school her son was attending in the Love Canal neighborhood of Niagara Falls, New York, was built on top of a toxic chemical dump containing twenty thousand tons of waste. Gibbs was alerted to the presence of the dump in December, 1977, when her son, Michael, began to experience asthma and seizures just four months after he entered kindergarten. She then went door-to-door and questioned other residents about their health in an attempt to determine the full extent of contamination. Two years later, Gibbs traveled to Washington, D.C., to testify on behalf of the Love Canal Homeowners Association at

Lois Gibbs, in December, 1978, makes adjustments to a Christmas tree trimmed with decorations that name some of the chemicals found in the Love Canal neighborhood. The sign at the top of the tree saying "Hooker is no angel" refers to the Hooker Electrochemical Corporation, which used the site as a chemical dump for ten years. (AP/Wide World Photos)

hearings on hazardous waste disposal held before the House Subcommittee on Oversight and Investigations.

Although homes immediately adjacent to the dump were evacuated, residents living outside the "inner ring" remained until January, 1980, when the federal government's Environmental Protection Agency (EPA) financed a controversial scientific investigation by Biogenics Corporation of Houston, Texas. On May 17, 1980, the EPA released a report that stated residents of the area were believed to have damaged chromosomes as a result of exposure to the dump and consequently were at increased risk of cancer, miscarriages, birth defects, and seizures. Gibbs described the report as "one more frightening, scary thing, and we couldn't take it any more. . . . People were very, very frightened, almost panicked." Shortly afterward, President Jimmy Carter declared a state of emergency at Love Canal and authorized relocation of the residents at a cost of $35 million to the federal government.

Lois Gibbs's fight to evacuate Love Canal captured the public's imagination and was depicted in *Lois Gibbs and the Love Canal,* a 1982 film made for television starring Marsha Mason. The most significant development following the Love Canal disaster was passage of the Comprehensive Environmental Response, Compensation, and Liability Act of 1980 (widely known as Superfund). This legislation created an industry-funded program for cleaning up toxic chemical dumps across the nation. In 1981 Gibbs formed the Citizens Clearinghouse for Hazardous Waste, which was later renamed the Center for Health, Environment, and Justice. This nonprofit organization began by helping community groups suffering from the effects of toxic dumps similar to Love Canal. It eventually expanded its programs to match the growing needs of grassroots environmental organizations, working with more than eight thousand community-based groups.

Peter Neushul

FURTHER READING

Blum, Elizabeth D. *Love Canal Revisited: Race, Class, and Gender in Environmental Activism.* Lawrence: University Press of Kansas, 2008.

Gibbs, Lois. "What Happened at Love Canal." 1982. In *So Glorious a Landscape: Nature and the Environment in American History and Culture,* edited by Chris J. Magoc. Wilmington, Del.: Scholarly Resources, 2002.

SEE ALSO: Center for Health, Environment, and Justice; Environmental justice and environmental racism; Hazardous and toxic substance regulation; Hazardous waste; Love Canal disaster; Superfund.

Gila Wilderness Area

CATEGORY: Preservation and wilderness issues

DATE: Established in June, 1924

DEFINITION: Area of protected wilderness within the Gila National Forest

SIGNIFICANCE: Establishment of the Gila Wilderness Area, the first designated wilderness area in the world, was a harbinger of the national wilderness system later established under the Wilderness Act of 1964.

The Gila Wilderness Area, located in southwestern New Mexico, comprises more than 202,000 hectares (500,000 acres) of rugged country containing highly diverse landforms, plants, and animals. It includes the termini of both the Rocky Mountains to the north and the Sierra Madre range to the south. The Chihuahuan and Sonoran deserts also reach into the Gila Wilderness Area, contributing to its high biodiversity.

The designation of this region as the world's first official wilderness area grew out of the nineteenth century conservation movement in the United States. As the United States became increasingly settled, some people grew concerned with the need to preserve the land's natural beauty. This led to the establishment of the first national park, Yellowstone, in 1872. At the same time, however, plans were being formulated to make the park and similar areas readily accessible to visitors. About thirty years later, a number of observers began to realize that some wildlands needed to be preserved in a manner devoid of roads and other amenities for visitors.

In 1899 President William McKinley set aside the Gila River Forest Reserve. Seven years later, the Gila National Forest was established. In the early 1920's threats to the area by developers led Aldo Leopold, then a young forester in the Southwest District of the U.S. Forest Service, to awaken the American public to the need to save wildlands. An avid hunter, Leopold observed that wildlands and wildlife habitats were shrinking, and he blamed these losses on road building. Upon his arrival in the Southwest in 1909, he had identified six roadless backcountry areas—each more than 202,000 hectares in size—in the region's national forests. By 1921, only one, at the headwaters of the Gila River, remained without roads.

Passage of the Federal Highway Act in 1922 threatened to bring more road building and tourism, leading Leopold to fear the loss of this last roadless area. As assistant district forester, he led a movement to establish the Gila headwaters area as the first officially designated wilderness, to keep it roadless and without buildings or artificial trails. Persuaded by his arguments, the district forester designated some 305,500 hectares (755,000 acres) as the Gila Wilderness Area in June, 1924—a harbinger of the national wilderness system later established under the Wilderness Act of 1964.

Under the 1924 U.S. Forest Service wilderness designation, hunting was encouraged in the Gila Wilderness Area. Predator control then led to deer overpopulation. In 1933 the Forest Service responded by rebuilding an abandoned wagon road through the wilderness to provide easier access for hunters. This resulted in division of the wilderness. After some modifications in the laws and name changes in the intervening years, the New Mexico Wilderness Act of 1980 designated approximately 226,000 hectares (558,000 acres) west of the road as the Gila National Wilderness and about 81,000 hectares (200,000 acres) east of the road as the Aldo Leopold National Wilderness. These areas together exceed the size of the original Gila Wilderness Area.

Jane F. Hill

FURTHER READING

Huggard, Christopher J. "America's First Wilderness Area: Aldo Leopold, the Forest Service, and the Gila of New Mexico, 1924-1980." In *Forests Under*

Fire: A Century of Ecosystem Mismanagement in the Southwest, edited by Christopher J. Huggard and Arthur R. Gómez. Tucson: University of Arizona Press, 2001.

Lewis, Michael, ed. *American Wilderness: A New History.* New York: Oxford University Press, 2007.

Lorbiecki, Marybeth. *Aldo Leopold: A Fierce Green Fire.* 1996. Reprint. Guilford, Conn.: Globe Pequot Press, 2005.

SEE ALSO: Forest Service, U.S.; Hunting; Leopold, Aldo; National forests; Predator management; Wilderness Act; Wilderness areas.

Gill nets and drift nets

CATEGORY: Animals and endangered species

DEFINITIONS: Gill nets are mesh nets that ensnare fish by their gills; drift nets are large vertical nets (often made up of multiple gill nets) that are allowed to drift free in a body of water

SIGNIFICANCE: Although they are made to snare particular types of fish, gill nets and drift nets can trap and kill untargeted types of marine life. Lost drift nets can float in the open sea, killing marine life indiscriminately.

Mesh fishing nets of the type known as gill nets have been used for centuries. Fish swim through the mesh openings of gill nets and become ensnared by their gills when they try to back out. Gill nets are manufactured with specific mesh opening sizes for use in selectively capturing certain fish species without affecting other, untargeted, species. Even if gill nets snare untargeted species, if fishers regularly tend their nets, they can return any entangled untargeted marine life (bycatch) to the sea and guarantee that all targeted fish are removed from the nets before they die so that they are brought to market in the freshest possible condition.

Not all fishers exercise such vigilance, however, and untended gill nets do kill untargeted marine life. Nylon gill nets, which are quite durable, are almost in-

Fishermen pull in a drift net loaded with salmon near Egegik, Alaska, on Bristol Bay. The United States permits drift-net fishing within U.S. waters. (AP/Wide World Photos)

visible in the water, and this increases the tendency of untargeted species to swim into them and become entangled. Gill-net bycatches may include sea turtles, seabirds, and marine mammals such as sea otters, seals, dolphins, porpoises, and whales.

Drift nets are very large nets, often made up of several gill nets, that hang vertically in the water; floats attached to the top and weights secured to the bottom keep the nets vertical. Because drift nets are not anchored to the seafloor, they drift with the currents. Drift nets have a high rate of bycatch. One trial use of drift nets to catch skipjack tuna in early 1989 showed that the nets killed one marine mammal for every ten tuna caught, compared with one dolphin per 70 tons of tuna in the eastern Pacific purse-seine tuna fisheries. The loss of drift nets during storms produces "ghost nets," which continue to catch marine life indiscriminately. It has been estimated that 300,000 dolphins, porpoises, and whales are killed by untended gill nets, drift nets, and ghost nets each year.

Some fishers attach small devices known as pingers, which emit audible signals, to their gill nets or drift nets to warn marine mammals that they are approaching the nets. Several studies have shown that the use of pingers can significantly decrease the bycatch of some marine mammals without affecting fish catches. The responses of different marine species to pingers vary, however. For example, sea lions have been known to become increasingly attracted to pingers over time, and such devices are not a reliable means of reducing bycatch of endangered species.

The United Nations General Assembly passed a resolution in December, 1989, prohibiting the expansion of drift netting and calling for a gradual phaseout of the practice by 1992. Illegal drift netting continues to occur, however. New Zealand, Australia, and South Africa have prohibited drift netting in their waters. Even Japan, which has one of the largest drift-netting operations in the world, banned large-scale drift netting in its own waters in response to the depletion of marine species that unregulated drift netting caused within Japan's own exclusive economic zone.

Michael A. Buratovich

FURTHER READING

Allsopp, Michelle, et al. *State of the World's Oceans.* New York: Springer, 2009.

Barlow, Jay, and Grant A. Cameron. "Field Experiments Show That Acoustic Pingers Reduce Marine Mammal Bycatch in the California Drift Gill Net Fishery." *Marine Mammal Science* 19, no. 2 (2003): 265-283.

Miller, Frederic P., Agnes F. Vandome, and John McBrewster, eds. *History of Fishing: Fishing, Gillnet, Trawling, Cod, Trepanging, Fly Fishing, History of Whaling.* Beau Bassin, Mauritius: Alphascript, 2009.

SEE ALSO: Commercial fishing; Dolphin-safe tuna; Environmental law, international; Environmental law, U.S.; Fisheries; Sea Shepherd Conservation Society; Sea turtles.

Glacial melting

CATEGORY: Weather and climate

DEFINITION: Gradual erosion of polar ice caps and mountain glaciers

SIGNIFICANCE: The most widespread indication that the earth has been warming over time has been the steady erosion of ice in the Arctic and Antarctic and on mountain glaciers. Although a few exceptions do exist, the worldwide erosion of ice leaves little doubt that that the earth has experienced steadily increasing warming for at least a century.

According to a report by Hamish D. Pritchard and his colleagues that appeared in the journal *Nature* late in 2009, accelerated ice flow (dynamic thinning), measured by high-resolution laser altimetry, reaches all latitudes in Greenland and has intensified on key Antarctic grounding lines, penetrating far into the interior of each ice sheet, spreading as ice shelves thin by ocean-driven melt. Mountain glaciers are in rapid retreat around the earth, with very few exceptions. The Intergovernmental Panel on Climate Change's models project that between one-third and one-half of existing mountain glacial mass could disappear over the next hundred years. Sometime during the twenty-first century, the last glacier may melt in Montana's Glacier National Park.

Mass loss from the mountain glaciers, as well as massive ice shelves and sheets along the coasts of Greenland and Antarctica, is accelerating faster than expected, contributing to sea-level rises. As seas gradually rise because of glacial melting and thermal expansion along some coasts, isostatic rebound, the gradual rising of land masses that have been compressed by glaciers in the past, has provoked the rise of Scotland's coast following the last ice age. The Great Lakes, remnants of melted glaciers, also experience some isostatic rebound.

MOUNTAIN GLACIERS IN RAPID RETREAT

Climbers have been plucked from the Matterhorn in the Swiss Alps as thawing mountainsides crumbled under them. During the summer of 2003, Mont Blanc, Europe's tallest mountain, was closed to hikers and climbers because its deteriorating snow and ice were too unstable to allow safe passage. The mountain was crumbling as ice that once held it together melted during a record-warm summer in Europe. Scientists have estimated that by 2025 glaciers in the Swiss Alps will have lost 90 percent of the volume they contained a century earlier.

The U.S. Geological Survey (USGS) has collected a digital library that describes the state of more than 67,000 glaciers around the world. Using historical photographs, images from space satellites, precision laser measurements, and other tools, the USGS has created an archive that tells a story of glacial retreat around the world. The rate of ice loss doubled from 1988 to 2002. Generally, the only glaciers gaining mass have been in wet maritime areas of the world, such as parts of Norway and Sweden, where melting has been offset by increased snowfall, another facet of climate change. Alaska's Hubbard Glacier is advancing so swiftly that it threatens to seal off the entrance to Russell Fjord near Yakutat.

GLOBAL EFFECTS OF ICE MELT

The snow-and-ice crown of Mount Kilimanjaro in equatorial Africa may vanish by the mid-twenty-first century. Kilimanjaro will no longer live up to its name, which in Swahili means "mountain that glitters." Nearby Mount Kenya's ice fields lost three-quarters of their mass during the twentieth century. By the end of the twenty-first century, Glacier National Park in Montana may lose the last of its permanent glaciers. The original 150 glaciers within Glacier National Park

Ice falls from Hubbard Glacier in Alaska into the water of the bay below. (©Linda Bair/Dreamstime.com)

had been reduced to 37 by 2002, and most of these were small remnants of once-mighty ice masses.

Millions of people around the world may face severe water shortages as glaciers around the world melt. Ecuador, Peru, and Bolivia, where major cities rely on glaciers as their main source of water during dry seasons, could be among the most intensely affected. Areas of the Himalayas face grave danger of flooding initially, as glaciers melt, followed by water shortages from the ebbing of glacier-fed rivers in the region that supply water to one-third of the world's population, mainly in India and China.

Evidence from Antarctica suggests that melting ice may flow into the sea much more easily than earlier believed, perhaps leading to an accelerating rise in sea levels. A study published in the journal *Science* in 2003 suggested that seas might rise as much as several meters during the next several centuries, given projected global warming based on continuing usage of fossil fuels at the levels of the early twenty-first century. Additional studies have provided further evidence of this trend, as the important Pine Island Glacier of the West Antarctic Ice Sheet has been receding into the ocean at a faster pace. This mass of ice could influence the entire region in years to come.

Bruce E. Johansen

FURTHER READING

Alley, Richard B., et al. "Ice-Sheet and Sea-Level Changes." *Science* 310 (October 21, 2005): 456-460.

Dutch, Steven I., ed. *Encyclopedia of Global Warming.* 3 vols. Pasadena, Calif.: Salem Press, 2010.

Hansen, James. "Defusing the Global Warming Time Bomb." *Scientific American*, March, 2004, 68-77.

Knight, Peter G., ed. *Glacier Science and Environmental Change.* Malden, Mass.: Blackwell, 2006.

Lynas, Mark. *High Tide: The Truth About Our Climate Crisis.* New York: Picador, 2004.

_____. *Six Degrees: Our Future on a Hotter Planet.* New York: HarperCollins, 2008.

Pritchard, Hamish D., et al. "Extensive Dynamic Thinning on the Margins of the Greenland and Antarctic Ice Sheets." *Nature* 461 (October 15, 2009): 971-975.

SEE ALSO: Alliance of Small Island States; Carbon dioxide; Climatology; Global warming; Greenhouse effect; Greenhouse gases; Polar bears; Sea-level changes.

Glen Canyon Dam

CATEGORY: Preservation and wilderness issues

IDENTIFICATION: Large hydroelectric dam built on the Colorado River in northeastern Arizona

DATES: Construction begun in 1956; reservoir filled in 1983

SIGNIFICANCE: Construction of the Glen Canyon Dam flooded the pristine, wild Glen Canyon and created Lake Powell. Downstream, the dam's effects on water temperature and sediment load affected the ecology of the Grand Canyon over time. Flow regulation experiments begun during the late 1990's have sought to reverse some of the dam's adverse effects.

The U.S. Congress authorized the building of the Glen Canyon Dam in 1956 to provide water storage by creating the second-largest human-made reservoir in the United States, Lake Powell. Construction of the dam began in 1956, and the lake reached its full mark in 1983. The dam and reservoir generate hydroelectric power, provide a recreation area, and decrease siltation in the downstream Lake Mead reservoir, which is formed by Hoover (Boulder) Dam. The combined storage behind the Glen Canyon and Hoover dams manages the flow of water in the Colorado River to California, Arizona, Nevada, and Mexico. Agriculture consumes approximately 85 percent of the Colorado River water, with most of the remainder going to urban areas in Southern California; Phoenix, Arizona; and Las Vegas, Nevada. Glen Canyon Dam, with power plants, cost $272 million to build. Although the plants were designed to generate up to 1,300 megawatts of electric power annually, environmental damage to the riparian (shoreline) ecosystem downstream in the Grand Canyon proved too great at this rate of electricity production, and the government limited production to fewer than 800 megawatts annually.

When construction on the dam began, Glen Canyon was considered extremely remote and was rarely visited. The Sierra Club initially fought the project but agreed to end its resistance in a compromise agreement with the U.S. Bureau of Reclamation that stopped construction of two other dams in Utah. The environmental organization's leaders soon regretted their decision when they learned of the spectacular beauty of Glen Canyon. Sierra Club executive director David Brower teamed with photographer Eliot

Porter to create a beautifully illustrated book titled *The Place No One Knew* (1963), which featured photographs of Glen Canyon before the dam. This book launched a Sierra Club policy of publishing illustrated books on outstanding natural areas to try to ensure that Americans would sacrifice no other spectacular areas because they did not know of their beauty. In 1996 the Sierra Club called on the federal government to drain Lake Powell and abandon the dam.

Many people view the Glen Canyon Dam as an icon of environmental destruction. The central theme of Edward Abbey's landmark novel on ecotage, *The Monkey Wrench Gang* (1975), involves plotting to destroy the dam. The radical preservationist group Earth First! leaped into national prominence during the early 1980's when members unfurled a 300-foot length of black plastic from the top of the dam that gave the appearance of a giant crack descending down the front of the structure. Brower suggested that the lake be drained but the dam left "as a tourist attraction, like the Pyramids, with passers-by wondering how humanity ever built it, and why."

IMPACTS

The most obvious postdam change was the flooding of a spectacular canyon and its tributaries, along with the accompanying loss of hiking and river-running experiences. The entire ecosystem of the inundated area, 653 square kilometers (252 square miles), was destroyed and replaced with a lake community. The reservoir water altered the color of the adjacent red sandstone rocks, resulting in a prominent white "bathtub ring" along the shore most of the year.

The reservoir behind the Glen Canyon Dam caused several other changes. Because the reservoir is located in the desert, an estimated 74,000 hectare-meters (1 hectare of water, 1 meter deep) of water (600,000 acre-feet) are lost to evaporation each year.

Glen Canyon Dam. (©Dreamstime.com)

Annually, more than 43,000 hectare-meters (350,000 acre-feet) of water seep into the surrounding rock strata. The river loses approximately 8 percent of its volume in Lake Powell. An additional concern is the rate of siltation. Silt and sand carried by the Colorado River settle onto the reservoir floor, and more than 123,000 hectare-meters (1 million acre-feet) of water storage have been loss to siltation.

Dramatic changes to the downstream riparian environment in the Grand Canyon are less apparent. Prior to dam construction, muddy river temperatures seasonally ranged from 26 degrees Celsius (80 degrees Fahrenheit) to nearly freezing. The dam releases water from the depths of the reservoir, and the river below the dam is approximately 9 degrees Celsius (48 degrees Fahrenheit). This water is also more saline than it was before the dam was erected, as evaporation allows the natural salts in the water to become more concentrated. In addition, the river below the dam bears only 15 percent of the sediment and nutrient load that it used to carry when it ran free. These changes have affected the aquatic fauna and flora of the Grand Canyon's Colorado River. Several species have become extinct, and many nonnative or previously sparse species have flourished.

Releasing water from the dam to meet hydroelectric needs eliminated large floods but caused a daily tide as more water was released during afternoon, high-power-demand times and less flowed through the dam during the low-demand nights. The resulting tide stressed aquatic organisms downstream and caused rapid beach and sandbar erosion. Lack of natural flooding prevented rebuilding of the beaches and sandbars. The rapid loss of sandy deposits along the river in the Grand Canyon brought about ecosystem changes and had negative impacts on the recreational experiences of customers of the multimillion-dollar rafting industry.

NEW APPROACHES

During the 1980's and 1990's, the U.S. Department of the Interior undertook an extensive review of the impacts of the Glen Canyon Dam that resulted in an environmental impact study. The major outcome of the $70 million study was a revised approach to the way water was released from the dam. Extreme and rapid daily fluctuations in river level to meet power needs were eliminated. Also, in the spring of 1996 the dam released an experimental flood with a slow rise to modest flood levels, followed by a gradual lowering of the river. The release was designed to scour the river channels and rebuild the beaches, as happens in a natural flood. This flow management measure succeeded in redistributing existing downstream sediments without contributing significant quantities of new sediment, but its benefits proved to be short-term. Other controlled flood experiments conducted in 2004 and 2008 also yielded transient benefits.

During the late 1990's, serious discussion began regarding the Sierra Club's call to decommission the dam, drain Lake Powell, and restore Glen Canyon. If such a plan were to be carried out, it would take the reservoir fifteen to twenty years to drain. According to an assessment issued in 2000 by the Glen Canyon Institute, the dam could continue to generate hydroelectric power for ten to fifteen years while the lake drained. Opponents of canyon restoration assert that decommissioning the dam would be too detrimental to water supplies, power generation, recreation, local economies dependent on tourism and the hydroelectric industry, and the ecosystems that have developed in the decades since the dam was built. Whether Lake Powell is maintained or drained, either choice would have a complex array of beneficial and adverse environmental effects on an extensive area upstream and downstream from Glen Canyon Dam.

Louise D. Hose
Updated by Karen N. Kähler

FURTHER READING

Carothers, Steven W., and Bryan T. Brown. *The Colorado River Through Grand Canyon: Natural History and Human Change.* Tucson: University of Arizona Press, 1991.

Farmer, Jared. *Glen Canyon Dammed: Inventing Lake Powell and the Canyon Country.* Tucson: University of Arizona Press, 1999.

Gloss, Steven P., Jeffrey E. Lovich, and Theodore S. Melis, eds. *The State of the Colorado River Ecosystem in Grand Canyon: A Report of the Grand Canyon Monitoring and Research Center, 1991-2004.* Reston, Va.: U.S. Geological Survey, 2005.

Lowry, William R. *Dam Politics: Restoring America's Rivers.* Washington, D.C.: Georgetown University Press, 2003.

McPhee, John. *Encounters with the Archdruid.* 1971. Reprint. New York: Farrar, Straus and Giroux, 2000.

Martin, Russell. *A Story That Stands Like a Dam: Glen Canyon and the Struggle for the Soul of the West.* Salt Lake City: University of Utah Press, 1999.

Parks, Timothy L. *Glen Canyon Dam.* Charleston, S.C.: Arcadia, 2004.

Porter, Eliot. *The Place No One Knew: Glen Canyon on the Colorado.* Edited by David Brower. Commemorative ed. Layton, Utah: Gibbs Smith, 2000.

Powell, James Lawrence. *Dead Pool: Lake Powell, Global Warming, and the Future of Water in the West.* Berkeley: University of California Press, 2008.

SEE ALSO: Abbey, Edward; Brower, David; Colorado River; Dams and reservoirs; Earth First!; Floods; Grand Canyon; Monkeywrenching; Sedimentation; *Sierra Club v. Morton.*

Global Biodiversity Assessment

CATEGORY: Ecology and ecosystems

IDENTIFICATION: Independent scientific analysis of all issues, theories, and views regarding biodiversity from a global perspective

DATES: 1993-1995

SIGNIFICANCE: The results of the Global Biodiversity Assessment made it clear that the degradation of the environment caused by expanding human populations had placed the earth's biological resources under serious threat. The assessment also revealed that much additional work needed to be done for scientists to reach consensus on all the various issues surrounding biodiversity.

From 1993 through 1995, a group of scientific experts helped develop, write, and review contributions to the Global Biodiversity Assessment. Funded by the Global Environment Facility (GEF) and the United Nations Environment Programme (UNEP), this independent, peer-reviewed assessment was the work of more than fifteen hundred scientists and other experts from all parts of the world. The final report focused on assessing the scientific understanding of biodiversity's various components—namely, ecosystems, species, and genes—and on identifying the gaps in the knowledge base that should be targeted for future research. It concluded that the preservation of biodiversity must include a blend of strategies, including programs to save species by creating controlled environments and policies to manage natural environments in ways that minimize adverse impacts on ecosystems and species.

While great advances in understanding the earth's biological diversity were made during the 1980's and 1990's, the results of the assessment demonstrated that this understanding continued to be incomplete. The researchers found a great range of opinion on many basic theoretical issues, and gaps in data were enormous, with estimates sometimes differing by several orders of magnitude. Decisions that rely on an understanding of ecosystem dynamics, such as knowledge about the optimal size of a nature reserve necessary to preserve species diversity, are greatly hindered by a lack of good information, as are decisions that rely on an understanding of how genetic diversity is distributed within populations and how species evolve and function.

The researchers who conducted the Global Biodiversity Assessment found ecosystems of all kinds around the world to be under great pressure. Lowland and coastal areas, native grasslands, wetlands, and many types of woodlands and forests had been adversely affected or destroyed by human activities. During the 1980's humid tropical forests were being lost at an annual rate of nearly 10.1 million hectares (25 million acres), dry tropical forests were being destroyed at an even faster rate, and 10 percent of the coral reefs of the world were eroded beyond recovery.

The researchers estimated that only about 13 percent of the total number of species on earth had been scientifically described. In addition, they found that the number of species reported as being threatened by extinction was far from the actual total. The researchers concluded that the root causes of the loss of biodiversity in general were closely tied to the ways in which human societies have used natural resources.

In conducting the Global Biodiversity Assessment, the researchers not only evaluated existing problems but also analyzed various options for ensuring that biodiversity would be conserved and protected wisely in the future. They concluded that biodiversity management must go far beyond steps such as establishing nature reserves and setting up agricultural seed banks. The management of biodiversity must be fully integrated into all aspects of landscape management.

Alvin K. Benson

FURTHER READING

Biermann, Frank. "Whose Experts? The Role of Geographic Representation in Global Environmental Assessments." In *Global Environmental Assessments: Information and Influence,* edited by Ronald B. Mitchell et al. Cambridge, Mass.: MIT Press, 2006.

National Research Council. *Analysis of Global Change Assessments: Lessons Learned.* Washington, D.C.: National Academies Press, 2007.

SEE ALSO: Biodiversity; Convention on Biological Diversity; Ecology as a concept; Ecosystems; Global Environment Facility.

Global Environment Facility

CATEGORIES: Organizations and agencies; ecology and ecosystems; weather and climate

IDENTIFICATION: Independent financial organization that provides grants and concessional funds to recipient countries for projects and activities that protect the global environment

DATE: Established in October, 1991

SIGNIFICANCE: The Global Environment Facility works to improve and protect the environment by funding developing nations' efforts to address problems related to such issues as climate change, biodiversity, and land degradation.

Jointly implemented by the United Nations Development Programme (UNDP), the United Nations Environment Programme (UNEP), and the World Bank, the Global Environment Facility (GEF) was launched as a pilot program in 1991. Since that time, seven more agencies have joined the GEF partnership: the United Nations Food and Agriculture Organization, the United Nations Industrial Development Organization, the African Development Bank, the Asian Development Bank, the European Bank for Reconstruction and Development, the Inter-American Development Bank, and the International Fund for Agricultural Development.

The GEF secretariat, which is functionally independent of the partner agencies, reports to the Council and Assembly of the GEF. The Council, the GEF's main governing body, consists of representatives from thirty-two countries; sixteen of these are from developing countries, fourteen are from developed countries, and two are from countries with transitional economies. Any country that is a member of the United Nations or one of its specialized agencies may become a GEF participant by filing a notification of participation with the GEF secretariat. By 2010, 181

countries were participating in the GEF program. Countries may be eligible for GEF project funds in one of two ways: if they are eligible to borrow from the World Bank or receive technical assistance funds from the UNDP, or if they are eligible for financial assistance through either the Convention on Biological Diversity or the Climate Change Convention. In either case, the country must be a participant in the Convention on Biological Diversity or the Climate Change Convention.

The organizing principle of the GEF is that no new bureaucracy will be created. The UNDP is responsible for technical assistance activities and helps identify projects and activities consistent with the GEF's Small Grants Programme for nongovernmental organizations (NGOs) and community groups around the world. Responsibility for initiating the development of scientific and technical analysis and for advancing environmental management of GEF-financed activities rests with UNEP, as does the management of the Scientific and Technical Advisory Panel, which provides scientific and technical guidance to the GEF. The World Bank serves as the repository of the GEF trust fund and is responsible for investment projects. It also seeks resources from the private sector in accordance with GEF objectives.

GEF provides financial assistance with the aim of facilitating projects with global environmental benefits. Funds are available for proposals dealing with six focal areas: biodiversity, climate change, international waters, land degradation, the ozone layer, and persistent organic pollutants. The projects and activities funded must be approved by the GEF Council. If at least four Council members request a review of a project's final documents, the Council must conduct a review prior to granting approval of GEF funds.

The GEF and the Montreal Protocol (1987) have a complementary relationship in funding projects concerned with the earth's ozone layer. Whereas the Montreal Protocol provides funds for ozone-related projects to countries where ozone-depleting substance (ODS) production or consumption is low, the GEF provides funding for countries where ODS production or consumption is high. This applies mainly to countries in Central and Eastern Europe.

Alvin K. Benson

FURTHER READING

Clapp, Jennifer, and Peter Dauvergne. "Global Financing and the Environment." In *Paths to a Green*

World: The Political Economy of the Global Environment. Cambridge, Mass.: MIT Press, 2005.

Miller, Alan S. "International Trade and Development." In *Global Climate Change and U.S. Law*, edited by Michael B. Gerrard. Chicago: ABA Publishing, 2008.

See also: Biodiversity; Debt-for-nature swaps; Development gap; Ecosystems; Montreal Protocol; Ozone layer.

Global ReLeaf

Categories: Organizations and agencies; activism and advocacy; forests and plants; preservation and wilderness issues

Identification: Conservation program that focuses on the planting of trees and the protection of forestlands from overdevelopment and pollution

Date: Initiated on October 12, 1988

Significance: Global ReLeaf has successfully planted millions of trees while also working to educate the public about the importance of the world's forests for the planet's environmental health.

In the United States, Global ReLeaf is the education and action program of the organization American Forests (formerly known as the American Forestry Association), which was founded in 1875 and is the nation's oldest national nonprofit conservation organization. Global ReLeaf sponsors educational programs to show the benefits of trees and forests for the environment and for the enhancement of people's lives. These programs highlight the value of trees for filtering air and water, sheltering and feeding wildlife, absorbing greenhouse gases, and reducing the runoff of polluted soil into rivers and streams. The program also provides funding, from private and corporate donations, for tree-planting projects across the United States.

Global ReLeaf's activities include ecosystem restoration projects called Global ReLeaf Forests, which involve the planting of trees, typically on public or private land that was once forested but has been cleared by wildfires, hurricanes, tornadoes, insects, or other natural occurrences; by developers; or by unintentional human interference, such as the spread of accidentally introduced exotic species. Program person-

The Goals of American Forests

American Forests, the organization that conducts the Global ReLeaf program, describes the organization's mission, vision, and strategy as follows:

Mission: Our mission is to grow a healthier world.

Vision: Our vision is to have healthy forest ecosystems for every community.

Strategy: Our strategy for achieving the mission is to provide action opportunities to targeted audiences to enable them to improve their environment with trees. We do this by using the best science to identify conservation issues, then develop and market practical solutions that individuals and groups can apply. American Forests' targeted audiences are individuals, community groups, government at all levels, educators, and businesses.

nel work with local groups to ensure that the new trees are native to the area and that they are properly planted and maintained.

From the time of its initiation in 1988 through 2010, Global ReLeaf planted more than thirty million trees during more than six hundred projects, aiming toward a goal of one hundred million trees planted by 2020. In 2010 alone, the organization planted more than four million trees in fourteen U.S. states and ten countries around the world.

Global ReLeaf has been successful in part because it works with governmental agencies, local organizations, and large corporations to make it easy for individual citizens to participate. Through extensive advertising and publicity, and through a colorful presence on the Internet, American Forests has encouraged donations of as little as ten dollars, with one tree being planted for each dollar received. In partnership with a major breakfast cereal, Global ReLeaf Kids supported a project to plant trees in rain forests in the Philippines and Hawaii.

Other organizations throughout the world have also used the name Global ReLeaf. One prominent group based in Slovakia in Eastern Europe was the former Slovak Union of Nature and Landscape Conservation, now called the Global ReLeaf Foundation. Like American Forests, this organization sponsors educational programs and prepares school curriculum materials, especially about the dangers of pollution and overdevelopment.

Cynthia A. Bily

FURTHER READING

Cohen, Shaul E. "American Forests: Planting the Future." In *Planting Nature: Trees and the Manipulation of Environmental Stewardship.* Berkeley: University of California Press, 2004.

Gray, Gerald J., Maia J. Enzer, and Jonathan Kusel, eds. *Understanding Community-Based Forest Ecosystem Management.* Binghamton, N.Y.: Haworth Press, 2001.

SEE ALSO: Coniferous forests; Deciduous forests; Deforestation; Forest management; Logging and clearcutting; National forests.

Global 2000 Report, The

CATEGORIES: Ecology and ecosystems; population issues

IDENTIFICATION: Report on the U.S. government's first attempt to use computer modeling to analyze the environmental impacts of global human activities

DATE: Published in 1980

SIGNIFICANCE: *The Global 2000 Report,* although criticized as both too pessimistic and too optimistic, helped to raise awareness of the impacts that the activities of human beings have on the worldwide environment and the need to establish plans aimed at minimizing negative environmental impacts.

As the use of computers became increasingly widespread in the early 1970's, the world was shocked by the publication of two books that reported on computer models that projected the eventual collapse of the world system if population growth and resource consumption were not curtailed: *The Limits to Growth* (1972), by Donella H. Meadows, Dennis L. Meadows, Jørgen Randers, and William W. Behrens III; and *Mankind at the Turning Point* (1974), by Mihajlo Mesarović and Eduard Pestel. The U.S. government was unprepared for the conclusions presented in these books, and in 1977 President Jimmy Carter focused the attention of the federal government on producing a computer model to serve as a planning tool for human activity. The simulation was completed in 1979, and the results were published in *The Global 2000 Report,* by the Council on Environmental Quality.

The Global 2000 Report concluded that, given the trends of the time, the world's population would increase by 55 percent and reach 6.35 billion in the year 2000 and that the number of people being fed on 1 hectare (2.5 acres) of land would increase from 2.6 in 1970 to 4 by 2000, requiring the increased use of biocides, fertilizers, and irrigation. The report also predicted that between 1970 and 2000, per capita consumption of food would increase by 15 percent, but the increase would be confined to the well-fed, industrialized nations. Finally, the report indicated that the number of malnourished people in the world would increase from 500 million in 1970 to 1.3 billion in 2000. The report also stated that if the predicted changes were to be avoided, change would be required, and "the needed changes go far beyond the capability and responsibility of this or any other single nation. An era of unprecedented cooperation and commitment is essential."

A number of serious flaws existed in the computer modeling that provided the data for *The Global 2000 Report.* Unlike the simulations developed for *The Limits to Growth* and *Mankind at the Turning Point, The Global 2000 Report* was not a single unified model. Individual government agencies produced data from their own simulations, and the models used were not combined into an integrated whole. As a result, some agencies assumed no interruptions in flows of necessary goods and services, whereas the agencies overseeing those goods and services were predicting interruptions. In addition, while other groups were using their computer models to extend predictions to the year 2100, the government's simulation stopped at the year 2000, severely limiting the usefulness and longevity of the results. Nevertheless, the projections that resulted from the U.S. simulation and reported in *The Global 2000 Report* were not significantly different from what other groups were reporting.

Although the report's projections were very pessimistic, many observers believed they were actually too optimistic, and a revised version of *The Global 2000 Report* was issued in 1988. In 1992, Senator Al Gore echoed the difficulties noted in *The Global 2000 Report* when he said in his book *Earth in the Balance,* "We find it difficult to imagine a realistic basis for hope that the environment can be saved, not only because we still lack widespread agreement on the need for this task, but also because we have never worked together globally on any problem even approaching this one in degree of difficulty."

Gary E. Dolph

FURTHER READING

Buell, Frederick. *From Apocalypse to Way of Life: Environmental Crisis in the American Century.* New York: Routledge, 2004.

Chasek, Pamela S., et al. *Global Environmental Politics.* 4th ed. Boulder, Colo.: Westview Press, 2006.

Davis, David Howard. "The Golden Age of Statistics: Planning as Prediction." In *Ignoring the Apocalypse: Why Planning to Prevent Environmental Catastrophe Goes Astray.* Westport, Conn.: Praeger, 2007.

SEE ALSO: Carter, Jimmy; Climate models; Club of Rome; *Limits to Growth, The*; Population-control movement; Population growth; Sustainable agriculture; Sustainable development; United Nations population conferences.

Global warming

CATEGORY: Weather and climate

DEFINITION: Increase in the average surface and ocean temperature of the earth since 1850 and the projected persistence of the trend

SIGNIFICANCE: The findings of scientists concerning the causes of global warming are extremely important in that they provide guidance for policy makers. Harmful consequences may result if the anthropogenic, catastrophic theory of global warming is correct and policy makers do not take the political and economic actions necessary to address the problem; conversely, harmful consequences may result if the theory is wrong but major political and economic decisions are made in the belief that it is correct.

According to the Intergovernmental Panel on Climate Change (IPCC), the overall global temperature during the twentieth century increased by a little less than 1 degree Celsius (1.8 degrees Fahrenheit). This involved an increase of about 0.5 degree Celsius (0.9 degree Fahrenheit) from 1910 to 1945 and a similar increase from about 1975 to 2000 or so (actually peaking in 1998), with a slight decrease in the intervening years. (The figures are approximate because of uncertain data and yearly fluctuations, occasionally as large as 0.25 degree Celsius up or down, and the complexity of adjusting the raw temperature data.) Explaining these increases and projecting future trends and their consequences are the key issues addressed by scientists who examine global warming.

CLIMATE CYCLES AND HUMAN CAUSES

Two basic theories have been posited regarding the source of the warming, both of which could easily be partially correct. Some see the warming as basically natural (as many scientists agree is probably the case for the pre-1945 warming, which predates the large increase in atmospheric carbon dioxide). In fact, short-term natural causes have been documented, such as volcanic eruptions (the Mount Pinatubo eruption in 1991 was followed by a strong temperature down-spike in 1992) and the El Niño/La Niña weather cycle (the very strong El Niño of 1998 resulted in a large temperature up-spike). In addition, some long-term fluctuations, such as the Pacific Decadal Oscillation and the Atlantic Multidecadal Oscillation, affect global as well as local temperatures. In addition, solar energy is not constant; there are slight cyclical variations that correlate with sunspot activity. These do not seem to be sufficient to explain the post-1975 warming; Patrick J. Michaels and Robert C. Balling, Jr., in their 2009 book *Climate of Extremes: Global Warming Science They Don't Want You to Know*, estimate that natural causes explain only 25 percent of the post-1975 warming (compared to 75 percent of the earlier warming). Some scientists, however, think natural answers can be found for the rest.

These scientists think the current warming is natural and cyclic, a Modern Warm Period to follow the Medieval Warm Period and Little Ice Age (which ended about 1850). A wide array of historical temperature proxies show that there is a roughly 1,500-year cycle, with shorter heating and warming subcycles. The proxies include ice cores dating back hundreds of thousands of years, six thousand boreholes (from all continents), seabed and lake-bed sediment cores, tree rings and tree lines, cave stalagmite cores, peat bogs, and records. These do show occasional remarkable shifts, global temperature increasing nearly 1 degree Celsius for about a decade at the end of the Younger Dryas (11,500 years ago) for reasons still unknown (there was an increase in greenhouse gases after the rise).

Environmental scientist S. Fred Singer has estimated that the Medieval Warm Period exceeded the current warming (so far). Others dispute this. Climatologist Michael E. Mann has argued that global temperatures changed only slightly during these periods, far less than in the twentieth century.

Among the many possible anthropogenic, or human-caused, influences on climate change is land use. The effects of land use are important, but they are for the most part local; cropland is warmer than forest, and urban areas are much warmer than cropland (the urban heat island effect). Overgrazing can lead to desertification, which makes the land warmer.

The strongest anthropogenic effect comes from the production of greenhouse gases such as carbon dioxide. Water vapor is an extremely important greenhouse gas, and methane (much of it from rice paddies and livestock raising) is far more powerful than carbon dioxide, but a trend toward increasing atmospheric methane halted during the mid-1990's. Carbon dioxide is the greenhouse gas that is the cause of greatest concern. Since the beginning of the twentieth century, the amount of carbon dioxide in the atmosphere has increased from about 290 parts per million to almost 390. The increase in the greenhouse effect is far smaller than the increase in greenhouse gases, however, particularly in areas of high humidity, owing to the atmospheric equivalent of the law of diminishing returns. Greenhouse gas warming is strongest at night and therefore in winter, and in upper latitudes; in the atmosphere it leads to a warmer troposphere and cooler stratosphere.

CLIMATE MODELS

Computer climate models can be used for historical research as well as future projections. Extremely complex general circulation models provide detailed information on how natural warming or greenhouse gas warming is likely to affect climate all over the planet and into the atmosphere, and the models' projections can be tested against observational data. Such testing is as necessary for the findings produced by computer models as for the findings produced by any other scientific experiments; results must be shown to be replicable by others, and the data must be freely available for examination by others. One problem with the testing of data from climate models is that observational data are often too recent (satellite tracking of hurricanes began in 1970, for example, and satellite measurement of Arctic sea ice in 1979) to allow scientists to determine reliably whether changes represent coincidental long-term oscillations or result from the current warming trend.

In the early twenty-first century, most climate models project a linear global surface temperature in-

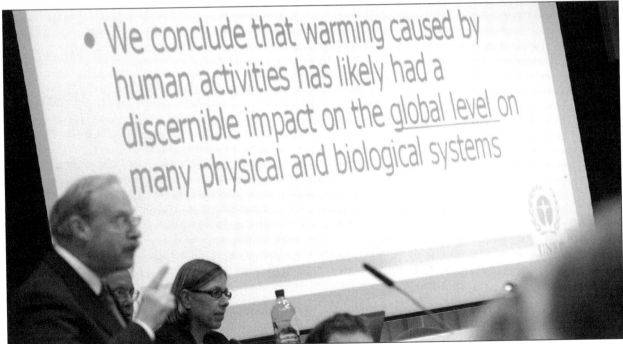

Martin Parry, left, of the Intergovernmental Panel on Climate Change, presents a report on global warming at an international conference in 2007. A conclusion of the IPCC report is displayed on the screen behind him. (AP/Wide World Photos)

crease from 2 to 3 degrees Celsius (3.6 to 5.4 degrees Fahrenheit) per century (occasionally much more owing to positive feedback effects, such as increased evaporation leading to increased humidity). Early models exaggerated the warming and could not match the previous history. Later models that added in sulfate aerosols were more accurate, but they failed to predict the absence of net warming since 1998 or the relative lack of warming in the Southern Hemisphere; the problem was that the models used one unknown to check another. Some of the early error may have resulted from negative feedbacks, such as clouds (low-level cumulus and stratus clouds reflect solar light, cooling the planet), which the models often ignored. Climate models predict different specific results from natural and greenhouse gas warming, and many observations (most notably the overall cooling of Antarctica) tend to support the latter, though not entirely.

Michael E. Mann and Lee R. Kump have praised three projections made by James E. Hansen, director of the National Aeronautics and Space Administration's Goddard Institute for Space Studies, when he presented testimony to the U.S. Congress on climate change in 1988. The most severe scenario (A) predicted an increase of just over 1 degree Celsius in the following thirty years, comparable to the high-end projection of 3 degrees per century, but started to diverge from reality within a few years as too high. The middle scenario (B) projected an increase of less than 1 degree, roughly comparable to an increase of 2 degrees per century, and the low scenario (C) projected an increase of about 0.25 degree in thirty years, probably less than 1 degree per century. Scenarios B and C tracked closely with each other, and with the actual observed data, up to 2005. Reports since then (including the Goddard Institute's own December, 2009, estimate) show that scenario C has been the most accurate.

The IPCC's 2007 Assessment of Climate Change

On its Web site, the Intergovernmental Panel on Climate Change summarized the following findings of its 2007 assessment report:

- Warming of the climate system is unequivocal.

- Most of (50% of) the observed increase in globally averaged temperatures since the mid-20th century is very likely (confidence level 90%) due to the observed increase in anthropogenic (human) greenhouse gas concentrations.

- Hotter temperatures and rises in sea level "would continue for centuries" even if greenhouse gas levels are stabilized, although the likely amount of temperature and sea level rise varies greatly depending on the fossil intensity of human activity during the next century.

- The probability that this is caused by natural climatic processes alone is less than 5%.

- World temperatures could rise by between 1.1 and 6.4° Celsius (2.0 and 11.5° Fahrenheit) during the twenty-first century. . . .

- Sea levels will probably rise by 18 to 59 centimeters (7.08 to 23.22 inches).

- There is a confidence level 90% that there will be more frequent warm spells, heat waves and heavy rainfall.

- There is a confidence level 66% that there will be an increase in droughts, tropical cyclones and extreme high tides.

- Both past and future anthropogenic carbon dioxide emissions will continue to contribute to warming and sea level rise for more than a millennium.

- Global atmospheric concentrations of carbon dioxide, methane, and nitrous oxide have increased markedly as a result of human activities since 1750 and now far exceed pre-industrial values over the last 650,000 years.

Consequences and Solutions

Global warming may have many possible effects. The Medieval Warm Period, though beneficial to European and Arctic agriculture, often led to drought elsewhere (including the drought suspected of having caused the collapse of the Native American Anasazi culture). Similar effects can be seen in the twenty-first century; the decline in the snowpack on Africa's Mount Kilimanjaro is apparently more a result of increased local aridity than of global warming.

Warming also leads to a sea-level rise of 1 to 2 centimeters (0.4 to 0.8 inch) per decade, which could increase if the vast Greenland and Antarctic ice packs melt significantly (most models predict more snow in the interiors and more meltwater on the edges of these ice packs, and observations confirm this), which could also seriously alter key ocean currents. Also with warming, warm-weather crops can be grown further north and warm-weather habitats invade cold-weather habitats. Some scientists fear that global warming will lead to more frequent or more severe extreme weather events (particularly tropical cyclones), but there has been no observational evidence of such a trend (for example, North Atlantic hurricanes declined after the severe 2005 season).

Suggested approaches to addressing global warming include both adapting to the heat (and the effects of the heat) when it occurs (and meanwhile devoting resources to solving other problems) and trying to reduce the increase in warming. The latter can have no effect on the natural component of temperature rise and will be unnecessary if the increase is small.

Some proposed solutions aimed at reducing the greenhouse gas emissions linked with global warming are questionable. In particular, the substitution of ethanol for fossil fuels has drawbacks: Growing the crops needed to produce ethanol can in some cases decrease food production or increase cropland at the expense of forestland. Given that forests help to remove carbon dioxide from the atmosphere, the net result may be an increase in greenhouse gases.

Among the most useful and affordable ways to reduce greenhouse gases may be to increase the numbers of hybrid and electric vehicles in relation to gasoline-fueled vehicles, to improve energy conservation by individuals and industries, and to reduce reliance on the burning of fossil fuels for electricity by developing alternative sources of power. Because stronger proposed changes would involve serious economic dislocations, calls for such changes generally include long time lines for achievement. Per-capita carbon dioxide emissions have declined slightly from a 1979 peak; if the trend continues, they will level off with global population around 2050.

Timothy Lane

FURTHER READING

Alley, Richard B. *The Two-Mile Time Machine: Ice Cores, Abrupt Climate Change, and Our Future.* Princeton, N.J.: Princeton University Press, 2000.

Dutch, Steven I., ed. *Encyclopedia of Global Warming.* 3 vols. Pasadena, Calif.: Salem Press, 2010.

Fagan, Brian. *The Great Warming: Climate Change and the Rise and Fall of Civilizations.* New York: Bloomsbury, 2008.

Houghton, John Theodore. *Global Warming: The Complete Briefing.* 4th ed. New York: Cambridge University Press, 2010.

Mann, Michael E., and Lee R. Kump. *Dire Predictions: Understanding Global Warming.* New York: DK, 2008.

Michaels, Patrick J., and Robert C. Balling, Jr. *Climate of Extremes: Global Warming Science They Don't Want You to Know.* Washington, D.C.: Cato Institute, 2009.

Singer, S. Fred, and Dennis T. Avery. *Unstoppable Global Warming: Every 1,500 Years.* Updated ed. Lanham, Md.: Rowman & Littlefield, 2008.

Weart, Spencer W. *The Discovery of Global Warming.* Rev. ed. Cambridge, Mass.: Harvard University Press, 2008.

SEE ALSO: Climate change and oceans; Climate change skeptics; Climate models; Climatology; El Niño and La Niña; Glacial melting; Greenhouse effect; Greenhouse gases; Hansen, James E.; Intergovernmental Panel on Climate Change; Kyoto Protocol; Sea-level changes.

Globalization

CATEGORIES: Resources and resource management; philosophy and ethics

DEFINITION: Intercontinental integration of regional economies, cultures, and political and financial systems, driven by the transnational exchange and circulation of labor, ideas, technologies, products, services, languages, and popular culture

SIGNIFICANCE: Globalization in the early twenty-first century rests on a free trade or neoliberal economic model that favors open markets and global competition among states and nonstate actors in the world economy. Intense competition among developing nations to secure investment and jobs from huge transnational corporations pushes ecological interests in those countries to the background of their political agendas. Corporate interests in the developed world tend to suppress movements for ecological reform that would cut into corporate profits.

The modern state emerged in western Europe in 1648, with the Peace of Westphalia, which brought to a close the Thirty Years' War among Europe's various princes and ended the political struggle between the Roman Catholic Church and the European state, with the secular state arising as the sovereign and independent actor on the world stage. In this state-centric model, each state was recognized by other states as having the exclusive right to determine domestic policy, and each was expected to address common issues and to resolve conflicted interests through negotiations with other states in a process referred to as international relations.

After World War II, the United Nations was established (in October, 1945) to usher in a new era of postinternational or global relations. The United Nations followed the model of the League of Nations, created in 1919 following World War I under the Treaty of Versailles, "to promote international cooperation and to achieve peace and security." The extent and appalling nature of the crimes of the Nazis against Jews and other victim groups forged a consensus among members of the global community regarding shared norms of behavior, while structural reconstruction of Europe and elsewhere after the war welded global economic networks; these factors together ultimately undermined the import of state sovereignty in favor of global cooperation. New global realities demanded that states cooperate with each other to deal with common threats, develop markets, exchange technologies, manage conflict, and share power with rising nonstate actors, nongovernmental organizations (NGOs). NGOs took the form of economic organizations such as transnational corporations, advocacy organizations such as Amnesty International and Greenpeace, and service organizations such as the Red Cross and Doctors Without Borders. However, the post-World War II era also saw the emergence of less benevolent nonstate actors networking globally. Terrorist organizations expanded globally, as did transnational criminal syndicates.

The twenty-first century model of globalization has many varied and interwoven aspects, and it has been evolving for a very long time, congealing most intensely since the mid-twentieth century. Globalization is primarily an economic system, characterized by globalized financial markets, trade networks, foreign investment, and capital flows, but it has many aspects, including political globalization (effected through the proliferation of international and regional coalitions of states and nonstate actors), military globalization (networks of military force and alliances), environmental globalization (global efforts to address environmental degradation and global warming at the supranational level), and cultural globalization (through an "acculturation" process whereby people's everyday lives are fundamentally altered by the exchange of foods, people, ideas, technologies, and other products).

The General Assembly of the United Nations is the forum where members of the global community meet to negotiate political interests and address common problems and challenges. Trade practices in the global economy are regulated by three international institutions: the World Trade Organization, which regulates global trade and rules on disputes in the global marketplace; the World Bank, which makes short-term, high-interest loans to economically struggling nations; and the International Monetary Fund, which intervenes in debt-bearing nations to reorient their trade practices and financial policies. Global cooperation has fostered trade and economic development across the planet, but economic prosperity—in the developing world as well as in the developed world—is often purchased at the cost of human rights and ecological devastation. Globalization also fosters international debate on these problems, however, and has sparked global movements for "fair trade" practices and Green movements that have culminated in ongoing cooperation and international agreements to restrict polluting practices.

CONTROVERSIES

Globalization is the subject of heated debate around the world, among politicians and economists as well as among scientists and environmental activists. From the standpoint of environmental ethics, globalizing trade practices have had devastating effects on the earth's natural environment. Regional neglect and the pollution of air, land, and ocean waters are driven by the "race to the bottom" phenomenon that pits developing countries against each other in efforts to lure global investors. Critics argue that the existing system is simply a broader-reaching, more profitable model of colonialism, a neocolonialism, whereby governments act as mere salespersons, promoting the profits of their corporations in a global marketplace.

Critics charge also that developing countries have no fighting chance in the global trade game, and so the rich get richer through the growing exploitation of the global poor and the devastation of the environment, in both the developed and the developing nations. Globalists, in contrast, assert that "free trade" promotes freedom and democracy, and that even as global inequality rises, poverty can be reduced through free trade. They argue that problems such as environmental degradation and global warming should be viewed as opportunities for entrepreneurial innovation and new economic ventures, and not as problems to be addressed through political intervention and legal restrictions.

Wendy C. Hamblet

FURTHER READING

Bhagwati, Jagdish. *In Defense of Globalization.* New York: Oxford University Press, 2004.

Braun, Joachim von, and Eugenio Diaz-Bonilla, eds. *Globalization of Food and Agriculture and the Poor.* New York: Oxford University Press, 2007.

Friedman, Thomas L. *The World Is Flat: A Brief History of the Twenty-first Century.* 2d rev. ed. New York: Farrar, Straus and Giroux, 2008.

Glyn, Andrew. *Capitalism Unleashed: Finance, Globalization, and Welfare.* New York: Oxford University Press, 2006.

McCulloch, Jock, and Geoffrey Tweedale. *Defending the Indefensible: The Global Asbestos Industry and Its Fight for Survival.* New York: Oxford University Press, 2008.

Mander, Jerry, and Edward Goldsmith. *The Case Against the Global Economy: And for a Turn Toward the Local.* San Francisco: Sierra Club Books, 1996.

SEE ALSO: Agenda 21; Environmental ethics; Environmental justice and environmental racism; Free market environmentalism; Global Environment Facility; Green movement and Green parties; Indigenous peoples and nature preservation; International Biological Program; International Institute for Environment and Development; International Institute for Sustainable Development; Johannesburg Declaration on Sustainable Development; Ocean pollution; Race-to-the-bottom hypothesis; Sustainable development; World Health Organization.

Gore, Al

CATEGORIES: Activism and advocacy; weather and climate

IDENTIFICATION: American environmental activist and politician who served in Congress and as vice president of the United States

BORN: March 31, 1948; Washington, D.C.

SIGNIFICANCE: Through his activism and particularly his participation in the documentary film *An Inconvenient Truth*, Gore has brought worldwide attention to the problem of global warming.

Before serving as the forty-fifth vice president of the United States (1993-2001) under President Bill Clinton, Al Gore was involved in American politics for more than two decades, serving in the U.S. House of Representatives (1977-1985) and then in the Senate (1985-1993) as a representative of Tennessee. Gore is less likely to be remembered primarily for his political service, however, than for his work as an environmental activist of international reputation and impact. From the time of his earliest days in Congress, Gore has been a leading advocate for confronting the threat of global warming. He addressed the issue in his best-selling book *Earth in the Balance: Ecology and the Human Spirit* (1992), and then, as vice president, he led the Clinton administration's efforts to develop economically profitable ways to protect the environment.

Environmental activist and former U.S. vice president Al Gore. (AP/Wide World Photos)

After Gore, a Democrat, ran for the U.S. presidency in 2000 and lost to Republican George W. Bush, he began traveling around the United States and internationally to present an educational slide show on global warming; the slide show became the basis of the documentary film *An Inconvenient Truth*, released in 2006. The film won two Academy Awards in 2007, and Gore's message, a warning of the disasters associated with the mounting threat of climate change, won a global audience for the issue. In 2007 Gore was awarded the Nobel Peace Prize, an honor he shared with the Intergovernmental Panel on Climate Change, for "efforts to build up and disseminate greater knowledge about man-made climate change, and to lay the foundations for the measures that are needed to counteract such change."

Gore has attributed the beginnings of his passion for environmental concerns to the influence of his Harvard professor of climate science, Roger Revelle, whose research on climate change ultimately steered Gore toward an interest in politics and government as a means of having a direct impact on environmental policy in the United States. Gore received his bachelor of arts degree in government from Harvard in 1969, then served in the U.S. Army (1969-1971), after which he returned to his studies, attending Vanderbilt University Divinity School (1971-1972) on a Rockefeller Foundation scholarship and then Vanderbilt University Law School (1974), while simultaneously working for the newspaper *The Tennessean* as an investigative reporter (1971-1976).

Gore holds a number of key positions on governing bodies that serve the cause of environmental protection. He is founder and chair of the Alliance for Climate Protection, cofounder and chair of Generation Investment Management, and a partner in the venture capital firm Kleiner Perkins Caufield & Byers, heading that firm's climate change solutions group. Gore has also held faculty positions at Middle Tennessee State University, Columbia University Graduate School of Journalism, Fisk University, and the University of California, Los Angeles.

Gore's critics have often raised four particular issues: They assert that a conflict of interest is suggested by Gore's discordant roles as a private investor in green technology and as a public advocate of taxpayer-funded green-technology subsidies; they point out that his personal level of energy consumption is high; they question the scientific basis of some of his claims; and they object to his refusal to join in open debate with opponents on the subject of global warming. In addition, some organizations, notably the animal rights group People for the Ethical Treatment of Animals (PETA), have criticized Gore for eating meat, a practice that they argue is environmentally unfriendly.

During the 1990's Gore spoke out in opposition to a number of situations that had negative implications for the environment. He opposed the ties of Ronald Reagan's and George H. W. Bush's presidential administrations to Iraqi leader Saddam Hussein, because of the latter's use of poisonous gas and his burgeoning nuclear program. After Saddam's Al-Anfal Campaign, which included nerve-gas attacks on Kurdish Iraqis, Gore cosponsored the Prevention of Genocide Act (1988), legislation that would have cut U.S. assistance to Iraq, but the bill was ultimately defeated.

Wendy C. Hamblet

FURTHER READING

Gore, Al. *Earth in the Balance: Ecology and the Human Spirit.* 1992. Reprint. New York: Rodale Books, 2006.

_____. *An Inconvenient Truth: The Planetary Emergency of Global Warming and What We Can Do About It.* New York: Rodale Books, 2006.

SEE ALSO: Carbon footprint; Climate change and oceans; Climatology; Conservation movement; Environmental ethics; Global warming; *Inconvenient Truth, An*; Intergovernmental Panel on Climate Change; Renewable energy; Renewable resources; Sea-level changes; United Nations Framework Convention on Climate Change; Volcanoes and weather.

Gorillas. *See* Mountain gorillas

Grand Canyon

CATEGORIES: Places; preservation and wilderness
issues

IDENTIFICATION: Deep gorge created by the
Colorado River in northern Arizona

SIGNIFICANCE: The popularity of the Grand Canyon
as a tourist destination has contributed to a number of environmental problems in and around the
canyon itself. These problems range from air and
noise pollution to issues related to the scarcity of
water resources.

The Grand Canyon is a deep, 450-kilometer-long
(280-mile) segment of the Colorado River and its
tributary canyons in northern Arizona. Grand Canyon National Park, one of the most heavily visited na-

tional parks, was established by President Theodore
Roosevelt in 1919 and has been designated a World
Heritage Site. The parts of the Grand Canyon that lie
outside the park's boundaries are managed by the
Hualaipai and Navajo tribal councils.

The arid climate of the Grand Canyon influences
every resource in the region. Well-exposed rock layers
reveal more than 1.8 billion years of the earth's history.
The arid climate preserves ancient human and animal
remains, including those of many extinct animals that
lived more than ten thousand years ago. Cliff dwell-
ings, human artifacts, and old adobe buildings represent habitation dating back more than four thousand
years. Grand Canyon tourists, however, often focus
most on the canyon's scenic grandeur and beautiful
views. People who hike into the canyon commonly
feel overwhelmed by the immensity of the chasm.

Visitors view the Grand Canyon at sunset from Mather Point on the South Rim after a dusting of snow. (NPS)

ECOSYSTEMS

Grand Canyon National Park, with more than 2,000 meters (6,500 feet) of relief, contains many ecosystems. The South Rim, a flat plateau, has an elevation of 2,100 meters (7,000 feet). The North Rim is an even higher plateau, 2,400 meters (8,000 feet) above sea level. Coniferous forests cover both rims and provide homes to deer, squirrels, and mountain lions. No streams cross the plateaus, as water from rain and snowmelt immediately flows underground in the karst terrains.

The coniferous forests extend down into the canyon, transitioning into an arid environment at lower elevations. Desert plants—such as cacti, acacia, mesquite, brittle bush, ocotillo, rabbitbrush, and agave—grow on the walls of the Grand Canyon. Desert bighorn sheep, lizards, snakes, skunks, and mice populate the slopes and side canyons. Water is scarce, particularly on the south side of the Colorado River. Cottonwood, ash, and redbud trees—as well as ferns, columbine, and other water-loving plants and animals—cluster around small seeps throughout the canyon and larger karst springs in some tributary canyons on the north side.

Another distinct ecological zone in the Grand Canyon is the riparian (riverside) habitat along the Colorado River at the bottom. Willow, arrowweed, and exotic tamarisk line the riverbanks. Otters, beavers, muskrats, fish, and other aquatic organisms call the Colorado River home.

ENVIRONMENTAL PROBLEMS

After visiting part of the Grand Canyon in 1858, Lieutenant Joseph C. Ives wrote, "The region is, of course, altogether valueless. . . . It seems intended by nature that the Colorado River . . . shall be forever unvisited and undisturbed." If his prediction had come true, the Grand Canyon would not be facing the many environmental problems that now threaten it.

More than five million people visit Grand Canyon National Park each year. Support facilities for these visitors (campgrounds, hotels, shops, toilets) require water and sewage treatment, yet no water is available on either rim. Drilling to the closest aquifer, several thousands of feet underground, would be extremely costly, and so all water used in the park comes from a cave about halfway up the north side of the canyon. A transcanyon pipeline and associated pumphouses lift the water to two places on the South Rim and one location on the North Rim. The purity and quantity of

Powell on the Great and Unknown Grand Canyon

John Wesley Powell combined into one narrative the story of his two trips into the Grand Canyon and down the Colorado River with his survey crew. In this excerpt from his 1875 book, Exploration of the Colorado River of the West and Its Tributaries, *Powell introduces the reader to the start of the journey "down the Great Unknown."*

We are now ready to start on our way down the Great Unknown. Our boats, tied to a common stake, chafe each other as they are tossed by the fretful river. They ride high and buoyant, for their loads are lighter than we could desire. We have but a month's rations remaining. The flour has been resifted through the mosquito-net sieve; the spoiled bacon has been dried and the worst of it boiled; the few pounds of dried apples have been spread in the sun and reshrunken to their normal bulk. The sugar has all melted and gone its way down the river. But we have a large sack of coffee. The lightening of the boats has this advantage: they will ride the waves better and we shall have but little to carry when we make a portage.

We are three quarters of a mile in the depths of the earth, and the great river shrinks into insignificance as it dashes its angry waves against the walls and cliffs that rise to the world above; the waves are but puny ripples, and we but pigmies, running up and down the sands or lost among the boulders.

We have an unknown distance yet to run, and unknown river to explore. . . .

Source: Excerpted in *The Wilderness Reader,* edited by Frank Bergon (Reno: University of Nevada Press, 1980), p. 152.

this modest stream are critical to keeping the Grand Canyon open to visitors. Occasionally, the water's impurities exceed state-set limits, and the park is forced to truck in water at tremendous expense. The numbers of visitors to the park and further development there will remain strictly limited unless other sources of water are developed.

The popularity of the Grand Canyon brings other problems as well. In an attempt to provide each visitor with a high-quality wilderness experience, the park requires overnight campers to register for permits and limits the number of visitors. In popular backcountry areas, campers are restricted to designated campgrounds that have solar-powered compost toilets. Because the desert recovers slowly from erosion, the park rangers enforce strict rules concerning vandalism and the cutting of switchbacks. They warn hikers to treat archaeological sites with respect. As in all national parks, visitors are forbidden to take any archaeological or historical materials, rocks, animals, or plants.

Because many visitors choose to see the park from the air, the presence of helicopters and low-flying airplanes has created a volatile issue. Many hikers have expressed dismay, and even anger, at the noise that accompanies the flights overhead, noting that it hinders their ability to enjoy the canyon's beauty and grandeur. The government struggles to balance the demands of these two different user groups by strictly controlling the routes and heights of over-flights.

Once known for being spectacularly clear, the air of the Grand Canyon is now occasionally marred by pollution from nearby coal-burning power plants. Since the 1970's, battles have raged between environmentalists concerned with preserving air quality in the area and those interested in developing the region's abundant coal supplies.

The most dramatic and contentious environmental issues in the Grand Canyon involve the Colorado River. The greatest change to the riparian zone resulted from construction of the Glen Canyon Dam, which has controlled the Colorado River flow through the Grand Canyon since 1966. The dam eliminated large floods, but the maintenance of more consistent flows throughout the year has had a severe impact on the riparian ecosystem, including elimination of many beaches and a general increase in vegetation (including nonnative species of plants) and wildlife. The dam has also altered water temperature and clarity. Predam water temperatures in the river fluctuated from 26 degrees Celsius (80 degrees Fahrenheit) in summer to nearly freezing in winter. Today, the river temperature remains constant at 7 degrees Celsius (46 degrees Fahrenheit). Once noted for its load of sediment, the river below the dam is now clear. The combined changes in water temperature and clarity have had dramatic effects on the canyon's aquatic ecology, resulting in the loss of many plant and fish species and the proliferation of nonnative carp and trout.

Louise D. Hose

FURTHER READING

Finkbine, Bob. *Smoke That Roars: The Grand Canyon and the Creation of a New Century.* Phoenix: Atwell, 2002.

Houk, Rose. *An Introduction to Grand Canyon Ecology.* Grand Canyon, Ariz.: Grand Canyon Association, 1996.

McPhee, John. *Encounters with the Archdruid.* 1971. Reprint. New York: Farrar, Straus and Giroux, 2000.

Webb, Robert. *A Century of Change: Rephotography of the 1889-90 Stanton Expedition.* Tucson: University of Arizona Press, 1996.

SEE ALSO: Coal-fired power plants; Glen Canyon Dam; Kaibab Plateau deer disaster; National parks; Powell, John Wesley; Roosevelt, Theodore.

Grand Coulee Dam

CATEGORIES: Preservation and wilderness issues; animals and endangered species

IDENTIFICATION: Dam located on the Columbia River west of Spokane, Washington

DATE: Completed in 1941

SIGNIFICANCE: The environmental impacts of the Grand Coulee Dam include a dramatic decrease in the salmon population in the Columbia River, once the largest natural salmon hatchery in the world.

The multipurpose, 168-meter-high (550-foot-high) Grand Coulee Dam, one of the largest concrete structures in the world, provides downstream flood control, irrigation, and hydroelectricity. The facility delivers irrigation water to more than 223,000 hectares (550,000 acres) of agricultural lands. Its associated electrical power production facilities are the largest in North America. President Franklin D. Roosevelt initiated construction of the dam in 1933 under the Public Works Administration of the National Industrial Recovery Act, which Congress authorized for emergency projects to relieve unemployment during the Great Depression. Electricity was first generated at the dam in 1941. Placing a dam across the fourth-largest river in North America created a 243-kilometer-long (151-mile-long) reservoir, Franklin D. Roosevelt Lake, which affords fishing and water sports to visitors.

Salmon ladders were built adjacent to downstream dams to allow the annual Columbia River salmon spawning migration to continue, but the Grand Coulee Dam is too high to accommodate this dam-

Grand Coulee Dam

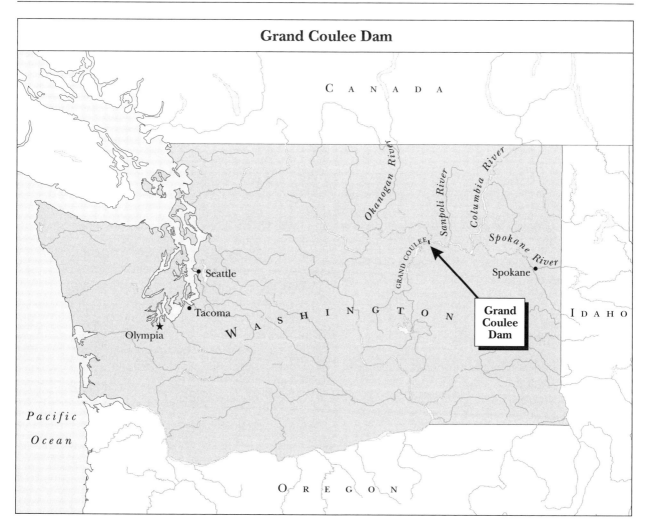

passing technique. At the time of construction, an elevator to lift the fish up and into the reservoir was considered but rejected because too many fish would die in the process. Instead, more than 1,600 kilometers (1,000 miles) of the world's most prolific salmon breeding river was eliminated upstream of the dam. Many miles of the affected river flow through Canada, but the Canadian government expressed a lack of concern about the potential loss, stating that no commercial salmon fisheries existed along the river in Canada.

In the spring of 1939 the U.S. Bureau of Reclamation captured chinook and blueback salmon downstream from the Grand Coulee Dam site and transported them to spawning grounds in both tributaries and the main river upstream from the dam. The fish that survived the truck ride and transplant spawned. They and their offspring returned over the dam spill-

ways and through the turbines, causing high mortality rates. The following year, captured fish were transferred to a hatchery for spawning. The young fish were released in downstream tributaries with the hope that they would return to the new spawning sites the following year. In 1943 the Bureau of Reclamation declared the salmon relocation program successful and turned the efforts over to the U.S. Fish and Wildlife Service.

The salmon population in the Columbia River, once the largest natural salmon hatchery in the world, dropped dramatically with construction of the Grand Coulee and other dams. At the downstream Rock Island Dam, more than fifty thousand salmon passed through ladders to bypass the dam in 1933. A 1942 census at the same site counted only a little more than seven thousand salmon. Improved understanding and accommodation of salmon diet and environmental needs has helped the Columbia River populace grow, but it remains far smaller than the predam population.

Louise D. Hose

FURTHER READING

McCully, Patrick. *Silenced Rivers: The Ecology and Politics of Large Dams.* Enlarged ed. London: Zed Books, 2001.

Moss, Brian. *Ecology of Fresh Waters: A View for the Twenty-first Century.* 4th ed. Hoboken, N.J.: Wiley-Blackwell, 2010.

Palmer, Tim. *Lifelines: The Case for River Conservation.* 2d ed. Lanham, Md.: Rowman & Littlefield, 2004.

SEE ALSO: Aquaculture; Dams and reservoirs; Floods; Habitat destruction; Hydroelectricity; Irrigation.

Grazing and grasslands

CATEGORIES: Resources and resource management; land and land use

DEFINITIONS: Grazing is the consumption of any plant species by any animal species; grasslands are ecosystems where grasses and other nonwoody vegetation predominate

SIGNIFICANCE: While grazing is of mutual benefit to plants and animals, overgrazing is ultimately detrimental to both the plant and animal populations, as well as to grassland ecosystems. Maintaining a balance between grazing animals and the plants on which they feed prevents deleterious consequences.

Grasslands are characterized by the presence of low plants, mostly grasses, and are distinguished from woodlands, tundra, and deserts. Grasslands experience sparse to moderate rainfall and are found in both temperate and tropical zones. Grassland plants coevolved over millions of years with the grazing animals that depended on them. Wild ancestors of cattle and horses, as well as antelope and deer, were found in Eurasian grasslands. On the North American prairie, bison and antelope prospered. Wildebeest, gazelle, zebra, and buffalo dominated African savannas, whereas the kangaroo was the preponderant grazer in Australia. Grasslands occupied vast areas of the world more than ten thousand years ago, before the development of agriculture and industrialization, and the subsequent explosive growth of the human population.

Grazing is a symbiotic relationship whereby animals gain their nourishment from plants, which in turn benefit from the activity. Grazing removes the vegetative matter required for grasses to grow, facilitates seed dispersal, and disrupts mature plants, permitting young plants to take hold. Urine and feces from grazing animals recycle nutrients to the plants. The grassland ecosystem also attracts other animals, including invertebrates, birds, rodents, and predators. The grasses, grazing animals, and grassland carnivores, such as wolves or cat species, constitute a food chain.

Grasses are generally well suited to periods of low rainfall because of their extensive root systems and can go dormant during periods of drought. Humans have been an increasing presence in grassland areas, where more than 90 percent of modern crop production occurs and much urbanization and industrialization have taken place. The remaining grasslands, unsuitable for crops because of inadequate rainfall or difficult terrain, are used for grazing by domesticated or wild herbivores. In addition, many woodland areas around the world have been cleared and converted to grasslands where animals can graze.

IMPACTS OF OVERGRAZING

Continued heavy grazing of a given area leads to deleterious environmental consequences. Even repeated removal of leaf tips will not adversely affect the regeneration of grasses, provided that the basal zone of the plant remains intact. Whereas animals can generally safely eat the upper half of the grass shoot, if they ingest the lower half, which sustains the roots and fuels regrowth, they will eventually kill the plants. Overgrazing leads to denuding of the land, invasion by less nutritious plant species, erosion caused by decreased absorption of rainwater by soil, and starvation of animal species. Because the loss of plant cover changes the reflectance of the land, climate changes can follow that make it virtually impossible for plants to return, with desertification an ultimate consequence.

The number of animals is not the only factor in overgrazing; the timing of the grazing can also be detrimental. Grasses require time to regenerate, and continuous grazing will inevitably kill them. Consumption too early in the spring can stunt their development. Semiarid regions are particularly prone to overgrazing because of low and often unpredictable rainfall; regrettably, these are the areas of the world to

A flock of sheep grazes in a grassland region. (©Serban Enache/Dreamstime.com)

which much livestock grazing has been relegated, because the moister grassland areas have been converted to cropland.

Overgrazing has contributed to environmental devastation worldwide. Excessive grazing by cattle, sheep, goats, and camels is partly responsible for the deserts of the Middle East. Uncontrolled livestock grazing during the late nineteenth century and early twentieth century negatively affected many areas in the American West, where sagebrush and juniper trees invaded the grasslands. Livestock overgrazing has similarly devastated areas of Africa and Asia. Feral horses in the American West and the Australian outback continue to damage those environments.

Overgrazing by wildlife can also be deleterious. The Kaibab Plateau deer disaster in Arizona is one such example, where removal of natural predators and livestock that competed with the deer for food led first to a deer population explosion, then to overgraz-ing by the deer, followed by starvation and large die-offs within the deer population. Protection of elk and bison in Yellowstone National Park has similarly led to high populations, excessive grazing, and changes to the environment. Only the provision of winter feed has prevented the die-offs that would otherwise naturally ensue. Ironically, winter feeding has perpetuated the problem by maintaining these populations at levels higher than grazing can sustain. Feeding has also encouraged the animals to congregate in unusually large numbers, which has contributed to the spread of disease. In 2010 a coalition of conservation groups lost a lawsuit to stop the supplemental feeding of elk and bison on the National Elk Refuge in Wyoming.

GRASSLANDS MANAGEMENT

Grassland areas need not deteriorate if they are properly managed, whether for livestock, wild animals, or both. The land's carrying capacity, or the

number of healthy animals that can be grazed indefinitely in a given area, must not be exceeded. Because
of year-to-year changes in weather conditions and
hence food availability, determining carrying capacity is not simple; worst-case estimates typically have
been used as guidelines to minimize the risk of exceeding carrying capacity. The goal should be a grassland rendered and kept healthy by optimizing, not
maximizing, the number of animals. For private land,
optimizing livestock numbers is in the long-term self-
interest of the landowner. For publicly held land,
managed in common or with unclear or disputed
ownership, restricting animals to the optimum level is
particularly difficult to achieve. Personal short-term
benefit often leads to long-term disaster, in a phenomenon known as the tragedy of the commons.

Managing grasslands involves controlling the numbers of animals and enhancing the habitat. Cattle and
sheep can be physically restricted through the use of
herding and fencing, although requiring such restrictions can be difficult to achieve through political
means. Much more problematic is controlling wildlife
when natural predators have been eliminated and
hunting is severely restricted. As for habitat improvement, the prudent use of chemical, fire, mechanical,
and biological approaches can increase carrying capacity for domesticated and wild herbivores. Removing woody vegetation by burning or mechanical
means can increase grass cover, fertilizing can stimulate grass growth, and reseeding with desirable species (plants native to the particular region) can enhance the habitat. Effective grassland management
also requires matching animals with the grasses on
which they graze.

An approach to grazing known as holistic management may have the potential not only to stave off ecosystem damage of grasslands but also to reverse desertification. This approach operates on the essential
principle that, because herbivores and perennial
grasses evolved together, the grasses will thrive only in
combination with herbivores grazing and roaming
naturally. Contrary to common wisdom regarding
best management practices for grazing, holistic management involves grazing livestock in ultradense, constantly moving herds that mimic big-game grazing
patterns. The livestock till the soil with their hooves
and fertilize it with their excrement. By grazing the
grasses, they allow sunlight to reach the grasses'
growth buds; by contrast, when grazing is so restricted
that the vegetation is able to die upright, the growth

buds are shielded from the sun and the entire plant
dies the following year. The common management
practice of allowing grazed land an extended period
to rest and recover, then, may not promote a resur-
gence of vegetative cover; rather, this practice may
cause the land to remain barren and dry.

In 1992 holistic management pioneer Allan Savory
began a program in Zimbabwe, Africa, in which livestock herds were increased by 400 percent on 2,630
hectares (6,500 acres) of land that had been barren
for hundreds of years. By 2010, after years of holistic
planned grazing, this area had become healthy grassland with open water. Other holistic management
practitioners around the globe have enjoyed similar
successes.

GRAZING IN THE UNITED STATES

There are roughly 312 million hectares (770 million acres) of rangelands (grasslands, forests, wetlands,
and other ecosystems that are suitable for grazing) in
the United States, more than half of which are privately owned. The federal government manages 43
percent, and the remainder is under state and local
government control.

Laws pertaining directly to grazing in the United
States include the Taylor Grazing Act of 1934, the Federal Land Policy and Management Act of 1976, and
the Public Rangelands Improvement Act of 1978. The
Taylor Grazing Act introduced measures to control
the unregulated grazing practices of homesteaders
that had led to overgrazing, enhanced erosion, damage to streams and springs, and the land's reduced
productivity; however, rancher needs still tended to
take precedence over range condition. Four decades
later, heightened environmental awareness led to passage of the Federal Land Policy and Management Act,
which established a multiple-use mandate for land
management agencies to serve present and future
generations in their practices. Not long after came
passage of the Public Rangelands Improvement Act,
which sought to improve the condition of public
rangelands so that they might meet their potential for
grazing and other uses. U.S. laws pertaining to environmental quality and endangered species also have
impacts on rangeland management.

The Bureau of Land Management (BLM) manages
livestock grazing on 64 million hectares (157 million
acres) of the 99 million hectares (245 million acres) of
public lands that it administers. The BLM administers
roughly eighteen hundred permits and leases, which

are held mostly by cattle and sheep ranchers. The U.S. Forest Service, which administers the 77 million hectares (191 million acres) of national forest system lands, manages some 39 million hectares (96 million acres) of rangelands. The Forest Service became the nation's first grazing control agency in the early 1900's. In 1934 the Department of the Interior's Division of Grazing Control (soon renamed the Division of Grazing) joined it; this division became the Grazing Service in 1939, which merged with the General Land Office in 1946 to form the BLM.

Both the Forest Service and the BLM implement a regulatory system of permits, rental fees, herd size limits, and grazing seasons. They must maintain a balance among several often-conflicting objectives: providing forage for grazing and browsing animals, ensuring the land's long-term health and productivity, protecting watersheds, managing wildlife habitat, administering permitted mineral and energy resource exploration and extraction, offering recreational opportunities, and preserving the land's distinctive character and aesthetic appeal. In order to meet the array of resource needs, rangeland management agencies inventory, classify, and monitor rangeland conditions. Where rangeland health needs improvement, they implement measures to restore ecosystem functions. Public land decision makers must take into account a variety of factors that affect rangelands, including severe and extensive wildfires, invasive plant species, rural residential development driven by population increases, and global climate change.

James L. Robinson
Updated by Karen N. Kähler

FURTHER READING

Arnalds, Olafur, and Steve Archer, eds. *Rangeland Desertification*. Norwell, Mass.: Kluwer Academic, 2000.

Cheeke, Peter R. *Contemporary Issues in Animal Agriculture*. 3d ed. Upper Saddle River, N.J.: Prentice Hall, 2004.

Chiras, Daniel D., and John P. Reganold. "Rangeland Management." In *Natural Resource Conservation: Management for a Sustainable Future*. 10th ed. Upper Saddle River, N.J.: Prentice Hall, 2010.

Gibson, David J. *Grasses and Grassland Ecology*. New York: Oxford University Press, 2009.

Gordon, Iain J., and Herbert H. T. Prins, eds. *The Ecology of Browsing and Grazing*. Berlin: Springer, 2008.

Lemaire, G., et al., eds. *Grassland Ecophysiology and Grazing Ecology*. New York: CABI, 2000.

Manning, Richard. *Grassland: The History, Biology, Politics, and Promise of the American Prairie*. New York: Penguin Books, 1995.

Vallentine, John F. *Grazing Management*. 2d ed. San Diego, Calif.: Academic Press, 2001.

Woodward, Susan L. *Grassland Biomes*. Westport, Conn.: Greenwood Press, 2008.

SEE ALSO: Bureau of Land Management, U.S.; Desertification; Dust Bowl; Erosion and erosion control; Federal Land Policy and Management Act; Overgrazing of livestock; Range management; Taylor Grazing Act; Tragedy of the commons.

Great Barrier Reef

CATEGORIES: Places; ecology and ecosystems; preservation and wilderness issues

IDENTIFICATION: Massive oceanic ecosystem off the northeast coast of Australia

SIGNIFICANCE: Despite ongoing efforts to protect and preserve the Great Barrier Reef, the survival of this intricate and delicate ecosystem is constantly threatened by both natural events and human activities.

The Great Barrier Reef is a biodiverse ecosystem of more than three thousand coral reefs and seven hundred individual islands (some barely a few yards across, but twenty-seven of them large enough to have tourist resorts) that follows the Australian coast off the state of Queensland. With its tremendous size (at more than 337,000 square kilometers, or 130,000 square miles, it is visible from space) and its compelling beauty, the reef enthralls the imagination apart from its value as an intricate ecosystem. It is, in a sense, a single living organism—although more precisely it is a colony of millions of tiny coral polyps (living creatures inside colored hard shells of aragonite, a calcium derivative that shapes the familiar fan and branch shapes of coral) that live atop the dead, bleached remains of earlier generations, building slowly, steadily, century after century, into an incredibly dense superorganism.

Within this vast construction of accumulated coral structures (most of it just feet below the ocean's surface) thrives a diverse ecosystem in the pristine tropical waters that includes a wide variety of animal and plant species, among them green sea turtles, sharks, porpoises, whales, crocodiles, dugongs, and snakes, as well as more than one thousand species of fish and more than two hundred species of both land and marine birds. The Great Barrier Reef was designated a World Heritage Site in 1981 by the United Nations Educational, Scientific, and Cultural Organization (UNESCO), a recognition reserved for natural and cultural sites deemed an irreplaceable part of humanity's heritage.

Protecting the natural integrity and rich biodiversity of the massive reef is the special mission of the Great Barrier Reef Marine Park Authority, an oversight committee of the Queensland state government. In addition to natural threats—including cyclones, disease, and periodic infestations of crown-of-thorns starfish that attack the coral polyps—the most prominent threat measured since the 1990's has come from slowly rising ocean temperatures from the effects of El Niño weather conditions. Warming waters result in bleaching, the loss of tiny plants in the polyps that provide the coral its nutrients and in turn its rich coloring. That loss kills the coral itself. In addition, the reef has been affected by changes in fish migrations that have resulted from overharvesting, by unchecked pollution from land-based industries, and, most directly, by fertilizers and pesticides carried into the ocean by a river system that collects irrigation runoff from the scores of farms in northeastern Australia.

The Great Barrier Reef Marine Park Authority also monitors the impacts of the more than two million visitors the reef attracts annually, an influx that accounts for close to $2 billion each year for the Australian tourism industry. More than five thousand commercial vessels cross the reef annually, ferrying scuba divers, snorkelers, and even people who want to walk the reef's formations that lie closest to the surface. The authority monitors every aspect of these invasive encounters, from the effects of suntan oil on the formations to the impacts of fuel dumped by boats. Despite such protection, the Great Barrier Reef, because

The Great Barrier Reef ecosystem includes branching coral and many species of fish. (©Pniesen/Dreamstime.com)

of the intricacy of its ecosystem and the fragility of its construction, is considered among the most threatened natural sites on the earth.

Joseph Dewey

FURTHER READING

Bowen, James, and Margarita Bowen. *The Great Barrier Reef: History, Science, Heritage.* New York: Cambridge University Press, 2003.

Sapp, Jan. *What Is Natural? Coral Reef Crisis.* New York: Oxford University Press, 2003.

Veron, J. E. N. *A Reef in Time: The Great Barrier Reef from Beginning to End.* Cambridge, Mass.: Harvard University Press, 2010.

SEE ALSO: Australia and New Zealand; Climate change and oceans; Coral reefs; Ocean currents; Oil spills; Runoff, agricultural; Water pollution.

Great Lakes International Joint Commission

CATEGORIES: Organizations and agencies; water and water pollution

IDENTIFICATION: Independent organization established by the United States and Canada to resolve transboundary water disputes between the two nations and to advise on issues related to shared water resources

DATE: Established in 1909

SIGNIFICANCE: The border between the United States and Canada stretches thousands of miles and crosses many major waterways crucial to both nations, including the Great Lakes-St. Lawrence River system, home to one-fifth of all surface freshwater in the world. The International Joint Commission ensures the cooperative and sustainable use of clean, safe water for both nations.

The Great Lakes International Joint Commission (IJC) was founded with the signing of the Boundary Waters Treaty of 1909. This treaty gives the organization jurisdiction over all rivers, lakes, and other waterways that cross the U.S.-Canada border. The IJC has two basic functions. First, in its quasi-judicial role, it is given the authority to approve projects that affect transboundary waters. This includes activities upstream that affect the natural flow of water downstream, although because of sovereignty issues the IJC does not have authority over tributaries that feed these bodies of water. Second, when requested to do so by the Canadian and U.S. governments, the commission investigates and provides recommendations on transboundary water issues.

The IJC is headed by six commissioners, three appointed by each country. These members are expected to act as impartial judges rather than as national representatives. Several boards of experts from the United States and Canada have been assembled to assist the IJC in its investigative and decision-making processes. The body investigates water issues only when requested to do so by the governments and offers only nonbinding resolutions. However, the IJC does hold the power to arbitrate agreements between governments, although it has never been called on to use this authority.

In 1991 Canada and the United States signed the Air Quality Agreement, an executive agreement aimed at addressing issues of transboundary air quality and reducing air pollution. The Air Quality Committee was established within the IJC to report on the progress of the agreement every two years, thus expanding the commission's role, although water issues continue to predominate.

As the largest transboundary body of water and the largest surface freshwater system in the world, the Great Lakes-St. Lawrence River system is of critical importance to the commission. The first step to monitor and control water quality in the system, the Great Lakes Water Quality Agreement, was signed in 1972. Amendments and a new protocol created in 1987 set specific objectives for water quality, and the governments were required to create procedures to meet these targets. The IJC's plans for individual severely degraded areas as well as larger "lakewide management plans" take a holistic ecosystem-based approach in order to reduce human impacts on the system and remediate damage already done. The IJC reviews these plans and the progress made periodically in required reports submitted by both governments.

The commission has had high levels of success in achieving its aims since its establishment, resolving more than one hundred disputes and fostering dialogue both between the governments and with the general public. As part of its mandate, the commission must ensure that all interested parties are given the opportunity to voice their opinions. Public participation in the process and the involvement of multiple levels of governments are thus of paramount importance. The commission holds public meetings every two years to discuss the cleanup of the Great Lakes and regularly arranges other forums for public participation.

Daniel J. Connell

FURTHER READING

Bakker, Karen J. *Eau Canada: The Future of Canada's Water.* Vancouver: University of British Columbia Press, 2007.

Garrido, Alberto, and Ariel Dinar, eds. *Managing Water Resources in a Time of Global Change: Mountains, Valleys, and Flood Plains.* New York: Routledge, 2009.

Thompson, Stephen A. *Water Use, Management, and Planning in the United States.* San Diego, Calif.: Academic Press, 1999.

SEE ALSO: Air pollution; Air-pollution policy; Environmental law, international; Lake Erie; Water pollution; Water-pollution policy; Water quality; Water use.

Great Swamp National Wildlife Refuge

CATEGORIES: Places; animals and endangered species; preservation and wilderness issues

IDENTIFICATION: Federally protected wildlife habitat in Morris County, New Jersey

DATE: Established on November 3, 1960

SIGNIFICANCE: The Great Swamp National Wildlife Refuge, which was established largely as the result of a grassroots effort, provides an important nesting and feeding habitat for migratory birds.

The Great Swamp National Wildlife Refuge in New Jersey occupies a 3,116-hectare (7,700-acre) region of bottomland hardwood swamps and mixed hardwood forests containing cattail marshes, grasslands, ponds, and streams. Its beginnings were established in 1960, and it was designated a National Natural Landmark in 1966. An estimated 300,000 people visit the refuge each year. The refuge supports approximately 240 species of birds, 39 species of reptiles and amphibians, 29 species of fish, 33 species of mammals, and 600 species of plants, 215 of which are wildflowers. Of those species, more than two dozen have been designated as threatened or endangered by the state of New Jersey, including the bog turtle, the wood turtle, and the blue-spotted salamander.

In 1959 the New York-New Jersey Port Authority identified a 4,047-hectare (10,000-acre) area in rural New Jersey as the site of a new airport to serve the New York City metropolitan area. The proposed site would cover twice the area of what was then known as Idlewild Airport (later renamed John F. Kennedy International). When the *Newark Evening News* broke the story about the proposed airport, citizens, politicians, and conservationists banded together to fight the project. Citizens objected to the Port Authority's expansion plans for a multitude of reasons: destruction of homes and businesses, unacceptable noise levels from the new airport, traffic, and contamination of underground water supplies.

Fourteen volunteer groups joined forces as the Jersey Jetport Site Association (JJSA). The JJSA fought the airport expansion on a variety of fronts: political, legal, and economic. Most important was the group's ability to influence public opinion. Some of the same people also became involved with the North American Wildlife Foundation (NAWF), which purchases threatened lands and holds them for future government purchase or donates the property outright. Through the efforts of the NAWF and the Bureau of Sport Fisheries and Wildlife, the U.S. Department of the Interior agreed to grant the area wildlife refuge status if the NAWF could raise the funds to purchase 1,214 hectares (3,000 acres). The NAWF acquired the first 405 hectares (1,000 acres) in 1960 and turned them over to the U.S. Fish and Wildlife Service later that same year, and on an November 3, 1960, an act of Congress established as a park the first part of what would eventually become the Great Swamp National Wildlife Refuge.

In 1962 public hearings began in New Jersey on a bill that would prohibit airport construction in seven northern counties—including the Port Authority site in Morris County. The bill passed by a wide margin but was vetoed by New Jersey governor Robert Meyner, who declared it unconstitutional. Governor Meyner lost his bid for reelection and was replaced by Richard Hughes, who supported the bill and the refuge. By 1963 the state had even provided $25,000 to purchase additional land, which was added in 1964. Additional parcels of land were appended to the existing refuge throughout the following years, until by 1990 the Great Swamp National Wildlife Refuge consisted of more than 2,800 hectares (7,000 acres) of land.

P. S. Ramsey

FURTHER READING

Dawson, Chad P., and John C. Hendee. *Wilderness Management: Stewardship and Protection of Resources and Values.* 4th ed. Boulder, Colo.: WILD Foundation, 2009.

Richman, Steven M. *The Great Swamp: New Jersey's Natural Treasure.* Atglen, Pa.: Schiffer, 2008.

SEE ALSO: Arctic National Wildlife Refuge; Fish and Wildlife Service, U.S.; National Wildlife Refuge System Administration Act; Nature preservation policy; Wildlife management; Wildlife refuges.

Green buildings

CATEGORIES: Atmosphere and air pollution; urban environments; pollutants and toxins; resources and resource management

DEFINITION: Structures designed and constructed to increase resource efficiency and reduce negative impacts on human health and the environment

SIGNIFICANCE: Residential and commercial buildings generate more than 30 percent of the world's emissions of carbon dioxide, a greenhouse gas that has been linked to global warming. Green buildings reduce carbon emissions substantially and provide significant environmental benefits by reducing solid waste, efficiently using energy and other resources, reducing air and water pollution, and conserving natural resources.

Although energy efficiency and sustainability were not major concerns at the time, the concept of green buildings originated during the mid-nineteenth century. The Galleria Vittorio Emanuele II in Milan, Italy, designed in 1861, and the Crystal Palace in London, England, built in 1851, both used underground air cooling and roof ventilators to reduce their negative impacts on the environment. New York City's Flatiron Building, constructed in 1902, used deep-set windows to control the interior temperature.

From the 1930's to the 1960's technological advances such as the inventions of reflective glass, structural steel, and air-conditioning resulted in the proliferation of high-rise buildings that consumed huge amounts of cheap fossil fuels. During the 1960's, however, environmental consciousness grew, and visionaries began defining green building. During this period

scientist James Lovelock formulated the Gaia hypothesis, a holistic concept of the earth as a single, complex organism. In 1969 landscape architect Ian L. McHarg published *Design with Nature*, which helped define green architecture.

BEGINNINGS OF THE MOVEMENT

On the first Earth Day, in April, 1970, millions of Americans showed their concern about the environment. The 1973 and 1979 oil crises demonstrated the need for the nation to seek energy from diversified sources and become less dependent on fossil fuels. The U.S. government and many corporations began investing in research into methods of energy conservation and alternative energy sources.

During the 1980's architect Malcolm Wells designed green underground and earth-sheltered buildings. In 1982, physicist Amory Lovins and his wife, environmentalist Hunter Lovins, emphasized the basic green principle of using regional resources in founding their Rocky Mountain Institute, a nonprofit resource policy center that promotes resource efficiency and global security. Beginning during the mid-1980's, popular environmental organizations—such as the Sierra Club, Greenpeace, the Nature Conservancy, and Friends of the Earth—became increasingly active. Growing awareness of the problem of sick building syndrome raised concerns regarding the indoor environments of some workplaces. In 1984 architect William McDonough designed a headquarters building for the Environmental Defense Fund in New York City using a high-performance building approach (the building was completed in 1985). During the late 1980's Pliny Fisk III designed Blueprint Farm—a green agricultural community—in Laredo, Texas, using recycled materials, wind power, and photovoltaic panels.

MILESTONES DURING THE 1990's

In 1992, the first local green building program began in Austin, Texas, and the U.S. Environmental Protection Agency (EPA) launched the Energy Star program, a voluntary energy-efficiency labeling program for consumer products. By 2009 Energy Star labels were appearing on more than sixty product categories and Energy Star ratings had become the standard for major appliances, homes, commercial buildings, and heating systems. By November, 2009, one million Energy Star-qualified homes had been built throughout the United States, resulting in an estimated reduction

This solar-powered building in Athens, Greece, named the Promitheus Pirforos, opened in 2007. It is the world's first energy-autonomous building, capable of operating for sixty days without connection to the electrical grid. (AP/Wide World Photos)

developing its green building certification program, known as LEED (for Leadership in Energy and Environmental Design), which became available for public use in 2000. This voluntary system provides third-party certification that certain standards have been met in the construction of high-performance, sustainable buildings, with an emphasis on reducing carbon dioxide emissions and increasing energy efficiency. LEED certification covers a wide range of existing and new commercial and residential buildings, including offices, schools, medical facilities, private homes, and stores.

ENVIRONMENTAL BENEFITS

The key areas measured in the LEED certification process reflect the environmental benefits of green building. Sustainable site development involves preserving natural resources for future generations and can include reusing existing buildings, planting around buildings, roof gardens, and underground or earth shelters. Building for water savings and efficiency involves monitoring water supplies and usage, recycling gray or previously used water, and constructing rainwater catchment systems. To improve energy and atmosphere efficiency, buildings can use geographically and climatically appropriate energy resources, including renewable energy. Eliminating chlorofluorocarbons in heating, ventilation, air-conditioning, and refrigeration systems helps reduce ozone depletion. Efforts to conserve materials and resources include using renewable, recycled, local, chemical-free, nonpolluting, and durable materials. Indoor environmental quality can be improved through the use of nontoxic materials, adequate ventilation, temperature controls, and materials that emit few or no volatile organic compounds.

In the twenty-first century, as environmental knowledge and building technologies continue to improve, the green building movement is gaining worldwide momentum. Given that buildings consume more than one-half of the world's resources and generate more than 30 percent of greenhouse gas emissions, the benefits of green building have become increasingly obvious. By 2010 forty-one countries had developed their own LEED initiatives, including Aus-

of greenhouse gas emissions equal to the emissions released by 370,000 gasoline-fueled motor vehicles. Many other countries adopted the Energy Star idea, including Japan, Taiwan, China, New Zealand, South Africa, and the nations of the European Union.

In 1993 Bill Clinton's presidential administration began the successful "Greening of the White House" initiative, and the nonprofit U.S. Green Building Council (USGBC) was created to promote the construction of environmentally responsible, healthy, and profitable buildings. USGBC is a national, voluntary, consensus coalition with members from all sectors of the building industry. In 1995, USGBC began tralia, Brazil, Canada, France, India, Israel, Mexico, the United Arab Emirates, and the United Kingdom.

Alice Myers

FURTHER READING

Fisanick, Christina, ed. *Eco-rchitecture*. Detroit: Greenhaven Press, 2008.

GreenSource. *Emerald Architecture: Case Studies in Green Building*. New York: McGraw-Hill, 2008.

Johnston, David, and Scott Gibson. *Green from the Ground Up: A Builder's Guide—Sustainable, Healthy, and Energy-Efficient Home Construction*. Newtown, Conn.: Taunton Press, 2008.

Yudelson, Jerry. *The Green Building Revolution*. Washington, D.C.: Island Press, 2008.

SEE ALSO: Air-conditioning; Earth-sheltered construction; Energy conservation; Energy-efficiency labeling; Green marketing; Heat pumps; Indoor air pollution; Renewable energy; Sick building syndrome; Solar water heaters; Water-saving toilets.

Green marketing

CATEGORY: Philosophy and ethics

DEFINITION: The touting of the environmental benefits of a product, service, or company to bolster its image and encourage sales

SIGNIFICANCE: As the environmental awareness of the general population has grown, companies have increasingly used production methods, product ingredients, and corporate policies to appeal to environmentally concerned consumers. Although some green marketing efforts have effected positive change, the lack of "green standards" has left this marketing approach subject to consumer confusion as well as corporate abuse in the form of greenwashing.

The term "green marketing" first arose during the 1970's during a period of increasing environmental awareness among consumers. New companies formed that emphasized environmentally friendly products, and existing corporations began to give more consideration to environmental issues, partly because of new governmental regulations but also because suddenly "being green" appeared to promise economic benefits. Corporations discovered that they could create loyal customer bases by exceeding regu-

latory compliance and becoming environmentally proactive. Whether businesses made ecologically friendly choices out of a sense of social responsibility or because they were forced to do so by environmental regulations, they found they could then use those choices as a means to appeal to consumers who were seeking to make healthy choices and reduce their impact on the planet.

Broadly defined, green marketing encompasses a wide variety of corporate practices, activities, and choices. In its simplest, most easily recognized form, it involves promoting or advertising a product that features an easily recognized environment-related attribute—for example, a toothbrush manufactured from recycled plastic yogurt cups, an energy-saving lightbulb, a biodegradable plastic bag, or pesticide-free produce. Green marketing is also used to promote product modifications that benefit the environment, such as the reformulation of a hairspray to eliminate chemicals that damage the ozone layer or contribute to greenhouse gas emissions. Reducing or otherwise changing packaging; altering a production process to minimize waste, consume fewer resources, or reduce dependence on hazardous and toxic raw materials; purchasing components from a local manufacturer rather than having them shipped from a source overseas—these and other corporate decisions that determine a company's overall environmental impact can serve as fodder for a green marketing campaign.

ENVIRONMENTAL RESPONSIBILITY AND PROFITABILITY

Green marketing represents a challenge to the traditional view that an inherent conflict exists between economic profits and environmental quality. By "going green" companies can make profits while producing environmentally friendly products or increase profits by changing production processes to use less energy or material, produce less pollution, or find new uses for onetime waste products. Corporate consideration of environmental consequences can increase efficiency and profitability by forcing executives to rethink normal operating procedures. For example, at the Atlantic Richfield Company's Los Angeles refinery, a series of relatively low-cost changes during the 1980's reduced waste volume by 8,600 tons per year. Because disposal costs were about $300 per ton, the company saved more than $2 million each year in disposal costs alone. Atlantic Richfield also found markets for much of its former waste—it began

FTC Guidance on Green Marketing

In its "Guides for the Use of Environmental Marketing Claims," first issued in 1992, the Federal Trade Commission includes helpful examples to clarify each of the guidelines presented, as this excerpt illustrates.

Overstatement of environmental attribute: An environmental marketing claim should not be presented in a manner that overstates the environmental attribute or benefit, expressly or by implication. Marketers should avoid implications of significant environmental benefits if the benefit is in fact negligible.

Example 1: A package is labeled, "50% more recycled content than before." The manufacturer increased the recycled content of its package from 2 percent recycled material to 3 percent recycled material. Although the claim is technically true, it is likely to convey the false impression that the advertiser has increased significantly the use of recycled material.

Example 2: A trash bag is labeled "recyclable" without qualification. Because trash bags will ordinarily not be separated out from other trash at the landfill or incinerator for recycling, they are highly unlikely to be used again for any purpose. Even if the bag is technically capable of being recycled, the claim is deceptive since it asserts an environmental benefit where no significant or meaningful benefit exists.

Example 3: A paper grocery sack is labeled "reusable." The sack can be brought back to the store and reused for carrying groceries but will fall apart after two or three reuses, on average. Because reasonable consumers are unlikely to assume that a paper grocery sack is durable, the unqualified claim does not overstate the environmental benefit conveyed to consumers. The claim is not deceptive and does not need to be qualified to indicate the limited reuse of the sack.

Example 4: A package of paper coffee filters is labeled "These filters were made with a chlorine-free bleaching process." The filters are bleached with a process that releases into the environment a reduced, but still significant, amount of the same harmful by-products associated with chlorine bleaching. The claim is likely to overstate the product's benefits because it is likely to be interpreted by consumers to mean that the product's manufacture does not cause any of the environmental risks posed by chlorine bleaching. A claim, however, that the filters were "bleached with a process that substantially reduces, but does not eliminate, harmful substances associated with chlorine bleaching" would not, if substantiated, overstate the product's benefits and is unlikely to be deceptive.

selling spent alumina catalyst to chemical companies, spent silica catalyst to cement makers, and alkaline carbonate sludge to a nearby sulfuric acid plant.

There are numerous other examples of corporate changes benefiting both profitability and environmental quality. The Pollution Prevention Pays (3P) program created by the giant 3M Company is the project perhaps most often cited. Initiated in 1975 by two environmental engineers and a communications specialist, the groundbreaking, companywide program cut 3M's generated pollution by 50 percent. By 2009, 3P had saved the company more than $1.2 billion and prevented more than 2.9 billion pounds of pollutants at the source. In addition to the savings, programs such as 3P benefit the public images of the corporations that implement them.

For some companies, green marketing is inextricably linked with corporate image; examples are Ben & Jerry's, the Body Shop, and Tom's of Maine, all of which were founded during the 1970's. These companies pioneered the concept of socially responsible, profitable business, and among their corporate values is the importance of minimizing environmental impact. Smaller firms are generally perceived as being "greener" and more socially responsible than huge multinational corporations, so some consumer confusion and alienation arose when during the early twenty-first century these companies were purchased by multinational giants—Ben & Jerry's by Unilever, the Body Shop by L'Oréal/Nestlé, and Tom's of Maine by Colgate-Palmolive.

While businesses may choose to be environmentally responsible without promoting the fact, many major corporations have come to consider their environmental efforts to be an important public relations tool. Market research has indicated that roughly one-half of the consumers in the United States and other industrialized nations are willing to pay a premium for environmentally friendly products. Even among those unwilling to pay higher prices, environmental attributes serve as a tiebreaker between competing brands, thus serving as a potential impetus for brand loyalty and product differentiation.

Not all publicized "greening" efforts result in corporate profits. For example, during the late 1980's McDonald's restaurants received widespread criticism for using polystyrene foam packaging for its sandwiches. As public pressure increased, McDonald's signed an agreement with the Environmental Defense Fund, which provided technical assistance in improving the company's environmental performance. Through these efforts, McDonald's switched to paper and cardboard packaging and made other adjustments to reduce its environmental impacts. These changes, which were highly publicized, ultimately did not alter McDonald's market position relative to its major competitors, and none of the competitors experienced enough economic pressure to respond with waste reduction programs to match that of McDonald's.

PROBLEMS

During the 1980's many companies began using product labels that touted the environmental friendliness of the contents and packaging. Many of these claims were dubious at best, however. For example, soon after dolphin-safe tuna became available, Greenpeace reported that some of the tuna being sold under this claim had been taken in the traditional manner, with no attempts made to prevent dolphins from dying in the nets.

The actual environmental impacts of products can be evaluated, and such evaluation can serve as a basis for product certification. In the European Union and Canada, government performs this function. In the United States, the federal government has no national program in place to evaluate products' environmental impacts, but two nonprofit organizations— Scientific Certification Systems (originally known as Green Cross) and Green Seal—systematically evaluate the environmental impact of products from production to disposal.

In 1992, the Federal Trade Commission (FTC) introduced guidelines governing the circumstances under which terms such as "biodegradable," "recyclable," and "ozone-safe" can be used in advertising in the United States. Firms that cannot substantiate their environmental claims are not allowed to use these terms in their advertising. Other terms, among them "all natural," remain unregulated by the FTC and are often used to promote environmentally unfriendly products.

The unethical practice of making exaggerated or false environmental claims to promote a product or business is known as greenwashing. That term is also sometimes applied to legitimate, environmentally beneficial actions taken by corporations that have flawed environmental records or reputations. The cynical consumer response of suspecting corporations of greenwashing has led some companies to practice what Joel Makower, author of *The Green Consumer* (1990), has dubbed "covert environmentalism," in which they shun green marketing to avoid having their environmentally responsible actions attacked and their motives questioned.

Although green marketing has grown in acceptance and prevalence in the decades since its inception, it remains problematic. Consumer concern reflected in market research surveys does not always translate to significant changes in purchasing and consumption habits. Not all consumers are willing or able to pay a premium for green products. Many are mistrustful of or confused by manufacturers' claims, and the challenge of actually determining the environmental impact of one good in comparison with another is considerable. In addition, the environmentally educated consumer is more likely to be overwhelmed by conflicting goals. Is it better for the environment, for example, for a meat eater to select grass-fed beef over feedlot-produced meats when grass-fed cattle are known to produce more of the greenhouse gas methane? Ultimately, if an eco-friendly product or service requires consumers to sacrifice too much in terms of price, quality, convenience, or availability, the majority will decide against purchasing it.

Allan Jenkins
Updated by Karen N. Kähler

FURTHER READING

Charter, Martin, and Michael Jay Polonsky. *Greener Marketing: A Global Perspective on Greening Marketing Practice.* 2d ed. Sheffield, England: Greenleaf, 1999.

Conroy, Michael E. *Branded! How the "Certification Revolution" Is Transforming Global Corporations.* Gabriola Island, B.C.: New Society, 2007.

Frause, Bob, and Julie Colehour. *The Environmental Marketing Imperative: Strategies for Transforming Environmental Commitment into a Competitive Advantage.* Chicago: Probus, 1994.

Grant, John. *The Green Marketing Manifesto.* Hoboken, N.J.: John Wiley & Sons, 2007.

Ottman, Jacquelyn A. *Green Marketing: Opportunity for Innovation.* 2d ed. Chicago: NTC Business Books, 1998.

Smith, Toby M. *The Myth of Green Marketing: Tending Our Goats at the Edge of Apocalypse.* Toronto: University of Toronto Press, 1998.

Wasik, John F. *Green Marketing and Management: A Global Perspective.* Cambridge, Mass.: Blackwell Business, 1996.

Welford, Richard, and Richard Starkey, eds. *Business and the Environment.* New York: Taylor & Francis, 1996.

SEE ALSO: Ceres; Certified wood; Dolphin-safe tuna; Eco-fashion; Ecotourism; Energy-efficiency labeling; Green buildings; Greenwashing; McToxics Campaign; Organic gardening and farming; Recycling.

Green movement and Green parties

CATEGORIES: Activism and advocacy; philosophy and ethics

DEFINITIONS: Movement seeking to influence society toward greater consciousness of environmental issues and the political parties established to bring about such change through existing democratic institutions

SIGNIFICANCE: The Green movement seeks to change certain fundamental values in Western society, particularly those that appear to create threats to humanity and the larger nonhuman environment, such as unrestricted technological progress and economic development. Although varied in their origins and interests, Green parties work for political change aimed at achieving the ideals of the Green movement.

The Green movement in Western nations evolved from the various protest movements of the 1960's. Specifically, Rachel Carson's book *Silent Spring* (1962) awakened people to the health hazards of industrial pollution and encouraged them to consider humanity and the environment as interdependent. In addition, the fear that world superpowers might resort to the use of nuclear weapons during the Cold War prompted antinuclear, propeace demonstrations. The peace movement in Europe produced the first formal Green parties in the 1970's, organized on the models of two precursors: New Zealand's New Values Party, started in 1972, and Great Britain's Ecology Party, started in 1973, both of which sought to formu-late new electoral strategies with environmental issues.

The Greens first organized in the United States in 1984 as the Green Committees of Correspondence, which took the German Green Party as their model. These organizations consisted of state parties and otherwise unaligned individual members. From the outset, however, Green parties were state controlled; their national organization was a loose confederation. The first individual Green candidate in the United States appeared on a ballot in 1986, and in 1990 the Alaska Green Party was the first to achieve ballot status, followed by the California party in 1992. The coalition of Green parties for the first time fielded a presidential candidate, Ralph Nader, in the 1996 election and attracted 1 percent of the votes. Despite interstate disputes over tactics during the early 1990's, twenty-five state parties formed the Association of State Green Parties (ASGP) in 1996 in order to prepare for national elections in 2000; in 2001, the ASGP was replaced by the Green Party of the United States (GPUS), a federation of forty-six state Green parties.

Some degree of international affiliation exists among national parties. The GPUS, for instance, is a member of the Federation of the Green Parties of the Americas, based in Mexico City, and is associated with the thirty-seven-member European Federation of Green Parties.

CORE PRINCIPLES

Central to the basic tenets of Green parties in all countries is the belief that the world order must be reshaped, with emphasis given to local governance. At the same time, the transformation must involve a shift in people's interest from the immediate future and self-centered satisfaction to the long-range future and sustainable production of material needs. The often-repeated slogan "Think globally, act locally" reflects the spirit of the movement. To this end, the European Greens published four basic principles in the 1970's, to which Greens in the United States added six more in the 1980's.

The "four pillars" of the early European Greens were ecology, social justice, grassroots democracy, and nonviolence. By "ecology" (or "ecological wisdom," the American term) is meant a redirection of mentality: People should consider themselves part of nature, not controllers of it, and so live in harmony with it. Practically, this entails devoting technology to achieving an energy-efficient economy and minimizing extraction of nonrenewable resources. Social justice encompasses universal equal rights, dignity, and social responsibility based on the values of simplicity and moderation. Greens advocate community-controlled, free education and programs to prevent crime. Grassroots democracy, in the American interpretation, would distribute most state and federal power to local elected officials and mediating institutions, such as neighborhood organizations, church groups, voluntary associations, and ethnic clubs. The goal is to restore civic vitality by involving as many people as possible in decision making and avoiding reliance on lawyers, legislators, and bureaucrats. The pillar of nonviolence involves not only seeking to end patterns of conflict in families and communities but also working to eliminate nuclear weapons. Worldwide, Greens opposed the 1991 Persian Gulf War and cited the environmental disaster of burning oil wells and oil spills in the Gulf as evidence that military action is counterproductive at best and likely to be ruinous.

The six key values added by American Green parties elaborate on issues inherent in the four pillars. First, respect for diversity would honor cultural, racial, sexual, and religious differences, while insisting that citizens bear individual responsibility to all beings for their actions. Second, feminism would replace traditions of dominance by a particular ethnic group, social class, or gender with ethics based on cooperation and respect for the contemplative, intuitive capacities of all people. Community-based economics calls for employee ownership of businesses, workplace democracy, and equal distribution of wealth to ensure basic economic security for all. Similar to grassroots democracy, decentralization would give power to economically defined localities and ecologically defined regions; it entails redesigning institutions so that control over regulations and money is greatest at the community level, rather than the national level, and control over environmental policy is greatest at the regional level. Personal and global responsibility encourages wealthy communities to assist grassroots groups in developing countries—directly rather than through their governments—in order to make them self-sufficient. To produce the means for such aid, the Greens want to decrease the national defense budget, although not to the point of compromising American security. Finally, future focus, or sustainability, requires all economic, scientific, and cultural policies to be formulated with careful attention to their long-range effects and not just to their immediate benefits. For this reason, European and American Greens have denounced such scientific developments as genetic engineering and nuclear power.

The European and American versions of the Green movement contain potential conflicts. For example, the European Greens call for internationalized security to prevent war; in particular they want to give control over military action to the United Nations after it is reformed so that each nation has equal voting power. They also want to replace regional trade treaties, such as the North American Free Trade Agreement (NAFTA), with treaties negotiated and monitored by the United Nations. The American goal of decentralization does not clearly accord with this vision of world order, nor is it clear how community-based economics would admit the European goal of planetwide economic solidarity.

POLITICAL POWER

In the 1980's and 1990's, doctrinal divisions that were already latent among the Greens produced contention and sometimes disaffection, especially in the European organizations. Divisions have tended to fall between moderates, who are open to compromises with other political parties in order to gain power, and radicals, for whom compromise is unacceptable. The moderates, sometimes referred to as "light green," tend to espouse the anthropocentric view: The Greens should help safeguard the human environment, although preferably not at the expense of other organisms. This faction seeks reform of existing social and economic institutions. The radicals, known as "dark green," are biocentric, holding that all creatures have equal natural rights to life and that humankind should not consider itself a favored species. They wish to end the affluent, technological, expansionist, service-providing orientation of society.

Green parties have achieved modest success, especially in Western Europe, in placing political candidates in office. Greens have constituted 10 percent of some parliaments and have entered into coalitions that have formed ruling governments. In nearly all countries, they have influenced environmental legislation. In the United States, by 1999 Green candidates had been elected to sixty-three local offices in fifteen states, mostly for such nonpartisan agencies as planning groups and school boards. Arcata in Northern California was the first municipality to have a Green majority on its city council, and in 2008 a Green was elected to the Arkansas House of Representatives (although he later changed party affiliation, becoming a Democrat).

The power of Green parties has manifested more in influencing policy and in education about environmental and social issues than in direct legislative or administrative action because opposition to the Greens' principles is formidable in most countries. Critics in the United States accuse the Green movement of elitism, asserting that it is the project of well-educated, middle-class white people. The Green reconception of society appears to leave little room for individualism, which also runs counter to mainstream American sensibilities, and the proposal to give international organizations control over security, particularly the United Nations, raises fears among nationalists in all countries that cultural identity and sovereignty will be lost. Perhaps most of all, the citizens of Western nations generally resist the ideas of the redistribution of wealth and sustainability, which would dismantle free market economies. Nonetheless, the Green movement has succeeded in a primary objective: to end the status of technological, economic expansion as an unquestioned value and place the burden of proof on its proponents to demonstrate that specific projects will not harm humanity or nature.

Some mainstream parties, such as the Democrats in the United States, complain that Green parties are spoilers, enticing away liberal votes for candidates who stand little chance of winning elections. For this reason, liberal parties have adopted some of the Green environmental goals in their own platforms, weakening the attraction of the Green parties' candidates.

Roger Smith

FURTHER READING

Audley, John J. *Green Politics and Global Trade: NAFTA and the Future of Environmental Politics.* Washington, D.C.: Georgetown University Press, 1997.

Carter, Neil. *The Politics of the Environment: Ideas, Activism, Policy.* 2d ed. New York: Cambridge University Press, 2007.

Dickerson, Mark O., Thomas Flanagan, and Brenda O'Neill. "Environmentalism." In *An Introduction to Government and Politics: A Conceptual Approach.* 8th ed. Toronto: Nelson Education, 2010.

Dobson, Andrew. *Green Political Thought.* New York: Routledge, 1995.

Johnson, Huey D. *Green Plans: Greenprint for Sustainability.* Lincoln: University of Nebraska Press, 1997.

Kline, Benjamin. *First Along the River: A Brief History of the U.S. Environmental Movement.* 3d ed. Lanham, Md.: Rowman & Littlefield, 2007.

Pepper, David. *Modern Environmentalism: An Introduction.* 1996. Reprint. New York: Routledge, 2003.

Switzer, Jacqueline Vaughn. *Green Backlash: The History and Politics of Environmental Opposition in the U.S.* Boulder, Colo.: Lynne Rienner, 1997.

SEE ALSO: Animal rights movement; Antinuclear movement; Environmental ethics; Environmental justice and environmental racism; European Green parties.

Green Plan

CATEGORY: Preservation and wilderness issues

IDENTIFICATION: Canadian national strategy to create a cleaner, safer, and healthier environment along with a sound economy

DATE: Initiated on December 11, 1990

SIGNIFICANCE: By establishing a series of sustainable development goals and then measuring progress toward those goals and mobilizing collective, nationwide efforts, Canada's Green Plan has provided a model for a national approach to environmental management.

After many years of extensive consultations with Canadians representing government, business, interest groups, and the public, Canada's Department of the Environment launched its internationally acclaimed Green Plan in December, 1990. The overall goal of the plan was to ensure that current and future generations would enjoy a safe, healthy environment and a sound economy. Although the Green Plan focused on a wide range of environmental issues, it also incorporated the fundamentals of sustainable development into all aspects of decision making at all levels of society.

Since the Green Plan was an umbrella document, many of the details were left to work out during implementation. Many programs were initiated that affected various aspects of the lives of all Canadians, including the air they breathed, the water they drank, and the food they ate. For example, numerous full assessments of priority toxic substances were performed, and in 1992 the number of full-time Canadian environmental inspectors and investigators was increased from forty-nine to seventy.

The Green Plan established a series of sustainable development goals for Canadians that now serve as benchmarks for measuring progress and mobilizing collective, nationwide efforts. The first goal is to ensure that current citizens and future generations have clean air, water, and land, all of which are essential to sustaining human and environmental health. Steps toward achieving this goal include the reduction of ground-level ozone (smog) to below the threshold of adverse health effects, and the reduction of Canada's generation of waste by 50 percent.

The second goal is the sustainable use of renewable resources, which involves shifting forest management from sustained yield to sustainable development. Some of the key areas for decision making are those of harvesting practices (particularly in old-growth forests), reforestation, and the use of forest pesticides. Answers are being provided through the creation of a network of model forests and the creation of Tree Plan Canada.

The third goal is the protection of special spaces and species. The Canadian government has set aside 12 percent of the country as protected space for parks, wildlife areas, and ecological reserves. Similarly, the fourth goal focuses on preserving and enhancing the integrity, health, biodiversity, and productivity of Canada's Arctic ecosystems. Waste cleanups and assessments have been carried out at numerous sites in the Yukon and the Northwest Territories.

The fifth goal of the Green Plan is to enhance Canada's commitment to global environmental security. For example, plans were implemented to phase out the use of human-made chlorofluorocarbons (CFCs), methyl chloroform, and other major ozone-depleting substances by the year 2000. In addition, national emissions of carbon dioxide and other greenhouse gases were to be stabilized at 1990 levels. The Green Plan also includes the goals of minimizing the impact of environmental emergencies and making environmentally responsible decisions. The Canadian government is accomplishing these goals by implementing plans for quick, effective responses to environmental emergencies and by providing accurate, accessible information about the environment to all Canadians.

Alvin K. Benson

Further Reading

Boyd, David R. *Unnatural Law: Rethinking Canadian Environmental Law and Policy.* Vancouver: University of British Columbia Press, 2003.

Dwivedi, O. P., et al. "The Canadian Political System and the Environment." In *Sustainable Development and Canada: National and International Perspectives.* Orchard Park, N.Y.: Broadview Press, 2001.

See also: Conservation; Environment Canada; Nature preservation policy; Nature reserves; Preservation; Wilderness areas.

Green Revolution

Category: Agriculture and food

Definition: Implementation of advances in agricultural science to raise food production levels, particularly in developing countries

Significance: The input-intensive agriculture associated with the Green Revolution has increased crop yields but has also created a number of environmental problems, such as rising nitrate levels in water supplies from the use of fertilizers, community health threats linked to pesticides, and damage to soil quality that includes compaction and salinization.

The Green Revolution can be traced back to a 1940 request from Mexico for the United States to provide technical assistance to increase Mexican wheat production. By 1944, with the financial support of the Rockefeller Foundation, a group of U.S. scientists began to research methods of adapting the new high-yield variety (HYV) wheat that had been successfully used on American farms in the 1930's to Mexico's varied environments. A major breakthrough in this effort is attributed to Norman Borlaug, who, by the late 1940's, was director of the research in Mexico. For his research and his work in the global dissemination of the Mexican HYV wheat, Borlaug won the 1970 Nobel Peace Prize.

From wheat, research efforts shifted to rice production. Through the work of the newly created International Rice Research Institute in the Philippines, researchers used advanced methods of plant breeding to develop an HYV rice. This so-called miracle rice was widely adopted in developing countries during the 1960's. Since that time, researchers have sought to

Agronomist Norman Borlaug looks at selected wheat stocks at the Rockefeller Agricultural Institute in Atizapán, Mexico, in October, 1970. (AP/Wide World Photos)

spread the success of the Green Revolution to other crops and to more countries.

Approximately one-half of the yield increases in food crops worldwide since the 1960's are attributable to the Green Revolution. Had there not been a Green Revolution, the amount of land used for agriculture would undoubtedly be higher today, as would the prices of wheat, rice, and maize, three species of plants that account for more than 50 percent of total human energy requirements. There is a concern, however, that the output benefits of the Green Revolution have had some negative equity and environmental effects.

In theory, a small farmer will get the same advantages from planting HYV seeds as will a large farmer. In practice, however, small farmers have had more difficulty in gaining access to the Green Revolution. To use the new seeds, farmers need adequate irrigation and the timely application of chemical fertilizers and pesticides. In many developing countries, small farmers' limited access to credit makes it difficult for them to obtain the variety of complementary inputs they need for success with HYV seeds.

The Green Revolution has promoted input-intensive agriculture, which has, in turn, created several problems: Greater usage of fertilizers is associated with rising nitrate levels in water supplies, pesticides have been linked to community health problems, and long-term, intensive production has resulted in compaction, salinization, and other soil-quality problems. Because agriculture is increasingly dependent on fossil fuels, food prices will become more strongly linked to energy supplies of this type, a fact that has raised concerns about the sustainability of the new agriculture. Biotechnological approaches to generating higher yields, the expected future path of the Green Revolution, will raise an additional set of equity and environmental concerns.

Bruce G. Brunton

FURTHER READING

Conkin, Paul K. *A Revolution Down on the Farm: The Transformation of American Agriculture Since 1929.* Lexington: University Press of Kentucky, 2008.
Federico, Giovanni. *Feeding the World: An Economic History of Agriculture, 1800-2000.* Princeton, N.J.: Princeton University Press, 2005.

SEE ALSO: Agricultural revolution; Borlaug, Norman; Genetically modified foods; High-yield wheat; Intensive farming; Irrigation; Pesticides and herbicides; Sustainable agriculture.

Greenbelts

CATEGORIES: Land and land use; preservation and wilderness issues
DEFINITION: Tracts of open space preserved to control urban growth patterns
SIGNIFICANCE: Greenbelts provide numerous environmental, social, and economic benefits to the areas that surround them.

Throughout the twentieth century and into the twenty-first, developed nations have urbanized at ever-increasing rates. As a result of the automobile, the United States and Canada in particular have been subjected to uncontrolled growth, resulting in the phenomenon referred to as urban sprawl. As development moves outward from a central city, prime agricultural and forested lands are converted to more intensive uses, resulting in a significant loss of wildlife and plant habitats. This decrease in natural areas also leads to a subsequent degradation of air and water quality.

The concept of creating greenbelts, or greenways, developed as a grassroots response to address these problems. With limited public funds for open-space preservation, greenbelt proponents have focused attention on "leftover" or abandoned lands. These parcels are often found along ridgelines and streams, areas that are too steep or too wet for development. Abandoned railroad and utility rights-of-way have become important potential resources as well. All of these areas have common physical characteristics: They are long, thin tracts of land that relate to the topography, threading through land more suitable for development.

Greenbelts, as linear open-space corridors, provide several important benefits. First, they enable urban areas to retain their biodiversity. This is important for maintaining plant and animal habitats, as well as for establishing sources of protection for air and water quality. The natural corridors provide migration routes for species interchange. This movement of plant and wildlife along natural pathways is particularly significant, since it may determine the ability of some species to survive in these areas. Second, the retention of undeveloped, vegetated lands allows surface water to be returned naturally to the water table, minimizing surface runoff, erosion, and subsequent stream sedimentation.

Greenbelts offer many recreational opportunities as well. Most greenbelts include systems of trails that may link larger, more intensive recreational facilities or provide people with access to natural amenities from urban areas. By connecting different sorts of facilities, in essence creating a system or network of urban parks, greenbelts increase the aggregate benefit to the community. Because of the linear nature of greenbelts, they have more edge area than do other kinds of parks or open spaces. This characteristic maximizes the available open space and provides potential access to greater numbers of people.

The economic benefits of greenbelts are also significant. As leftover or derelict lands, suitable parcels may be purchased relatively inexpensively; thus mini-

mum expenditure is often required for the development of a greenbelt system. In addition, the aesthetic improvement of the green edge provided by a greenbelt often enhances the value of adjacent properties.

Steven B. McBride

FURTHER READING
Amati, Marco, ed. *Urban Green Belts in the Twenty-first Century.* Burlington, Vt.: Ashgate, 2008.
Hough, Michael. *Cities and Natural Process: A Basis for Sustainability.* 2d ed. New York: Routledge, 2004.

SEE ALSO: Biodiversity; Olmsted, Frederick Law; Open spaces; Urban ecology; Urban parks; Urban planning; Urban sprawl.

Greenhouse effect

CATEGORY: Weather and climate
DEFINITION: Natural process of atmospheric warming in which solar energy absorbed by the earth's surface is reradiated and absorbed by certain atmospheric gases, primarily carbon dioxide and water vapor
SIGNIFICANCE: Without the natural warming process provided by the greenhouse effect, the earth's atmosphere would be too cold for the planet to support life. However, an intensification of the greenhouse effect resulting from human-generated greenhouse gas emissions could significantly alter surface atmospheric temperatures on a global scale, which would have major environmental and societal implications.

Since 1880, at which point historical measurement records become reliable enough and of sufficient spatial distribution to provide an overview of recent global climate trends, the earth's surface atmospheric temperatures have on average become warmer. During this period—and since the mid-eighteenth century, when the Industrial Revolution began—human activity has released increasing quantities of what are known as greenhouse gases into the atmosphere. These gases include naturally occurring substances such as carbon dioxide, methane, nitrous oxide, and ozone, as well as synthetic chemicals such as chlorofluorocarbons (CFCs), hydrofluorocarbons (HFCs), perfluorocarbons (PFCs), and sulfur hexafluoride. (Water vapor is the most abundant of the naturally occurring greenhouse gases, but human activity has an insignificant influence on its atmospheric concentrations.) Anthropogenic (human-caused) greenhouse gases have been identified as likely contributors to the rise in global surface temperature.

The temperature increase may lead to drastic changes in climate and food production, as well as widespread coastal flooding. As a result, many scientists, organizations, and governments have called for curbs on greenhouse gas emissions. Because their predictions are not definite, however, debate continues about the financial and societal costs of reducing the production of these gases given the lack of certainty regarding the benefits.

GLOBAL WARMING AND HUMAN INTERFERENCE

The naturally occurring greenhouse effect takes place because the gases that make up the atmosphere are able to absorb only particular wavelengths of energy. The atmosphere is largely transparent to shortwave solar radiation, so sunlight basically passes through the atmosphere to the earth's surface. Some is reflected or absorbed by clouds, some is reflected from the earth's surface, and some is absorbed by dust or the earth's surface. Only small amounts are actually absorbed by the atmosphere. Sunlight therefore contributes very little to the direct heating of the atmosphere. On the other hand, the greenhouse gases are able to absorb longwave, or infrared, radiation from the earth, thereby heating the earth's atmosphere.

Discussion of the greenhouse effect has been confused by terms that are imprecise and even inaccurate. For example, when the term "greenhouse effect" was coined during the early nineteenth century, the atmosphere was believed to operate in a manner similar to that of a greenhouse, in which glass lets visible solar energy in but is also a barrier preventing the heat energy from leaving. In actuality, the reason that the air remains warmer inside a greenhouse is probably because the glass prevents the warm air from mixing with the cooler outside air. Therefore the greenhouse effect could more accurately be called the "atmospheric effect," but the term "greenhouse effect" continues to be used.

Even though the greenhouse effect is necessary for life on earth, the term gained harmful connotations during the late twentieth century with the discovery of apparently increasing atmospheric temperatures

The Greenhouse Effect

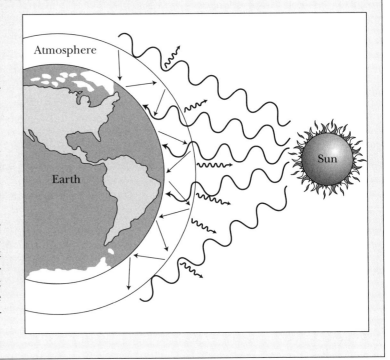

Clouds and atmospheric gases such as water vapor, carbon dioxide, methane, and nitrous oxide absorb part of the infrared radiation emitted by the earth's surface and reradiate part of it back to the earth. This process effectively reduces the amount of energy escaping to space and is popularly called the "greenhouse effect" because of its role in warming the lower atmosphere. The greenhouse effect has drawn worldwide attention because increasing concentrations of carbon dioxide from the burning of fossil fuels may result in a global warming of the atmosphere.

Scientists know that the greenhouse analogy is incorrect. A greenhouse traps warm air within a glass building where it cannot mix with cooler air outside. In a real greenhouse, the trapping of air is more important in maintaining the temperature than is the trapping of infrared energy. In the atmosphere, air is free to mix and move about.

and growing concentrations of greenhouse gases. The concern, however, is not with the greenhouse effect itself, but rather with the intensification or enhancement of the greenhouse effect, presumably caused by increases in the level of gases in the atmosphere resulting from human activity, especially industrialization. Thus the term "global warming" or "global climate change" is a more precise description of this presumed phenomenon.

A variety of human activities have contributed to greenhouse gas emissions. Industry emits a host of natural and synthetic greenhouse gases. Fossil-fuel combustion releases massive quantities of carbon dioxide. Agriculture also affects atmospheric carbon dioxide concentrations, because once natural vegetation and forests have been cleared for agriculture, the crops that replace them may not be as efficient in absorbing carbon dioxide. Increased numbers of ruminant livestock such as cattle and sheep have led to growing levels of methane, which is produced in the animals' digestive systems.

Several greenhouse gases, including CFCs and nitrous oxides, have also been implicated in ozone depletion. Stratospheric ozone shields the earth from solar ultraviolet (shortwave) radiation; therefore, if the concentration of these ozone-depleting gases continues to increase and the ozone shield is depleted, the amount of solar radiation reaching the earth's surface should increase. More solar energy would thus be intercepted by the earth's surface to be reradiated as longwave radiation, which would presumably increase the temperature of the atmosphere.

However, whether there is a direct cause-and-effect relationship between increases in greenhouse gases and surface temperature may be impossible to determine because the atmosphere's temperature has fluctuated widely over millions of years. Over the past 800,000 years, the earth has had several long periods of cold temperatures—during which thick ice sheets covered large portions of the planet's surface—interspersed with shorter warm periods. Since the most recent retreat of the glaciers around 10,000 years ago, the earth has been relatively warm.

PROBLEMS OF PREDICTION

How much the temperature of the earth might rise as a result of an intensified greenhouse effect is not clear. So far, the temperature increase of around half a degree Celsius (a single degree Fahrenheit) since the late nineteenth century is within the range of nor-mal (historical) trends. The possibility of global warming became a serious concern during the late twentieth century because the decades of the 1980's and the 1990's included some of the hottest years recorded for more than one hundred years. This trend continued into the twenty-first century, with the decade from 2000 to 2009 establishing itself as the warmest on record. On the other hand, warming has not been consistent since 1880. For example, there was a slight global cooling from the 1940's to the 1970's, possibly a result of the increase of another product of fossil-fuel combustion, sulfur dioxide aerosols, which reflect sunlight and thus lessen the amount of solar energy entering the atmosphere. The reduction in sulfur dioxide emissions following the implementation of pollution controls after 1970 may account for the subsequent observed rise in temperature. Similarly, during the early 1990's temperatures declined, most likely because of the ash and sulfur dioxide ejected into the atmosphere during the 1991 eruption of Mount Pinatubo (the second-largest terrestrial volcanic eruption of the twentieth century). By the late 1990's temperatures were on the rise again, suggesting that products of volcanic explosions may have masked the process of global warming.

Proper analysis of global warming is dependent on the collection of accurate temperature records from many locations around the world and over many years. Because human error is always possible, "official" temperature data may not be accurate. This possibility of inaccuracy compromises examination of past trends and predictions for the future. However, the use of satellites to monitor temperatures has served to increase the reliability, comparability, and spatial distribution of the data.

Predictions for the future are hampered in various ways, including lack of knowledge about all the factors and complex interactions affecting atmospheric temperature. Therefore, computer programs cannot be made sufficiently precise to yield accurate predictions. A prime example is the relationship between ocean temperature and the atmosphere. As the temperature of the atmosphere increases, the oceans should absorb much of that heat. Therefore, the atmosphere might not warm as quickly as predicted. However, the carbon dioxide absorption capacity of oceans declines with temperature. Therefore, the oceans would be unable to absorb as much carbon dioxide as before, but exactly how much is unknown. Increased ocean temperatures might also lead to more plant growth, including phytoplankton. These plants would be expected to absorb carbon dioxide through photosynthesis. A warmer atmosphere could hold more water vapor, resulting in the potential for more clouds and more precipitation. Whether that precipitation would fall as snow or rain and where it would fall could also affect air temperatures. Air temperature could drop, as more clouds might reflect more sunlight; on the other hand, more clouds might absorb more infrared radiation.

To complicate matters, any change in temperature would probably not be uniform over the globe. Because land heats up more quickly than water, the Northern Hemisphere, with its much larger landmasses, would be likely to experience greater temperature increases than the Southern Hemisphere. Similarly, ocean currents might change in both direction and temperature. These changes would affect air temperatures as well. In reflection of these complications, computer models of temperature change range widely in their estimates. The Intergovernmental Panel on Climate Change (IPCC) has projected a global temperature increase by the year 2100 of 2 to 3 degrees Celsius (3.6 to 5.4 degrees Fahrenheit).

MITIGATION ATTEMPTS

International conferences have been held and international organizations have been established to research and minimize the potential detriments of global warming. In 1988 the United Nations Environment Programme and the World Meteorological Organization established the IPCC to assess and compile climate change information for use by policy makers. The IPCC issued assessment reports in 1990, 1995, 2001, and 2007; its fifth assessment report is due in 2013 or 2014. The first of the IPCC's assessment reports concluded that global warming was sure to result if human emissions of greenhouse gases were not brought under control.

In June, 1992, the United Nations Conference on Environment and Development, also known as the Earth Summit, was held in Brazil. Participants devised the United Nations Framework Convention on Climate Change, considered a landmark international treaty. It required signatories to reduce and monitor their greenhouse gas emissions. Developed nations agreed on a voluntary year 2000 target of stabilizing their emissions at 1990 levels, a goal that many ratifying governments failed to meet.

A binding agreement, the Kyoto Protocol, was adopted in December, 1997, and entered into force in February, 2005. It set compulsory targets for reducing emissions of carbon dioxide, methane, nitrous oxide, sulfur hexafluoride, HFCs, and PFCs over the five-year period of 2008 through 2012 for thirty-seven industrialized countries and the European Community; no emissions targets were established for developing countries. The treaty allows signatory nations to use afforestation (the creation of forests where none existed before) to offset emissions and includes the economic incentive of allowing the trading of emissions permits—that is, countries that meet their targets can sell their excess permits to other nations. The United States, which produces roughly one-fourth of the world's greenhouse gas emissions, has not ratified the treaty. President George W. Bush formally withdrew the nation from the protocol in 2001, citing the damage it would do to the American economy and stating his objection to the exemption of developing nations such as India and China—both fast-growing polluters—from obligations to make emissions cuts before 2012.

The first commitment period of the Kyoto Protocol ends in 2012, by which time a follow-up international framework must be in place. Conferences on post-Kyoto climate policy have proved contentious. Attempts to negotiate a successor to the protocol, notably at the United Nations Climate Change Conference held in Copenhagen, Denmark, in December, 2009, generated more than one hundred domestic policies and government initiatives by participating nations to reduce greenhouse gas emissions but failed to produce a new international treaty.

Margaret F. Boorstein
Updated by Karen N. Kähler

FURTHER READING

Gore, Al. *An Inconvenient Truth: The Planetary Emergency of Global Warming and What We Can Do About It.* New York: Rodale Books, 2006.

Houghton, John Theodore. *Global Warming: The Complete Briefing.* 4th ed. New York: Cambridge University Press, 2010.

Lankford, Ronald D., ed. *Greenhouse Gases.* Detroit: Greenhaven Press, 2009.

Schneider, Stephen Henry, et al., eds. *Climate Change Science and Policy.* Washington, D.C.: Island Press, 2010.

Shulk, Bernard F., ed. *Greenhouse Gases and Their Impact.* New York: Nova Science, 2007.

Weart, Spencer W. *The Discovery of Global Warming.* Rev. ed. Cambridge, Mass.: Harvard University Press, 2008.

SEE ALSO: Air pollution; Air-pollution policy; Carbon dioxide; Climate change and human health; Climate change and oceans; Global warming; Greenhouse gases; Intergovernmental Panel on Climate Change; Kyoto Protocol; United Nations Framework Convention on Climate Change.

Greenhouse gases

CATEGORIES: Atmosphere and air pollution; pollutants and toxins

DEFINITION: Atmospheric gases that allow sunlight to reach the earth's surface but at least partially block infrared from radiating back into space

SIGNIFICANCE: Greenhouse gases raise the earth's average temperature 33 degrees Celsius (59 degrees Fahrenheit) above what it would be if these gases were not present in the atmosphere. Most scientists agree that human activities are contributing to increased concentrations of greenhouse gases and thus to increases in the average surface temperature of the earth.

The majority of scientists accept that a small global warming has taken place—that is, the earth's surface temperature has warmed about 0.7 degree Celsius (1.3 degrees Fahrenheit)—since the end of the nineteenth century. At least half of this temperature rise is attributed to the release of greenhouse gases into the atmosphere by human beings. It is thought that if greenhouse gases were returned to their 1990 levels, the temperature would still rise another 0.5 degree Celsius (0.9 degree Fahrenheit). Since greenhouse gases are transparent to visible light, sunlight passes through the atmosphere and warms the earth's surface. The warmed surface radiates infrared into the sky, where greenhouse gases absorb infrared and then reradiate it. They radiate about half of the infrared upward into space and half of it back down to the earth's surface. Since the earth absorbs more energy than it radiates back into space, it heats up until the energies entering and leaving are equal. Balance is possible because a hotter earth radiates with greater intensity and at shorter wavelengths where

greenhouse gases allow more infrared to escape into space.

Water vapor strongly absorbs infrared of about 3 microns wavelength (3,000 nanometers), and so does carbon dioxide. Adding more carbon dioxide will not change the amount of energy passing out into space, since water vapor absorbs all of the energy near that wavelength. However, more carbon dioxide would make a difference at 4.5 microns wavelength (4,500 nanometers) because water vapor does not absorb at that wavelength. This means that doubling the amount of carbon dioxide in the atmosphere does not necessarily double the effect of carbon dioxide. In fact, most climate scientists believe that increasing carbon dioxide will increase the surface temperature, which will cause more water to evaporate. More water vapor in the atmosphere should further warm the earth's surface, but it will also cause more clouds that reflect sunlight back into space. More clouds will cool the earth. Under these competing processes it is believed that temperature will increase until a new equilibrium is reached.

EARTH'S MAJOR GREENHOUSE GASES

Water vapor, carbon dioxide, methane, nitrous oxide, fluorocarbons, and ozone are the major greenhouse gases. The effect of a particular gas depends on which other gases are present, how much of the gas is present, and how likely a gas molecule is to absorb infrared radiation.

Water vapor is the most important greenhouse gas, but human activities have no direct effect on the global average amount of water vapor, which is fixed mainly by evaporation from the earth's oceans. Other greenhouse gases are significantly affected by human activities. Burning fossil fuels and deforestation in the Tropics increase the amount of carbon dioxide in the atmosphere. Rice paddy farming and the digestive processes of livestock produce large amounts of methane. Fertilizers used in farming produce nitrous oxide, and refrigeration systems and some manufacturing processes release chlorofluorocarbons, other perfluorocarbons, and sulfur hexafluoride.

Since the mid-nineteenth century, carbon dioxide concentration in the air has increased from 280 parts per million to almost 380 parts per million. Normal carbon is carbon 12 (6 protons and 6 neutrons in the nucleus), but about 1 percent of carbon is carbon 13 (6 protons and 7 neutrons). Plants prefer carbon 12, so plant carbon (and fossil fuels from plants) has a

U.S. Greenhouse Gas Emissions
(millions of metric tons)

	1990	2000	2002	2003	2004	2005	2006
Carbon dioxide	5,017.5	5,890.5	5,875.9	5,940.4	6,019.9	6,045.0	5,934.4
Methane gas	708.4	608.0	598.6	603.7	605.9	607.3	605.1
Nitrous oxide	333.7	341.9	332.5	331.7	358.3	368.0	378.9
High GWP gases	87.1	138.0	137.8	136.6	149.4	161.2	157.6

Source: U.S. Energy Information Administration, Emissions of Greenhouse Gases in the United States, 2006, 2006.
Note: High GWP (global warming potential) gases are hydrofluorocarbons, perfluorcarbons, and sulfur hexafluoride.

smaller ratio of carbon 13 to carbon 12 than the atmosphere does. Analysis of air bubbles in ice cores shows that the ratio of carbon 13 to carbon 12 in the atmosphere has been decreasing since the mid-nineteenth century, presumably because of increased burning of fossil fuels and the practice of setting fire to forests to clear land.

POLITICS

Under the sponsorship of the United Nations, the Kyoto Protocol was adopted in December, 1997, and went into force in February, 2005. Seeing the protocol as flawed because it places no limits on China (the world's largest polluter) or India, the United States opted out of the protocol. Under the protocol, industrialized nations set goals for greenhouse gas reductions, and developing nations negotiated for money from those nations to help them industrialize with fewer greenhouse emissions. By 2007 only Germany, Norway, France, and the United Kingdom were meeting their goals. Several nations that had been part of the former Soviet Union reduced emissions, in large part because their economies floundered.

In December, 2009, the U.S. Environmental Protection Agency (EPA) released a finding that greenhouse gases threaten the public health and welfare of the American people. The finding will allow the EPA to regulate greenhouse gas emissions in the United States if the Congress fails to pass legislation to do so.

In November, 2009, a large number of e-mails were stolen from the Climatic Research Unit at the University of East Anglia, United Kingdom. Some climate change skeptics alleged that the e-mails provided evidence that scientists had doctored data on global warming. Subsequent investigations found evidence of frustration on the part of the climate researchers and a lack of willingness among them to share raw data, but no evidence of wrongdoing. This incident nevertheless cast some doubt on the East Anglia data on global warming, despite the fact that the findings of the Climatic Research Unit are supported by a great deal of data from other sources. This doubt probably contributed to the weakness of the agreements reached in regard to greenhouse gas emissions at the United Nations Climate Change Conference held in Copenhagen, Denmark, in December, 2009.

Charles W. Rogers

FURTHER READING

Houghton, John. *Global Warming: The Complete Briefing.* New York: Cambridge University Press, 2009.

Singer, S. Fred, and Dennis T. Avery. *Unstoppable Global Warming: Every 1,500 Years.* Updated ed. Lanham, Md.: Rowman & Littlefield, 2008.

Weart, Spencer W. *The Discovery of Global Warming.* Rev. ed. Cambridge, Mass.: Harvard University Press, 2008.

SEE ALSO: Carbon dioxide; Chlorofluorocarbons; Climate change skeptics; Climate models; Climatology; Global warming; Greenhouse effect; Methane; Nitrogen oxides; Rain forests.

Greenpeace

CATEGORIES: Organizations and agencies; activism and advocacy; animals and endangered species; nuclear power and radiation

IDENTIFICATION: International watchdog organization dedicated to protecting the environment

DATE: Established in 1971

SIGNIFICANCE: Through its persistent, nonviolent confrontational tactics, Greenpeace applies pressure to governments, organizations, and private corporations that often results in positive changes for the environment while greatly raising public awareness of particular environmental issues.

Greenpeace grew out of the small organization known as the Don't Make a Wave Committee, which was formed in 1971. The members of the group, which was based in Vancouver, British Columbia, had joined together because of their mutual opposition to nuclear testing; they aimed to use nonviolent, creative confrontation to raise public awareness of dangers to the global environment. In September, 1971, twelve volunteers from the organization, now known as Greenpeace, sailed a small boat into the U.S. atomic test zone on the island of Amchitka, off the coast of Alaska, in an effort to stop the United States from conducting a test of a nuclear weapon. The activists believed the bomb's blast could destroy the wildlife haven, cause an earthquake, or create a tidal wave.

Since that first voyage, Greenpeace protesters have often taken their boats into dangerous zones, placing themselves at risk to save whales from harpoons or to block the dumping of toxic or radioactive wastes into the ocean. Within twenty years of its founding, Greenpeace had supporters in thirty countries; by 2010, it had offices in forty-one countries and 2.8 million supporters around the world.

The work of Greenpeace is based on a belief in the importance of the role of individuals and of the visibility of high-profile actions. The group's funding comes mainly through small, individual donations. The stated "core values" of Greenpeace include "'bear[ing] witness' to environmental destruction in a peaceful, non-violent manner" and using "non-violent confrontation to raise the level and quality of public debate" on environmental issues.

Greenpeace's ships travel the world to bring the attention of the news media to local and regional environmental problems, to raise public awareness. The ships have been positioned between whaling fleets and the whales they were trying to catch and have been used to stop the killing of seals. In 1985 the Greenpeace ship *Rainbow Warrior* was bombed and sunk in Auckland Harbor, New Zealand, by French intelligence agents because it was shadowing French nuclear vessels to protest nuclear testing in French Polynesia.

The main governing body of Greenpeace is Greenpeace International, which is based in Amsterdam. Greenpeace International develops and coordinates global strategies and policies with input from Greenpeace regional offices. Over the years the organization has undergone changes and disputes within its leadership, and in 1996 its articles of association were changed to allow for more efficient allocation of resources. Because of the complexity of the issues that Greenpeace addresses, the organization has devoted increasing resources to scientific research, using its ships as mobile laboratories.

Colleen M. Driscoll

Raising Awareness by Making Waves

In his memoir Making Waves: The Origins and Future of Greenpeace *(2001), Greenpeace cofounder Jim Bohlen relates a realization he had following his return from the ultimately aborted but highly publicized first voyage of the Greenpeace:*

The next Monday morning, November 1st, I returned to work at the UBC Forest Products Laboratory. I walked in the front door, checked in with the receptionist, and walked down the glassed-in connecting corridor towards my office. On the way I met one of my colleagues. He was all smiles, congratulating me on the success of the voyage, and ended with, "Well Jim, what are you going to do next?" Suddenly overcome with a great weariness, I replied, "Bob, what are *you* going to do next?"

Those few words summed up the essence of the campaign. Our job had been to make the people of North America aware of the real, and potential, danger of testing nuclear weapons. We imagined massive public opinion raging against the . . . underground blast and bringing about, once and for all, an end to the mad escalation of nuclear weapons. The *Greenpeace* and its crew had done its job.

FURTHER READING

Bohlen, Jim. *Making Waves: The Origins and Future of Greenpeace*. Tonawanda, N.Y.: Black Rose Books, 2001.

Weyler, Rex. *Greenpeace: How a Group of Ecologists, Journalists, and Visionaries Changed the World*. Emmaus, Pa.: Rodale Press, 2004.

SEE ALSO: Antinuclear movement; *Brent Spar* occupation; International whaling ban; Nuclear accidents; Nuclear and radioactive waste; Nuclear testing; Nuclear weapons production; Save the Whales Campaign; Whaling.

Greenwashing

CATEGORY: Philosophy and ethics

DEFINITION: Public relations and marketing practice of exaggerating or misrepresenting the environmental benefits and friendliness of products, services, policies, or practices

SIGNIFICANCE: With the rapid growth of the environmental movement and consumer spending on green products since the 1960's, greenwashing has also proliferated, resulting in consumer mistrust that is detrimental to the entire market. Greenwashing has also hurt the environment by persuading consumers to purchase products that are not actually green instead of truly green products. Valuable economic resources have been funneled into greenwashing instead of into environmental problem solving.

In 1986, environmentalist Jay Westerveld coined the word "greenwashing" in an essay criticizing the hotel industry's money-saving practice of asking guests to reuse their towels as a way to reduce impacts on the environment. (The term combines the concepts of green, or environmentally friendly, and whitewashing, in the sense of covering something up.) The *Oxford English Dictionary* added the term "greenwashing" in 1999. The actual phenomenon of greenwashing began during the mid-1960's, however, as, in response to the growing environmental movement, automobile, chemical, oil, and utility companies began to develop green corporate images and increasingly sought to promote their products and activities as environmentally friendly. By the time of the first Earth Day in 1970, companies were spending about $1 billion annually on green advertising, much more than they were spending on actual environmental research and development.

GLOBAL GREENWASHING

During the 1980's, as the antinuclear and environmental movements gained momentum, greenwashing became both more prevalent and more sophisticated. By the twentieth anniversary of the first Earth Day in 1990, 25 percent of new household products were being advertised as organic, recyclable, natural, ozone-friendly, compostable, toxic free, or biodegradable.

Greenpeace, which was established in 1971 as an independent organization dedicated to protecting the environment, published *The Greenpeace Book of Greenwash* in 1992. This book described how greenwashing had become global, with transnational corporations (TNCs) assuming a significant role in the 1992 United Nations Conference on Environment and Development, or Earth Summit. The book profiled nine TNCs that had spent huge amounts of money to develop a green image, among them the Du Pont chemical company, inventor and leading producer of chlorofluorocarbons (CFCs), the primary chemicals destroying the earth's protective ozone layer. Du Pont projected a green company image in a television advertisement that featured Ludwig van Beethoven's "Ode to Joy" playing over images of seals, flamingos, and dolphins. Greenpeace also cited Westinghouse Electrical Corporation for suggesting that nuclear plants, like trees, are good for the environment, and described how the Mitsubishi Corporation distributed a comic book in high schools that justified the corporation's commercial logging in tropical rain forests by blaming deforestation on poverty and shifting cultivation by indigenous peoples.

PROTECTING CONSUMERS

In 1992 the U.S. Federal Trade Commission (FTC) established its first guidelines for environmental marketing. In 2010 the FTC announced plans to develop new, more enforceable standards to deal with escalating greenwashing.

The environmental marketing company TerraChoice, located in Ottawa, Ontario, Canada, has published commentary on greenwashing that includes a list of the "sins" of such marketing, such as the hidden trade-off (promoting a product's single green attribute while ignoring its harmful qualities), vagueness (for example, "all natural" could mean that a product contains naturally occurring poisons), and irrelevance (providing truthful but not helpful information). Futerra Sustainability Communications, a communications agency based in London, England, produced its *Greenwash Guide* in 2008 as an aid to consumers. According to Futerra, consumers should be alert for signs of greenwashing such as the following: "fluffy" language using terms with no clear meaning, supposedly green products made in polluting factories, irrelevant claims emphasizing insignificant green attributes, "best in the class" suggestions that a product is greener than others, jargon that only scientists can understand, and imaginary third-party endorsements.

In efforts to protect consumers from greenwashing, numerous groups have formed to rate, label, or certify green products. In 1989 the nonprofit National Fenestration Rating Council was founded in the United States to provide independent verification of the energy performance of windows and doors. Also in 1989 the U.S. nonprofit Green Seal was created to provide scientific certification of the green claims made by home, construction, and office products. In 1992, the U.S. Environmental Protection Agency began its Energy Star program, a voluntary rating and labeling system concerned with the energy-efficiency of consumer products.

Alice Myers

FURTHER READING

Clegg, Brian. *Eco-Logic: Cutting Through the Greenwash—Truth, Lies, and Saving the Planet.* London: Eden Project, 2009.

Esty, Daniel, and Andrew Winston. *Green to Gold: How Smart Companies Use Environmental Strategy to Innovate, Create Value, and Build Competitive Advantage.* Hoboken, N.J.: John Wiley & Sons, 2009.

Grant, John. *The Green Marketing Manifesto.* Hoboken, N.J.: John Wiley & Sons, 2007.

Greer, Jed, and K. Bruno. *Greenwash: The Reality Behind Corporate Environmentalism.* New York: Apex Press, 1997.

Lubbers, Eveline. *Battling Big Business: Countering Greenwash, Infiltration, and Other Forms of Corporate Bullying.* Monroe, Me.: Common Courage Press, 2002.

MacDonald, Christine C. *Green, Inc.: An Environmental Insider Reveals How a Good Cause Has Gone Bad.* Guilford, Conn.: Lyons Press, 2008.

Tokar, Brian. *Earth for Sale: Reclaiming Ecology in the Age of Corporate Greenwash.* Boston: South End Press, 1997.

SEE ALSO: Alternative energy sources; BP *Deepwater Horizon* oil spill; Chlorofluorocarbons; Eco-fashion; Ecological economics; Energy-efficiency labeling; Environmental economics; Green marketing; Nader, Ralph.

Groundwater pollution

CATEGORY: Water and water pollution

DEFINITION: Degradation, by chemicals and other substances, of the water found below the surface of the earth

SIGNIFICANCE: Many public and private water supplies rely on wells that tap important groundwater reserves. Pollution of groundwater leads to changes in water quality that can affect groundwater use for a given purpose.

Humans require vast amounts of fresh water for use in homes, livestock operations, agriculture, and industrial processes. Groundwater is an important source of fresh water. The pollution of groundwater by human activity can contaminate water-supply wells, making the water they provide unacceptable for drinking and other purposes. This can lead to a need for new water supplies that may not be readily available or easily accessible. In some instances polluted groundwater interacts with surface water, thus contaminating the surface-water environment as well.

Groundwater constitutes a small but significant portion of the world's overall water supply. Much of the earth's surface is covered by water, but an estimated 97.2 percent of it exists as salt water. Since fresh surface water may account for as little as 0.009 percent of the earth's water, groundwater is a significant source of readily available fresh water. Groundwater occurs in the saturated zone of the earth, which is the area below the surface where pores between particles—void spaces in the soil or rock—are filled with water. In some places groundwater may be encountered near the surface, but in other areas, such as arid regions, it can be quite deep below the surface. Groundwater flows from areas of high hydrostatic head to areas of low hydrostatic head. Shallow

The Freshwater Crisis

Toxic chemicals are contaminating groundwater on every inhabited continent, endangering the world's most valuable supplies of fresh water, reports a study published in 2000 from the Worldwatch Institute, a Washington, D.C.-based research organization on the environment. This worldwide survey of groundwater quality shows that pesticides, nitrogen fertilizers, industrial chemicals, and heavy metals are contaminating groundwater everywhere, and that the damage is often worst in places where people most need water.

There are at least three essential roles of groundwater: providing drinking water, irrigating farmland, and replenishing rivers, streams, and wetlands. About one-third of the world population relies almost exclusively on groundwater for drinking. Groundwater provides irrigation for some of the world's most productive farmland. Over 50 percent of irrigated croplands in India, and 40 percent in the United States, is watered by groundwater. Groundwater plays a crucial role in replenishing rivers, streams, and wetlands. It provides much of the flow for great rivers such as the Mississippi, the Niger, the Yangtze, and many more.

The range of groundwater contamination is stunning. Groundwater in all twenty-two major industrial zones surveyed by the Indian government in the late 1990's was unfit for drinking. In a 1995 study of four northern Chinese provinces, groundwater was contaminated in more than 50 percent of surveyed locations. One-third of the wells tested in a California region in 1988 contained the pesticide 1,2-dibromo-3-chloropropane (DBCP) at levels ten times higher than the maximum allowed for drinking. The list goes on. There is a compelling urgency to prevent groundwater contamination.

groundwater often mimics topography, flowing downhill toward streams and lakes.

The soil and rock through which groundwater flows consist of particles of varying size, which help determine the classification of the soil or rock and how well water will move through the material. Sand-sized particles are seen in unconsolidated sandy soils or sandstones. Smaller particles may form silty or clayey soils or their bedrock equivalents of siltstones and shales. In the saturated zone, groundwater saturates the pores and voids between the particles. The size of the pores and the degree to which they are interconnected affect hydraulic conductivity—a measure of the ability of water to move through the rock or soil.

Transmissivity is the measure of the ability of an aquifer to transmit water and is a measure of the hydraulic conductivity multiplied by the saturated thickness of the aquifer. Therefore, a thick aquifer with relatively poor hydraulic conductivity might be able to transmit as much water as a thinner aquifer composed of materials with greater hydraulic conductivity. Groundwater is recharged by rainwater percolating through the soil, snowmelt, and rivers and streams.

THREATS TO GROUNDWATER

Humans produce a wide array of pollutants and combinations of pollutants. The degree and extent to which individual pollutants can affect groundwater quality is dependent on a large number of variables, which can include the amount of contaminant introduced into the environment, the time frame in which it is introduced, its toxicity, its mobility, whether it will readily degrade in the environment, and the chemical and physical characteristics of the soil or rock through which it will pass.

Even something as common as nitrogen can lead to pollution in groundwater. Nitrogen can be mobile in the environment in the form of dissolved nitrates and nitrites. Sources for pollution include septic tanks, leaks from sewage treatment plants and lagoons, and animal wastes. Nitrogen is also an important component of many fertilizers used in agriculture, and such fertilizers may become dissolved by rainwater and percolate down into groundwater. In high enough concentrations, nitrates can make water unacceptable for human consumption. At even higher concentrations, the water can become unacceptable for livestock and other animals.

Gasoline spills and leaks from underground storage tanks are relatively common sources of groundwater pollution. Some of the dissolved-phase components of gasoline are quite mobile in the environment; however, many are also susceptible to biological degradation. Gasoline and other substances less dense than water can float on the surface of groundwater, but seasonal fluctuations in the water table can smear such contaminants in the soil, potentially making them more difficult to remove. Other contaminants, such as chlorinated solvents, can be denser than water and have the capacity to sink into aquifers.

Although metals as a group are generally not considered very mobile and tend to be adsorbed onto soils, some are quite mobile, and contamination by heavy metals can be a relatively common form of groundwater contamination. Although less common, radiological contamination of groundwater can be a concern. Groundwater often moves slowly, but radioactive half-lives can be quite long.

Raymond U. Roberts

FURTHER READING

Appelo, C. A. J., and D. Postma. *Geochemistry, Groundwater, and Pollution*. 2d ed. New York: Balkema, 2005.

Chiras, Daniel D. "Water Pollution: Sustainably Managing a Renewable Resource." In *Environmental Science*. 8th ed. Sudbury, Mass.: Jones and Bartlett, 2010.

Heath, Ralph C. *Basic Groundwater Hydrology*. Reston, Va.: U.S. Geological Survey, 2004.

Sampat, Payal. *Deep Trouble: The Hidden Threat of Groundwater Pollution*. Washington, D.C.: Worldwatch Institute, 2000.

Todd, David Keith, and Larry W. Mays. *Groundwater Hydrology*. 3d ed. Hoboken, N.J.: John Wiley & Sons, 2005.

Younger, Paul L. *Groundwater in the Environment: An Introduction*. Malden, Mass.: Blackwell, 2007.

SEE ALSO: Acid mine drainage; Aquifers; Drinking water; Sewage treatment and disposal; Soil contamination; Stormwater management; Water pollution; Water quality; Wells.

Group of Ten

CATEGORIES: Activism and advocacy; ecology and ecosystems

IDENTIFICATION: Coalition of representatives of the ten leading environmental groups in the United States

SIGNIFICANCE: Although the existence of the Group of Ten was short-lived, the formation of the coalition demonstrated widespread recognition of the essential interrelatedness of all environmental concerns and of the need for environmental advocates to engage corporations and political leaders

On January 21, 1981, leaders of nine U.S. environmental organizations met at a Washington, D.C., restaurant to discuss common goals. In attendance were J. Michael McCloskey, the chief executive officer of the Sierra Club; John Hamilton Adams of the Natural Resources Defense Council; Janet Brown of the Environmental Defense Fund; Thomas Kimball of the National Wildlife Federation; Jack Lorenz of the Izaak Walton League; Russell Wilbur Peterson, the former governor of Delaware, of the National Audubon Society; William Turnage of the Wilderness Society; Louise C. Dunlap of the Environmental Policy Center; and Rafe Pomerance of Friends of the Earth. Although he was not in attendance at the original meeting, Paul Pritchard of the National Parks and Conservation Association was later invited to join the group, which thus became known as the Group of Ten.

The activists who formed the group hoped to send a signal to newly inaugurated President Ronald Reagan and his incoming administration that American environmental organizations would fight to preserve the environmental legislation of the 1960's and 1970's. In addition, the attendees hoped to protect and enhance environmental quality by promoting public awareness and corporate social responsibility. The group's leaders thus conducted meetings with the heads of major corporations, including Du Pont, Exxon Chemical, Union Carbide, Dow, American Cyanamid, and Monsanto. A number of these companies were sensitive to the unfavorable publicity they had been receiving for the environmental consequences of their businesses. In the meetings, the members of the Group of Ten urged the corporate leaders to assume greater social responsibility for the protection of the environment and for the welfare of future generations.

In 1985, the Group of Ten published *An Environmental Agenda for the Future*, a detailed plan for the worldwide environment's preservation and renewal. With the waning of the Reagan administration's assault on domestic environmental legislation in the late 1980's, the Group of Ten disbanded as a formal organization, but its legacy of united action in defense of the environment continued to prove influential.

Glenn Canyon

FURTHER READING

Adams, John H., et al. *An Environmental Agenda for the Future*. Edited by Robert Cahn. Washington, D.C.: Agenda Press, 1985.

Bevington, Douglas. *The Rebirth of Environmentalism: Grassroots Activism from the Spotted Owl to the Polar Bear*. Washington, D.C.: Island Press, 2009.

Gottlieb, Robert. *Forcing the Spring: The Transformation*

of the American Environmental Movement. Rev. ed. Washington, D.C.: Island Press, 2005.

Lash, Jonathan, Katherine Gillman, and David Sheridan. *A Season of Spoils: The Reagan Administration's Attack on the Environment.* New York: Pantheon Books, 1984.

SEE ALSO: Friends of the Earth International; Green movement and Green parties; National Audubon Society; Natural Resources Defense Council; Sierra Club; Wilderness Society.

Gulf War oil burning

CATEGORIES: Disasters; atmosphere and air pollution; water and water pollution

THE EVENT: Fires started in Kuwait oil fields by retreating Iraqi soldiers

DATES: January-November, 1991

SIGNIFICANCE: When Iraqi armed forces damaged an oil pipeline and set fire to more than five hundred oil wells, they began what became the worst oilfield disaster in history.

The Persian Gulf is a shallow, northwest-trending body of water that covers an area of about 260,000 square kilometers (100,400 square miles). The Gulf is actually a large bay about 800 kilometers (500 miles) long, 200 kilometers (125 miles) wide, and 90 meters (300 feet) deep at the deepest point. It is bordered by Iran, Iraq, Kuwait, Saudi Arabia, Bahrain, Qatar, Oman, and the United Arab Emirates. The Shatt-al-Arab, a river formed by the merging of the Tigris and Euphrates rivers, has created a combined river floodplain and delta region at the head of the Persian Gulf that covers more than 3,200 square kilometers (1,235 square miles).

The Persian Gulf is teeming with wildlife. The coastal mangrove swamps and coral reefs provide habitats for birds and fish. Hundreds of species of fish, including mackerel, snapper, and mullet, live in the region and feed on the abundant algae. Wading birds such as shanks and sand plovers feed along the coastal mudflats. Also, valuable crustaceans such as prawn and shrimp are farmed along the Persian Gulf's shores.

The region holds more than one-half of the world's proven reserves of oil and natural gas. The Persian Gulf also provides the world's major shipping lanes for oil tanker traffic. It has been estimated that some 40 million liters (10 million gallons) of oil spills or leaks into Persian Gulf waters each year. The results of these oil releases are usually absorbed by the environment without significant ecological damage. However, in January, 1991, the area was hit by an environmental disaster in the form of the Persian Gulf War.

On August 2, 1990, Iraqi forces invaded the small, oil-rich neighboring nation of Kuwait. On November 29, 1990, the United Nations voted to permit the use of force to expel Iraqi forces from Kuwait. On January 16, 1991, air attacks on Iraqi targets began, followed on February 23 with a ground attack against Iraq by a coalition of forces from twenty-eight countries, including the United States, Saudi Arabia, Egypt, and Great Britain.

A Kuwaiti oil field worker kneels for midday prayers near a burning oil field outside Kuwait City in March, 1991. (AP/Wide World Photos)

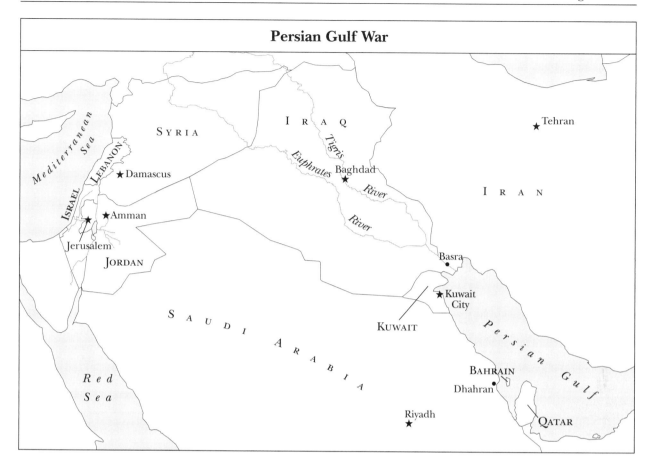

Persian Gulf War

SOIL AND WATER POLLUTION

During the ground war, which lasted only one hundred hours, Iraqi soldiers departing from Kuwait damaged an oil pipeline and set fire to more than five hundred Kuwaiti oil wells. These damaged wells poured hundreds of thousands of liters of oil into the surrounding countryside, forming large lakes up to 1.6 kilometers (1 mile) long and nearly 1 meter (3.28 feet) deep. More than 800 million liters (211 million gallons) of oil issued from the damaged wells each day. The spilled oil threatened to pollute the water supply of Kuwait City and other inhabited areas, and the contaminated soil and vegetation threatened to harm wildlife. Soil is adversely affected by the oily mist from such spills, as the mist forms a thin film over the topsoil and reduces the amount of oxygen and water that can penetrate the soil profile. This seal reduces the activity of a number of microbes and earthworms that help to keep the soil fertile.

Because currents in the upper Persian Gulf generally move counterclockwise around the bay, a thick oil slick spread northwest to southeast along the western coast from disrupted oil terminals, pipelines, and individual wells in the vicinity of Kuwait City to the lower part of Saudi Arabia; a thin oil film spread from the coastal region to the middle of the upper Persian Gulf. The colossal oil slick threatened the entire ecosystem of the Persian Gulf. It resulted in the deaths of some twenty thousand birds, including flamingos, as the oil soaked their feathers and caused them to drown or die from exposure. The Persian Gulf's populations of sea cows (dugongs), dolphins, and green turtles were at risk, as were the endangered hawksbill turtles that lay eggs on the local islands. Although many scientists believe that most marine life requires only three years to recover from the effects of exposure to crude oil, marine life exposed to refined oil, especially in enclosed areas such as the Persian Gulf, may require ten years or longer to recover.

AIR POLLUTION

A number of the wells that were ignited released poisonous gases into the air that endangered people (including military personnel) and wildlife. The smoke released by the fires contained noxious gases such as deadly carbon monoxide and sulfuric acid. The sky was darkened by clouds of toxic black smoke, causing temperatures to drop 5.5 degrees Celsius (10 degrees Fahrenheit) lower than normal.

The Burgan Oil Refinery complex south of Kuwait City was devastated. Most of the damage occurred within an 800-kilometer (500-mile) range that included most of Iran, Iraq, and Kuwait. Soot and acid rain clouds extended nearly 1,920 kilometers (1,200 miles) away from the sabotaged oil fields. In March, 1991, *The New York Times* reported that the pollution "rained on Turkey and reached the western shore of the Black Sea, touching Bulgaria, Rumania, and the southern Soviet Union, becoming more prevalent over Afghanistan and Pakistan." The newspaper also indicated that the toxic clouds resulted in a significant increase in respiratory diseases among the elderly and the very young. In *Oil Spills* (1993) Jane Walker reports that many cattle and sheep "died in Kuwait either from breathing oil droplets in the air, or from eating oil-covered grass."

CLEANUP

The cleanup operation was undertaken in two phases: extinguishing the burning wells and controlling the oil slick. Several professional firefighting groups were called upon to extinguish the oil-field fires; the process took less than one year. Among the techniques used to combat the fires were cooling the well equipment with water, removing the well debris, cutting off oxygen to the fires by blowing flames out with explosives (usually dynamite), capping well heads with stingers (plugs), and attaching new valve assemblies to shut off the flow of oil.

The main technique used to confine and recover oil from the waters of the Persian Gulf was skimming. Oil-skimming ships scraped the greasy layer off the water's surface, recovering between 20,000 and 30,000 barrels (about 3 to 5 million liters) of oil per day. The Saudi Arabian national oil company, Saudi Aramco, placed about 40 kilometers (25 miles) of floating booms along the periphery of the spill and sent more than twenty oil-recovery craft to the area.

Donald F. Reaser

FURTHER READING

Browning, K. A., et al. "Environmental Effects from Burning Oil Wells in Kuwait." *Nature*, May 30, 1991, 363-367.

Falola, Toyin, and Ann Genova. "Environmental Concerns." In *The Politics of the Global Oil Industry: An Introduction*. Westport, Conn.: Praeger, 2005.

Hawley, T. M. *Against the Fires of Hell: The Environmental Disaster of the Gulf War*. Orlando, Fla.: Harcourt Brace Jovanovich, 1992.

Hobbs, Peter V., and Lawrence F. Radke. "Airborne Studies of the Smoke from the Kuwait Oil Fires." *Science* 256 (March 26, 1993): 987-991.

Horgan, John. "Burning Questions: Scientists Launch Studies of Kuwait's Oil Fires." *Scientific American*, July, 1991, 17-22.

SEE ALSO: Air pollution; Nuclear winter; Oil crises and oil embargoes; Oil spills; PEMEX oil well leak; Soil contamination.

H

Habitat destruction

CATEGORIES: Preservation and wilderness issues; land and land use

DEFINITION: Degradation of a natural landscape so that it becomes functionally incapable of supporting its native species

SIGNIFICANCE: The destruction of habitat represents a pressing threat to global biodiversity, as habitat loss leads to species extinctions. In the twenty-first century most habitat destruction is occurring in developing nations, where overpopulation and poverty contribute to the need to convert forest-land to agriculture.

Habitat destruction occurs when human beings remove or significantly alter the land or aquatic communities in which other animal species dwell. Certainly, habitat destruction is not always a negative practice; human civilization was built on the practice of altering land for human purposes, such as the burning of forest for pasture, logging for construction materials, the draining of marshland for development, and mining for resource extraction. In the twenty-first century, however, the density of human population on the earth, combined with modern industrial technology, means that humans' alterations of the natural landscape greatly exceed the ability to render land productive for human needs. Further, the most biodiverse ecoregions of the world, tropical rain forests and coral reefs, have seen rapid increases in habitat destruction. Tropical forests are down to about 1 billion hectares (2.5 billion acres) from the nearly 1.6 billion hectares (4 billion acres) they occupied approximately two hundred years ago. One-fifth of coral reefs have been destroyed, and another one-fifth have been severely degraded.

Habitat destruction is most often caused by expansion of agriculture. Planting crops or raising livestock requires wide expanses of bare soil or grassy plain. When suitable land is unavailable naturally, many land types can be converted to agricultural uses: Wetlands can be drained, forests can be logged, deserts can be irrigated. Although the productivity of such converted land may be relatively high, biodiver-sity and ecosystem functionality drop harshly. Modern agriculture is often monocultural (that is, devoted to a single type of crop), without topographical complexity, and devoid of any plants not being grown for either sale or consumption. Therefore, the available niches for native species are nearly all removed.

Other causes of habitat destruction include mining, ocean trawling, oil prospecting, urban sprawl, and infrastructure development. These practices directly destroy ecosystems, but other practices indirectly degrade surrounding ecosystems. Desertification is caused by overgrazing of livestock and excessive extraction of groundwater, which render the land unusable, usually affecting communities that already live in resource-impoverished landscapes. Deforestation on a small scale can cause ecosystem collapse by dividing a forest into fragments, rendering the land unfit to support animals with large ranges and plants with wide dispersal needs. Coral degradation is rarely caused by direct destruction; rather, coral is negatively affected by increases in water temperature and changes in water chemistry resulting from climate change and industrial pollution.

Depending on terrain characteristics and climate patterns, landscapes naturally acquire ecosystem types that are functional to their locations. When humans alter these land types for unnatural purposes, the consequences can be deleterious. Modern examples include the levy system in New Orleans, which replaced an extensive natural wetland buffer; when Hurricane Katrina hit the city in 2005, the levy system failed and the city was flooded. The devastation that resulted when Haiti was struck by a massive earthquake in 2010 was magnified by the high rate of deforestation, and therefore high rate of soil erosion, in that nation. Maintaining functional natural landscapes is increasingly seen as a priority, and restoration efforts are being developed to restore native habitat types even in urban areas.

DEVELOPING COUNTRIES

Since the mid-twentieth century, most of the world's habitat destruction has taken place in developing countries, as developed countries have already exhausted their most accessible resources and altered much of their land for development. For example, nearly 50 percent of wetlands in the United States and 60-70 percent of European wetlands have been destroyed.

The main factors contributing to land alteration in the developing world are poverty, overpopulation, lack of sustainable technology, and adherence to cultural practices. For example, many communities in the developing world frequently cook with charcoal, as electricity and natural gas are in short supply. Charcoal is acquired through the burning of forestland, and the results are mass deforestation and air pollution. New technologies that do not require large monetary investments, such as solar ovens and permaculture techniques, are increasingly being offered as solutions to poor communities that rely on habitat destruction to survive.

Many farmers in the developing world are wary of foreign technology and are unwilling to risk the failure of a crop to adopt a new technology, even if it could result in increased production. Because of this wariness, education is needed to help farmers ease into the use of new technologies that can benefit them, and the environment, in the long term. For ecological conservation to be sustainable, it must be beneficial for both the farmers and the environment.

FUTURE SOLUTIONS

The ability of humans to alter the land for their own gain is one of the main reasons humans have been able to expand over the planet and live in nearly every climate. Since the dawn of large civilizations, humans have increasingly acquired the means to reap more and more from the environment, sometimes altering it so severely that they have eradicated whole communities of species. Mass extinctions occurred in prehistoric times during the early years of humanity, but the species that died off at that time were mostly those targeted for food. In contrast, in modern times entire ecosystems are destroyed for resources and agriculture, and this can only have increasingly negative effects on both the natural world and human habitations.

In order to measure the direct importance for humans of many landscapes, scientists have defined the ecosystem services provided by the landscapes. These services, such as erosion prevention, storm buffering, soil productivity, and wildfire prevention, are often given monetary equivalents to introduce them into a system of economics.

Other approaches to reducing habitat destruction include the use of modern techniques to increase agricultural productivity that can reduce the need to clear more land for farming. Planting native species, creating tree-shaded spaces, and increasing topographical heterogeneity on agricultural land can increase biodiversity without having a large impact on production. In addition, the incorporation of natural areas around cities to buffer weather, prevent erosion, and safeguard watersheds may save much money in repairs and provide protection against disasters.

Jamie Michael Kass

FURTHER READING

Barbault, R., and S. D. Sastrapradja. "Generation, Maintenance, and Loss of Biodiversity." In *Global Biodiversity Assessment*, edited by V. H. Heywood. New York: Cambridge University Press, 1995.

Cincotta, Richard P., and Robert Engelman. *Nature's Place: Human Population Density and the Future of Biological Diversity*. Washington, D.C.: Population Action International, 2000.

Pullin, Andrew S. "Effects of Habitat Destruction." In *Conservation Biology*. New York: Cambridge University Press, 2002.

Tibbetts, John. "Louisiana's Wetlands: A Lesson in Nature Appreciation." *Environmental Health Perspectives* 114 (January, 2006): A40-A43.

SEE ALSO: Agricultural revolution; Coral reefs; Deforestation; Overgrazing of livestock; Rain forests; Strip and surface mining.

Hanford Nuclear Reservation

CATEGORIES: Nuclear power and radiation; waste and waste management

IDENTIFICATION: Site in Washington State that was formerly used in the production of plutonium for nuclear weapons

SIGNIFICANCE: Procedures used at the Hanford Nuclear Reservation when its reactors were operational allowed radioactive wastes to be discharged directly into the soil, slightly radioactive cooling water to be discharged into the Columbia River, and radioactive iodine to be released into the atmosphere. Environmental cleanup at the site continues into the twenty-first century.

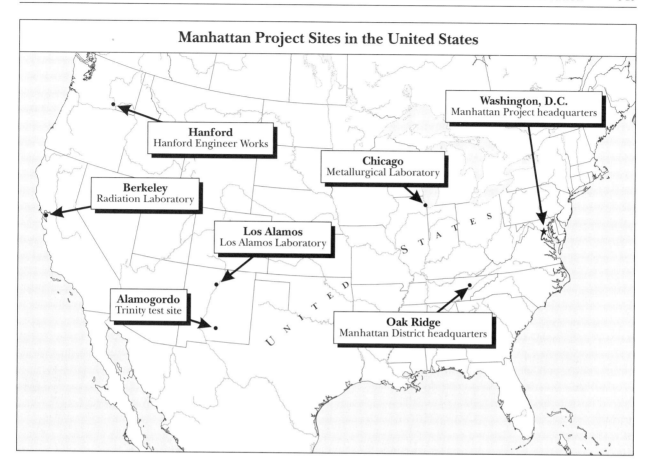

Manhattan Project Sites in the United States

The Hanford Nuclear Reservation is a roughly circular area 40 kilometers (25 miles) in diameter lying on the Columbia River in south-central Washington State. As a key part of the Manhattan Project to build the atomic bomb, several nuclear reactors were constructed at Hanford beginning in 1943. Plutonium produced in these reactors was extracted from spent nuclear fuel and then shipped off-site for fabrication into nuclear weapons. Although the river water used for cooling spent only one or two seconds in the reactor core, the intense neutron bombardment made some trace elements in the water radioactive. Radioactivity was also picked up from uranium fuel elements on which the aluminum cladding had corroded. After leaving the reactor, the cooling water was held in basins to allow short-lived radioactive elements to decay, after which it was returned to the river.

Radioactive elements that were eventually detected downstream from Hanford include sodium 24, phosphorus 32, and zinc 65. The latter two were found to be concentrated in the flesh of fish and waterfowl. People who ate fish and drank water from the Columbia River were estimated to experience a maximum increased dose from this radioactivity of about 10 percent above natural background radiation, a value believed to be safe.

The chemical procedures used to extract plutonium from spent fuel at Hanford released significant amounts of radioactive iodine 131 into the air, particularly in 1945. This was believed to be safe because iodine 131 has a short half-life (eight days), and its concentration is greatly diluted as it disperses across the countryside. It has since been learned that cows and goats concentrate iodine in their milk if they are fed grass contaminated with iodine and that the iodine is concentrated further in the thyroids of people who drink this milk. Infants and small children who drank a great deal of milk from grass-fed cows or goats near Hanford may have received harmful doses in 1945, but no such individuals have yet been identified.

The reactors at Hanford were shut down in the 1970's. During the years of reactor operation, Hanford stored some 256,000 cubic meters (9 million cubic feet) of highly radioactive liquid waste in 177 concrete-encased steel tanks. Some tanks contained chemical mixtures that could become explosive under certain conditions, so great vigilance was required. Approximately one-half of the waste was stored in tanks with only a single steel liner, and over the years some of these tanks leaked an estimated 4 million liters (1 million gallons) of highly radioactive waste into the soil. In addition, 800 billion liters (200 billion gallons) of low-level radioactive wastes were discharged directly into the soil, as it was believed that they would not migrate far. In fact, some elements migrated farther than anticipated, but only low levels of radioactivity have been found to have migrated off-site.

In 1989, the Washington Department of Ecology, the U.S. Environmental Protection Agency, and the U.S. Department of Energy initiated the legal steps needed to begin environmental cleanup at the Hanford Nuclear Reservation. The cleanup project, which has continued into the twenty-first century, is the largest such cleanup ever undertaken. It includes stabilization or removal of contaminated soil and the placement of restrictions on groundwater usage. The liquid waste stored at Hanford is to be vitrified (incorporated into glass) and taken to a permanent storage site; construction of the plant to be used in the vitrification process is expected to be completed in 2018.

Charles W. Rogers

FURTHER READING

Gephart, Roy E. *Hanford: A Conversation About Nuclear Waste and Cleanup.* Columbus, Ohio: Battelle Press, 2003.

Gerber, Michele Stenehjem. *On the Home Front: The Cold War Legacy of the Hanford Nuclear Site.* 2d ed. Lincoln: University of Nebraska Press, 2002.

McCutcheon, Chuck. *Nuclear Reactions: The Politics of Opening a Radioactive Waste Disposal Site.* Albuquerque: University of New Mexico Press, 2002.

SEE ALSO: Groundwater pollution; Nuclear accidents; Nuclear and radioactive waste; Nuclear power; Nuclear weapons production; Radioactive pollution and fallout.

Hansen, James E.

CATEGORIES: Activism and advocacy; weather and climate

IDENTIFICATION: American climate change scientist

BORN: March 29, 1941; Charter Oak Township, Iowa

SIGNIFICANCE: As a prominent climate scientist and activist, Hansen has been an important contributor to increased public awareness of global warming.

James E. Hansen was born in a farmhouse and reared in Denison, Iowa, about 97 kilometers (60 miles) northeast of Omaha, Nebraska. Hansen's family moved to Denison when James was four years of age; his father, who had been a tenant farmer, took up work as a bartender, and his mother worked as a waitress. With a scholarship and money saved from his job delivering newspapers, Hansen attended the University of Iowa, where he majored in mathematics and physics; he graduated summa cum laude in 1963.

Hansen went on to earn a master's degree in astronomy and a doctoral degree in physics at the University of Iowa; he studied for his Ph.D. with the chairman of the physics department, James Van Allen, who discovered the earth-girdling radiation belts named for him. Hansen decided to specialize in the study of the atmosphere of Venus at a time when scientists

Climate change scientist James E. Hansen. (AP/Wide World Photos)

were discovering that the planet's super-hothouse atmosphere (with temperatures above 454 degrees Celsius, or 850 degrees Fahrenheit) was 95 percent carbon dioxide.

His doctorate completed, Hansen went to work at the National Aeronautics and Space Administration's Goddard Institute for Space Studies (GISS), which is affiliated with Columbia University in New York City. During 1976 Hansen was working as principal investigator on the Pioneer Venus Orbiter when a Harvard postdoctoral researcher asked him to help calculate the greenhouse effect of human-generated emissions in the earth's atmosphere, and soon Hansen was captivated by the subject of global warming. He was appointed director of GISS in 1981, and that same year, with several coauthors, he was the first to use the term "global warming" in a scientific context. During 1988, Hansen went public with warnings about global warming in testimony before the U.S. Senate.

Several times over many years, transcripts of Hansen's warnings were edited before they were released to the public by the government. Each time, Hansen publicly objected to the edits, provoking a public furor. Soon Hansen was making headlines in *The Washington Post* and *The New York Times* as he compared censorship of science to the tactics of Nazi Germany and the old Soviet Union.

Every month, Hansen's laboratory takes the earth's temperature, monitoring ten thousand temperature gauges around the planet. Hansen has stated that he anticipates that the earth will not experience another ice age unless the human race becomes extinct—and even then it would take several thousand years to restore equilibrium to natural cycles that have been disrupted by the burning of fossil fuels. He has advanced an argument that the earth will not be safe until the proportion of carbon dioxide in the atmosphere is reduced to 350 parts per million (it was 387 parts per million in 2009).

Over time, Hansen has become increasingly adamant in his opposition to coal mining and to coal-fired electrical power plants, which are significant contributors to the carbon dioxide emissions linked with human-caused global warming. He has advocated civil disobedience as an oppositional strategy, and he was among thirty-one protesters arrested for obstructing officers and impeding traffic during a West Virginia demonstration against mountaintop coal mining in June, 2009.

Bruce E. Johansen

FURTHER READING

Hansen, James E. *Storms of My Grandchildren*. New York: Bloomsbury, 2009.

_____, et al. "Climate Impact of Increasing Atmospheric Carbon Dioxide." *Science* 213 (1981): 957-966.

Johansen, Bruce E. "The Paul Revere of Global Warming." *The Progressive*, August, 2006, 26-28.

SEE ALSO: Carbon dioxide; Climate change and oceans; Climate models; Climatology; Coal; Coal-fired power plants; Global warming; Greenhouse effect; Greenhouse gases; Manabe, Syukuro.

Hardin, Garrett

CATEGORIES: Activism and advocacy; population issues; resources and resource management

IDENTIFICATION: American ecologist

BORN: April 21, 1915; Dallas, Texas

DIED: September 14, 2003; Santa Barbara, California

SIGNIFICANCE: Through his writings—in particular his widely read 1968 article "The Tragedy of the Commons"—Hardin raised awareness of the environmental problems caused by human overpopulation and overexploitation of resources.

Garrett Hardin was an ecologist who argued strongly for the need to control human population growth. He first received public attention for his ideas in 1968 with the publication of his article "The Tragedy of the Commons" in the journal *Science*. The title of the article came from the concept of the English commons. In a typical medieval English community, all inhabitants could graze their animals on pasture that was held in common by all. The nature of the commons made it possible for users to overexploit the pasture, because any extra animals that one person raised were available for that user's benefit alone, while the grazing resources would be lost to all users of the commons. When the total population is low, any misuse of a commons can be compensated for by a move to a new area. When the total population has increased to the point where no new land is available, the resource held in common will be depleted. The results can be seen in acid rain and ozone depletion (in which the air is treated as a commons), overgrazing (in which the land is treated as a commons), and the whaling and commercial fishing industries (in which the resources of the sea are treated as a commons).

Hardin noted that each problem could be treated as though it were separate and distinct from every other, but he believed that this approach would simply hide the underlying cause of these problems. The pivotal commons is the freedom to breed, particularly when the costs of having children are spread over the entire population and are not restricted to the parents. According to Hardin, human overpopulation is the single root cause that lies behind problems such as excessive resource consumption, war, starvation, so-called ethnic cleansing, poverty, noise pollution, foreign aid, and simple traffic jams.

As explained in his book *The Limits of Altruism* (1977), Hardin believed that every nation has the responsibility to reduce its population to the level it can support using its own resources. Special privileges enjoyed by a few people today should be tolerated if they contribute to national self-reliance in the future. If all nations do not become self-reliant, the resulting population crash will be so devastating that civilization may never recover even if small enclaves of humans survive.

In one of his last books, *Living Within Limits: Ecology, Economics, and Population Taboos* (1993), Hardin compared the United States to a lifeboat in a sea of human misery and concluded that the nation would be able to survive the coming era of resource scarcity only by banning immigration and stabilizing its population. Hardin denied the workability of the "global village" and offered no apology for his apparent lack of compassion. In the absence of human moderation, he believed, nature will once more assert dominance in controlling human population size through disease, starvation, and war.

Gary E. Dolph

FURTHER READING

Hardin, Garrett. "The Tragedy of the Commons." *Science* 162 (December 13, 1968): 1243-1248.

Pepper, David, ed., with Frank Webster and George Revill. *Environmentalism: Critical Concepts.* New York: Routledge, 2003.

Sooros, Marvin S. "Garrett Hardin and Tragedies of Global Commons." In *Handbook of Global Environmental Politics*, edited by Peter Dauvergne. Northampton, Mass.: Edward Elgar, 2005.

SEE ALSO: Ehrlich, Paul R.; Malthus, Thomas Robert; Population-control and one-child policies; Population-control movement; Population growth; Tragedy of the commons.

Hawk's Nest Tunnel disaster

CATEGORIES: Disasters; human health and the environment

THE EVENT: Construction of a tunnel in West Virginia that resulted in the deaths of hundreds of workers from acute silicosis

DATES: March 31, 1930-November, 1931

SIGNIFICANCE: The willingness of the Union Carbide and Carbon Corporation to rush construction of the Hawk's Nest Tunnel despite possible danger to the workers from the inhalation of silica particles resulted in one of the largest workplace tragedies in the history of the United States.

In 1917 the Electro-Metallurgical Company of Fayette County, West Virginia, merged with three other companies and became the Union Carbide and Carbon Corporation. Interested in developing the water resources of West Virginia, Union Carbide, through the New Kanawha Power Company, filed its plans to build dams, tunnels, and power stations with the federal government. The Federal Power Commission took no action, the climate of the times being that government hesitated to interfere with the wishes of industry. Union Carbide rushed the building of the Hawk's Nest Tunnel out of fear that the government's attitude would change. Built near Gauley Bridge, West Virginia, in 1930-1931, the purpose of the 5-kilometer-long (3-mile-long) tunnel was to divert a portion of the New River underground to produce hydroelectric power.

The rushed construction of the tunnel resulted in the deaths of hundreds of workers. Although accidents, such as rock falls, accounted for some of these deaths, the majority resulted from silicosis, a lung disease that occurs when silica particles are inhaled. When granite and sandstone are pulverized, silica particles small enough to inhale are released; the entire length of the Hawk's Nest Tunnel was dug through granite and sandstone, and much silica dust was produced. Although techniques to reduce such dust were in use in some mining operations at the time, these techniques were not employed in the building of the tunnel because they would have slowed productivity. Although Union Carbide admitted that 109 men died as a result of working on the tunnel, some estimates place the death toll at closer to 700 workers if the deaths from silicosis that occurred over the next five years are included.

The Hawk's Nest Tunnel disaster served to bring the plight of mine workers and others working underground to national attention. Dozens of personal lawsuits were filed against Union Carbide. The lawsuits claimed that the company's managers knew the dangers of silica dust and that they knew of alternate, but slower, digging methods that would have protected the workers. Although Union Carbide won most of these lawsuits, a congressional subcommittee that convened in 1936 to study the issue did raise awareness about the dangers of silica.

The disaster also highlighted the environmental issue of water rights. In 1934 the Federal Power Commission sued Union Carbide and claimed full jurisdiction over the New River and all of the artifacts that affected its flow. The attorney general of the state of Virginia, siding with industry, argued the case of Union Carbide before the U.S. Supreme Court. The federal government decided that the case was critical to its future jurisdiction over water rights and continued to battle Union Carbide for the next thirty years. In 1964 the federal government finally won and forced Union Carbide to install stream gauges and fishways at the Hawk's Nest Dam. Throughout the legal battle, the government was concerned only with the cost of water. The human costs of disease and the damage to the waterways and the surrounding environment were not issues of interest.

Louise Magoon

FURTHER READING

Spangler, Patricia. *The Hawk's Nest Tunnel: An Unabridged History.* Proctorville, Ohio: Wythe-North, 2008.
Waite, Donald E. *Environmental Health Hazards: Recognition and Avoidance.* Rev. ed. Columbus, Ohio: Environmental Health Consultants, 2002.
Williams, John Alexander. "Hawks Nest." In *West Virginia: A History.* 2d ed. Morgantown: West Virginia University Press, 2001.

SEE ALSO: Asbestosis; Black lung; Environmental illnesses; Silicosis.

Hazardous and toxic substance regulation

CATEGORIES: Human health and the environment; pollutants and toxins

DEFINITION: Legally enforceable rules pertaining to substances that pose a threat to people, animals, or the environment because of their chemical, physical, or biological properties

SIGNIFICANCE: The regulation of hazardous and toxic substances is necessary in order to protect people and the environment from the by-products of industrialization. Policy makers must find a balance between safeguarding against undesirable health or environmental effects and reaping the benefits of the activities that produce such hazards.

The survival of the human species is inseparable from the preservation of the environment. This concern may be divided into three categories: the health of the general population, occupational hazards, and the availability of food and water supplies that are pure and free from pollutants. Within these categories, industrial development has created new problems and environmental pressures, even as technological advances improve the ability to detect, study, and analyze environmental changes.

Since ancient times humans have been concerned with issues surrounding public health. The most common concerns before the Industrial Revolution of the eighteenth and nineteenth centuries were related to food and water supplies. Tasks involving occupational hazards were left for slaves and the lower classes and did not, therefore, receive much recognition. After the Industrial Revolution, contact with occupational hazards and toxic substances became much more commonplace.

By the early twentieth century, conditions had worsened for the average working person, and those not faced with occupational hazards were beginning to feel the effects of an industrialized society that gave rise to hazardous by-products and toxic waste. Consciousness of environmental problems began to increase, but the rate of industrial development outstripped societal awareness of the hazards such development was creating.

Although Congress made attempts to regulate the hazardous and toxic wastes being generated in the United States from mining, agriculture, and industry, progress was slow because there was little public interest in the situation. Awareness of the negative impact of industrialization substantially increased during the 1960's as environmental activists such as Barry Commoner and Rachel Carson succeeded in publicizing issues relating to the deterioration of the environment. Public interest reached a peak with the celebration of the first Earth Day in 1970, and Congress began to respond to public pressure by strengthening the laws that had previously been passed to regulate air and water pollution.

Efforts to decrease hazardous and toxic waste levels in the environment were relatively ineffective until August 2, 1978, when New York State officials ordered the emergency evacuation of 239 families living within two blocks of the Love Canal chemical waste dump in Niagara Falls. Headlines throughout the nation declared Love Canal the largest human-made environmental disaster in decades, and Americans began paying attention to the issue of hazardous and toxic wastes. The general public held policy makers accountable for the disaster and the prevention of similar occurrences in the future.

DEVELOPMENT OF U.S. LEGISLATION

The U.S. regulatory style is considered to be one of open conflict, with interest groups, various media, legislators, and the courts all playing important roles in the development of laws. Policy makers must also rely on the consensus of scientists in understanding the risks and magnitude of problems involving hazardous and toxic materials. This reliance complicates and considerably slows the regulation process. Policy makers must also make tough decisions on the health benefits versus economic costs of controlling hazardous and toxic wastes.

This difficult task of creating legislation that works to protect the environment and the health of citizens in a hazardous society can be broken down into a four-step process: creating a law, putting the law to work, creating a regulation, and enforcing the law. The collaboration of Congress, government agencies, and societal awareness and responsibility is needed to create effective legislation.

To create new legislation, a member of Congress must propose a bill, which, if approved by both the House of Representatives and the Senate and signed by the president, becomes a new law. The act is codified by the House of Representatives and published in the United States Code. To put the law to work, regulations for the law must be created by government agencies authorized by Congress. Regulations are rules about the law, specifying what is legal and what is not. To create a regulation, the authorized government agency, usually the U.S. Environmental Protection Agency (EPA) in the case of hazardous and toxic substances, determines the need to form a regulation. The regulation is proposed on the Federal Register, and members of the public are allowed to provide input in the form of comments and suggested modifications. Revisions to the regulation may be made accordingly. Once a completed regulation is finished, it is published in the Code of Federal Regulations (CFR). Twice a year each agency publishes a comprehensive report that describes all the regulations it is working on or has recently finished. Laws and regulations are enforced by the government agency that put them into effect.

IMPORTANT U.S. LEGISLATION

The responsibility for regulating and enforcing hazardous materials laws in the United States is primarily split among a few federal regulatory agencies. There is some overlap between the fields that regulate hazardous and toxic materials. Federal governmental agencies that are involved in regulating hazardous and toxic substances are the EPA, the Occupational Safety and Health Administration (OSHA), and the Department of Transportation (DOT).

Several environmental laws affect hazardous and toxic materials policy in the United States. The Occupational Safety and Health Act, passed in 1970, provides standards of allowable exposure to toxic chemicals in the workplace. This law establishes occupational exposure limits for hundreds of toxic and hazardous substances. The act also establishes labeling standards for equipment, standards for personal protection, and monitoring requirements for the health of workers. With passage of the act, OSHA and the National Institute of Occupational Safety and Health (NIOSH) were created. The 1972 amendments to the Federal Insecticide, Fungicide, and Rodenticide Act (FIFRA) established a regulatory program for the EPA to control the manufacture of potentially harmful pesticides. This legislation was created to prevent the adverse environmental effects posed by new pesticides and to ensure safety standards for people using pesticides.

The Safe Drinking Water Act (SDWA) of 1974 was established to protect groundwater and drinking-water sources from contamination by hazardous chemicals. The act sets two levels of standards to limit the amount of contamination that might be found in drinking water: primary standards with a maximum contaminant level (MCL) to protect human health and secondary standards that relate to color, taste, smell, and other physical characteristics. The Hazardous Materials Transportation Act (HMTA) of 1975 provides a high level of environmental protection during hazardous waste transportation managed by the DOT. By requiring special packing and routing, the act ensures the careful shipment of hazardous substances.

The 1976 Resource Conservation and Recovery Act (RCRA) and its amendments deal with the ongoing management of solid wastes throughout the United States. With RCRA, a "cradle-to-grave" approach to hazardous waste management was introduced that was designed to protect groundwater supplies by focusing on the treatment, storage, and disposal of such wastes. RCRA focuses on five main areas for hazardous waste management: identification and classification of hazardous waste; requirements for generators of hazardous waste to identify themselves so that hazardous waste activities can be tracked and standards of operation for generators can be established; adoption of standards for the transportation of hazardous wastes; standardization of treatment, storage, and disposal facilities; and provisions for enforcement of the standards with a program of legal penalties for noncompliance. RCRA classifies waste materials based on four characteristic properties: ignitability, corrosivity, reactivity, and toxicity. The need to clean up abandoned toxic waste sites such as Love Canal, which RCRA did not address, gave rise to the Comprehensive Environmental Response, Compensation, and Liability Act (CERCLA) of 1980, widely known as Superfund, and its amendments, including the Superfund

Amendments and Reauthorization Act (SARA) of 1986.

The Toxic Substances Control Act (TSCA) of 1976 requires that all chemicals produced in or imported into the United States be tested, regulated, and screened for toxic effects prior to commercial manufacture. This law bans the manufacture of polychlorinated biphenyls (PCBs) and regulates asbestos. The EPA works with other federal agencies under this law to fill in the gaps of the other acts that attempt to manage hazardous materials. Additional laws under which

---

Ford Signs the Toxic Substances Control Act

On October 12, 1976, President Gerald R. Ford made this statement before signing the Toxic Substances Control Act into law:

I believe this legislation may be one of the most important pieces of environmental legislation that has been enacted by the Congress.

This toxic substances control legislation provides broad authority to regulate any of the tens of thousands of chemicals in commerce. Only a few of these chemicals have been tested for their long-term effects on human health or the environment. Through the testing and reporting requirements of the law, our understanding of these chemicals should be greatly enhanced. If a chemical is found to present a danger to health or the environment, appropriate regulatory action can be taken before it is too late to undo the damage.

The legislation provides that the Federal Government, through the Environmental Protection Agency, may require the testing of selected new chemicals prior to their production to determine if they will pose a risk to health or the environment. Manufacturers of all selected new chemicals will be required to notify the Agency at least 90 days before commencing commercial production. The Agency may promulgate regulations or go into court to restrict the production or use of a chemical or to even ban it if such drastic action is necessary.

The bill closes a gap in our current array of laws to protect the health of our people and the environment. The Clean Air Act and the Water Pollution Control Act protect the air and water from toxic contaminants. The Food and Drug Act and the Safe Drinking Water Act are used to protect the food we eat and the water we drink against hazardous contaminants. Other provisions of existing laws protect the health and the environment against other polluting contaminants such as pesticides and radiation. However, none of the existing statutes provide comprehensive protection. . . .

The administration, the majority and minority members of the Congress, the chemical industry, labor, consumer, environmental, and other groups all have contributed to the bill as it has finally been enacted. It is a strong bill and will be administered in a way which focuses on the most critical environmental problems not covered by existing legislation while not overburdening either the regulatory agency, the regulated industry, or the American people.

the EPA acts include the Clean Air Act (CAA) and amendments and the Clean Water Act (CWA) and amendments.

INTERNATIONAL REGULATORY EFFORTS

A host of international agreements and guidelines shape national and regional hazardous and toxic substances regulations around the world. The Basel Convention on the Control of Transboundary Movements of Hazardous Wastes and Their Disposal, for example, was adopted in 1989 to keep developed countries from dumping their hazardous wastes in developing countries. It addresses the transport of hazardous wastes among countries and illegalizes such transport without prior informed consent. Similarly, the Waigani Convention to Ban the Importation into Forum Countries of Hazardous and Radioactive Wastes and to Control the Transboundary Movements and Management of Hazardous Wastes Within the South Pacific Region was adopted in 1995 to halt the practice of waste traders of using the South Pacific region as a dump for hazardous and nuclear wastes. The convention also provides for the environmentally responsible management and disposal of existing wastes in this region.

The Rotterdam Convention on the Prior Informed Consent Procedure for Certain Hazardous Chemicals and Pesticides in International Trade, adopted in 1998, facilitates the environmentally sound management of severely restricted pesticide formulations and other extremely hazardous chemicals by requiring countries to share information regarding these substances. Exporters must supply information to importing countries and must abide by the importers' wishes. The Stockholm Convention on Persistent Organic Pollutants, adopted in 2001, restricts and ultimately prohibits the production, use, import, and export of bioaccumulative and environmentally persistent chemicals such as PCBs, furans, dioxins, and dichloro-diphenyl-trichloroethane (DDT). The Strategic Approach to International Chemicals Management, a policy framework designed to foster safe practices in chemical production and use worldwide, was adopted in 2006.

To minimize the likelihood of hazardous and toxic materials transport accidents that could harm people or property or damage the environment, the United Nations Economic and Social Council developed the U.N. Recommendations on the Transport of Danger-

ous Goods. The council administers regional agreements regarding hazardous and toxic materials transport and works to keep various nations' regulatory systems from impeding the flow of trade. Following the 1992 Earth Summit, at which the harmonization of chemical classification and labeling was identified as an international priority, the Globally Harmonized System of Classification and Labeling of Chemicals (GHS) was created to facilitate international trade, transport, and management of hazardous and toxic substances. Various modes of transport have their own regulatory schemes, notably the International Air Transport Association Dangerous Goods Regulations, the International Maritime Dangerous Goods Code, and the Regulations Concerning the International Carriage of Dangerous Goods by Rail.

Marcie L. Wingfield and Massimo D. Bezoari
Updated by Karen N. Kähler

FURTHER READING

Bergeson, Lynn L. *TSCA: The Toxic Substances Control Act.* Chicago: American Bar Association, 2000.

Cranor, Carl F. *Toxic Torts: Science, Law, and the Possibility of Justice.* New York: Cambridge University Press, 2006.

Griffin, Roger D. *Principles of Hazardous Materials Management.* 2d ed. Boca Raton, Fla.: CRC Press, 2009.

Hill, Marquita K. "Hazardous Waste." In *Understanding Environmental Pollution.* 3d ed. New York: Cambridge University Press, 2010.

LaGrega, Michael D., Philip L. Buckingham, and Jeffrey C. Evans. *Hazardous Waste Management.* 2d ed. New York: McGraw-Hill, 2001.

Sprankling, John G., and Gregory S. Weber. *The Law of Hazardous Wastes and Toxic Substances in a Nutshell.* 2d ed. St. Paul, Minn.: Thomson/West, 2007.

Switzer, Carole Stern, and Peter L. Gray. *CERCLA: Comprehensive Environmental Response, Compensation, and Liability Act (Superfund).* 2d ed. Chicago: American Bar Association, 2008.

SEE ALSO: Basel Convention on the Control of Transboundary Movements of Hazardous Wastes; Federal Insecticide, Fungicide, and Rodenticide Act; Hazardous waste; Nuclear and radioactive waste; Resource Conservation and Recovery Act; Right-to-know legislation; Solid waste management policy; Stockholm Convention on Persistent Organic Pollutants; Superfund.

Hazardous waste

CATEGORY: Waste and waste management

DEFINITION: Waste products of industrial society that pose dangers to human health and the environment

SIGNIFICANCE: Although many national governments have taken steps to regulate the disposal of hazardous wastes, some have not, and many such wastes continue to be produced all over the world. Improper disposal of hazardous wastes creates serious problems for the environment.

In the United States, hazardous wastes are legally defined as materials that have ignitable, corrosive, reactive, or toxic properties. In the early 1990's approximately 97 percent of all hazardous waste in the United States was produced by 2 percent of the waste generators. Remediation and cleanup of these wastes involve substantial economic cost. Since the 1970's the United States and other Western democracies have tried to regulate hazardous waste disposal. Hazardous wastes are also a serious problem in the former Soviet Union and other Eastern European nations.

ENVIRONMENTAL PROBLEMS

Improper disposal of hazardous waste can lead to the release of chemicals into the air, surface water, groundwater, and soil. High-risk wastes are those known to contain significant concentrations of constituents that are highly toxic, persistent, highly mobile, or bioaccumulative. Examples include dioxin-based wastes, polychlorinated biphenyls (PCBs), and cyanide wastes. Intermediate-risk wastes may include metal hydroxide sludges, while low-risk wastes are generally high-volume, low-hazard materials. Radioactive waste is a special category of hazardous waste, often presenting extremely high risks, as do biomedical and mining wastes.

Hazardous waste presents varying degrees of health and environmental hazards. When combined, two relatively low-risk materials may pose a high risk. Factors that affect the health risk of hazardous waste include dosage received; age, gender, and body weight of those exposed; and weather conditions. The health effects posed by hazardous waste include cancer, genetic defects, reproductive abnormalities, and central nervous system disorders.

Environmental degradation resulting from hazardous waste can render various natural resources, such as croplands and forests, useless and can harm animal life. For example, chemicals can leach out of improperly stored waste and into groundwater. Hazardous wastes may also generate long-lasting air pollution, water pollution, or soil contamination. In the past, before standards were in place for managing hazardous wastes, such materials were often buried or stored in unattended drums or other containers. This situation created threats to the environment and human health when the original containers began to leak and the materials leached into the soil and the water supply.

METHODS FOR HANDLING WASTES

The technologies and methods used in dealing with hazardous solid and liquid wastes continue to evolve. Several approaches have had positive impacts on the environment and the consumption of natural resources. One is the reduction of the volume of waste material through efforts to generate less of it. Another approach is to recycle hazardous materials as much as possible. A third means of dealing with hazardous waste is to treat it to render it less harmful; often, such treatment also reduces its volume. Least desirable among methods of addressing the problem is the storage of hazardous wastes in landfills. The Environmental Protection Agency (EPA) has established standards for the responsibility and tracking of hazardous wastes, based on the principle that waste generators are responsible for their waste "from cradle to grave." This principle requires that waste generators and disposal sites keep extensive records.

The costs for the cleanup and remediation of hazardous waste are substantial and are likely to continue to grow. This situation is particularly true in Eastern Europe and the nations of the former Soviet Union, where the magnitude of past dumping of hazardous materials is slowly becoming apparent. Meanwhile, less industrialized nations generally are ignoring the issue of hazardous waste, thereby setting themselves up for future difficulties.

U.S. LEGISLATION

The 1984 amendments to the federal Resource Conservation and Recovery Act (RCRA) included a thorough overhaul of hazardous waste legislation. Previously exempt sources that generated between 100 and 1,000 kilograms (220 and 2,200 pounds) of hazardous waste per month were brought under

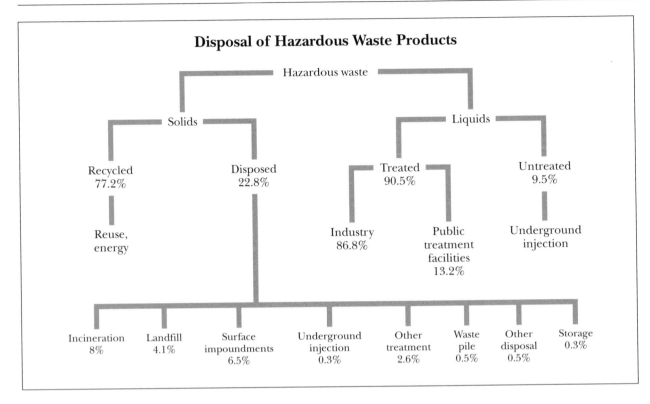

Disposal of Hazardous Waste Products

RCRA provisions. Congress further tried to force the EPA to adopt a bias against the landfilling of hazardous waste with a "no land disposal unless proven safe" provision. The amendments also added underground storage tanks for gasoline, petroleum, pesticides, and solvents to the list of sources to be regulated and remediated. In addition to RCRA, the Comprehensive Environmental Response, Compensation, and Liability Act of 1980 (known as Superfund) provides for the cleanup of all categories of abandoned hazardous waste sites except for radioactive waste sites. Several other statutes (and ensuing EPA regulations) have dealt with these aspects of the hazardous waste problem. The cleanup of existing sites will continue to be a troubling problem, while the cleanup and disposal of radioactive waste will be a major issue for the future.

The waste-minimization philosophy expressed in RCRA is a sound long-range strategy for dealing with hazardous waste. However, some materials will continue to be deposited in landfills. Incineration offers one solution to the problem of volume of material but poses issues of air quality and disposal of the highly toxic ash remaining. As some firms have found, minimizing their waste stream affords them economic benefits while conserving natural resources.

Household waste, which is not regulated by RCRA, often includes small quantities of hazardous materials such as pesticides. Many of these materials are still being landfilled in the early twenty-first century, as individual consumers remain ignorant of the proper ways to dispose of such wastes.

John M. Theilmann

FURTHER READING

Fletcher, Thomas H. *From Love Canal to Environmental Justice: The Politics of Hazardous Waste on the Canada-U.S. Border.* Orchard Park, N.Y.: Broadview Press, 2003.

Grisham, Joe. *Health Aspects of the Disposal of Waste Chemicals.* New York: Pergamon, 1986.

Hill, Marquita K. "Hazardous Waste." In *Understanding Environmental Pollution.* 3d ed. New York: Cambridge University Press, 2010.

LaGrega, Michael D., Philip L. Buckingham, and Jeffrey C. Evans. *Hazardous Waste Management.* 2d ed. New York: McGraw-Hill, 2001.

McKinney, Michael L., Robert M. Schoch, and Logan Yonavjak. "Municipal Solid Waste and Hazardous Waste." In *Environmental Science: Systems and Solutions.* 4th ed. Sudbury, Mass.: Jones and Bartlett, 2007.

SEE ALSO: Côte d'Ivoire toxic waste incident; Electronic waste; Hazardous and toxic substance regulation; Incineration of waste products; Landfills; Nuclear and radioactive waste; Solid waste management policy; Superfund; Waste management; Waste treatment.

Heat islands

CATEGORIES: Weather and climate; urban environments

DEFINITION: Dome-shaped areas of warm air that form over cities

SIGNIFICANCE: Heat islands increase temperatures, concentrate pollution, and may affect precipitation in urban environments.

Five factors are responsible for the formation of urban heat islands: urban fabric, or the rocklike material of a city's buildings and streets, which conducts heat three times faster than nonurban sandy soil; city structure, a complex web of multiple reflections and energy exchanges; artificial heat production, which occurs mainly in winter when artificial heating is highest; urban water balance, marked by rapid drainage and reduced humidity; and urban air pollution, which retains heat within the dome.

The impacts of heat islands on urban environments are many. The retention of heat leads to temperatures that are several degrees higher than those in surrounding nonurban outskirts. These temperature differences tend to be greatest in winter and at night, but heat islands also exist during summer. Higher temperatures lead to longer freeze-free seasons in cities, with first freezes in the fall occurring about one month later and last freezes in the spring occurring about one month earlier on average than in the surrounding rural areas.

Pollution concentration is greater in urban areas because of the local circulation produced within cities and the polluting industries situated in them. The local circulation of a city's heat island results from warm air rising near the central business district, spreading out at the top of the dome, descending at the outer edges, and flowing back into the city center. Pollution tends to attenuate incoming shortwave radiation and reduce visibility. Pollution levels tend to be highest in winter, when the sun angle is low, and on workdays. Heat islands are associated with pollution disasters such as those that occurred in London, England, in 1952; in Donora, Pennsylvania, in 1948; and in the Meuse Valley of Belgium in 1930. Each of these events occurred during the cold season.

The impacts of heat islands on precipitation remain in question. Some researchers have found that greater precipitation amounts over cities are largely the result of air rising in the city center and the greater abundance of condensation nuclei over the city. Others argue that the number of condensation nuclei is so vast that cloud drops are too tiny to grow to raindrop size, thus reducing the amount of rain. However, it is generally accepted that the number of rainy days in cities is slightly higher than the number in rural areas. Most urban climatologists have reported higher frequencies of hail and thunderstorms in cities and have noted that fog is more prevalent there than in outlying areas. The heat island effect generally produces less snow and more rain during storms. Humidity, both relative and absolute, tends to be slightly lower in urban environments than in areas outside cities.

Ralph D. Cross

FURTHER READING

Aguado, Edward, and James E. Burt. "Human Effects: Air Pollution and Heat Islands." In *Understanding Weather and Climate.* 5th ed. Upper Saddle River, N.J.: Pearson Education, 2010.

Alberti, Marina. *Advances in Urban Ecology: Integrating Humans and Ecological Processes in Urban Ecosystems.* New York: Springer Science, 2008.

Gartland, Lisa. *Heat Islands: Understanding and Mitigating Heat in Urban Areas.* Sterling, Va.: Earthscan, 2008.

SEE ALSO: Air pollution; Atmospheric inversions; Black Wednesday; Donora, Pennsylvania, temperature inversion; London smog disaster; Smog; Weather modification.

Heat pumps

CATEGORY: Resources and resource management
DEFINITION: Devices that transfer heat from air, earth, or water to heat or cool buildings
SIGNIFICANCE: Heat pumps use heat sources that are naturally occurring, plentiful, and environmentally sustainable, and because they replace the use of fossil fuels for heating, they substantially reduce the production of greenhouse gases, which are linked to global warming. Heat pump systems are among the most reliable green heating technologies.

All types of heat pumps operate on the basic principle of heat transfer from an existing source rather than a source that requires the burning of fuel to create heat. A heat pump uses a small amount of energy to pull heat out of a relatively low-temperature area and then pumps it into a higher-temperature area. Heat is moved from a heat source (ground, air, or water) into a heat sink, a building. When switched to the reverse mode, a heat pump can cool a building. Heat pumps operate most efficiently in moderate climates.

The four major components of any heat pump system are a compressor, a condenser, an evaporator, and a metering device. Modern heat pumps use hydrofluorocarbons or other refrigerants that have little to no ozone-depleting effect. The main environmental impacts associated with using heat pumps are the pollution produced by the use of the electricity to power the pumps (most electricity is generated from the burning of fossil fuels, which contributes to pollution) and the possibility of toxic refrigerants leaking into the environment.

The efficiency rating of a heating system, known as the coefficient of performance (COP), is the ratio of the benefit obtained from the pump (heat moved) divided by the work that must be paid for to move heat from the source to the sink. For an electric furnace, the COP is 1; for a high-efficiency natural gas furnace, the COP can exceed 2. For heat pumps, the COP ranges from 1 to greater than 4. The COP of a heat pump used in the cooling mode is often referred to as the energy-efficiency ratio (EER) or seasonal energy-efficiency ratio (SEER).

TYPES OF HEAT PUMPS

The most common type of heat pump is an air-source device that takes heat from the air outside a building and pumps it inside through refrigerant-filled coils. The basic constituents of this kind of heat pump are two fans, refrigerator coils, a reversing valve, and a compressor. The reversing valve reverses the flow of the refrigerant so that the pump can work in the opposite direction. In the reverse mode, the refrigerant absorbs heat from inside the building and releases it to the outside environment as an air conditioner would. The refrigerant cools and flows back inside to pick up more heat. Although air-source heat pumps work well, the operating costs are relatively high because of the electricity needed to run them. Typical COP ranges are between 1.0 and 2.9.

A water-source heat pump is an open-loop system that takes water drawn from a well directly to a heat exchanger, where heat is obtained. The water is then discharged back to an aboveground body of water or into a separate well. It uses water as its source of heat in the winter and water to move heat from a building in the summer. Typical open-loop systems move approximately 45 liters (12 gallons) of water per minute. Water-source heat pumps have the highest COP, but they require more maintenance than do other heat pump types. The COP typically ranges between 3.2 and 4.2. Potential environmental concerns with open-loop systems include possible contamination of surface water by accidental release of refrigerant and temperature changes in surface water that can affect aquatic habitats.

A ground-source heat pump is a closed-loop system. In the winter, it collects heat from a continuous loop of piping that is buried in the ground. During the summer, the pipes absorb heat to cool a building. Ground-source heat pumps are very dependable, but they are more costly to install than other types of heat pumps, and they also pose landscaping concerns. Typical COP ranges are between 2.6 and 3.4. The carbon dioxide output of both water-source and ground-source heat pumps is less than half of that produced by electric or gas furnaces or air conditioners.

ADVANCES IN TECHNOLOGY

Electric-powered heat pumps are the most common kinds in use, but a 5-horsepower, single-cylinder, glycol-cooled motor that runs on natural gas has been developed to power heat pumps. Pumps using the motor have been shown to have operating costs about 50 percent lower than electric heat pumps. With such advancements, applications of heat pumps have expanded to include heating swimming pools and heating water for household use.

For large-scale applications, absorption heat pumps have been developed. They differ from standard air-source heat pumps in that rather than compressing a refrigerant, an absorption pump absorbs ammonia into water, which is pressurized by a low-power pump. The heat source then boils the ammonia out of the water, and the process is repeated.

New types and blends of refrigerants are being developed for heat pumps that will produce minimal negative impacts on the environment. Environmentally friendly gases, such as hydrogen, helium, nitrogen, and plain air, are being used in Stirling-cycle heat pumps. Isobutane, which does not deplete ozone and is friendly to the environment, is also being used in heat pumps.

Alvin K. Benson

FURTHER READING

Banks, David. *An Introduction to Thermogeology: Ground Source Heating and Cooling.* Malden, Mass.: Blackwell, 2008.

Langley, Billy C. *Heat Pump Technology.* 3d ed. Upper Saddle River, N.J.: Prentice Hall, 2001.

Silberstein, Eugene. *Heat Pumps.* Clifton Park, N.Y.: Delmar Cengage Learning, 2002.

SEE ALSO: Air-conditioning; Air pollution; Alternative energy sources; Carbon dioxide; Coal-fired power plants; Earth-sheltered construction; Geothermal energy; Renewable energy.

Heavy metals

CATEGORY: Pollutants and toxins

DEFINITION: Dense metallic chemical elements

SIGNIFICANCE: Heavy metals released into the environment by mining, industry, and disposal of wastes can poison living organisms by interfering with metabolic functions. Such poisoning can cause impaired nervous system functioning, birth defects, and death.

The chemical elements lead, mercury, cadmium, and thallium are located together at the central bottom portion of the periodic table. Although mercury, thallium, and cadmium are fairly rare, lead is an abundant element on earth. All are dense, soft metals (mercury is a liquid at room temperature) that have a high affinity for chemically bonding with the element sulfur. They occur on or near the earth's surface as sulfur-containing minerals (sulfides) that are insoluble in water. This lack of water-solubility kept the heavy metals isolated from life-forms as they were evolving prior to the onset of the technological era, at which point humans began to mine and purify these useful elements. Once the heavy metals and their ions became more abundant in the environment, human beings—particularly mine workers and workers in certain industries—were at risk for toxic exposure to elements their bodies were not well equipped to process.

The ions of heavy metals resemble beneficial lighter metals such as zinc, calcium, magnesium, and iron in terms of their diameters and charges. This enables heavy metals to substitute for the beneficial elements and thus reside in the body over time. These lighter metals are much more abundant in the environment, and so life evolved with them, even employing them in critical roles.

The strong ability of heavy metals to bond to sulfur is significant because body biochemicals known as metalloproteins are composed of lighter metals such as zinc bonded to sulfur. (Sulfur is a normal constituent element of proteins.) In cases of heavy metal intoxication, defective metalloproteins are biosynthesized, employing an incorrect toxic metal that is strongly bound to sulfur. Although the bonding ability and even the size and charge of the heavy metal are appropriate for fitting into the structure of the biochemical, the biochemical fails to function as required, leaving the organism deficient in some vital way.

The degree of poisoning depends on the level and duration of exposure to the toxic element. Organisms have limited detoxification defenses, including proteins called metallothioneins, which scavenge metal atoms by virtue of having a large number of sulfur atoms in their structures. Medical intervention is possible in certain cases; for example, chelating agents are drugs designed to scavenge metal atoms and make them easier to eliminate from the body in the urine.

LEAD

Lead has been mined since antiquity and occurs in one form as the common mineral galena. Lead is a soft, easily worked metal, and it was often used to make food containers in ancient Roman times. Historians believe that acidic wine and other food leached lead from these vessels to cause lead poisoning, resulting in neurological damage that contributed to the decline of the Roman Empire. In modern times, the use of lead in paints and as an antiknock agent in gasoline helped to spread lead throughout the environment, although its use in these capacities has been largely phased out in the United States. Lead was also once used extensively in plumbing, and some older structures still have plumbing systems that deliver drinking water through lead pipes. Lead is currently used on a vast scale in batteries.

Lead in soil is absorbed by crops and works its way up the food chain. Lead dust from crumbling paint has been implicated in learning impairments and violent behavior in children. The nature of this neurotoxicity is not well understood. Other effects of chronic lead exposure include kidney damage and anemia, which results from lead's interference with the production of hemoglobin, the iron-containing component of blood. Lead is able to displace calcium in bone, where it can remain in the body for years.

CADMIUM, MERCURY, AND THALLIUM

Cadmium, a fairly rare metal, occurs in zinc-containing ores and is found in high concentrations near zinc smelters. It is used industrially in batteries and metal coatings. Airborne cadmium gets into soil, from which, like lead, it is absorbed into crops, including tobacco. Smokers consume about twice the cadmium each day that nonsmokers receive from the environment. Results of cadmium poisoning include lung and kidney damage and painful damage to the joints known in Japan (where cadmium pollution has been severe) as *itai-itai* (ouch-ouch) disease. A diet high in calcium can help to limit cadmium poisoning.

Mercury is used in industry in many ways. In the past, the improper disposal of mercury caused poisonings, such as what took place at Minamata Bay in Japan in the 1950's. Mercury use is being phased out where other compounds can safely be used, such as in pesticides. Many mercury compounds are volatile and easily absorbed by the body. Several can cross the blood-brain barrier; therefore, key symptoms of mercury poisoning involve vision and hearing distur-

bances, loss of coordination, and tremors. Mercury poisoning can also cause miscarriage and birth defects.

The effects of thallium poisoning are similar to those of poisoning involving the other heavy metals. Thallium, a by-product of zinc and lead production, can behave chemically similar in some ways to the vital element potassium. Unlike potassium, however, once it has been absorbed from the environment, it binds to sulfur, from which it is slowly released, effectively exposing the individual over long periods. Thallium was once used as a rat poison, but owing to its toxic nature such use is now forbidden in the United States.

Wendy Halpin Hallows

FURTHER READING
Cox, P. A. *The Elements on Earth: Inorganic Chemistry in the Environment.* New York: Oxford University Press, 1995.
Kaim, Wolfgang, and Brigitte Schwederski. *Bioinorganic Chemistry.* 2d ed. Hoboken, N.J.: John Wiley & Sons, 2006.
Mirsal, Ibrahim A. "Major Types of Soil Pollutants." In *Soil Pollution: Origin, Monitoring, and Remediation.* 2d ed. New York: Springer, 2004.
Silva, J. J. R. Fraústo da, and R. J. P. Williams. *The Biological Chemistry of the Elements: The Inorganic Chemistry of Life.* 2d ed. New York: Oxford University Press, 2001.
Timbrell, John. "Environmental Pollutants." In *Introduction to Toxicology.* 3d ed. New York: Taylor & Francis, 2002.
Wang, Lawrence K., et al., eds. *Heavy Metals in the Environment.* Boca Raton, Fla.: CRC Press, 2009.

SEE ALSO: Environmental illnesses; Lead; Mercury; Minamata Bay mercury poisoning; Oak Ridge, Tennessee, mercury releases; Soil contamination.

Herbicides. *See* Pesticides and herbicides

Hetch Hetchy Dam

CATEGORY: Preservation and wilderness issues

IDENTIFICATION: Part of a hydroelectric system built in Yosemite National Park to provide water and power to the city of San Francisco

DATE: Completed in 1923

SIGNIFICANCE: The controversy over O'Shaughnessy Dam's construction in Yosemite National Park during the early twentieth century was the first environmental issue argued on the national stage in the United States. Although environmentalists were unsuccessful in stopping the construction, they developed strategies and gathered support that became useful in later battles—including twenty-first century efforts to have the dam removed.

In 1890, eighteen years after the U.S. Congress named Yellowstone the first national park in the United States, Yosemite was named the second. Occu-

pying some 3,100 square kilometers (1,200 square miles) in the Sierra Nevada in eastern California, the new park featured giant redwoods and sequoias and two great scenic mountain valleys less than 32 kilometers (20 miles) apart: Yosemite Valley (which technically remained under state control for several more years) and Hetch Hetchy Valley. Both valleys offered breathtaking wilderness: flowering meadows surrounded by sheer cliffs of colorful granite punctuated by dramatic waterfalls. Writers such as John Muir, John Burroughs, and Mary Austin tramped through both valleys and the surrounding glacier-scoured mountains, bringing back descriptions of awe-inspiring beauty. Many thought that of the two, the oddly named Hetch Hetchy was the more beautiful. The 5.6 kilometers (3.5 miles) of the flat valley floor were traversed by a clear, clean river, and its granite walls were straight and steep. Because Yosemite's state control was less stringent than Hetch Hetchy's federal control, concessions and tourist businesses sprang up around Yosemite Valley, making it the more popular attraction.

View of the Hetch Hetchy Valley before it was dammed. (Sierra Club Bulletin, 1908)

Muir Opposes the Dam

In his 1912 book The Yosemite, *John Muir argued eloquently against the damming of the Tuolumne River in the Hetch Hetchy Valley:*

Hetch Hetchy Valley, far from being a plain, common, rock-bound meadow, as many who have not seen it seem to suppose, is a grand landscape garden, one of Nature's rarest and most precious mountain temples. As in Yosemite, the sublime rocks of its walls seem to glow with life, whether leaning back in repose or standing erect in thoughtful attitudes, giving welcome to storms and calms alike, their brows in the sky, their feet set in the groves and gay flowery meadows, while birds, bees, and butterflies help the river and waterfalls to stir all the air into music— things frail and fleeting and types of permanence meeting here and blending, just as they do in Yosemite, to draw her lovers into close and confiding communion with her.

Sad to say, this most precious and sublime feature of the Yosemite National Park, one of the greatest of all our natural resources for the uplifting joy and peace and health of the people, is in danger of being dammed and made into a reservoir to help supply San Francisco with water and light, thus flooding it from wall to wall and burying its gardens and groves one or two hundred feet deep. This grossly destructive commercial scheme has long been planned and urged (though water as pure and abundant can be got from sources outside of the peoples park, in a dozen different places), because of the comparative cheapness of the dam and of the territory which it is sought to divert from the great uses to which it was dedicated in the Act of 1890 establishing the Yosemite National Park.

The making of gardens and parks goes on with civilization all over the world, and they increase both in size and number as their value is recognized. Everybody needs beauty as well as bread, places to play in and pray in, where Nature may heal and cheer and give strength to body and soul alike. This natural beauty-hunger is made manifest in the little window-sill gardens of the poor, though perhaps only a geranium slip in a broken cup, as well as in the carefully tended rose and lily gardens of the rich, the thousands of spacious city parks and botanical gardens, and in our magnificent National parks—the Yellowstone, Yosemite, Sequoia, etc.— Nature's sublime wonderlands, the admiration and joy of the world.

Source: John Muir, *The Yosemite* (New York: Century, 1912).

San Francisco, 240 kilometers (150 miles) to the west, was one of the fastest-growing cities in the United States. As its population increased, so did its need for fresh water and electricity. At the time there was no public water supply, and San Franciscans were at the mercy of private companies that were not always responsive or responsible. By the beginning of the twentieth century, San Francisco's need for water was desperate. In 1901, under pressure from California legislators, the U.S. Congress passed the Right of Way Act, giving local governments the right to use national park lands for water projects such as dams and reservoirs if the projects were in the public interest. San Francisco officials wasted no time in declaring their intention to build a dam on the Tuolumne River at the narrow end of Hetch Hetchy Valley, flood the valley, and create a reservoir to supply the city with water. The very things that contributed to the valley's natural beauty—the flat valley floor, the steep cliffs, and the purity of the river—also made it an ideal spot for a reservoir.

When the city first applied for a right-of-way in 1903, the request was denied by U.S. secretary of the interior Ethan A. Hitchcock, who felt a dam was not in the public interest. After San Francisco's devastating earthquake and fire of 1906, the city's efforts intensified. Gifford Pinchot, chief of the U.S. Forest Service, was sympathetic to San Francisco's plan, and he urged the city to reapply for a right-of-way. It appeared that the new secretary of the interior, James Rudolph Garfield, was inclined to grant permission for the dam.

Immediately John Muir and the Sierra Club sprang into action to prevent the project. Muir wrote a personal letter to his old hiking companion, President Theodore Roosevelt, asking that other locations be developed instead. Roosevelt agreed that other rivers and dams should be exploited first but did not completely rule out the eventual flooding of Hetch Hetchy. Sierra Club members and others wrote letters to the editors of major newspapers on both coasts and garnered enough support for preserving Hetch Hetchy to stall the project in Congress.

PASSAGE OF THE RAKER ACT

For the next six years the issue was debated by the Congress and by the public in newspapers, newsletters, and public addresses. San Francisco's need for water was real, but many people argued that the need should be met in some way that would not destroy irreplaceable wilderness. Some argued that a reservoir in the valley would actually be more beautiful and attract more tourists to the area than the wild valley. For others, the issue was a matter of private versus public control of the city's water supply. Still others expressed the debate in terms of Pinchot's preservationism and the Sierra Club's conservationism.

Several congressional hearings on the right-of-way were held between 1908 and 1913. A brochure written by Muir in 1911 titled *Let Everyone Help to Save the Famous Hetch-Hetchy Valley and Stop the Commercial Destruction Which Threatens Our National Parks* included tips on lobbying Congress—a strategy that remains common for environmentalists today. Although conservationists attracted an impressive amount of support across the nation, they were ultimately defeated. In 1913 Congress passed the Raker Act, giving permission for the construction of a dam at Hetch Hetchy Valley.

O'Shaughnessy Dam—at the time the nation's largest concrete dam—was finished in 1923, and the associated water and power system was completed in 1934. Additional construction in 1938 raised the dam's height. The flooded valley never did attract many tourists, even during the late 1990's, when Yosemite Valley was so crowded that access by car was restricted. Water from the reservoir did help meet the needs of San Francisco, but only after control over its distribution was turned over to a private utility.

In 1987 Secretary of the Interior Donald P. Hodel proposed to study the removal of O'Shaughnessy Dam and the restoration of Hetch Hetchy Valley. State and federal studies concluded that the valley was more valuable as a water source than as a restored environment. During the early twenty-first century public interest in dam removal and valley restoration increased after the release of major reports by the Environmental Defense Fund (2004) and the organization Restore Hetch Hetchy (2005), along with two associated master's theses (2003 and 2004), all of which generally supported the concept of returning Hetch Hetchy to its original state. A 2006 California Resources Agency restoration study found the concept feasible, and the report on the research called for additional study. Although hundreds of small dams have been removed in the United States, no removal has ever been conducted for a dam as large as O'Shaughnessy. Restore Hetch Hetchy stated its hope of winning congressional approval for dam removal

by December, 2014, the month marking the one hundredth anniversary of Muir's death.

Cynthia A. Bily
Updated by Karen N. Kähler

FURTHER READING

Jones, Holway R. *John Muir and the Sierra Club: The Battle for Yosemite.* San Francisco: Sierra Club Books, 1965.

Righter, Robert W. *The Battle Over Hetch Hetchy: America's Most Controversial Dam and the Birth of Modern Environmentalism.* New York: Oxford University Press, 2005.

Simpson, John W. *Dam! Water, Power, Politics, and Preservation in Hetch Hetchy and Yosemite National Park.* New York: Pantheon Books, 2005.

SEE ALSO: Dams and reservoirs; Muir, John; National Park Service, U.S.; Pinchot, Gifford; Roosevelt, Theodore; *Sierra Club v. Morton*; Yosemite Valley.

High-yield wheat

CATEGORY: Agriculture and food

DEFINITION: Varieties of wheat that have been bred or otherwise genetically modified to produce maximum crop yields

SIGNIFICANCE: High-yield wheat has helped to increase food production worldwide, but the development of such wheat and other high-yield crops has also changed the structure of agriculture, with some negative effects on the environment.

Wheat (*Triticum sativum*) is the world's most important grain crop because it serves as a major food source for most of the world's population. Large portions of agricultural land are devoted to the production of wheat worldwide. Wheat constitutes a large part of the domestic economy of the United States, contributes a large part to the nation's exports, and serves as the national bread crop. Wheat is the national food staple for forty-three countries and about 35 percent of the people of the world, and it provides 20 percent of the total food calories for the world's population.

No one knows for certain when wheat was first cultivated, but by six thousand years ago, humans had discovered that seeds from wheat plants could be collected, planted in land that could be controlled, and later gathered for food. As human populations continued to grow, farmers found it necessary to select and produce increasingly high-yielding wheat. The Green Revolution of the twentieth century helped to make this possible. Agricultural scientists developed new, higher-yielding varieties of numerous crops, particularly the seed grains such as wheat that supply most of the calories necessary for maintenance of the world's population.

Wheat, like other major crops, originated from a low-yield native plant, but it has been converted into one of the highest-yielding crops in the world. There are two major ways to improve yield in seed grains such as wheat. One way is to produce more seed per seed head, and the second way is to produce larger seed heads. Both of these approaches have been utilized to produce high-yield wheat.

Numerous agricultural practices are required to produce higher yields, but one of the most important is the selection and breeding of genetically superior cultivars. When a grower observes a plant with a potentially desirable gene mutation that produces a change that improves a yield characteristic, the grower collects its seed and grows additional plants, which produces higher yields. This selection process remains one of the major means of improving yield in agricultural crops. Advances in the understanding of genetics have made it possible to breed some of the desirable characteristics that have resulted from mutation into plants that lacked those characteristics. In addition, the advent of recombinant deoxyribonucleic acid (DNA) technology has made it possible for scientists to transfer genetic characteristics from one plant to any other plant.

While tremendous increases in the world's food supply have resulted from the planting of crops such as high-yield wheat, the changes in agriculture accompanying the production of these crops have had some negative impacts on the environment. The production of high-yield wheat in modern agricultural units is highly mechanized and thus uses large amounts of energy; in addition, it is highly reliant on agricultural chemicals such as fertilizers and pesticides. The development of the new crop varieties also has led to an increased reliance on monoculture, the practice of growing only one crop over a vast number of hectares.

D. R. Gossett

FURTHER READING

Chiras, Daniel D. "Creating a Sustainable System of Agriculture to Feed the World's People." In *Environmental Science.* 8th ed. Sudbury, Mass.: Jones and Bartlett, 2010.

McNeill, J. R. "The Biosphere: Eat and Be Eaten." In *Something New Under the Sun: An Environmental History of the Twentieth-Century World.* New York: W. W. Norton, 2000.

Nelson, Gerald C., ed. *Genetically Modified Organisms in Agriculture: Economics and Politics.* San Diego, Calif.: Academic Press, 2001.

SEE ALSO: Agricultural revolution; Biotechnology and genetic engineering; Borlaug, Norman; Genetically modified foods; Green Revolution; Pesticides and herbicides; Selective breeding; Sustainable agriculture.

Hiroshima and Nagasaki bombings

CATEGORIES: Disasters; nuclear power and radiation

THE EVENTS: The dropping of two atomic bombs on cities in Japan during World War II—a uranium bomb on Hiroshima and a plutonium bomb on Nagasaki

DATES: August 6 and 9, 1945

SIGNIFICANCE: The use of atomic weapons by the United States is credited with ending the war with Japan, but debates continue regarding the necessity and moral justification of the bombings. The bomb blasts and the radiation emitted immediately on detonation were responsible for widespread devastation and the deaths of hundreds of thousands of people, but the effects of radiation on the surrounding areas were less long-lasting than had been expected.

By the summer of 1945, Japan had lost the war against the United States and the other Allied Powers, but the ruling Japanese war cabinet was deadlocked—three for and three against—on a decision on surrender. With the tie vote, the previous policy of war prevailed. The Japanese devised a strategy to end the war that would force the Allies to invade the Japanese homeland, where so many would be killed that Japan would receive more favorable surrender terms to end the war. The Allies did not invade Japan, however; on August 6 and 9, 1945, the United States dropped atomic bombs on the cities of Hiroshima and Nagasaki. Even after the devastation caused by the bombings, the war cabinet remained deadlocked, but Emperor Hirohito survived an attempted coup and gained enough support that he was able to lead Japan's surrender.

THE BOMBS

The bomb code-named Little Boy exploded 580 meters (1,900 feet) over Hiroshima. It weighed 4,400 kilograms (9,700 pounds). Its great power came from 64 kilograms (141 pounds) of uranium 235. The uranium was assembled into critical mass (the amount needed to sustain a chain reaction) in 1 millisecond by the firing of a 25-kilogram (55-pound) uranium bullet from a short cannon into 39 kilograms (86 pounds) of uranium shaped into three target rings. Although the bomb was only 1.3 percent efficient, it exploded with the power of 13 kilotons of trinitrotoluene (TNT).

A three-second thermal pulse from the blast burned the exposed skin of people as far away as 3.5 kilometers (2.2 miles) from ground zero (the point directly below the burst) and started fires throughout the city. The blast crushed buildings around ground zero, leaving only skeletons of reinforced concrete or steel. Numerous fires spread, joined, and formed a violent firestorm that consumed everything combustible within 13 square kilometers (5 square miles). It is estimated that some 70,000 people were killed, and another 100,000 died over the next five years from bomb-related injuries. An estimated 60 percent of the deaths were from burns, 30 percent from the blast, and 10 percent from radiation.

Because the Japanese did not surrender immediately, and to demonstrate that the Hiroshima bomb was not a fluke, the United States detonated a plutonium bomb, code-named Fat Man, 595 meters (1,952 feet) over Nagasaki three days later. (The bombs' code names, Little Boy and Fat Man, referred to U.S. president Harry Truman and British prime minister Winston Churchill, respectively.) Fat Man weighed 4,900 kilograms (10,800 pounds), half of which was high explosives used to compress the softball-sized plutonium core into critical mass. The 22-kiloton blast and subsequent fire destroyed 6.7 square kilometers (2.6 square miles). The destruction was not greater because part of the city was protected by a hill. Approximately 70,000 people were killed, and another 70,000 died over the next five years from bomb-

related injuries. An estimated 77 percent of these deaths were from burns, 16 percent from the blast, and 7 percent from radiation.

RADIATION

Prompt radiation consists of the neutrons and gamma rays emitted as a nuclear bomb explodes and during the first minute thereafter. About 85 percent of the people within 1 kilometer (0.6 mile) of ground zero of both blasts were killed outright by blast or heat, or they died within the first year of radiation or injuries. People up to 2 kilometers (1.2 miles) away received significant but nonlethal prompt doses of radiation. A study of thirty women who had been pregnant when they were exposed to radiation within 2 kilometers of ground zero and showed signs of radiation damage found that only sixteen of the children they subsequently bore lived more than one year, and four of those children were developmentally disabled. Deaths from prompt radiation began during the first week and peaked three to four weeks later. Be-

A mushroom cloud forms over the Japanese city of Nagasaki on August 9, 1945, moments after the United States dropped the second atomic bomb ever used in warfare. (National Archives)

tween 1950 and 1990 an estimated 850 people who had been in the vicinities of the Hiroshima and Nagasaki bombings died of radiation-caused cancer, and another 400 to 800 died of radiation-induced non-cancer injuries.

Radiation after the first minute of a nuclear bomb blast is called delayed radiation. Near ground zero, neutrons make elements such as sodium, aluminum, and manganese radioactive, but these newborn isotopes have short half-lives and disappear within minutes to hours. The other source of delayed radiation is fallout. It consists of radioactive bomb vapor that has condensed into small, solid particles. Had the Hiroshima and Nagasaki bombs been ground bursts, dirt and rocks would have been sucked up into the fireball and made radioactive by neutrons, which would have made the fallout much worse. Since the bombs were air bursts, this did not happen. Within an hour of each burst, soot from the fires helped precipitate black rain. In spite of reports to the contrary, the rain was only mildly radioactive. Direct measurements found the highest levels to be in the Nishiyama district of Nagasaki, but even there radiation levels should not have caused lasting harm. People who came into the city right after the explosion and were exposed to the fallout and residual radiation showed no symptoms attributable to radiation.

Although some scientists expected that Hiroshima and Nagasaki would be uninhabitable for generations because of the radiation caused by the atomic weapons, in reality radiation in the cities had returned to near background level one month after the bombings. Both cities were rebuilt and achieved their prebombing populations within ten years. In the twenty-first century they are modern and prosperous.

Charles W. Rogers

FURTHER READING

Pike, Frances. *Empires at War: A Short History of Modern Asia.* London: I. B. Tauris, 2010.

Rotter, Andrew J. *Hiroshima: The World's Bomb.* New York: Oxford University Press, 2008.

Schull, William J. *Effects of Atomic Radiation: A Half-Century of Studies from Hiroshima and Nagasaki.* New York: Wiley-Liss, 1995.

SEE ALSO: Antinuclear movement; Bikini Atoll bombing; Chernobyl nuclear accident; Nuclear weapons production; Radioactive pollution and fallout; Union of Concerned Scientists.

Homestead Act

CATEGORIES: Treaties, laws, and court cases; land and land use; resources and resource management

THE LAW: U.S. federal legislation that granted large sections of the West to settlers and farmers

DATE: Enacted on May 20, 1862

SIGNIFICANCE: The Homestead Act opened much U.S. territory to development, but a lack of protections in the act also led to rampant speculation and to land and resource abuse.

The Homestead Act of 1862 granted 65 hectares (160 acres) of land to individual potential farmers free of charge. To obtain full ownership, settlers were required to farm their land for five years. By 1867, at least 1 million hectares (2.5 million acres) had been allotted under the act.

When the U.S. Congress created the Homestead Act, it failed to classify the available land parcels in terms of use, whether farming, mining, cattle raising, or logging. All plots were open to interested settlers, regardless of their intentions, and the government employed too few people to monitor the uses of the land. The Homestead Act, in theory, was designed to deal only with farmland, which left the other lands unprotected. The General Reform Act of 1891 and the Desert Land Act of 1877 repaired the omissions of the Homestead Act and accounted for the diversity of the western United States.

The Homestead Act's commutation clause, which stated that after six months the land could be purchased for $1.25 per acre, helped lead to speculation in natural resources. Between 1881 and 1904, settlers gathered 23 percent of the Homestead lands by commutation; companies and speculators, in particular, used the clause to buy large plots. In Aberdeen in the Dakota Territory, for example, speculators claimed 75 percent of the land. With thousands of settlers moving west, the demand for firewood and building materials greatly increased. Because of speculation to meet the settlers' needs, forests in Minnesota, Wisconsin, and Michigan were almost destroyed. Congress hired special agents to investigate timber misuse and illegal land claims, but these officials were responsible for large regions and had little effect.

By the 1880's, timber misuse had become a problem. Congress was forced to enact the 1873 Timber Culture Act, granting an additional 65 hectares to interested settlers. These farmers were required to

Homesteader Qualifications

The first section of the Homestead Act of 1862 spelled out the qualifications for claiming land.

Be it enacted by the Senate and House of Representatives of the United States of America in Congress assembled, That any person who is the head of a family, or who has arrived at the age of twenty-one years, and is a citizen of the United States, or who shall have filed his declaration of intention to become such, as required by the naturalization laws of the United States, and who has never borne arms against the United States Government or given aid and comfort to its enemies, shall, from and after the first January, eighteen hundred and sixty-three, be entitled to enter one quarter section or a less quantity of unappropriated public lands, upon which said person may have filed a preemption claim, or which may at the time the application is made, be subject to preemption at one dollar and twenty-five cents, or less, per acre; or eighty acres or less of such unappropriated lands, at two dollars and fifty cents per acre, to be located in a body, in conformity to the legal subdivisions of the public lands, and after the same shall have been surveyed: Provided, That any person owning and residing on land may, under the provisions of this act, enter other land lying contiguous to his or her said land, which shall not, with the land so already owned and occupied, exceed in the aggregate one hundred and sixty acres.

plant at least 16.2 hectares (40 acres) of timber for ten years, but 90 percent of the land was used to produce timber for sale, not to replace the misused forest regions. In Kansas, Nebraska, and the Dakota Territory, however, the Timber Culture Act replaced a small number of trees. In addition to being a response to the loss of forestlands caused by the Homestead Act, the Timber Culture Act was designed to encourage the planting of trees to reduce winds, increase rainfall, and attract new settlers.

Although speculators used the Homestead Act to their advantage, the act's opening of the frontier also prompted settlers to make use of lands that were previously vacant. During the Civil War, for example, the farmers who had claimed their land under the act helped to produce enough food for 1.5 million soldiers. Allied with improvements in farming methods such as irrigation, the settlement of areas opened by the Homestead Act made the West a new source of food production for the nation.

Keith E. Rolfe

FURTHER READING

Greenberg, Michael R. *Environmental Policy Analysis and Practice.* Piscataway, N.J.: Rutgers University Press, 2007.

Koontz, Tomas M. *Federalism in the Forest: National Versus State Natural Resource Policy.* Washington, D.C.: Georgetown University Press, 2002.

Platt, Rutherford H. *Land Use and Society: Geography, Law, and Public Policy.* Rev. ed. Washington, D.C.: Island Press, 2004.

SEE ALSO: Dust Bowl; Forest and range policy; Land-use policy; Range management.

Hoover Dam

CATEGORY: Preservation and wilderness issues
IDENTIFICATION: Dam on the lower Colorado River, on the border between Arizona and Nevada
DATE: Completed in 1936
SIGNIFICANCE: Hoover Dam, which was built to regulate the flow of the lower Colorado River in order to prevent floods, provide consistent water levels necessary for irrigation, and produce hydroelectric power, also altered the environment of the southwestern United States, often in unanticipated ways.

At the time of its construction, Hoover Dam, then known as Boulder Dam (it was officially renamed in 1947), was rightfully hailed as an engineering marvel, with its concrete wall 221 meters (726 feet) high and 379 meters (1,244 feet) long at its crest. Spanish explorers named the Colorado River for its reddish-brown color, a result of the sediment that the river carries. The new dam captured the sand and silt, which settled to the bottom of the water collected behind the dam, in the reservoir named Lake Mead. Experts knew that the silt would eventually fill the reservoir, but they estimated that this process would take several hundred years. In the meantime, they sought ways to reduce the amount of silt deposited.

Arguing that overgrazing by Navajo sheep herds was causing soil erosion, which was the major source of silt in the Little Colorado and San Juan rivers (two tributaries of the Colorado), government officials successfully forced the Navajo to accept a stock reduction program. Nonetheless, silt continued to build up behind the dam, and by 1949 the sediment was more than 82 meters (270 feet) deep in some areas. This changed the flow of the river upstream, slowing rapids in the Grand Canyon more than 160 kilometers (100 miles) north of the dam. Silt buildup in Lake Mead slowed after the completion of the Glen Canyon Dam north of the Grand Canyon in 1963.

Ironically, the reduction in the Colorado's flow downstream of the dam caused severe floods in Needles, California, during the 1940's. The water released from the newly finished dam picked up large deposits of sediment that stood at the base of the dam, carrying it downriver. Because the river's flow had been reduced, it could not carry the silt great distances, and sand quickly built up where the river widened near Needles. Plants soon filled the river bottom, providing yet another obstacle to water flow. The magnitude of the problem became apparent in 1941, when water releases from the dam increased in size and frequency. The sand, silt, and plant life diverted the rushing waters into the nearby town, forcing the evacuation of several families. During the 1940's the region became a veritable swamp. Flooded cesspools polluted the groundwater, creating a health hazard. In response, the federal government authorized levee construction and river dredging, expensive solutions to problems caused by a dam that was intended to prevent flooding.

One of the benefits to which the government pointed in publications praising the dam was the creation of new wildlife habitats along the shores of Lake Mead. A 1985 brochure noted that more than 250 species of birds had been identified in the region, including migratory species that use the lake as a stopover. It also claimed that the lake provides water for native animals, including desert bighorn sheep. In addition, several species of game fish live in the lake, a boon to the area's reputation as a recreational site and a plus for the region's economy.

The brochure neglected to mention the negative impacts of the dam on wildlife. For centuries the Colorado had deposited its sediment in a delta located at the river's mouth at the Gulf of California. This delta was a haven for wildlife, such as deer, birds, bobcats, and numerous other species. The reduced flow of the river and the reduction in sediment altered the delta, and wildlife began to disappear; some observers likened the changed area to a desert. Changes also occurred upriver from the dam. The game fish in Lake Mead competed with native fish upriver, possibly contributing to the extinction of some species in the lower reaches of the Grand Canyon.

The construction of the dam may have also increased the possibility of earthquakes in the region.

Aerial view of Hoover Dam. (©Photoquest/Dreamstime.com)

When Lake Mead began to fill during the late 1930's, scientists recorded several seismic shocks, a phenomenon that had not been noted in the area prior to the dam's construction. Arguing that several faults exist in the region, some scientists dismissed any notion of a general link between earthquakes and the weight of water in reservoirs. However, earthquakes in the regions of several major dams around the world during the 1960's led engineers to reconsider the connection between seismic activity and dams. Two major shocks in the Lake Mead area in 1972 confirmed suspicions that the reservoir contributes to earthquake activity.

Perhaps the two most important environmental impacts of Hoover Dam are the precedent its construction created and the population growth that it fostered in the American Southwest. The success of the dam prompted the construction of several dams on the Colorado River, including the Glen Canyon Dam, which significantly altered the ecosystem within the Grand Canyon. In addition, the electricity generated at Hoover Dam and the lure of Lake Mead increased human migration to the area, which in turn led to increased pollution and greater demand for water from the already beleaguered Colorado River.

Thomas Clarkin

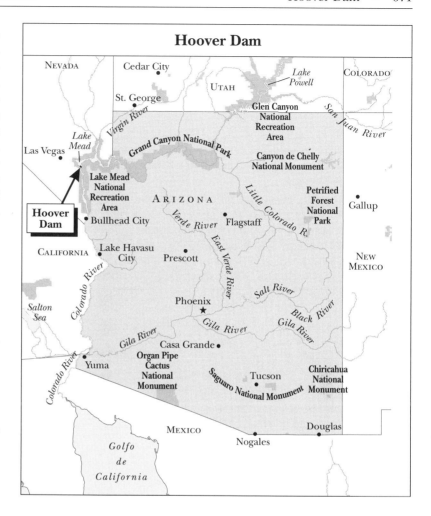

FURTHER READING

Berkman, Richard L., and W. Kip Viscusi. *Damming the West.* New York: Grossman, 1973.

Billington, David P., and Donald C. Jackson. *Big Dams of the New Deal Era: A Confluence of Engineering and Politics.* Norman: University of Oklahoma Press, 2006.

Goldsmith, Edward, and Nicholas Hildyard. *The Social and Environmental Effects of Large Dams.* New York: Random House, 1984.

Kleinsorge, Paul Lincoln. *The Boulder Canyon Project: Historical and Economic Aspects.* Palo Alto, Calif.: Stanford University Press, 1940.

McCully, Patrick. *Silenced Rivers: The Ecology and Politics of Large Dams.* Enlarged ed. London: Zed Books, 2001.

Mann, Elizabeth, and Alan Witschonke. *The Hoover Dam: The Story of Tough Times, Tough People, and the Taming of a Wild River.* New York: Mikaya Press, 2001.

Reisner, Marc. *Cadillac Desert: The American West and Its Disappearing Water.* Rev. ed. New York: Penguin Books, 1993.

Stevens, Joseph E. *Hoover Dam: An American Adventure.* Norman: University of Oklahoma Press, 1988.

Watkins, T. H., et al. *The Grand Colorado: The Story of a River and Its Canyons.* Palo Alto, Calif.: American West, 1969.

SEE ALSO: Aswan High Dam; Colorado River; Dams and reservoirs; Floods; Glen Canyon Dam; Grand Canyon; Hydroelectricity.

Horizontal gene transfer

CATEGORY: Biotechnology and genetic engineering
DEFINITION: Transfer of deoxyribonucleic acid
between organisms by means other than
reproduction
SIGNIFICANCE: Horizontal gene transfer is widespread
among bacteria and viruses but relatively rare between multicellular organisms or between multicellular organisms and bacteria. Environmentalists
have expressed concerns that genetically engineered organisms could potentially pass their artificially introduced genes to other nonrelated
organisms and thus change local and global ecologies, but research has shown that there are limits to
the extent of such transfer.

Deoxyribonucleic acid (DNA) is used by all cell-
based life on the earth to store genetic information. Parents pass their DNA to their offspring during
reproduction. DNA transfer by means of reproduction is referred to as vertical gene transfer. DNA transfer between organisms that is not the result of reproduction is known as horizontal gene transfer (HGT)
or lateral gene transfer.

Bacterial genomic studies have demonstrated that
HGT is rather common between bacteria. HGT occurs by three different mechanisms: transduction, or
DNA transfer between cells by viruses; conjugation, in
which cells physically touch each other and transfer
DNA; and transformation, or the direct uptake of
DNA from the environment by cells. The rate of HGT
between the common intestinal bacterium *Escherichia
coli* and its close relative *Salmonella* is approximately
one gene every 50,000 years, and 18 percent of the ge-
nome of *E. coli* was acquired through HGT. Genomic
studies have also shown that not all genes are transferred at the same rate and not all groups of organisms experience HGT to the same extent.

HGT between bacteria can imbue bacterial species
with new genes and opportunities to invade new environments. HGT can give bacteria antibiotic resistance, the ability to cause diseases, or the capacity to
utilize new food sources. Genetically modified bacteria, which contain novel genes, can potentially move
their new genes into large populations of bacteria.
The introduction of new genetic material into bacteria has the potential to alter seriously the ecological
and pathogenic character of bacterial species.

HGT from bacteria to multicellular organisms occurs at a lower rate. The plant pathogen *Agrobacterium*
transfers DNA to plants by means of a conjugation-
type mechanism. Likewise, intracellular parasites can
also transfer genes into their multicellular plant or animal hosts. With respect to HGT from multicellular
organisms to bacteria, genomic analyses have confirmed that bacteria have acquired genes from multicellular organisms. A consequence of these data is
that genetically modified (GM) plants, which are
commonly used in modern agriculture, might be able
to transfer genes to soil bacteria. However, even
though GM plants can transfer to soil bacteria under
laboratory conditions, such transfer has not been de-
finitively demonstrated under natural conditions in
field studies. Studies have addressed the transfer of
genes from GM plants to bacteria in the digestive
tracts of those animals that consume foods made from
GM plants, and the research has shown that such
transfer is highly inefficient. Other studies on animals
and humans have established that the transfer of
genes from GM plants to animal species that consume
GM foods does not occur at any significant rate.

HGT to viruses from plants and animals is somewhat common. For example, viral infections of GM
plants cause the production of novel plant viruses that
have exchanged genes with the host plant. This generates new viruses with new genes and novel properties.
Similarly, when animal viruses infect animal cells they
can acquire altered versions of animal genes that
transform them into tumor-causing viruses.

Michael A. Buratovich

FURTHER READING
Bushman, Frederic. *Lateral DNA Transfer: Mechanisms
and Consequences.* Cold Spring Harbor, N.Y.: Cold
Spring Harbor Laboratory Press, 2002.
Fedoroff, Nina V., and Nancy Marie Brown. *Mendel in
the Kitchen: A Scientist's View of Genetically Modified
Food.* Washington, D.C.: Joseph Henry Press, 2006.
Stewart, C. Neal, Jr. *Genetically Modified Planet: Environ-
mental Impacts of Genetically Engineered Plants.* New
York: Oxford University Press, 2004.

SEE ALSO: *Bacillus thuringiensis*; Bacterial resistance;
Biotechnology and genetic engineering; Cloning; Genetically altered bacteria; Genetically modified foods;
Genetically modified organisms; Pesticides and herbicides.

Hunting

CATEGORY: Animals and endangered species
DEFINITION: Pursuit and killing of game animals by
   humans for sport or for needed food
SIGNIFICANCE: Hunting persists as a sport and a tool
   for wildlife or environmental regulation, but the
   assessment of hunting from the standpoint of values is politically and philosophically controversial.

Hunting originated as a means of subsistence, but
it seldom has that status today. When hunting
was necessary for human survival, no real environmental issues surrounded the practice, merely expe-
diencies. Native Americans burned land to make it
more attractive to buffalo and deer, but this was done
because such conditions favored human subsistence.
When European colonizers traveled to North America between the sixteenth and eighteenth centuries,
American Indians were able to collaborate in the fur
trade and rise above a subsistence economy, but at a
price that they could not foresee. Commercially desirable species were soon trapped out of settled regions
and on the frontiers, forcing Native Americans who
relied on such species to move
to new areas.

The British colonists began
to regulate hunting by enforcing laws that limited the taking
of desirable species of mammals, birds, and fish and also by
offering bounties on predators
that threatened livestock. Laws
seldom prevented, and sometimes were not intended to prevent, market hunters from using guns, nets, and traps to kill
wildlife and then sell their take
in public markets. Initially, settlers had little time to contem-
plate hunting as a sport, even if
they enjoyed doing it, because it
was an important supplement
to farming.

In Europe, from which the
settlers came, hunting was often
the exclusive right of royalty
and the aristocracy, who hunted
exclusively for sport. Such a formal class system did not take

hold in the Americas, and class restriction of hunting
never developed. However, property owners often felt
that they had special rights to the wildlife on their
lands and in their waters. Therefore colonies, and
later states, began to legislate the hunting, fishing,
and trapping rights of property owners and others. Although some species were depleted in the East, they
seemed abundant both in the West and in northern
Canada, and therefore most people did not worry
about depletions.

BEGINNINGS OF OPPOSITION

Eventually, naturalists, such as Pehr Kalm in the
1740's and Alexander Wilson in the early nineteenth
century, realized that Americans were too compla-
cent about their wildlife heritage and began writing
against the reckless killing of animals. By the mid-
nineteenth century nature writers were awakening
concerns for wildlife, and after the Civil War, when
sports hunters rode the transcontinental railroad west
to kill buffalo, the carnage was reported in newspapers and magazines. However, commercial hunters
also used the railroads to take buffalo hides and
tongues to eastern markets, and settlers on the plains

*A photograph from the mid-1870's shows a huge pile of American bison skulls, which were to
be ground up to make fertilizer. The American bison had been hunted almost to extinction by
the late nineteenth century.* (Courtesy, Burton Historical Collection, Detroit Public
Library)

believed that farms and ranches were incompatible with freely roaming buffalo and American Indians. Thus, although several bills protecting buffalo were introduced in the U.S. Congress during the 1870's, none became law. By the 1880's the buffalo was nearly extinct, and in 1889 Congress reacted by establishing the National Zoological Park to help breed endangered species.

By 1900 the ethics of hunting and the preservation of game species were serious public concerns. However, the ways these matters have been understood and evaluated have changed over time. Initially, abundance was thought to depend on the passing and enforcement of hunting restrictions. This approach was later supplemented by the establishment of federal wildlife refuges and national parks, which preserved habitats.

Ethical Debates

The ethical issue that dominated the late nineteenth and early twentieth centuries was whether market hunting of mammals and birds should continue. In 1907 the White House Conference on Conservation emphasized the need for state wildlife or conservation departments to study and manage hunting and fishing in each state. In that new climate, market hunting was outlawed, except for species trapped for their fur.

Nonhunting nature lovers asked whether sport hunting was any more defensible on moral grounds than market hunting, but opposition to sport hunting was muted by the fact that many sport hunters were active in habitat preservation and supported enforcement of wildlife laws. Poachers were enemies of both groups. Aldo Leopold, an avid hunter and pioneer wildlife manager, also helped bridge the gap between these groups by developing a land ethic that became influential after his death in 1948. Wildlife managers also discovered that in environments transformed by civilization, hunting can become one of the tools for wildlife management.

After World War II the numbers of nonhunting nature hobbyists greatly increased, whereas the numbers of hunters declined as a percentage of the total U.S. population. Hunters constitute only about 10 percent of the population, but nature hobbyists are several times as many. Along with these changes came increased interest in environmental ethics.

At the practical level, deer hunters argue that their hunting serves to control deer populations now that wolves and American Indians no longer do so, because deer will degrade their own environment if their numbers are not limited to the environment's carrying capacity. In contrast, at the philosophical level, some environmental ethicists respond that no one imagines that hunters go to all the trouble they do in order to do the deer a favor. Many environmentalists have suggested that deer populations could be better controlled if their natural predators, which were removed by hunters in the first place, were reintroduced into their native habitats. During the 1990's programs to reintroduce predator species were implemented in Yellowstone National Park and the southwestern United States.

Frank N. Egerton

Further Reading

Bronner, Simon J. *Killing Tradition: Inside Hunting and Animal Rights Controversies.* Lexington: University Press of Kentucky, 2008.

Cartmill, Matt. *A View to a Death in the Morning: Hunting and Nature Through History.* Cambridge, Mass.: Harvard University Press, 1993.

Dizard, Jan E. *Mortal Stakes: Hunters and Hunting in Contemporary America.* Amherst: University of Massachusetts Press, 2003.

Light, Andrew. "Methodological Pragmatism, Animal Welfare, and Hunting." In *Animal Pragmatism: Rethinking Human-Nonhuman Relationships*, edited by Erin McKenna and Andrew Light. Bloomington: Indiana University Press, 2004.

Trefethen, James B. *An American Crusade for Wildlife.* 1975. Reprint. Alexandria, Va.: Boone and Crockett Club, 1985.

SEE ALSO: American bison; Animal rights movement; Leopold, Aldo; Poaching; Predator management; Wolves.

Hybrid vehicles

CATEGORY: Energy and energy use

DEFINITION: Motorized vehicles that combine electric motors and internal combustion engines so that they can be powered by gasoline, batteries, or a combination of the two

SIGNIFICANCE: Because hybrid vehicles use less gasoline than do vehicles powered by gasoline only, they produce substantially lower amounts of toxic carbon emissions than do conventional vehicles. Thus at the same time they help conserve fossil-fuel resources, they reduce air pollution.

The gasoline-powered automobiles of the twentieth century replaced the steam-powered automobiles of the late nineteenth and early twentieth centuries. As early as 1889, when Karl Benz and Gottlieb Daimler built the first automobiles, steam had already been used extensively to power locomotives. Among the earliest automobiles was the Stanley Steamer,

which was fueled exclusively by steam. In 1900, car manufacturers in the United States sold 1,681 steam vehicles, 1,575 electric cars, and a mere 936 vehicles powered by gasoline. In 1906, twin brothers Francis E. Stanley and Freelan O. Stanley raced one of their steam-powered automobiles to a top speed of 127.7 miles per hour at Daytona Beach, Florida. Stanley Steamers were still being sold as late as 1920.

The development of gasoline-powered internal combustion engines greatly increased the range that automobiles could travel, and increasing numbers of such cars were manufactured as the twentieth century progressed. Gasoline was inexpensive and readily available, and the development of hybrid vehicles was seen as unnecessary and economically infeasible. The fuel shortages of the 1970's, however, awakened the public and automobile companies to the urgent need to produce automobiles that could offer drastically increased fuel economy in such places as the United States, much of Europe, and many Asian countries. Much of the world had become increasingly depen-

Journalists inspect the first commercially produced hybrid automobile, the Toyota Prius, in October, 1997, during a Tokyo press preview of the car. (AP/Wide World Photos)

dent on automobiles for transportation, especially in countries such as India and China, where rapidly emerging middle classes provided a market for automobiles.

Nevertheless, American automobile companies, especially Detroit's Big Three—Chrysler, Ford, and General Motors—were slow to move toward the production of compact cars and hybrids because the profit margin on small cars is considerably less than that on the full-size sedans and trucks that American companies had promoted strenuously. Eventually, however, public demand forced struggling American automobile companies to move quite vigorously into the development, manufacture, and sale of hybrids.

FUEL ECONOMY

The electrical power component of most hybrid vehicles is provided by lithium-ion batteries. In a typical hybrid car, every time the driver depresses the brake, the friction caused by braking automatically begins recharging the batteries the car carries. When the vehicle stops, such as for a traffic light, the gasoline-powered engine automatically shuts off to save fuel; it is then restarted immediately when the accelerator is pressed. The batteries in some hybrids are recharged through connection to the electrical grid (that is, by being plugged into electrical outlets); these plug-in hybrids are very expensive, owing to the cost of the batteries they use. Also, although plug-ins are fuel-efficient, they shift energy consumption to the electrical grid, which often uses electricity produced by the burning of fossil fuels.

Contrary to what many drivers expect, hybrid vehicles deliver better gas mileage in stop-and-start situations than they achieve on the open road at highway speeds. As is true of gasoline-powered vehicles, the mileage that hybrids deliver depends to some extent on the drivers' habits. Mileage is reduced by fast starts, for example. Even drivers who do not achieve optimal gas mileage in modern-day hybrid vehicles, however, can generally depend on getting more than 40 miles per gallon of gasoline on a regular basis.

Although experimental hybrid vehicles were produced throughout much of the twentieth century, the first commercially produced hybrid automobile, the Toyota Prius, was not available to the general public until October, 1997, when it was marketed in Japan. Honda marketed a hybrid, the Insight, in the United States in December, 1999, and seven months later, in July, 2000, Toyota's Prius, produced in limited quanti-

ties, was made available to American buyers. The Prius and the Insight have regularly accomplished fuel economies approaching 50 miles per gallon, and experimental versions of these hybrids have achieved double that fuel economy. Hybrid vehicles are generally more expensive than comparable conventional cars, but for consumers the higher prices are offset by the fuel savings that hybrids provide, along with incentives such as the tax credits the U.S. government allows hybrid buyers to claim under specified conditions.

Since the first hybrid vehicles were introduced, automakers have added many additional models to their hybrid lines, offering sport utility vehicles (SUVs), trucks, minivans, and mid- and full-size cars along with compacts. Some manufacturers that had not previously pursued the hybrid market, including the American companies Ford and General Motors, have developed or are planning to develop hybrid lines of their own. European automobile companies, notably Audi, BMW, Mercedes-Benz, Porsche, Volvo, and Volkswagen, have also shown concern for fuel economy and the use of alternative fuels, but rather than developing hybrids they have concentrated their efforts on developing efficient diesel-powered vehicles.

R. Baird Shuman

FURTHER READING

Chan, C. C., and K. T. Chau. *Modern Electric Vehicle Technology*. New York: Oxford University Press, 2001.

Ehsani, Mehrdad, et al. *Modern Electric, Hybrid Electric, and Fuel Cell Vehicles: Fundamentals, Theory, and Design*. Boca Raton, Fla.: CRC Press, 2005.

Erjavec, Jack, and Jeff Arias. *Hybrid, Electric, and Fuel Cell Vehicles*. Clifton Park, N.J.: Thomson Delmar Learning, 2007.

Halderman, James D., and Tony Martin. *Hybrid and Alternative Fuel Vehicles*. Upper Saddle River, N.J.: Pearson/Prentice Hall, 2009.

Husain, Iqbal. *Electric and Hybrid Vehicles: Design Fundamentals*. Boca Raton, Fla.: CRC Press, 2003.

Nerad, Jack R. *Hybrid and Alternative Fuel Vehicles: Get the Lowdown on Today's Green Machines*. New York: Alpha Books, 2007.

SEE ALSO: Alternative fuels; Alternatively fueled vehicles; Automobile emissions; Carbon dioxide; Cash for Clunkers; Electric vehicles; Environmental engineering; Solar automobile racing; Synthetic fuels.

Hydroelectricity

CATEGORY: Energy and energy use

DEFINITION: Electricity produced through hydropower, which is energy from water in motion, such as flowing or falling water

SIGNIFICANCE: Hydropower is the leading renewable energy source, producing 24 percent of the world's electricity for more than one billion people. Hydropower is a natural, inexpensive, clean, and replenishable energy source, but the building and operation of hydroelectric plants often have adverse environmental impacts on local natural habitats.

As early as 4000 B.C.E., hydropower was used in the waterwheel, a device that used the power of running water striking wooden paddles mounted around a wheel to drive machinery to grind grain. Eventually, waterwheels were used to irrigate crops, drive pumps, provide drinking water, and power the machinery used in textile mills and sawmills. During the nineteenth century, the waterwheel gave way to the water turbine, which was then replaced by the steam engine in mills. The development of the electric generator and an improved hydraulic turbine, combined with a growing need for electricity, led to the first commercial hydroelectric power plant in 1882 in Appleton, Wisconsin. Typical large modern hydroelectric plants have dams with reservoirs for storing water, which flows through turbines that activate generators. Electrical substations or grids transmit the electricity to consumers.

ADVANTAGES AND DISADVANTAGES

Compared with other energy sources, hydropower has numerous advantages. Water is a naturally free, inexpensive, clean, and renewable power source. Because water can be stored in reservoirs and the amount of electricity generated through the turbines can be increased or decreased rapidly in response to fluctuations in demand, hydropower is economical and efficient. Hydroelectric plants have relatively low maintenance and operational costs, and hydropower is more flexible than limited resources such as coal and oil or intermittent renewable energy sources such as wind or solar energy.

Another advantage to the use of hydroelectricity is that it does not require the combustion of fuels that produce air pollution. Unlike oil, coal, and gas, water is a nonpolluting fuel source that does not create toxic by-products. Hydroelectric plants do not release carbon dioxide, a greenhouse gas emitted with the burning of fossil fuels, and they have no problems with waste disposal, in contrast to nuclear power plants.

As a domestic energy source, hydropower decreases American dependence on foreign oil and thus reduces the nation's vulnerability to fluctuating world economic and political conditions. Other significant benefits of hydropower include the contributions of dams and reservoirs to the management of water supplies (for both municipal needs and irrigation) and flood control. In addition, the reservoirs and lakes formed by dams are often used by the public for recreational activities. For example, the Grand Coulee Dam, completed in 1942, is the largest U.S. hydroelectric producer. Located in Washington State, the dam provides irrigation water for more than 243,000 hectares (600,000 acres) of land and controls floods on the Columbia River. The dam also forms Franklin D. Roosevelt Lake. The Lake Roosevelt National Recreation Area was established in 1946 for activities such as swimming, boating, hunting, camping, and fishing.

Hydroelectric power generation does raise some negative environmental concerns, however. Among these is that methane, a potent greenhouse gas, can form in reservoirs and be released into the air. In addition, the building of dams, along with the reservoirs that dams create, changes the environment in the areas in which the dams are located. Often the natural habitats of local fish, animal, and plant species are dis-

Leading World Producers of Hydroelectricity, 2006

RANK	COUNTRY	KILOWATT-HOURS (BILLIONS)
1	China	431.43
2	Canada	351.85
3	Brazil	345.32
4	United States	289.25
5	Russia	173.65
6	Norway	118.21
7	India	112.46
8	Japan	84.90
9	Venezuela	81.29
10	Sweden	61.11

Source: Energy Information Administration, 2008.

turbed, either modified or destroyed, and local human populations may be displaced by the flooding of reservoirs. Further, hydroelectric plant structures can restrict fish migration upstream or downstream. The operation of hydropower plants can also change water quality and the flow rates of rivers and streams.

WORLDWIDE USAGE OF HYDROELECTRICITY

Hydropower accounts for some 70 percent of the renewable energy market and more electricity-generating capacity than any other renewable energy source. China is the leading producer of hydroelectricity, with more than half of the world's hydropower dams and plans to build more. Some nations have maximized the development of their available waterways for generating hydroelectricity; these include Canada, Norway, and Switzerland, which received 70 percent, 95 percent, and 74 percent, respectively, of their electricity from hydropower in 2010. Other major users of hydroelectricity include Brazil and Paraguay, which jointly built the Itaipu Dam on the Paraná River, which forms the border between the two countries. The Itaipu Dam is the world's largest hydroelectric plant in terms of generating capacity. In the United States, 10 percent of the electricity generated nationwide comes from hydropower plants at only three hundred dams (some eighty thousand dams exist in the United States, but most are used primarily for irrigation).

With increasing demand for clean, renewable energy to replace finite fossil fuels, interest has grown in using untapped water resources in Central Africa, China, India, and Latin America for the construction of more hydroelectric plants. Electrical power could improve the economies and the standards of living in many of the world's developing areas. Environmental issues must be addressed, however, such as how to develop hydroelectrical power while protecting the planet's remaining pristine rivers and forests and protecting needed agricultural lands. Emerging technologies may be able to mitigate some of the negative environmental impacts of hydroelectric power by eliminating the need to build more dams. Alternatives for generating hydropower without building dams include technologies that capture the power of waves or tides. Such hydrokinetic hydroelectricity would not require changes to the watercourse. Small systems, known as micro hydroelectric or micro hydel systems, producing 100 kilowatts or less can generate enough power for individual farms or homes. Small-scale hydro projects using small canals instead of dams to move water through turbines would also help preserve ecosystems.

Alice Myers

FURTHER READING

Craddock, David. *Renewable Energy Made Easy: Free Energy From Solar, Wind, Hydropower, and Other Alternative Energy Sources.* Ocala, Fla.: Atlantic, 2008.

Førsund, Finn R. *Hydropower Economics.* New York: Springer, 2007.

Miller, Frederic P., Agnes F. Vandome, and John McBrewster. *Hydroelectricity.* Beau Bassin, Mauritius: Alphascript, 2009.

Nersesian, Roy L. *Energy for the Twenty-first Century: A Comprehensive Guide to Conventional and Alternative Sources.* Armonk, N.Y.: M. E. Sharpe, 2010.

Surhone, Lambert M. et al., ed. *Wave Farm: Wave Power, Wave Hub, Electricity, Generation, Renewable Energy, Hydroelectricity, Ocean Thermal Energy Conversion.* Beau Bassin, Mauritius: Betascript, 2010.

Tiwar, Ramakant. *Hydroelectricity, Environment, and Quality of Life.* Scottsdale, Ariz.: Regal, 2010.

SEE ALSO: Alternative energy sources; Aswan High Dam; Dams and reservoirs; Echo Park Dam opposition; Franklin Dam opposition; Grand Coulee Dam; Hoover Dam; Renewable energy; Tellico Dam; Three Gorges Dam.

Hydrogen economy

CATEGORY: Energy and energy use

DEFINITION: An economy in which energy needs are met by molecular hydrogen produced predominantly from water

SIGNIFICANCE: Conversion to a hydrogen economy would drastically reduce pollution because using molecular hydrogen as an energy source generates only water, in contrast to the greenhouse gases produced by the burning of fossil fuels. In addition, the hydrogen gas used for fuel can be made from renewable resources.

The term "hydrogen economy" first appeared during the energy crisis of the 1970's. The use of hydrogen as a fuel source was proposed as a way to avoid energy crises resulting from the use of nonrenewable fuels. Interest in a hydrogen economy was resurrected in the 1990's when increasing numbers of people started to understand that the burning of fossil fuels generates carbon dioxide (CO_2), a greenhouse gas that has been linked to global warming; the use of molecular hydrogen (H_2), or hydrogen gas, as a fuel does not generate CO_2. It is interesting to note that nineteenth century science-fiction author Jules Verne imagined the use of hydrogen as a fuel in 1874, in his novel *Le Secret de l'île* (*The Mysterious Island*, 1875).

Hydrogen as a Fuel

H_2 is an ideal fuel for transportation, because the energy content of hydrogen is three times greater than that of gasoline and four times higher than that of ethanol. By the early years of the twenty-first century, growing numbers of automobile manufacturers around the world had begun making prototypes of hydrogen-powered vehicles.

H_2 can be used as a fuel in vehicles with internal combustion engines, but a more environmentally friendly way to use hydrogen power in motor vehicles is to replace their internal combustion engines with fuel cells that generate no greenhouse gases. Hydrogen is used in such fuel cells to produce electricity that powers the vehicle. Fuel cells are like batteries—that is, they generate electricity through a chemical reaction, in this case, between H_2 and oxygen (O_2). The resulting emissions consist of just water and heat with no CO_2 or other greenhouse gases. In addition, a fuel cell is two and one-half to three times more efficient than an internal combustion engine in the conversion of H_2 energy.

One problem with creating a hydrogen economy is that H_2 is not abundant on the earth. Although many microorganisms produce H_2 during fermentation, it is used almost immediately by other microbes because it is an excellent source of energy. If H_2 is to be used as a primary fuel source, it must be generated from other energy sources. Hydrogen as a chemical element (H) is the most plentiful element in the universe, and it is a part of the most abundant chemical compound on the earth—water. Therefore, to make H_2 widely available for fuel, cost-effective and environmentally friendly ways must be found to generate H_2 from water or other chemical compounds. H_2 has been obtained mainly from natural gas (methane and propane) through steam reforming. Although this approach is practically attractive, it is clearly not sustainable.

Molecular hydrogen can be also produced by electrolysis. In this case, electrical energy is used to split water into H_2 and O_2. However, the process is not efficient, and it requires significant expenditure of energy and purified water. One promising sustainable method of H_2 production is a biological approach. A great number of microorganisms produce H_2 from inorganic materials (for example, water) or from organic materials (for example, sugar) in reactions catalyzed by enzymes—hydrogenase, nitrogenase, or both. Hydrogen produced by microorganisms is called biohydrogen. For industrial applications, the most attractive method of H_2 production is one using photosynthetic microbes. These microorganisms, such as microscopic algae, cyanobacteria, and photosynthetic bacteria, use sunlight as an energy source and water to generate hydrogen. Hydrogen production based on photosynthetic microbes holds the promise of generating a renewable hydrogen fuel, as large amounts of solar light and water are available.

Obstacles

Several technological and economic problems have so far hindered progress toward a hydrogen economy. These problems include difficulties in storing and distributing H_2, as well as in convincing the general public of its safety. Hydrogen has gained an unwarranted reputation as a highly dangerous substance. Like other fuels, H_2 may produce explosions, but it has been used for years in industry and has earned an excellent safety record when handled properly.

Hydrogen has much lower energy density by volume than other fuels, and as a gas it requires three thousand times more space for storage than gasoline. Hydrogen storage, especially in motor vehicles, represents a challenge for scientists and engineers. For storage, H_2 is generally pressurized in cylinders or liquefied in cryostatic containers at −253 degrees Celsius (−423 degrees Fahrenheit). Both processes require a significant expenditure of energy and generate large quantities of waste CO_2. In most hydrogen-powered vehicles, H_2 is stored as compressed gas.

Another problem hindering the growth of a hydrogen economy has been the scarcity of refueling stations for hydrogen-powered cars. Gasoline stations cannot be converted into hydrogen stations, because H_2 stations require different pump technologies. Considerable monetary investment will be required to build and operate sufficient H$_2$ fueling stations to increase the attractiveness of owning hydrogen-powered vehicles.

Sergei A. Markov

FURTHER READING

Ball, Michael, ed. *The Hydrogen Economy. Opportunities and Challenges.* New York: Cambridge University Press, 2009.

Cammack, Richard, Michel Frey, and Robert L. Robson. *Hydrogen as Fuel: Learning from Nature.* London: Taylor & Francis, 2001.

Ogden, Joan. "High Hopes for Hydrogen." *Scientific American,* September, 2006, 94-99.

Rifkin, Jeremy. *The Hydrogen Economy: The Creation of the Worldwide Energy Web and the Redistribution of Power on Earth.* New York: J. P. Tarcher/Putnam, 2002.

Service, Robert F. "The Hydrogen Backlash." *Science* 305 (August 13, 2004): 958-961.

SEE ALSO: Alternative fuels; Alternatively fueled vehicles; Global warming; Renewable energy.

I

Incineration of waste products

CATEGORY: Waste and waste management

DEFINITION: Burning of waste materials under controlled conditions

SIGNIFICANCE: Incineration has several benefits as a method of waste disposal: It reduces the volume of waste by about 95 percent while producing useful amounts of heat, and it can be used to sterilize medical waste and to neutralize dangerous chemicals. Waste incineration can also produce environmentally harmful by-products if it is not conducted carefully, and for this reason the practice has not achieved widespread acceptance in the United States.

The incinerators used in burning waste products vary in type depending on the kinds and amounts of wastes to be processed. Solid household waste usually generates a lot of ash; if large amounts of such waste are incinerated, ash must be continuously extracted. This may be done with a moving grate incinerator, in which the grate is a conveyor belt. Waste is dumped onto the grate's front end, and ash and clinkers (unburned solids) are removed at the back end. Air is forced up through the grate to cool the grate and to aid combustion. If necessary, the grate can also be water-cooled. Air is also injected above the grate to ensure complete combustion of the gases. European law concerning waste incineration requires that the gases reach at least 850 degrees Celsius (1,560 degrees Fahrenheit) for at least 2 seconds to guarantee the breakdown of toxic organic material. If the gases are not hot enough, an oil burner is used, so wastes with relatively low fuel value can be treated in a moving grate incinerator.

If heat from the incinerator is to be used, the combustion gases pass through a heat exchanger. The heated working fluid from the exchanger may be used to produce steam that powers a turbine and produces electricity, or the hot fluid may carry energy elsewhere. Among the nations that have found ways to use the heat created by incinerators is Denmark; in 2005 waste incinerators produced 4.8 percent of the electricity and 13.7 percent of the space heating used in that country. A typical moving grate incinerator can handle 38.5 tons of municipal waste per hour, and 1.1 tons of such waste may produce 0.67 megawatt-hours of electricity and 2 megawatt-hours of space heating.

Pollutants not destroyed by heat or burning during incineration may be confined to the ash, which must be disposed of properly. Pollutants in the flue gas must be neutralized. Ash mixed with flue gas, called fly ash, must be captured. The cheapest capture method works best on larger particles: Flue gas is made to swirl in a cyclone chamber so that centrifugal force drives dust particles toward the outer chamber wall, from which they then drop into a hopper. In electrostatic precipitation, dust particles are given a negative charge and then the flue gas is passed between large, highly positively charged plates. The charged dust particles migrate to the plates and stick to them. Periodically shaking the plates allows the dust to fall down into a hopper for later removal. An excellent precipitator can remove 99.9 percent of fly ash.

Another capture method involves forcing flue gas through filter bags in a bag room. A final method involves spraying water droplets into the gas and letting the drops gather the dust as they fall. If there are pollutants in the flue gas, chemicals can be added to the water spray, or they may be blown in as a dry powder. Acids are neutralized with sodium bicarbonate, and bases and chlorides are treated with sodium hydroxide.

OTHER TYPES OF INCINERATORS

Medical wastes must be incinerated at more than 1,000 degrees Celsius (1,832 degrees Fahrenheit) to ensure the destruction of all pathogens. Such wastes are burned in a rotary kiln incinerator, which is completely lined with firebricks and has no moving parts in the heated region, so it can withstand such temperatures. The incinerator consists of a rotating horizontal cylinder with one end higher than the other. Waste is put in at the high end, and the burning residue automatically migrates to the lower end for ash removal.

When the waste particles and their composition are relatively homogeneous, as they are in liquid municipal sludge waste, a fluidized bed incinerator is used. The bed is a layer of sand about 1 meter (3.3

A rotary furnace at a waste incineration facility in Oftringen, Switzerland. (Alessandro Della Bella/KEYSTONE/LANDOV)

feet) thick. Air blown from beneath the sand bed lifts the sand particles and keeps them suspended. Fuel is blown in and ignited at start-up, then waste is sprayed onto the fluidized bed. Fluid waste droplets are quickly reduced to particles that are burned. They are given sufficient oxygen and uniformly heated from all sides so that burning is complete. Flue gases pass through a heat exchanger, preheating the air blown from under the sand bed. The gases are treated as necessary before they are vented to the atmosphere.

In a plasma arc incinerator, a plasma arc (which looks like a very big spark) is created by the passing of a high-voltage, high-amperage current between two electrodes. A gas passing between the electrodes is heated to 13,900 degrees Celsius (25,000 degrees Fahrenheit). The gas temperature drops to 2,800 to 4,400 degrees Celsius (5,100 to 8,000 degrees Fahrenheit) as it circulates into the incinerator's waste-containing chamber. The temperature is hot enough to break chemical compounds into their constituent atoms, so with proper flue gas treatment a plasma arc incinerator can handle difficult wastes such as batteries and asbestos.

Incineration provides one of the best methods for destroying dangerous chemicals such as those used in chemical weapons. Since the United Nations Convention on the Prohibition of the Development, Production, Stockpiling, and Use of Chemical Weapons and on Their Destruction went into effect in 1997, the United States has been using incineration to destroy its stockpile of some 33,000 tons of nerve gas and mustard gas. The poison chemicals are broken down by extremely high temperatures; they are heated in an incinerator to 1,500 degrees Celsius (2,700 degrees Fahrenheit), and their former containers are heated to 900 degrees Celsius (1,650 degrees Fahrenheit) for 2.5 hours. By 2010 the United States had destroyed 75 percent of its chemical weapons stockpile.

Charles W. Rogers

FURTHER READING

Pichtel, John. "Incineration of MSW." In *Waste Management Practices: Municipal, Hazardous, and Industrial.* Boca Raton, Fla.: CRC Press, 2005.

Royte, Elizabeth. *Garbage Land: On the Secret Trail of Trash.* New York: Little, Brown, 2005.

Tammemagi, Hans. "Incineration: The Burning Issue." In *The Waste Crisis: Landfills, Incinerators, and the Search for a Sustainable Future.* New York: Oxford University Press, 1999.

SEE ALSO: Hazardous waste; Landfills; Plastics; Refuse-derived fuel; Sludge treatment and disposal; Solid waste management policy.

Inconvenient Truth, An

CATEGORIES: Activism and advocacy; weather and climate
IDENTIFICATION: Documentary film on climate change
DATE: Released in 2006
SIGNIFICANCE: Winner of the Academy Award for best feature-length documentary, *An Inconvenient Truth* helped to increase public awareness of the issue of global warming.

An *Inconvenient Truth* is an influential documentary film directed by Davis Guggenheim; it centers on the environmental advocacy work of former U.S. vice president Al Gore, particularly his efforts to inform the public about the dangers posed by climate change. The film documents a lecture and slide show on global climate change presented by Gore in various locations around the world in the aftermath of his failed 2000 campaign for the U.S. presidency. The slide show includes a well-constructed combination of graphs and charts that present the leading edge of climate science in terms easily understood by nonexperts. In his presentation, Gore also shows maps projecting the extent of possible climate change and photos documenting receding glaciers, melting polar ice, softening tundra, and drying lakes.

An effective narrative strategy deployed in the film is the interspersion of autobiographical vignettes that contextualize Gore's motivations for taking on his environmental agenda. The vignettes help to personalize the sometimes cold scientific data, so that when Gore concludes with a call for individual, community, and governmental action on climate change, he is able to make a more personal human appeal than would have been possible if the film had documented the lecture and slide show alone. The topics of the vignettes range from idyllic childhood remembrances

of Gore's emotional connection to his family's farm in Tennessee to the near-fatal traffic accident of his youngest son in 1989 and his experience of losing the presidential election in 2000. Critics of Gore and his message regarding climate change have objected to the film's vignettes, especially the one discussing the lost presidential election, as self-aggrandizement; they have also asserted that some of the graphic extrapolations in the slide show and film are exaggerated for propaganda purposes, although the climate science Gore shares in the film has been substantiated by the peer review process.

Michael Mooradian Lupro

SEE ALSO: Climate change and oceans; Climate change skeptics; Droughts; Global warming; Gore, Al; Greenhouse effect; Greenhouse gases; Intergovernmental Panel on Climate Change; Kyoto Protocol; Sea-level changes.

Indicator species

CATEGORIES: Animals and endangered species; forests and plants
DEFINITION: Animal and plant species whose presence, relative abundance, or conditions are diagnostic for some factor in the environment
SIGNIFICANCE: Indicator species are useful for monitoring the impacts of human activities on the environment, particularly in assessing cumulative effects for which more direct measures are not available. Sometimes the investigation of population fluctuations in species not known to be indicators leads to recognition of previously unrecognized environmental problems. Indicator species are also used to monitor the progress of environmental remediation efforts.

Certain species of plants and animals exhibit strong responses to particular environmental factors, which are not necessarily human-made or deleterious. Observing these species in the field provides a convenient method for initial detection of factors of interest. Usually direct measurements are necessary if the information is to be used for determining environmental policy. Rather than using single species, many environmental surveys employ groups of species, defined either by taxonomic categories or by form and function, that respond similarly to a given environmental stressor.

CHARACTERISTICS

The most useful indicator species in environmental monitoring are those that are cosmopolitan (that is, occur in a variety of habitat types over a wide area), are fairly common, and are easily recognized. Rare and endangered species, and those that are narrowly endemic, make poor indicators. One of the strengths of using indicator species is that it allows a person without extensive training or specialized equipment to survey a large number of sites rapidly and identify those that merit more detailed monitoring. This advantage is lost if a species is rare enough to be absent from sites where no pollution or other degradation is present, or if the species is difficult to recognize in the field.

The more specific the response, the better the indicator. A combination of pollution, physical disturbance, and climate change may be causing a general decline of plants and animals. Under those conditions, an epiphytic lichen that concentrates pollutants from the air would be a good indicator of atmospheric pollution, while an introduced weedy species of herb might be a better indicator of disturbance, and a common native insect would be a better indicator of the overall effect of environmental degradation on food webs. If the mechanism of a pollutant's action is known, scientists may look for specific metabolic changes in a variety of species.

If a dominant or keystone species also has sensitivities making it a useful indicator species, its value in survey work is strengthened. A dominant species (in terrestrial ecosystems, a plant) is the one with the largest biomass. A keystone species is one whose removal would profoundly affect other members of the community—for example, a predator that keeps the most common herbivore in check. Changes in the health or relative abundance of a dominant or keystone species have disproportionate effects on the functioning of an ecosystem as a whole.

EXAMPLES IN ENVIRONMENTAL MONITORING

One of the earliest biological responses to environmental degradation to be recognized was that of epiphytic lichens to industrial air pollution. Lichens, a symbiotic association of a fungus, an alga, and in some cases a cyanobacterium, absorb water and nutrients directly from the air or rainwater. Acid rain, high levels of nitrogen, and heavy metals all affect lichen growth in ways that are quite species-specific. In forested areas such as Central Europe and the northwestern United States, where unpolluted mature forests support a diverse lichen flora, total lichen cover, relative abundance of certain species, and the chemical makeup of lichen thalli all provide a cumulative picture of air quality over a number of years. The cumulative effect is helpful to investigators because continual monitoring of air quality can be prohibitively expensive and pollution is often episodic in nature.

Lichenologists recognized in the nineteenth century that members of the lichen family Stictaceae, which contain cyanobacteria, were very sensitive to air pollution. Only in the early twenty-first century did forest management biologists discover that these lichens are important sources of nitrogen in coniferous forests, and that their conservation is a matter of concern in its own right.

Planktonic organisms, both plants and animals, are useful for monitoring pollution and temperature changes in the oceans. Some species concentrate particular pollutants. Relative and ab-

A California clapper rail in the Baylands Nature Preserve in Palo Alto, California. The clapper rail often serves as an indicator species in wetlands restoration projects in California. (©Rob Pavey/iStockphoto.com)

solute abundance can be determined from dragnet samples. If a pollutant interferes with a particular metabolic function, analysis of enzyme levels and metabolic by-products in a mass sample of many different species can provide a direct measure of that pollutant's impact on the biosphere, including animals much farther up the food chain. Such bioassays detect levels of toxic compounds high enough to be of biological concern but too low to detect from direct analysis of seawater samples.

In many areas of the world attempts are under way to remediate environmental damage, restoring, as much as possible, original natural environments. The presence of indicator species is one measure of whether such efforts, which are never complete, are considered successful. In wetlands restoration projects in California, biologists have used the presence of thriving breeding populations of clapper rail as an indication that a healthy wetland ecosystem has been reestablished. This native bird is fairly common and tolerates moderate disturbance but had disappeared from large areas because of draining and severe pollution.

Mapping extensions and contractions in the ranges of individual species of plants and of vegetation types has been useful in reconstructing past climate fluctuations and in rounding out the picture of progressive global warming since the mid-twentieth century. In the early twenty-first century, broad-leaved evergreens, which are characteristic of regions with mild winters and overall drier climate, began extending their ranges both in Europe and in western North America. Both the alpine and Arctic tree lines are slowly advancing. These effects are reminders that change is not necessarily negative: People living near the Arctic tree line welcome the milder winters and more vigorous forest growth.

Martha A. Sherwood

FURTHER READING

Conti, M. E., ed. *Biological Monitoring: Theory and Applications—Bioindicators and Biomarkers for Environmental Quality and Human Exposure Assessment.* Billarica, Mass.: WIT Press, 2008.

Dunne, Niall. "Global Warming. Tracking the Effects of Climate Change on Plants." *Plant and Garden News* 18, no. 3 (2003): 1-4.

Jovan, Sarah. *Lichen Bioindication of Biodiversity, Air Quality, and Climate: Baseline Results from Monitoring in Washington, Oregon, and California.* Portland, Oreg.: Pacific Northwest Forest Experiment Station, 2008.

Spellerberg, Ian F. *Monitoring Ecological Change.* 2d ed. New York: Cambridge University Press, 2005.

SEE ALSO: Air pollution; Bioassays; Bioremediation; Climatology; Conservation; Forest management; Ocean pollution.

Indigenous peoples and nature preservation

CATEGORIES: Philosophy and ethics; preservation and wilderness issues

DEFINITION: Involvement of the native inhabitants of a region in the protection of the region's natural resources

SIGNIFICANCE: The role of indigenous peoples in nature preservation is a growing side of environmental awareness. While some people are practicing preservation by learning from indigenous peoples' traditional ecological knowledge, indigenous peoples elsewhere are fighting for recognition and political power to preserve their lands and cultures.

Nature preservation encompasses the protection of biodiversity, land, and culture. Nature reserves are areas set aside by governments to protect the species that inhabit those areas from harm or loss. Such reserves generally seek to protect natural areas from human disturbance to reduce biodiversity loss and prevent extinction of species and destruction of habitats. Protecting nature through preservation in this way is often not the best approach for managing natural resources, however; the reality is that ecosystems, and the biodiversity within them, move through natural changes, and trying to prevent these changes can have negative impacts on ecosystem health.

Many human activities threaten the biodiversity of the natural world, including logging, drilling, and industries that cause pollution. Another threat to the natural world is the fading of indigenous cultures that have long traditional relationships with the biodiversity of the areas in which they live. Indigenous peoples have evolved not with the practice of preservation but instead with the practices of adaptation and healthy use of natural resources. An important part of preserving biodiversity is to preserve the lands and cultures of the indigenous peoples who best recognize how to maintain healthy relationships with the environment.

TRADITIONAL ECOLOGICAL KNOWLEDGE

The accumulated knowledge of indigenous peoples regarding the environments they inhabit and their relationships with those environments is often referred to as traditional ecological knowledge, or TEK. Indigenous cultures around the world have a great accumulation of TEK through generations of familiarizing themselves with their home areas, and many scientists recognize the value of TEK in nature preservation and conservation. Although the approaches of science and TEK differ greatly, the combination of the two approaches strengthens both, as well as nature preservation as a whole. In many different parts of the world, indigenous cultures have developed through relationships with the land that are based on survival and respect. Indigenous peoples in nature-based cultures use their TEK to preserve their cultures and manage natural resources by protecting the land and the biodiversity with which they are familiar. Because the ecological knowledge of indigenous peoples is developed through experience, it can be preserved only if the cultures of the peoples are also preserved.

Around the world, about sixty million indigenous persons rely on forests for survival and for cultural identity. In the Amazon, indigenous peoples have a strong connection to and extensive knowledge of the rain forests. Three-quarters of the national parks in Latin America include land that is inhabited by indigenous peoples, and globally, up to 85 percent of protected areas are inhabited by indigenous peoples.

Most TEK is passed orally in cultures that have no written history, and scientists who recognize the value of TEK attempt to learn from indigenous peoples how they live in their natural environments, how they relate to the land, and how they effectively conserve the resources in their environments. An example of such efforts is a study that was conducted from 1995 to 1998 by scientists seeking to understand TEK on beluga whales. This research included interviews with members of indigenous communities in Alaska and Russia to learn from their experiences with and knowledge of beluga whales. The researchers found that through indirect, conversation-style interviews with indigenous persons, both individually and in groups, they were able to gather unexpected and insightful infor-mation that would not have been revealed through question-by-question interviews.

CONFLICTS AND CONTROVERSIES

TEK is gradually becoming an accepted supplement to Western science and methods of nature preservation, but progress in accepting the importance of TEK is slow because of the many differences between TEK and mainstream scientific approaches to environmental issues. In contrast with the quantitative and mechanistic character of mainstream science, for example, TEK is qualitative and spiritual.

Because of differences in cultures, values, and approaches to knowledge, conflicts arise between the interests of indigenous peoples and those of Western science and the modern world. The indigenous peoples of the Amazon rain forests, for example, have interests that are in opposition to the modern world's commercial interests in the oil, minerals, and timber found in regions where indigenous communities live. In a world of commercially interested societies, indigenous cultures exist under great threat, as reports have suggested that indigenous peoples' lands hold the majority of the earth's remaining natural resources, including minerals, oil, fresh water, and medicinal plants.

Another conflict between indigenous peoples and Western scientists concerns the commercial use of TEK. Indigenous peoples have profound ecological knowledge based on generations of firsthand experience, much of which includes natural remedies, access to unique plants, and TEK that could prove useful in bioscience. For example, Brazilian indigenous communities traditionally use a poison produced by a certain species of frog as a strong painkiller, and this substance has become part of numerous patent requests in both Europe and the United States. The exploitation by outsiders of the biodiverse resources of the lands where indigenous peoples live (referred to as bioprospecting or biopiracy, depending on the viewpoint of the speaker) is the source of ongoing controversy.

Many countries have made treaties with the indigenous peoples located within their borders, often regarding land rights and federal recognition. The United States officially recognizes approximately 560 Native American tribes, and hundreds of other tribes are in the process of trying to attain federal recognition. The status of U.S. federal tribal recognition establishes a government-to-government relationship in which the tribe is recognized as a sovereign nation. Federal recognition is beneficial for Native American tribes that are active in environmental issues and preservation because the tribes are treated as separate governments with their own lands to preserve or manage. In the United States, federal funding and other benefits aid tribes in establishing their own natural resources departments, which are necessary for nature preservation as well as for preserving the tribes' cultures.

CLIMATE CHANGE

Climate change is a well-recognized environmental issue that affects indigenous peoples across the globe. Some indigenous populations live in societies with modern cultures, such as the indigenous tribes of the United States, but for many of the world's indigenous peoples climate change poses risks to their ancestral lands, to the sustainability of their food sources, and to the traditional ways of their cultures.

The potential effects of climate change on indigenous peoples' agricultural practices are particularly important because many indigenous cultures rely on farming for survival. Changes in climate conditions that result in droughts, floods, hurricanes, or frosts can have serious consequences for traditionally agricultural cultures, displacing populations and changing their relationship to the land.

Rising sea levels pose a climate change-related risk to island and coastal indigenous cultures, which may be directly affected through coastal land inundation that forces people to relocate and increases population density. Residents of the Alaskan village of Newtok, the majority of whom are Yupik Alaska Natives, began being driven from their homes in 2007 by the melting of the permafrost on which the village was built. Many other communities of Alaska Natives are likely to face similar situations in the future if climate change progresses as scientists have projected it will. Such displacements of indigenous populations put cultures as well as communities at risk.

In April, 2009, the Inuit Circumpolar Council hosted the Indigenous Peoples' Global Summit on Climate Change in Anchorage, Alaska, which focused on the effects of climate change on indigenous peoples. Among the topics discussed by the four hundred participants in the summit was the role that indigenous peoples might play in developing the successor agreement to the Kyoto Protocol, which was to be addressed at the United Nations Climate Change Conference held in Copenhagen, Denmark, in Decem-

ber, 2009. The participants also discussed the role that TEK could play in the mitigation of and adaptation to climate change.

Allyson Leigh Hughes

FURTHER READING
Colchester, Marcus. "Self-Determination or Environmental Determinism for Indigenous Peoples in Tropical Forest Conservation." *Conservation Biology* 14, no. 5 (2000): 1365-1367.
Mauro, Francesco, and Preston D. Hardison. "Traditional Knowledge of Indigenous and Local Communities: International Debate and Policy Initiatives." *Ecological Applications* 10, no. 5 (2000): 1263-1269.
Menzies, Charles R., ed. *Traditional Ecological Knowledge and Natural Resource Management.* Lincoln: University of Nebraska Press, 2006.
Posey, Darrell Addison, and Michael J. Balick, eds. *Human Impacts on Amazonia: The Role of Traditional Ecological Knowledge in Conservation and Development.* New York: Columbia University Press, 2006.
Selin, Helaine, ed. *Nature Across Cultures: Views of Nature and the Environment in Non-Western Cultures.* Norwell, Mass.: Kluwer Academic, 2003.
Westra, Laura. *Environmental Justice and the Rights of Indigenous Peoples: International and Domestic Legal Perspectives.* Sterling, Va.: Earthscan, 2008.

SEE ALSO: Biodiversity; Biopiracy and bioprospecting; Conservation biology; Cultural ecology; Deforestation; Environmental ethics; Forest management; Nature preservation policy; Nature reserves; Preservation; Rain forests.

Indoor air pollution

CATEGORIES: Atmosphere and air pollution; human health and the environment

DEFINITION: Contamination of the air contained within buildings

SIGNIFICANCE: Pollutants in indoor air are believed to cause thousands of deaths each year—mainly from lung cancer caused by radon and carbon monoxide poisoning—as well as a considerable amount of illness and discomfort.

Because most people in developed nations spend more time indoors than outdoors, the air quality in homes, offices, stores, and other buildings can have a greater effect on human health than the quality of outdoor air. The term "sick building syndrome" is used when a majority of a building's occupants experience health and comfort problems caused by a variety of indoor pollutants that are difficult to identify.

Outdoor air is one source of indoor air pollution, because the ventilation systems in buildings bring in air from the outside. Fortunately, some pollutants are trapped as they enter buildings; particulate matter, for example, may stick to walls and pass no further than entryways.

CARBON MONOXIDE AND CARBON DIOXIDE

Far more important to indoor air are activities that occur inside buildings. One significant contributor to indoor air pollution is combustion; such pollution may come from appliances or from tobacco-smoking building occupants. An unvented or improperly vented furnace or water heater, or a cracked heat exchanger in a furnace, may allow combustion products such as carbon monoxide to enter the indoor space. Carbon monoxide, which can also enter buildings from garages in which motor vehicle engines are running, causes hundreds of accidental deaths annually in the United States and produces a large amount of often-unrecognized illness. Tobacco smoke is known to be harmful not only to smokers but also to nonsmokers exposed to a significant amount of secondhand smoke; tobacco smoke is also a major source of annoyance and discomfort to most people.

Carbon dioxide, normally regarded as nontoxic, causes nausea and headaches at elevated levels and should not exceed 5,000 parts per million (ppm). Even at 1,000 ppm, however, a buildup of carbon dioxide will make a room seem stuffy. Unvented gas or kerosene space heaters should never be used indoors because they necessarily lead to high carbon dioxide levels and often produce high levels of carbon monoxide, nitrogen oxide, and sulfur oxide (the last of these is a particularly severe problem with kerosene heaters, because kerosene contains sulfur).

BUILDING MATERIALS AND VOLATILE COMPOUNDS

Building materials can also contribute to indoor air pollution. Asbestos, a known cause of lung cancer, may be present in indoor air if insulation or other materials containing asbestos have broken down; these can be especially dangerous when the materials are being removed or otherwise disturbed, such as during building renovation. Formaldehyde, a major component of urea-formaldehyde foam insulation, particle board, and some packaging materials, can produce acute eye, nose, and throat irritation at levels below 1 ppm. Formaldehyde is mainly a concern during the first few years after building construction or renovation, after which it eventually disappears.

A variety of other volatile organic compounds may contribute to poor indoor air quality. These include acetone and other ketones, alcohols, aromatic hydrocarbons (such as benzene and toluene), and halogenated hydrocarbons (such as methylene chloride) found in adhesives, household cleaners, enamels, glues, paints, solvents, and varnishes. Indoor hobbies or renovations involving such compounds should be undertaken only in areas that are well ventilated.

Microorganisms such as bacteria, fungi, molds, and viruses can be dangerous in buildings that are not kept clean. They appear to be most troublesome in buildings with low relative humidity, but some thrive in moist areas. The organism responsible for Legionnaires' disease has been found to be capable of growing in improperly maintained cooling and ventilation systems.

VENTILATION

The ventilation rate of a building plays an important role in its indoor air quality. The concentration of air pollutants can rise at a rapid rate in a poorly ventilated building because the pollutants generated inside the building are not being removed quickly enough. This problem has been aggravated since the energy crisis of the 1970's by the practice of making buildings airtight in order to reduce energy costs associated with the heating or cooling of outdoor air that has entered a building. It is possible for a building to be ventilated well and yet have low energy costs if an air-to-air heat exchanger is used; this device allows the incoming and outgoing airstreams to pass near each other across a thin conducting barrier, so that a large fraction of the heat from the outgoing air in winter is transferred to the incoming air. On the other hand, a building in which there is little indoor generation of air pollutants can be airtight and still have superior air quality; increasing the ventilation rate in such a building may actually decrease indoor air quality by bringing in outdoor pollution.

Radon levels are particularly and subtly dependent on the way a building is ventilated. Radon is normally present in underground air in concentrations sufficient to cause concern if even a small percentage of the air in the building comes from underground; the ventilation of the lower level of a building may actually increase this percentage.

Laurent Hodges

FURTHER READING

Burroughs, H. E., and Shirley Hansen. *Managing Indoor Air Quality.* 3d ed. Lilburn, Ga.: Fairmont Press, 2004.

Godish, Thad. *Indoor Environmental Quality.* Boca Raton, Fla.: CRC Press, 2001.

Hines, Anthony L., et al. *Indoor Air: Quality and Control.* Englewood Cliffs, N.J.: Prentice Hall, 1993.

Moffat, Donald W. *Handbook of Indoor Air Quality Management.* Englewood Cliffs, N.J.: Prentice Hall, 1997.

Vallero, Daniel. "Indoor Air Quality." In *Fundamentals of Air Pollution.* 4th ed. Boston: Elsevier, 2008.

SEE ALSO: Air pollution; Carbon dioxide; Carbon monoxide; Radon; Secondhand smoke; Sick building syndrome.

Industrial Revolution

CATEGORY: Urban environments

THE EVENT: Historical transformation of agricultural economies and societies into industrial economies and societies

DATES: Mid-eighteenth through late nineteenth centuries

SIGNIFICANCE: The Industrial Revolution, beginning around the middle of the eighteenth century in England and followed by similar developments on the European continent and the United States, transformed ecologies, class structures, economies, and politics. Modern industry and new technologies thrived at the expense of agriculture and saw the formation of an urban industrial working class. Among the negative effects for the environment were increased air and water pollution, habitat destruction, and resource depletion.

Eighteenth century England was particularly suited to be the birthplace of the Industrial Revolution. The nation was blessed with important natural resources ranging from coal to copper, salt, stone, and water—used for both energy and transportation by river. Self-sufficiency in food left much land and labor available for manufacturing, and the English countryside possessed many experienced weavers and millers. The enclosure movement, which began in earnest under King Henry VIII during the early sixteenth century, saw peasants expelled from their lands by law or rising rents, providing cheap labor for wool manufacturing in cottage industry. The rural unemployed flocked to the large urban centers looking for work, furnishing cities such as London and Manchester with industrial labor for the next three centuries.

INDUSTRY AND LABOR

The political stability that came with the end of the Napoleonic Wars in 1815 made it safe for the aristocracy, along with bankers and merchants, to invest in urban industry; it was safe also for families and even lone individuals to start up workshops with little capital, while the Bank of England guaranteed long-term loans that stabilized the national currency. English naval power secured markets for industrial exports and agricultural imports ranging from cotton—key to the manufacturing of textiles—to wheat for sale to the growing urban working class. England's and Wales's abundance in coal made energy available to manufacturers at low prices, and new inventions—the steam engine, the spinning machine, and the railroad—increased labor productivity while lowering the cost of capital investment, as did new methods of producing iron and steel. London became the self-proclaimed workshop of the world, simultaneously a great urban manufacturing center and national capital, exercising political and police control over the working class.

The social and physical environment of England changed profoundly during the Industrial Revolution. Workers toiled by the hundreds in cramped, wretched quarters dubbed "Satanic Mills" by the poet William Blake (1757-1827). In the countryside, the extension of the railroad brought the demise of the village pub and chapel. Urban workers—men, women, and children—were subject to harsh factory discipline, including work by piece rate, corporal punishment for disobedience, and imprisonment, exile, and even the death penalty for forming trade unions or striking. Many guildsmen, including tailors, carpenters, and master mechanics, were displaced by both the new technologies and the influx of less skilled workers into the factories. By the end of the nineteenth century Prime Minister Benjamin Disraeli

(1804-1881) would speak of England as two nations separated by wealth, class, and social conditions.

THE UNITED STATES

The United States was destined for its own industrial revolution and set to outstrip England by the end of the nineteenth century. Without a history of feudalism or a homegrown aristocracy, but with a high level of literacy and a large percentage of the labor force possessing entrepreneurial or managerial skills, the former British colonies lacked only land for expansion and an influx of labor to make the leap from agriculture to industry. The Louisiana Purchase (1803), war with Mexico (1846-1848), and the forced expulsion of the indigenous population from the Midwest to the Far West to make way for cattle, sheep, cotton, and wheat, combined with new inventions such as the mechanical harvester, fomented a revolution in agricultural productivity in a few decades.

The Civil War (1861-1865) eliminated slavery and the last vestige of noncapitalist production methods. The war also led to a relaxation of immigration laws, flooding cities such as New York and Chicago with millions of both skilled and unskilled laborers, mostly from Europe. Government investment in infrastructure, including the Erie Canal and other waterways, bridges, and roads, made the transportation of goods less expensive. The Bessemer process for making steel furnished steel plows and railroad tracks. The completion of the transcontinental railroad in 1869 symbolized the triumph of industry and commerce over small-scale agriculture. By 1890, when the national census declared the end of the American frontier, the United States was already the world's foremost agricultural and industrial producer, but at the cost of the lives and lands of the indigenous peoples and a national government hostile to labor unions and socialism.

CONTINENTAL EUROPE

Germany led the way in the second wave of the Industrial Revolution, starting around 1870. German unification in 1871, and victory over France in the Franco-Prussian War a year earlier, with the capture of Alsace-Lorraine and its coal fields, made Germany a leading industrial contender in iron and textiles, focused primarily in the Rhine River region, and also in relatively new industries such as electrical power and chemicals. Electricity, along with the internal com-

bustion engine, allowed Germany after 1880 to assume the lead in automobile manufacturing, while the German chemical industry supplied Europeans in everything from explosives to aspirins. At the end of the nineteenth century Germany could boast of being the most powerful industrial country on the continent. Germany's archrival to the east, czarist Russia, developed its own cotton industry, mostly for domestic consumption, while important iron- and steelworks surged in the Donets River basin. Oil deposits in the Caspian Sea region, and the railroad construction necessary to exploit them, pumped government investment in Russian industry.

As the new century approached, only England could claim to have made a full transition from an agricultural to an industrial society in Europe. Germany was still playing catch-up to the United Kingdom, while France, Italy, and the rest of the Mediterranean and Eastern European countries had developed only small-scale industries and had overwhelmingly agrarian populations. The locus of the Industrial Revolution had shifted to the other side of the Atlantic, and American-made industrial goods began to push the British out of important trade zones from South America to Japan.

Julio César Pino

FURTHER READING

Allen, Robert C. *The British Industrial Revolution in Global Perspective.* New York: Cambridge University Press, 2009.

Hobsbawm, Eric. *Industry and Empire: The Birth of the Industrial Revolution.* Rev. ed. New York: New Press, 1999.

Mokyr, Joel. *The Enlightened Economy: An Economic History of Britain, 1700-1850.* New Haven, Conn.: Yale University Press, 2010.

More, Charles. *Understanding the Industrial Revolution.* London: Routledge, 2000.

Stearns, Peter N. "Global Industry and the Environment." In *The Industrial Revolution in World History.* 3d ed. Boulder, Colo.: Westview Press, 2007.

Vallero, Daniel. "Air Pollution and the Industrial Revolution." In *Fundamentals of Air Pollution.* 4th ed. Boston: Elsevier, 2008.

SEE ALSO: Coal; Ecology in history; Enclosure movement; Europe; Nature preservation policy; North America; Rhine River; Urban sprawl.

Integrated pest management

CATEGORY: Agriculture and food

DEFINITION: The use of coordinated tactics aimed at maintaining populations of insect, animal, or plant pests below damaging levels in an economical and environmentally compatible manner

SIGNIFICANCE: The practice of integrated pest management helps to protect the environment by reducing the need for the development and use of chemical pesticides, which can pose dangers to human health if they enter surface water or groundwater.

In the past, pest management strategies in agriculture focused primarily on eliminating all of a particular pest organism from a given field or area. These strategies depended on the use of chemical pesticides to kill all of the pest organisms. Prior to the twentieth century, farmers used naturally occurring compounds such as kerosene or pyrethrum for this purpose. During the latter half of the twentieth century, synthetic pesticides began to play a prominent role in controlling crop pests. After 1939 the use of pesticides such as dichloro-diphenyl-trichloroethane (DDT) was so successful in terms of controlling pest populations that farmers began to substitute a heavy dependence on pesticides for sound pest management strategies. The more pesticides the farmers used, the more dependent they became. Soon, pests in high-value crops became resistant to one pesticide after another. In addition, outbreaks of secondary pests occurred because either they developed resistance to the pesticides or the pesticides killed their natural enemies. This supplied the impetus for chemical companies to develop new pesticides, to which the pests also eventually developed resistance.

Eventually, certain pests developed resistance to all U.S.-government-approved materials designed to con-

Jerry Knight of Knight Orchards in Burnt Hills, New York, looks at a sticky bug trap hung in an orchard tree. The trap is designed to enable monitoring of insect populations as part of an integrated pest management program funded by the U.S. Department of Agriculture. (AP/ Wide World Photos)

trol them. In addition, many pesticides are toxic to humans, wildlife, and other nontarget organisms and therefore contribute to environmental pollution. Also, it became very expensive for chemical companies to put new pesticides on the market. For these reasons, many producers in the late twentieth century began looking at alternative strategies for managing pests; integrated pest management (IPM) is one such strategy. The driving force behind the development of IPM programs has been concern regarding various negative effects of the use of pesticides: the contamination of groundwater and other nontarget sites, adverse effects on nontarget organisms, and development of pesticide resistance. Pesticides will probably continue to play a vital role in pest management, even in IPM, but it is believed that their role will be greatly diminished over time.

An agricultural ecosystem consists of the crop environment and its surrounding habitat. The interactions among soil, water, weather, plants, and animals in this ecosystem are rarely constant enough to provide the ecological stability seen in nonagricultural ecosystems. Nevertheless, it is possible to utilize IPM to manage most pests in an economically efficient and environmentally friendly manner. IPM programs have been successfully implemented in the cropping of cotton and potatoes, and they are being developed for other crops.

Three Stages of IPM Development

There are generally three stages of development associated with IPM programs, and the speed at which a program progresses through these stages is dependent on the existing knowledge of the agricultural ecosystem and the level of sophistication desired. The first phase is the pesticide management phase. The implementation of this phase requires that the farmer know the relationship between pest densities and the resulting damage to crops so that the pesticide is not applied unnecessarily. In other words, farmers do not have to kill all of the pests all of the time. They must use pesticides only when the economic damage caused by a number of pest organisms present on a given crop exceeds the cost of using a pesticide. This practice alone can reduce the number of chemical applications by as much as one-half.

The implementation of the next stage, the cultural management phase, requires knowledge of the pest's biology and its relationship to the cropping system. Cultural management includes such practices as de-laying planting times, utilizing crop rotation, altering harvest dates, and planting resistance cultivars. It is necessary that farmers understand pest responses to other species as well as abiotic factors, such as temperature and humidity, in the environment. If farmers know the factors that control the population growth of particular pests, they may be able to reduce the impacts of those pests on their crops. For example, if particular pests require short days to complete their development, farmers might be able to harvest their crops before the pests have a chance to develop.

The third stage of IPM, the biological control phase, involves the use of biological organisms rather than chemicals to control pests. This is the most difficult phase to implement because farmers must understand not only the pests' biologies but also the biologies of the pests' natural enemies and the degree of effectiveness with which these agents control the pests. In general, it is not possible for farmers to rely completely on biological control methods. A major requirement in using a biological agent is that sufficient numbers of the control agent must be present at the same time the pest population is at its peak. It is sometimes possible to change planting dates so that the populations of the pests and the biological control agents are synchronized. Also, multiple pest species are often present at the same time within a given crop, and it is extremely difficult to control more than one pest at a time with biological agents.

D. R. Gossett

Further Reading

Altieri, Miguel Angel, and Clara Ines Nicholls. *Biodiversity and Pest Management in Agroecosystems.* Binghamton, N.Y.: Haworth Press, 2004.

Pedigo, Larry P., and Marlin E. Rice. *Entomology and Pest Management.* 6th ed. Upper Saddle River, N.J.: Pearson Prentice Hall, 2009.

Radcliffe, Edward B., William D. Hutchison, and Rafael E. Cancelado, eds. *Integrated Pest Management: Concepts, Tactics, Strategies, and Case Studies.* New York: Cambridge University Press, 2009.

Rechcigl, Jack E., and Nancy A. Rechcigl. *Insect Pest Management: Techniques for Environmental Protection.* Boca Raton, Fla.: CRC Press, 2000.

See also: Agricultural chemicals; Agricultural revolution; Biopesticides; Organic gardening and farming; Pesticides and herbicides; Sustainable agriculture.

Intensive farming

CATEGORY: Agriculture and food

DEFINITION: Large-scale, high-production agriculture that relies heavily on the use of mechanization and chemical pesticides, herbicides, and fertilizers

SIGNIFICANCE: The world's growing population requires ever-increasing amounts of food, and intensive farming provides high levels of crop production on the land available. The greatest challenge to such farming is its long-term sustainability; intensive farming's dependence on fossil-fuel-burning heavy equipment and its reliance on chemicals to eliminate pests and replace nutrients in soil are damaging to the environment.

During the nineteenth century, the economy of the United States was largely dependent on the agricultural production of thousands of small farms. As the country industrialized, and particularly after World War II, small farmers began to consolidate their lands, with the result that there were fewer farm operators and much larger farms. Mechanization, in the form of tractors, seed application implements, and other heavy equipment, as well as technological advancements in chemical pesticides, herbicides, and fertilizers contributed to supporting the infrastructure of these large-scale farming operations and lowered the farmers' labor costs.

As world population continues to rise, demand for food also rises, and food crops require land for production. With the spread of urbanization, productive cropland has become increasingly scarce. Intensive farming, also known as industrial farming, maximizes crop production on the land available by relying heavily on the mechanization of many processes, by treating even marginal soils with chemical fertilizers to increase yield, and by using chemical pesticides and herbicides. Such large-scale commercial agriculture is very capital-intensive in nature, and most intensive farming operations are thus found in the world's economically developed regions.

Industrial farming operations make extensive use of technology to run systems and monitor resources, both to keep labor costs low and to increase productivity. For example, many such vast farms have switched from labor-intensive manually operated irrigation systems to computer-controlled center-pivot irrigation systems. Some farms use geographic information sys-

tem (GIS) technology to facilitate the monitoring and control of their land resources.

BENEFITS OF LARGE-SCALE FARMING

From the 1940's into the 1980's, the so-called Green Revolution in agriculture did much to transform regions that had been dependent on more traditional forms of farming into areas that could produce greater amounts of food for their populations through more intensive agriculture. The impact of the Green Revolution was especially strong in some developing nations, such as India, where the shift to larger farms and greater mechanization was highly successful.

Increased food production is the largest benefit of intensive commercial farming. In the United States, the level of crop production on farms has doubled since the nineteenth century. It has been estimated that in the 1960's one U.S. farmworker produced enough fiber and food for about thirty people; by 2010, the number of people whose food needs could be met by a single U.S. farmworker had increased to more than one hundred.

ENVIRONMENTAL IMPACTS

The practices associated with industrial farming have a number of negative impacts on the environment. Running heavy machinery, producing fertilizers and pesticides, and processing, storing, marketing, selling, and transporting massive amounts of crops require a great deal of energy, most of which comes from natural gas and oil. For example, grain grown in the midwestern United States might be shipped to Michigan for milling and processing into cereal; the finished product, a breakfast cereal, thus might travel thousands of miles before it reaches a consumer's table. This intense dependence on petroleum for agricultural production has long-term negative impacts for the environment, as fossil-fuel supplies are finite and the burning of fossil fuels contributes to carbon dioxide emissions, which have been linked to global warming.

The pesticides used in intensive farming have come under scrutiny for health reasons, with many critics pointing out the danger of pesticide exposure to farmworkers and questioning how residues that may remain on crops could affect consumers of the resulting food product. In addition, pesticides and fertilizers used across vast fields are carried into water sources by irrigation and rainfall runoff. Furthermore, the soil of industrial farms is degraded over time and requires large amounts of fertilizers to rebuild nutrients; yet another impact on the soil is the erosion that can be caused by the use of heavy machinery and intensive irrigation.

The water requirements of huge industrial farms are another factor with impacts on the environment. Many dams have been built specifically to direct water resources to enable crop irrigation, and other sources have also been affected. For example, it has been projected that the deep wells being used as water sources for industrial farms in the American Midwest could eventually run low given the demands of these farms and the continual encroachment of cities into formerly rural areas.

In recognition of the negative impacts that traditional industrial farming practices can have on the environment, some farm operators have instituted conservation methods such as minimal tillage, which reduces erosion, and integrated pest management, which reduces pesticide use. Additionally, some farms have invested in fuel-efficient heavy equipment and in technologies, such as GIS, that can increase efficiency of land management while keeping crop production high.

M. Marian Mustoe

FURTHER READING

Avery, Alex, and Dennis Avery. "High-Yield Conservation: More Food and Environmental Quality Through Intensive Agriculture." In *Agricultural Policy and the Environment*, edited by Roger E. Meiners and Bruce Yandle. Lanham, Md.: Rowman & Littlefield, 2003.

Conkin, Paul K. *A Revolution Down on the Farm: The Transformation of American Agriculture Since 1929.* Lexington: University Press of Kentucky, 2008.

Filson, Glen C., ed. *Intensive Agriculture and Sustainability: A Farming Systems Analysis.* Vancouver: University of British Columbia Press, 2004.

Gliessman, Stephen R., and Martha Rosemeyer, eds. *The Conversion to Sustainable Agriculture: Principles, Processes, and Practices.* Boca Raton, Fla.: CRC Press, 2010.

Laidlaw, Stuart. *Secret Ingredients: The Brave New World of Industrial Farming.* Toronto: McClelland & Stewart, 2004.

SEE ALSO: Agricultural chemicals; Agricultural revolution; Erosion and erosion control; Green Revolution; Integrated pest management; Irrigation; Monoculture; Organic gardening and farming; Runoff, agricultural; Soil conservation; Strip farming; Sustainable agriculture.

Intergenerational justice

CATEGORY: Philosophy and ethics
DEFINITION: The sense of obligation or fair play that one generation of humanity holds toward the generations that follow and precede it
SIGNIFICANCE: The laws and policies that are created in relation to environmental issues are in part influenced by legislators' and policy makers' views about the necessity of pursuing intergenerational justice.

In addition to societal issues such as how the young should treat the elderly or whether one generation should pay for the education of the next, the concept of intergenerational justice encompasses numerous questions regarding the environment. Is it fair or just for the current generation to exploit natural resources to the point where those resources may become exhausted? Is it fair or just for today's society to fill landfills with garbage or the atmosphere with pollutants that tomorrow's citizens will have to clean up? Although the answers to such questions regarding an implicit social contract reaching across generations may seem self-evident, not everyone agrees that the members of each generation have a moral obligation to leave the world a better place than they found it.

Some economists and policy analysts have argued in favor of what might appear to be shortsighted selfishness on the part of the current generation of humanity. They point to past ecologically unsound practices, such as overreliance on fossil fuels, and assert that the technological progress that humans have made can be attributed to their need to respond to problems created by the selfish behavior of past generations. Using this line of reasoning, they claim that it is unnecessary for current generations to preserve natural resources, curb population growth, or reduce industrial pollution. Frequently coupled to this argument is the statement that past generations showed no restraint or consideration for intergenerational justice, and current generations should be equally free to engage in selfish behavior. This latter argument is sometimes referred to as "mutual unconcern" between generations.

The flaw in pursuing a policy of mutual unconcern is that it is based on an assumption of continual technological and scientific progress. While it may be historically true that technological advances allowed past generations to substitute new resources for depleted ones, such as the substitution of coal for fuel when deforestation rendered charcoal scarce in Great Britain during the Industrial Revolution, humans cannot presume that science will always provide technical solutions to environmental problems. The historical record is rife with examples of technical solutions that, in the long run, generated more problems than they solved.

Further, engaging in unsound or damaging practices while arguing that the next generation will find a way to clean up the resulting mess fails on moral grounds. People should recognize that current actions do have significant impacts on the future. The fact that current generations may live to see the consequences of their actions should not release them from the moral obligations implicit in the social contract. The idea of distributive justice within a generation suggests, for example, that it is immoral for the wealthy to exploit the poor; that same concept of distributive justice suggests that rather than pursuing a policy of mutual unconcern, intergenerational justice mandates mutual concern, particularly regarding the environment.

Nancy Farm Männikkö

FURTHER READING

Gosseries, Axel, and Lukas H. Meyer, eds. *Intergenerational Justice*. New York: Oxford University Press, 2009.

Hiskes, Richard P. *The Human Right to a Green Future: Environmental Rights and Intergenerational Justice*. New York: Cambridge University Press, 2009.

SEE ALSO: Environmental ethics; Environmental justice and environmental racism; NIMBYism; Overconsumption; Public opinion and the environment; Public trust doctrine; Resource depletion; Sustainable development.

Intergovernmental Panel on Climate Change

CATEGORIES: Organizations and agencies; weather and climate

IDENTIFICATION: International body of scientific experts that evaluates humanity's impact on global climate

DATE: Established in 1988

SIGNIFICANCE: Through the Intergovernmental Panel on Climate Change the international scientific community and the world's governments come together to work cooperatively to compile climate change information for use in formulating policy.

During the 1980's scientists, governments, policy experts, and others began to consider seriously the possibility that human activities might affect the earth's climate in ways that could have broad impacts on vital natural and human-managed systems. In 1988 the World Meteorological Organization (WMO) and the United Nations Environment Programme (UNEP) established the Intergovernmental Panel on Climate Change (IPCC) to conduct an ongoing assessment of scientific and technical information and policy alternatives related to climate change and its possible impacts. Scientists from many nations were invited to participate, with the objective of evaluating the best available scientific research relevant to national and international policy.

The IPCC's membership includes representatives of the world's national governments, who work in cooperation with the organization's elected leaders and the international scientific community. The organization does not conduct original research; rather, it assesses the massive body of climate science literature contained in peer-reviewed scientific journals. So-called gray literature—sources that are unpublished or that have not undergone peer review, such as industry journals, workshop proceedings, internal organizational publications, and reports by governmental agencies and nongovernmental organizations—may also be assessed after the literature's quality and validity have been carefully considered. The IPCC's assessments are intended to inform and support, but not prescribe, climate change policy.

The IPCC has three working groups that carry out its mandate. Working Group I focuses on the science of the earth's climate system. Working Group II as-

Former U.S. vice president Al Gore, right, and Dr. Rajendra K. Pachauri, chairman of the Intergovernmental Panel on Climate Change, at the Nobel Peace Prize ceremony in Oslo, Norway, in December, 2007. Gore and the IPCC shared the prize for their work in building and disseminating knowledge about anthropogenic climate change. (AP/Wide World Photos)

sesses scientific, environmental, economic, and social impacts of climate change—both negative and positive—on ecological and socioeconomic systems and human health, emphasizing regional and sectoral analyses. Working Group III examines all aspects of climate change mitigation strategies and response alternatives.

In general, the IPCC's analyses have provided strong support for the theories and projected broad impacts of global warming. The IPCC published its First Assessment Report in 1990, which was followed by a supplementary report in 1992. These publications provided useful summaries of current scientific information in advance of the negotiation of the 1992 United Nations Framework Convention on Climate Change. The IPCC's Second Assessment Report was published in 1995. One sentence from the 530-page

Working Group I report has been widely quoted: "The balance of evidence suggests that there is a discernible human influence on global climate." The IPCC released its Third Assessment Report in 2001 and its Fourth Assessment Report in 2007. The Fifth Assessment Report is planned for 2013-2014. The IPCC has also published several supporting documents and technical reports addressing specific scientific and policy issues.

The IPCC's rules of procedure require that each assessment undergo extensive scientific peer review and governmental review processes prior to publication. For the four-volume Fourth Assessment Report, 450 scientists from 130 countries served as lead authors, another 800 were contributing authors, and more than 2,500 experts participated in the peer-review process. An assessment summary for policy makers, which includes an assessment report's key messages, is released only after representatives from participating governments have scrutinized the assessment report thoroughly and approved it by consensus. IPCC authors and reviewers all work on the assessments as unpaid volunteers, as do the panel's chair and elected leaders.

In 2007, the IPCC and Al Gore were jointly awarded the Nobel Peace Prize for their work in building and disseminating knowledge about anthropogenic (human-caused) climate change. Two years later, the organization made headlines for very different reasons. In November, 2009, less than one month before the United Nations Climate Change Conference was to be held in Copenhagen, Denmark, a computer server used by the Climatic Research Unit at the University of East Anglia in the United Kingdom was hacked, and hundreds of e-mail messages, thousands of documents, and source code stolen from the server were posted on various Internet sites. Climate change skeptics cited the contents of some of the correspondence between IPCC contributing researchers as evidence of deliberate manipulation and suppression of data. The scientists who were involved countered that the correspondence excerpts were taken out of context or were deliberately misinterpreted in the worst possible light, and that they revealed no unethical behavior (although some angrily worded e-mails reflected the frustrations of working in such a politically contentious field). The IPCC stood behind its scientific community and defended the Fourth Assessment Report as being comprehensive, unbiased, open, and transparent. The IPCC emphasized that its thorough-going assessment procedures made it impossible for any individual or small group to omit or distort findings or change conclusions. In March and July, 2010, three official inquiries all found that the scientists at the Climatic Research Unit had not manipulated any data.

The IPCC's credibility suffered another blow in January, 2010, after news stories reported that Working Group II's contribution to the 2007 assessment report had referenced poorly substantiated and overly pessimistic estimates regarding the fate of Himalaya's glaciers—information from a gray-literature source. The IPCC expressed regret over the improper application of its well-established standards of evidence and reaffirmed its commitment to those standards.

Phillip A. Greenberg
Updated by Karen N. Kähler

FURTHER READING

Bolin, Bert. *A History of the Science and Politics of Climate Change: The Role of the Intergovernmental Panel on Climate Change.* New York: Cambridge University Press, 2007.

Dessler, Andrew Emory, and Edward Parson. *The Science and Politics of Global Climate Change: A Guide to the Debate.* New York: Cambridge University Press, 2006.

Intergovernmental Panel on Climate Change. *Climate Change 2007: Synthesis Report.* Geneva: Author, 2008.

SEE ALSO: Climate change and oceans; Climate change skeptics; Climate models; Global warming; Greenhouse effect; Kyoto Protocol; United Nations Environment Programme; United Nations Framework Convention on Climate Change.

Internal combustion engines

CATEGORIES: Energy and energy use; atmosphere and air pollution

DEFINITION: Engines containing chambers in which the fuel that powers the engines is burned

SIGNIFICANCE: Internal combustion engines are the most commonly used engines in vehicles and in other applications where portability is important. Most internal combustion engines burn fossil fuels, so they are significant sources of greenhouse gases and other air pollutants.

Technically, any engine powered by combustion inside the engine is an internal combustion engine, including rockets and jet engines. Standard practice, however, is to limit the term to engines that convert the thermodynamic energy of the burned fuel directly into mechanical work inside the engine rather than relying on the expulsion of the combustion by-products to provide work, as in jet engines and rockets.

Most modern internal combustion engines are of a reciprocating piston design that relies on the four-stroke cycle described by German inventor Nikolaus August Otto in 1876. This four-stroke Otto cycle takes in the fuel and air mixture in the first step of the process. The mixture is then compressed and ignited. The release of energy in the combustion process pushes on the piston, providing the work of the en-

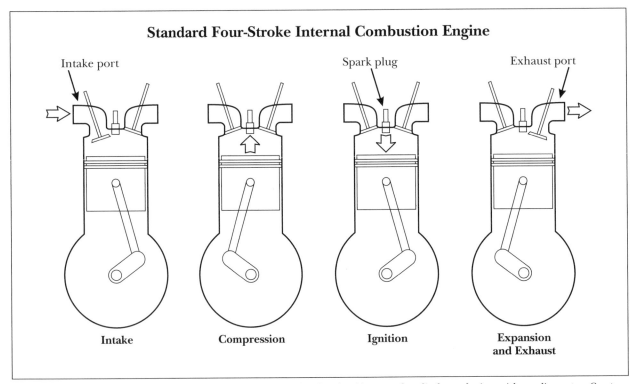

Standard Four-Stroke Internal Combustion Engine

Intake port   Spark plug   Exhaust port

Intake   Compression   Ignition   Expansion and Exhaust

A generalized depiction of the four-stroke internal combustion engine. Intake: Air enters the cylinder and mixes with gasoline vapor. Compression: The cylinder is sealed, and the piston moves upward to compress the air-fuel mixture. Ignition: The spark plug ignites the mixture, creating pressure that drives the piston downward. Expansion (exhaust): The burned gases exit the cylinder.

gine. The piston then pushes the waste gas out of the engine. Valves are carefully timed to admit gas when the piston is moving down and to permit the exhaust of gas when the piston is moving up. A related cycle of steps is performed in the rotary engine devised by German inventor Felix Wankel in 1957.

The Otto cycle produces work only once for every two times that the pistons move up and down. A somewhat more complicated engine, called a two-stroke engine, combines some of the steps of the Otto cycle. Since they produce work with every cycle of the piston, two-stroke engines provide more power for their size than do four-stroke engines. In a two-stroke engine, fuel and air are admitted into the engine as the exhaust gas is being expelled. Since both the inlet and the outlet valves are open at the same time, some mixing is inevitable, so two-stroke engines almost always cause more pollution than four-stroke engines through unburned fuel being released in the exhaust. Two-stroke engines, however, produce more power for their size and are thus often used when high portability is desired, such as on lawn maintenance equipment.

Because fuel must be admitted into the engine, gas or liquid fuels are most suited for internal combustion engines. While almost any combustible fuel may be used, most modern internal combustion engines use either diesel oil or gasoline as fuel. These fuels often contain contaminants that are released into the atmosphere when they are burned. Early engines also relied on lead additives in gasoline to control ignition and to prevent damage to the engines; however, these additives can release toxins into the environment. Modern engines are designed to operate with different additives performing the same function, thus reducing pollution from gasoline. Historically, engines have been designed to produce power with little concern over any air pollution they may create. Many engines thus do not completely burn their fuel, releasing the partially burned fuel as more pollution.

Because for many years gasoline and diesel fuel were comparatively inexpensive and plentiful, most engines were designed with little regard for fuel efficiency. The energy shortages of the 1970's, however, encouraged automobile manufacturers to build more efficient engines. Higher fuel efficiency also reduces tailpipe emissions of air pollutants, including greenhouse gases. The linking of these gases to global warming is another factor in the push for greater fuel efficiency of internal combustion engines.

Raymond D. Benge, Jr.

FURTHER READING

Hoag, Kevin L. *Vehicular Engine Design*. Warrendale, Pa.: SAE International, 2006.

Knight, Ben. "Better Mileage Now." *Scientific American*, February, 2010, 50-55.

Parker, Barry. *The Isaac Newton School of Driving: Physics and Your Car*. Baltimore: The Johns Hopkins University Press, 2003.

SEE ALSO: Air pollution; Alternatively fueled vehicles; Automobile emissions; Carbon dioxide; Corporate average fuel economy standards; Gasoline and gasoline additives.

International Atomic Energy Agency

CATEGORIES: Organizations and agencies; nuclear power and radiation; energy and energy use

IDENTIFICATION: International body that promotes the peaceful use and deters the military use of nuclear energy

DATE: Established on July 29, 1957

SIGNIFICANCE: The International Atomic Energy Agency serves the important functions of providing nuclear safety and security information, monitoring possible sources of radioactive pollution, and conducting research aimed at developing radiation technology.

The International Atomic Energy Agency (IAEA) was established by the International Atomic Energy Agency Statute, an international treaty. Although it reports to the United Nations General Assembly and the United Nations Security Council, the IAEA operates independent of the United Nations. The purposes of the IAEA are to develop, expand, and promote the peaceful use of nuclear technology, to provide education regarding nuclear safety procedures, and to guard against the military use of nuclear energy. The IAEA has no power to force the cooperation of any of its member states, however; each state can choose to follow or ignore the guidelines set by the IAEA.

The IAEA is headquartered in Vienna, Austria, and has regional and liaison offices in Toronto, Tokyo, New York, and Geneva. The agency is staffed by approximately 2,200 professional and support personnel from more than ninety countries. The IAEA also operates scientific laboratories in Vienna and Seibersdorf, Austria; Monaco; and Trieste, Italy. These include the International Seismic Safety Centre, which examines seismic conditions related to nuclear facilities, and the Marine Environment Laboratory, which studies the amounts and the effects of radiation in the world's oceans. Radiation technology developed in IAEA laboratories has shown success in fighting the screwworm fly and some types of grain fungi. The laboratories have also engaged in research into nuclear techniques for eradicating the olive fruit fly and the tsetse fly.

Critics of the IAEA assert that the agency cannot simultaneously push for greater use of nuclear energy and protect against the proliferation of nuclear weapons, because more widespread use of nuclear techniques will necessarily lead to the production of radioactive materials suitable for weapons use. Even given the IAEA's limited power to force compliance, however, most nations have agreed to cooperate with the agency's guidelines. After the accidents that took place at the Three Mile Island nuclear power plant in Pennsylvania in 1979 and at the Chernobyl nuclear plant in the Soviet Union in 1986, the IAEA bolstered its efforts to promote nuclear safety.

When the Comprehensive Nuclear-Test-Ban Treaty was approved by the U.N. General Assembly in 1996, much of the monitoring of residual nuclear materials remaining from the production of weapons was delegated to the IAEA. The agency monitors the storage of radioactive materials from dismantled weapons, measures pollution from nuclear ships that have sunk in the Arctic, and monitors the pollution of nuclear test sites. Another responsibility of the IAEA is to develop protections against the procurement and use of nuclear materials by terrorists.

In 2005 the IAEA and its general director at the time, Mohamed ElBaradei, were awarded the Nobel Peace Prize "for their efforts to prevent nuclear energy from being used for military purposes and to ensure that nuclear energy for peaceful purposes is used in the safest possible way."

C. Alton Hassell

FURTHER READING

Fischer, David. *History of the International Atomic Energy Agency: The First Forty Years.* Vienna: IAEA, 1997.

Olwell, Russell B. *The International Atomic Energy Agency.* New York: Chelsea House, 2009.

Pilat, Joseph F, ed. *Atoms for Peace: A Future After Fifty Years?* Baltimore: The Johns Hopkins University Press, 2007.

SEE ALSO: Atomic Energy Commission; Comprehensive Nuclear-Test-Ban Treaty; Limited Test Ban Treaty; Nuclear and radioactive waste; Nuclear power; Nuclear Regulatory Commission; Nuclear regulatory policy; Nuclear testing; Nuclear weapons production; Radioactive pollution and fallout.

International Biological Program

CATEGORIES: Organizations and agencies; ecology and ecosystems

IDENTIFICATION: International coalition of scientists who studied ecosystems and evaluated the impacts of natural and human-made changes

DATES: September, 1964-June 30, 1974

SIGNIFICANCE: The work of the International Biological Program increased cooperation among scientists around the world and helped lay the groundwork for further international efforts to study and preserve ecosystems.

The planning phase of the International Biological Program (IBP) took place from September, 1964, through June, 1967; the research stage lasted from July 1, 1967, to June 30, 1974. The program had two foci: environmental management and human adaptability. The environmental half was much better funded; in the United States, funding came largely from the National Science Foundation (NSF). W. Frank Blair, an American animal ecologist and administrator for the U.S. IBP team from 1968 to 1974, helped set up multidisciplinary teams that studied five biomes: grasslands (the largest and most successful biome project), eastern deciduous forest, coniferous forest, desert, and tundra. Joseph S. Weiner, a British biological anthropologist, organized the planning for the human adaptability section. American scientists studied five populations that were experiencing dramatic change: South American Indian, circumpolar, high altitude, migrant, and nutritionally stressed peoples.

Modeling complete ecosystems was a primary goal of the biome studies, but modeling subsystems was more practical. Many of the researchers gained new skills over the course of the program. The biome studies also provided more balanced coverage than did nonbiome research in the four areas traditionally examined by ecologists—abiotic, biotic producers (green plants), biotic consumers (animals), and biotic decomposers (bacteria, fungi and some animals)—in that more attention was paid to nutrient cycling and decomposition. Multidisciplinary cooperation resulted in standardized methodologies and techniques, and the human studies provided baseline information on child development, nutrition, and inherited variation that was later used by international agencies for planning. However, interactions were minimal between investigators working on the environmental management side and those on the human adaptability side of the IBP, and so the study of natural systems was essentially separated from the study of human impacts such as agriculture. The human adaptationists were also criticized for neglecting biocultural interactions.

The IBP officially ceased its studies in 1974, but in the United States some funding was transferred to a new NSF program called Ecosystem Studies. In addition, the human adaptability program served as the framework for and was superseded by the Man and the Biosphere Programme (MAB) established during the General Conference of the United Nations Educational, Scientific, and Cultural Organization (UNESCO) in 1970. The MAB differs from the IBP in its intergovernmental structure and because its primary objective is to use interdisciplinary teams to determine how ecosystems function under different levels of human impact. The MAB's activities include training, informational exchange, and support for research in developing areas. An important element is the incorporation of community input.

The U.S. Department of State established the U.S. counterpart to the MAB in 1974. Six program directorates were established: a biosphere reserve program and five research foci on ecosystems (high latitude, human-dominated, marine and coastal, temperate, and tropical). Although the IBP no longer exists, its vision is being carried forward.

Joan C. Stevenson

FURTHER READING

Appel, Toby A. *Shaping Biology: The National Science Foundation and American Biological Research, 1945-1975.* Baltimore: The Johns Hopkins University Press, 2000.

Newman, Michael C., and William H. Clements. "Introduction to Ecosystem Ecology and Ecotoxicology." In *Ecotoxicology: A Comprehensive Treatment.* Boca Raton, Fla.: CRC Press, 2008

SEE ALSO: Biodiversity; Biosphere; Biosphere reserves; Ecology as a concept; Ecosystems; Global Biodiversity Assessment.

International Convention for the Regulation of Whaling

CATEGORIES: Treaties, laws, and court cases; animals and endangered species

THE CONVENTION: International agreement setting limits on the hunting of whales

DATE: Opened for signature on December 2, 1946

SIGNIFICANCE: Intended to balance the conservation of whales with benefits from their exploitation, the International Convention for the Regulation of Whaling failed to prevent the collapse of many of the world's whale populations. Since its adoption of a ban on all commercial whaling in 1986, the convention has been a focal point for controversy.

The International Convention for the Regulation of Whaling (ICRW) was established to consolidate several previous whaling agreements and to encourage rapid development of the whaling industry of countries ravaged by World War II. The ICRW created the International Whaling Commission (IWC), which has power to amend the ICRW's schedule (a document that limits which whales can be taken and under what conditions), although the IWC cannot amend the ICRW itself.

The foremost controversial issues facing the ICRW have been its inability to prevent the collapse of the world's great whale stocks during the 1950's and 1960's and its inability to force compliance by its signatory states—these problems have continued to serve as backdrops to debates over whaling. While Japan, Norway, and some other nations argue that a sustainable harvest of some whale stocks has robust scien-

tific support, others—led by the United Kingdom, the United States, Australia, and New Zealand—contend that historical overexploitation counsels a more precautionary approach and that the killing of whales is unethical. Dissatisfaction with the ICRW and the whaling moratorium implemented by the IWC in 1986 has encouraged the creation of the rival North American Marine Mammal Commission, which Iceland, Norway, Greenland, and the Faroe Islands established in 1992 so that they might regulate their own whaling operations.

The ICRW is also challenged by noncompliance occurring either explicitly outside IWC guidelines or within the guidelines but against the spirit of the regulations. The most explicit noncompliance occurred when the Soviet Union engaged in illegal whaling for banned stocks during the Cold War. More recently meat from whale species that are banned from harvest by the IWC was allegedly found in the markets of Japan and South Korea.

The IWC allows for three types of exemptions to the whaling moratorium, the uses of which are sometimes interpreted to be against the intent of the moratorium, though within IWC guidelines. First, a member state can file a general reservation to the moratorium, an exemption Norway has used to engage in limited commercial whaling. Second, a scientific exemption allows for the taking of whales for research purposes. Only Japan practices scientific whaling, and it has been accused of using this exemption as a cover for commercial operations. Finally, an exemption for aboriginal subsistence whaling allows harvesting by indigenous peoples with strong historical whaling cultures. Japan, Norway, and Iceland have objected to the exemption for aboriginal subsistence whaling, asserting that it is inconsistent with restrictions placed on their own whaling communities.

Because the ICRW lacks effective enforcement mechanisms, compliance must often be negotiated externally. Notably, Japan agreed to comply with the whaling ban after the United States threatened to bar Japan from lucrative American fishing grounds, and Norway has received similar threats regarding its whaling expeditions. Despite these controversies, the ICRW remains the premier global whaling treaty, and the IWC has extended its scope of responsibility to other kinds of threats to whale populations, such as competition with fisheries, chemical and acoustic pollution, ozone depletion, and global warming.

Adam B. Smith

FURTHER READING

Burns, William C. G., and Alexander Gillespie, eds. *The Future of Cetaceans in a Changing World.* Ardsley, N.Y.: Transnational, 2003.

Gillespie, Alexander. *Whaling Diplomacy: Defining Issues in International Environmental Law.* Northampton, Mass.: Edward Elgar, 2005.

Tønnessen, J. N., and A. O. Johnsen. *The History of Modern Whaling.* Berkeley: University of California Press, 1982.

SEE ALSO: Animal rights; Animal rights movement; Convention on International Trade in Endangered Species; Endangered species and species protection policy; Environmental law, international; Extinctions and species loss; Greenpeace; International whaling ban; International Whaling Commission; Marine Mammal Protection Act; Save the Whales Campaign; Sea Shepherd Conservation Society; Watson, Paul; Whaling.

International Institute for Environment and Development

CATEGORIES: Organizations and agencies; activism and advocacy; resources and resource management

IDENTIFICATION: Nonprofit international environmental research and advocacy organization

DATE: Established in 1971

SIGNIFICANCE: The International Institute for Environment and Development works to address global poverty and to promote the sustainable management of the world's natural resources.

The International Institute for Environment and Development (IIED), which is based in London, was launched in 1971 by Barbara Ward, a world-renowned economist and policy adviser. IIED is considered to be one of the first nonprofit organizations to link the environment with development. The stated mission of the IIED is "to build a fairer, more sustainable world, using evidence, action and influence in partnership with others." Since its creation, IIED has played a key role in shaping major environ-ment-related international conferences, including the 1972 United Nations Conference on the Human Environment, the 1992 Earth Summit, and the 2002 World Summit on Sustainable Development.

IIED's research and development work is centered on addressing the local needs of some of the world's most vulnerable people. The organization's aim is to ensure that these people, the disadvantaged, have a say in the environmental policies that affect them. In collaboration with grassroots partner organizations, IIED works to develop programs that are relevant to the needs of disadvantaged populations. The five major program areas on which IIED focuses are as follows: natural resources, including sustainable agriculture, biodiversity, drylands, and forestry; climate change, including mitigation, adaptation, and vulnerability; human settlements and related areas, such as urban poverty, urban environment, and rural-urban interactions; sustainable markets, especially environmental economics, corporate responsibility, market reorganization, and trade; and governance, local and global, in the areas of law and planning.

Over the years, IIED has developed core concepts and methods for addressing the relationship between sustainable development and social issues. Some of the first ecotaxation and green accounting methodologies, now used by many governments and businesses, were developed by economists at IIED. Consultation with IIED has led a number of companies to improve how they address the environmental problems created by their processes and to develop best management practices to protect and conserve natural resources. IIED has also helped develop methodologies that promote public participation in environment-related decisions; these visual techniques are now widely used by the international and local development communities to encourage the airing of stakeholder ideas and views.

IIED has been very involved in shaping the global debate on climate change. The organization's climate change programs focus on the least developed countries and on small island developing states, with particular attention to the economics of climate change adaptations. Many small island countries are vulnerable to the sea-level rises projected as a result of climate change. IIED seeks to provide decision makers with the tools and options they need to take appropriate and effective measures in adapting to climate change.

Lakhdar Boukerrou

See also: Earth Summit; Johannesburg Declaration on Sustainable Development; Sustainable development; United Nations Commission on Sustainable Development; United Nations Environment Programme; World Summit on Sustainable Development.

International Institute for Sustainable Development

Categories: Organizations and agencies; resources and resource management

Identification: Nonprofit organization devoted to promoting human innovation to achieve sustainable development

Date: Founded in March, 1990

Significance: The International Institute for Sustainable Development plays an important role in environmental protection by working with other nongovernmental organizations and with government agencies to conduct research and disseminate information about the benefits of sustainable development.

In 1983, the United Nations convened the World Commission on Environment and Development to address rising international concerns about a degraded global environment and endangered natural resources. The findings of this commission (commonly known as the Brundtland Commission, for its chair, Gro Harlem Brundtland) were compiled and published in 1987 under the title *Our Common Future.* This report introduced the rubric "sustainable development," a long-term view of development that incorporates concerns for the needs of future generations in efforts to meet present-time goals. Sustainable development emphasizes direct assistance in meeting the needs of the world's impoverished communities and clear-eyed assessment of the influences of technology and social organization on efforts to meet the present and future needs of the global community. *Our Common Future,* also known as the Brundtland Report, was the first document of its kind to connect economic activity to the state of the global environment.

In 1988, Canadian prime minister Brian Mulroney announced to the United Nations General Assembly that Canada was planning to build a center for sustainable development in Winnipeg, Manitoba. Its mission would be to work on issues related to sustainable development with the United Nations Environment Programme and other international institutions dedicated to environmental affairs. The funding agreement to create the International Institute for Sustainable Development (IISD) was signed in March, 1990.

IISD was conceived as an institution independent of government and business entities. This independent stature enabled its founders to attract highly skilled directors and to create funding structures that would not create conflicts of interest within the intergovernmental network. The stated mission of IISD is to promote human innovation to achieve sustainable development as defined and measured by environmental health, economic prosperity, and better living standards for all peoples. The institute focuses on six key programs: trade and investment, sustainable natural resources management, measurement and assessment, climate change and energy, global connectivity, and reporting services. IISD is headquartered in Winnipeg and has branch offices in Ottawa, Ontario, New York City, and Geneva, Switzerland. More than two hundred organizations around the world have worked with IISD on its various projects and programs.

IISD was one of the first institutions to recognize the potential of the Internet to disseminate timely information about sustainable development. It created the online *Earth Summit Bulletin* in 1992 to report on the proceedings of the United Nations Conference on Environment and Development (also known as the Earth Summit) held in Rio de Janeiro, Brazil, in June of that year. IISD continues to publish the bulletin online under the title *Earth Negotiations Bulletin*; it covers all U.N. conferences and summits related to the environment. Representatives of IISD attended the 2009 United Nations Climate Change Conference, held in Copenhagen, Denmark, to promote the mitigation of and adaptation to climate change.

Victoria M. Breting-García

Further Reading

Hayward, Lillian. "A Successful Institution in a Struggling System: The Story of the International Institute for Sustainable Development and Sustainable Development in Canada." In *The New Humanitarians: Inspiration, Innovations, and Blueprints for Visionaries,* edited by Chris E. Stout. Westport, Conn.: Praeger, 2009.

International Institute for Sustainable Development. *Sustaining Excellence: The 2008-2009 Annual Report of the International Institute for Sustainable Development.* Winnipeg: Author, 2009.

World Commission on Environment and Development. *Our Common Future.* Washington, D.C.: Author, 1987.

SEE ALSO: Agenda 21; Globalization; International Institute for Environment and Development; Johannesburg Declaration on Sustainable Development; *Limits to Growth, The*; *Our Common Future*; Rio Declaration on Environment and Development; Sustainable agriculture; Sustainable development; United Nations Commission on Sustainable Development; World Summit on Sustainable Development.

International Union for Conservation of Nature

CATEGORIES: Organizations and agencies; preservation and wilderness issues; animals and endangered species; resources and resource management

IDENTIFICATION: Environmental organization devoted to addressing issues of conservation and sustainable development

DATE: Founded on October 5, 1948

SIGNIFICANCE: The International Union for Conservation of Nature brings together governments, nongovernmental organizations, scientists, communities, and commercial interests in its efforts to protect species and ecosystems and promote sustainable use of natural resources. It has been instrumental in drafting and gaining signatories for a number of important international treaties designed to protect the environment.

The International Union for Conservation of Nature (IUCN) is the world's oldest global environmental organization and its largest professional global conservation network. Scientists working within IUCN gather data, identify species and wildlife areas in need of protection, and determine ways to sustain the earth's resources for present and future generations.

This democratic membership union, headquartered in Gland, Switzerland, serves as a neutral forum where governments, scientists, nongovernmental or-

ganizations, commercial interests, and communities come together to find solutions for conservation and development concerns. Its membership spans 140 countries and includes more than two hundred governmental organizations and more than eight hundred nongovernmental organizations. IUCN employs the voluntary expertise of almost eleven thousand scientists and specialists to provide policy advice on conservation issues. A leading authority on sustainability and the environment, the organization develops and supports conservation science and runs thousands of field projects and activities around the world.

The idea for an international conservation organization emerged as early as 1910. Swiss naturalist Paul Sarasin advocated for one and succeeded in establishing it, but it lost momentum during World War I. Sarasin was unable to reestablish a viable organization before his death in 1929, and none of the other persons who had been inspired by his vision were able to make significant headway over the next two decades. The establishment of the United Nations after World War II, however, meant new hope for Sarasin's dream. In 1948, an international conference on nature conservation was held at Fontainebleu, France, and a majority of the delegates in attendance—representatives from twenty-three governments, 126 national institutions, and eight international organizations—signed a formal act that established the International Union for the Protection of Nature (IUPN) on October 5.

Originally headquartered in Brussels, Belgium, IUPN pledged to protect endangered species of wildlife on a worldwide scale. It created an international network of conservationists who used the union and its periodic conferences as means for sharing information. In 1950 IUPN launched the Survival Service (later the Species Survival Commission), a group of scientists who worked on a volunteer basis to document the plights of the world's endangered species. IUPN began to interact directly with government representatives from many nations to promote conservation of their endangered species.

In 1956 IUPN changed its named to the International Union for Conservation of Nature and Natural Resources, or IUCN. Between 1990 and 2008, the name World Conservation Union was also used in conjunction with the IUCN name.

THE ORGANIZATION MATURES

In 1961, IUCN moved its headquarters from Belgium to Switzerland. That same year it welcomed the fledgling World Wildlife Fund (WWF) to share its new headquarters. WWF was created to raise funds for IUCN and other worldwide conservation efforts, to conduct public relations campaigns, and to gain public support for conservation efforts. Ornithologist Peter Scott, then vice president of IUCN, became WWF's first chairperson. IUCN was among the earliest recipients of WWF grants, which funded projects such as creating a footpath in a reserved forest of Madagascar, transporting eight endangered white rhinos from South Africa to what is now Zimbabwe for breeding, and saving the rare Arabian oryx from extinction. The two organizations shared accommodations until the burgeoning WWF moved its offices in 1979. (WWF changed its name to the World Wide Fund for Nature in 1986 but retained use of the acronym by which it has been known since its founding.)

In 1962, the United Nations General Assembly adopted the first World List of National Parks and Equivalent Reserves, a list more than three hundred pages long compiled by IUCN's newly formed Commission on National Parks and Protected Areas (later its World Commission on Protected Areas). The commission subsequently worked to develop detailed criteria for what constitutes a protected area. The U.N. List of Protected Areas, a listing of the world's national parks, scientific reserves, and natural monuments, later became a regular publication of IUCN.

In 1963 IUCN began its Red List system for identifying the world's threatened species and assessing conservation efforts. The Red List has since become recognized as the authoritative global listing of plants and animals facing possible extinction.

IUCN and the United Nations Educational, Scientific, and Cultural Organization (UNESCO) codrafted the text of the Convention Concerning the Protection of the World Cultural and Natural Heritage in 1972. This convention established the World Heritage List, which names valued natural and cultural sites to be protected. IUCN continues to serve as a technical advisory body to UNESCO's World Heritage Committee.

In 1975 the Convention on International Trade in Endangered Species of Wild Fauna and Flora (CITES) went into force. This international agreement, created to ensure that trade in specimens of wild animals and plants does not jeopardize the survival of species, emerged from a 1963 IUCN member resolution. IUCN teamed with WWF in 1976 to set up a monitoring network called the Trade Records Analysis of Flora and Fauna in Commerce (TRAFFIC). TRAFFIC investigators found that the smuggling and poaching of animal and plant species was occurring throughout the world, even in countries that had signed treaties opposing such activities. Increased awareness of the problem encouraged local governments to work with TRAFFIC to stop illegal commerce in exotic animals and plants.

In 1980 IUCN, WWF, and the United Nations Environment Programme issued a joint publication titled *World Conservation Strategy: Living Resource Conservation for Sustainable Development*, which emphasized the need for a holistic approach to conservation and the sustainable use of natural resources. Fifty countries soon created their own national conservation strategies based on the recommendations presented in *World Conservation Strategy*. The three organizations copublished a follow-up strategy document, *Caring for the Earth: A Strategy for Sustainable Living*, in 1993, in which they recommended 132 actions that individuals at all social and political levels can take to protect the environment and improve quality of life.

Other major IUCN contributions include preparation of the World Charter for Nature (1982), the Convention on Biological Diversity (1992), and the draft Covenant on Environment and Development (1995; updated 2000, 2004, and 2010). IUCN also sponsors the World Conservation Congress, which convenes every four years.

Lisa A. Wroble
Updated by Karen N. Kähler

FURTHER READING

Bräutigam, Amie, and Martin Jenkins. *The Red Book: The Extinction Crisis Face to Face*. Gland, Switzerland: IUCN, 2001.

Holdgate, Martin W. *From Care to Action: Making a Sustainable World*. Washington, D.C.: Taylor & Francis, 1996.

International Union for Conservation of Nature, United Nations Environment Programme, and World Wide Fund for Nature. *Caring for the Earth: A Strategy for Sustainable Living*. Gland, Switzerland: Authors, 1993.

_____. *World Conservation Strategy: Living Resource Conservation for Sustainable Development*. Gland, Switzerland: Authors, 1980.

Morphet, Sally. "NGOs and the Environment." In *The Conscience of the World: The Influence of Non-governmental Organizations in the U.N. System*, edited by

Pete Willetts. Washington, D.C.: Brookings Institution Press, 1996.

Van Dyke, Fred. *Conservation Biology: Foundations, Concepts, Applications.* 2d ed. New York: Springer, 2008.

Vié, Jean-Christophe, Craig Hilton-Taylor, and Simon N. Stuart, eds. *Wildlife in a Changing World: An Analysis of the 2008 IUCN Red List of Threatened Species.* Gland, Switzerland: IUCN, 2009.

SEE ALSO: Convention on Biological Diversity; Convention on International Trade in Endangered Species; Endangered species and species protection policy; International Institute for Sustainable Development; Poaching; Sustainable development; United Nations Educational, Scientific, and Cultural Organization; United Nations Environment Programme; World Heritage Convention; World Wide Fund for Nature.

International whaling ban

CATEGORY: Animals and endangered species

IDENTIFICATION: Prohibition on commercial whaling approved by the International Whaling Commission

DATE: Implemented on January 1, 1986

SIGNIFICANCE: The International Whaling Commission's ban on commercial whaling was a response to the dramatic decline in the world's whale populations that resulted from the excesses of the whaling industry. The ban was successful in helping some whale populations rebound, despite the fact that several nations continued their whaling activities.

By the 1920's the whaling industry had decimated the populations of some whale species. Fearing a collapse of the industry, which was essential to the economies of a number of nations, some countries implemented laws banning the catch of certain species or of females with calves. However, international cooperation proved to be difficult, if not impossible. The International Whaling Commission (IWC), which was founded on December 2, 1946, in Washington, D.C., could do little to slow the continuing destruction of the world's whale population.

During the 1960's growing interest in protecting the environment led many people to turn their attention to the problems facing whales, especially the impact of the whaling industry. Public interest led participants at the United Nations Conference on the Human Environment, held in Stockholm, Sweden, in 1972, to call for a ten-year moratorium on commercial whaling. That same year the U.S. Congress passed the Marine Mammal Protection Act, which ended U.S. involvement in commercial whaling and allowed the United States to press for a global ban at IWC meetings.

Despite public pressure and the advocacy of the United States, the IWC resisted efforts to adopt a moratorium. However, by the early 1980's it had become abundantly clear that the whale population had diminished to the point where commercial whaling was fast becoming an industry unable to make profits. The IWC responded to the realities of the situation on July 23, 1982, when it approved a ban on commercial whaling to commence during the 1986 whaling season.

Although environmentalists greeted the passage of the international whaling ban with enthusiasm, several whaling nations, including Japan, Norway, and the Soviet Union, all of which had voted against the ban, sought to create loopholes that would allow them to continue their whaling activities. By 1987 declining profits led the Soviet Union to support the ban, but Japan requested exemptions that allowed whaling near the Japanese coastline. When the United States threatened economic sanctions, the Japanese agreed to end commercial whaling. Because the IWC did permit "research" whaling for scientific purposes, Japan was able to continue whaling on a limited basis. While the ostensible reason for this whaling was scientific study, the Japanese used the whale meat as a food resource.

The IWC planned to examine studies of whale populations for a period following implementation of the ban, to determine whether the ban needed to be continued. Although some species did increase in number during the first years after the ban was in place, in 1993 most IWC nations agreed to continue the prohibition. Japan and Norway vigorously protested, and Norway, which had abandoned commercial whaling for economic reasons in 1967, announced its intention to begin hunting minke whales. The Norwegians claimed that minkes had populations large enough to permit whaling to take place without any negative environmental impact. Environmentalists condemned the move, but the IWC had no mechanism for enforcing the ban; the nations that complied did so voluntarily.

The continuation of the ban threatened the existence of the IWC. Iceland quit the commission in 1992 (it rejoined in 2002), and other members expressed growing dissatisfaction with the prohibition when evidence existed that many whale species had undergone significant increases in population. During the 1997 IWC meeting, members debated implementation of the Revised Management Procedure (RMP), which would end research whaling but would allow Norway and Japan to conduct coastal whaling for domestic consumption with strict quotas. While some observers praised the RMP because it would close the research whaling loophole and keep Japan and Norway in the IWC, environmental groups condemned it as the first step toward a return to full-scale commercial whaling. Work toward implementing the RMP continued into the early years of the twenty-first century.

Another issue that raised objections from environmentalists was that of aboriginal whaling. The 1986 ban permitted indigenous peoples to continue whaling for subsistence and ceremonial purposes. For example, the Inuit were allotted an annual quota of bowhead whales. Opponents argued that because the Inuit were employing modern whaling techniques, their activities no longer met the standards of aboriginal whaling. The issue came to the fore again in 1998, when the Makah in Washington State received permission to hunt five California gray whales, a species that had been removed from the endangered species list in 1993. Although the Makah had not hunted whales since 1928, a nineteenth century treaty with the U.S. government guaranteed their whaling rights. Environmentalists contended that the U.S. government's support for Makah whaling was intended as a signal to Japan that the United States would support efforts to resume commercial whaling.

Although the ban had symbolic significance, its importance as an environmental measure was limited. It was enacted only after commercial whalers experienced declining profits, a result of decades of hunts that had depleted whale populations. As such, it merely acknowledged the reality that whaling had been devastating for whales. The ban's loopholes regarding scientific and aboriginal whaling allowed hunts to continue, and whalers killed an estimated

Japanese schoolchildren on a school tour watch as a Baird's beaked whale is slaughtered in port by workers for a whaling company that hunts Baird's beaked and pilot whales, species not protected by the international whaling ban. (AP/Wide World Photos)

eighteen thousand whales during the first twelve years of the ban. Lacking any enforcement mechanisms, the IWC depended on the voluntary cooperation of its member states, many of which adamantly opposed the continuation of the ban during the 1990's. One reason the ban remained in place during this period was the United States' use of economic sanctions against nations that violated the ban; however, the United States failed to impose sanctions on Norway when it resumed minke whaling in 1993. While the United States officially supported a continuation of the ban, it nonetheless participated in efforts to implement the RMP.

Thomas Clarkin

FURTHER READING

Day, David. *The Whale War.* San Francisco: Sierra Club Books, 1987.

Ellis, Richard. *Men and Whales.* 1991. Reprint. New York: Lyons Press, 1999.

Friedheim, Robert L., ed. *Toward a Sustainable Whaling Regime.* Seattle: University of Washington Press, 2001.

Heazle, Michael. *Scientific Uncertainty and the Politics of Whaling.* Seattle: University of Washington Press, 2006.

Kalland, Arne. *Unveiling the Whale: Discourses on Whales and Whaling.* New York: Berghahn Books, 2009.

Stoett, Peter J. *The International Politics of Whaling.* Vancouver: University of British Columbia Press, 1997.

Tønnessen, J. N., and A. O. Johnsen. *The History of Modern Whaling.* Berkeley: University of California Press, 1982.

SEE ALSO: International Convention for the Regulation of Whaling; International Whaling Commission; Marine Mammal Protection Act; Save the Whales Campaign; Whaling.

International Whaling Commission

CATEGORIES: Organizations and agencies; animals and endangered species

IDENTIFICATION: International body established to regulate whaling

DATE: Established on December 2, 1946

SIGNIFICANCE: Although the International Whaling Commission was created to ensure the conservation of whales through the regulation of whaling, it eventually evolved into a forum serving nations primarily interested in weakening restrictions on whaling.

During the 1930's, concerns regarding declining whale stocks and the near extinction of some species prompted efforts to regulate the whaling industries of the world's nations. These efforts at international cooperation met with little success, but as World War II drew to a close, proponents of regulation pressed for a regulatory framework. In November, 1945, representatives of several whaling nations gathered in Washington, D.C., for the International Whaling Conference. Although some delegates ar-

gued for an agency under the direction of the United Nations, the conference led to the creation of an autonomous organization called the International Whaling Commission (IWC) in 1946. Great Britain, Norway, the Soviet Union, and the United States were among the fifteen charter nations. Japan joined in 1950, and by 1982 membership had climbed to thirty-seven nations; as of 2009, the number was eighty-eight.

The stated mission of the IWC at its establishment was to regulate international whaling to promote the increase of whale stocks, thereby ensuring the continued existence of the lucrative whaling industry. The commission set a whaling season and created sanctuary zones in which whaling was not permitted. Defined in terms of blue whale units (BWUs), the whaling season ended when that year's quota had been met. However, the IWC was virtually powerless to enforce the regulations it implemented, and it allowed member nations a multitude of opportunities to bypass those regulations. As a result, the IWC's activities did little to slow the destruction of whale populations during the first several decades of the commission's existence.

Despite its ineffectiveness, the IWC became an arena of conflict and disagreement among the member states. Because the quota system created a free-for-all environment in which each nation attempted to catch as many whales as possible before the season ended, many nations argued that they did not have an opportunity to catch their fair share of the total number of BWUs available. However, efforts to assign quotas to individual nations proved impossible, as each nation jockeyed for a high percentage of the total at the expense of other nations. Debates grew so acrimonious that the IWC closed its meetings to the public and reporters for a time during the 1960's.

As a regulatory body, the IWC failed to live up to its promise. Quotas were based on estimates of whale populations that were too high. In addition, the commission proved both unwilling and unable to rein in the whaling industries of member nations. In 1982 the IWC voted for an international ban on whaling, which was implemented in 1986. This act reflected the declining economic vitality of the whaling industry more than a concern for the environment or the fate of the whales. As such, the ban came far too late to ensure the survival of many species. After the implementation of the ban, the IWC became more of a forum for nations attempting to circumvent or weaken

International Whaling Commission Member Countries

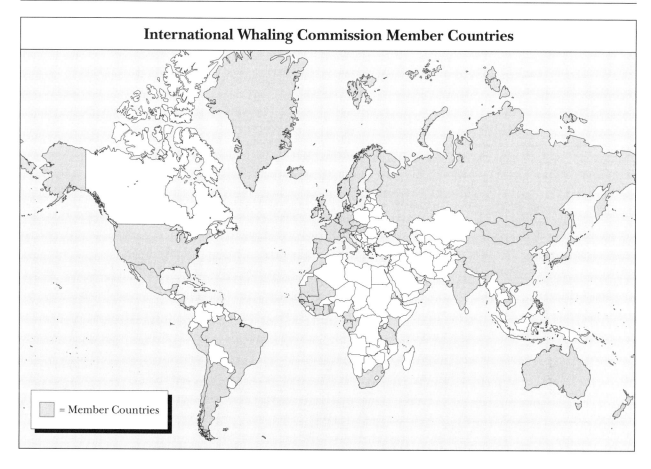

= Member Countries

restrictions on whaling than an organization committed to the preservation of whales.

Thomas Clarkin

FURTHER READING

Andresen, Steinar. "The International Whaling Commission (IWC): More Failure than Success?" In *Environmental Regime Effectiveness: Confronting Theory with Evidence*, by Edward L. Miles et al. Cambridge, Mass.: MIT Press, 2002.

Friedheim, Robert L., ed. *Toward a Sustainable Whaling Regime.* Seattle: University of Washington Press, 2001.

Heazle, Michael. *Scientific Uncertainty and the Politics of Whaling.* Seattle: University of Washington Press, 2006.

Stoett, Peter J. *The International Politics of Whaling.* Vancouver: University of British Columbia Press, 1997.

SEE ALSO: International Convention for the Regulation of Whaling; International whaling ban; Marine Mammal Protection Act; Save the Whales Campaign; Whaling.

Introduced and exotic species

CATEGORIES: Animals and endangered species; ecology and ecosystems

DEFINITION: Species of plants and animals that have been inadvertently displaced or purposefully transplanted from their native geographic ranges to other areas

SIGNIFICANCE: Exotic species often disrupt the functioning of the ecosystems into which they have been displaced or introduced.

An exotic species is one that has been displaced from its native or original geographic range to another area, generally by human activity. An introduced species is one that has been transplanted for some intended purpose. All introduced species may be considered exotic, but not all exotic species are introduced. Natural range extensions, although they may have been assisted indirectly by human activities, do not qualify for either category, and a species transported by a natural event (such as a storm or flood) may be considered exotic for a time while its fate is being determined by natural selection. Transplants may be accidental or purposeful. Accidental transport is usually suspected to occur by ship or other means of commercial conveyance.

Since most ecological communities have evolved containing unique assemblages of species with their own unique systems of interactions and interdependencies, seldom can new species be added to a community without causing one or more significant disruptions in some community pattern (such as competition, the food web, or predator-prey relationships). While some introduced species have been beneficial to their new communities, many have had one or more detrimental effects.

ASIATIC CLAM

The Asiatic clam (*Corbicula fluminea*) is an example of an exotic species that has had damaging impacts. The clam was apparently accidentally deposited from a ship's bilge near the upper West Coast of the United States in the mid-1930's. From there it spread rapidly into virtually all the river systems of the West Coast. By the late 1960's it had spread over all major portions of the lower Mississippi River drainage, and by 1983 the Asiatic clam had infested nearly all rivers in the western and southern states and most of the eastern half of the United States.

The Asiatic clam has exhibited two significant detrimental effects. Its high reproductive capacity, its somewhat mysterious ability to disperse rapidly (even

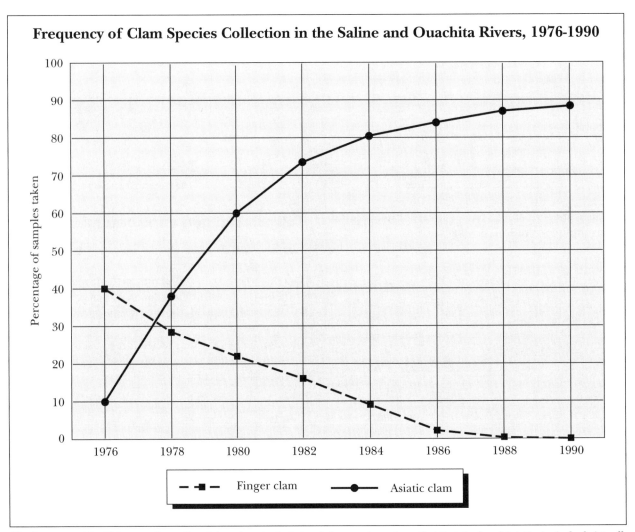

Frequency of Clam Species Collection in the Saline and Ouachita Rivers, 1976-1990

The frequency with which researchers collected these clam species in two Arkansas rivers shows that the introduced Asiatic species had virtually eliminated the native species by 1990.

upstream), and its rapid particle-filtering rate give the Asiatic clam the ability to compete with native species of fingernail clams (*Sphaerium* spp.). In the Saline and Ouachita rivers of south Arkansas between 1976 and 1988, the Asiatic clam virtually eliminated the fingernail clam.

The second effect arises from the clam's affinity for hiding in dark places, particularly water-intake structures. If eggs or small juveniles are drawn into water systems having some minimal flow, they may attach, filter particulates, mature, reproduce, and cause occlusion of the pipes. During the 1970's and 1980's, this characteristic of the clam caused considerable consternation among industrial plant managers and necessitated significant monetary investment in the control and subsequent prevention of Asiatic clam reproduction.

ZEBRA MUSSEL

Another problematic exotic species is the zebra mussel (*Dreissena polymorpha*). Larvae of the mussel were apparently flushed from a ship's ballast in St. Clair Lake (on the border between Michigan and Ontario, Canada) in the mid-1980's. Subsequent studies have revealed one of the most phenomenal dispersal rates of all time. By 1991 the zebra mussel had been found in the St. Lawrence River and the Hudson River, and by 1992 it appeared in the Mississippi River system. By early 1994 it had spread to virtually all tributaries of the Mississippi system navigable by commercial ships or barges.

Whereas the Asiatic clam is a more passive filterer and prefers smaller streams with sandy or gravelly substrate into which it can burrow, the zebra mussel is a more positive, pumping filterer that prefers large rivers and lakes with hard, rocky substrate to which it can attach. The zebra mussel's dispersal pattern coincides with previously unoccupied freshwater niches in North America. Consequently, the zebra mussel disperses more rapidly than and successfully competes with native species of freshwater mussels, which are more passive filterers. Since zebra mussels must attach to hard surfaces, they often attach to each other, producing two or more layers of mussels. Adult densities up to 100,000 per square meter and veliger densities up to 350,000 per cubic meter have been reported.

Several concerns have arisen regarding the zebra mussel. Because of the species' positive filtering mechanism and high densities, zebra mussels may re-move a significant portion of the plankton and particulates from a water column, thus robbing other species of the resource. The zebra mussels' need to settle and attach to hard surfaces has made the mussels that are native to much of the eastern two-thirds of the United States particularly vulnerable to being covered by zebras, because the partially exposed shells of mussels are often the only hard substrate in the lower reaches of most of the rivers of the region. When a native mussel is settled upon by zebra mussels, it is partially smothered, so it pulls itself farther out of the substrate, only to be covered more extensively by the zebras. Finally the zebra mussels may completely cover the larger mussel and smother it. This is a source of grave concern in all aspects of the shelling industry and involves several nations.

The zebra mussels' need to settle on hard surfaces also increases the probability that they will be transported on the hulls of sporting boats from large rivers to recreational lakes and reservoirs. Resource managers in North America are concerned that zebra mussels will significantly alter additional biological communities, reducing their value for sport fishing and possibly other types of water recreation.

EXOTIC FISHES

More than fifty fish species of foreign origin have been introduced into North American waters. Most of the introductions have been done intentionally to provide additional food or sporting opportunities. Among the most notable are the common carp (*Cyprinus carpio*) and the grass carp (*Ctenopharyngodon idella*). The common carp possesses such an omnivorous talent and high reproductive rate that it quickly became well established in all types of aquatic habitats, preferring larger rivers and lakes. In its European homeland it is prized for food, but in North America people are more interested in other, more desirable native species as food sources. On the negative side, the common carp's tendency to disturb sediments and destroy vegetation while feeding causes problems with high turbidity and community disruption. For a time, resource managers feared the carp would overpopulate and outcompete some native species.

Great caution accompanied the grass carp's introduction in the late 1960's. There were still concerns about the common carp's potential, and many feared that the grass carp would adapt and reproduce as well if not better. This concern was grave enough to stimulate the legislatures of several states in the Mississippi

River system to ban the grass carp's import. The Arkansas Game and Fish Commission sponsored research to learn how to control the propagation and dispersal of the species; as a result, sterile grass carp are now produced and widely marketed as vegetation control agents. As it turned out, most of the concerns about overpopulation were unfounded.

Perhaps the best-known example of exotic fish introduction occurred when the Welland Canal was built to bypass Niagara Falls for shipping into Lake Erie and points beyond. The sea lamprey (*Petromyzon marinus*) discovered the canal and invaded Lake Erie, displacing lake trout (*Salvelinus namaycush*) and whitefish (*Coregonus* spp.) populations. Canada and the United States have invested millions of dollars to discover a control for

Fears that the invasive Asian carp could overrun native species in the Great Lakes have led to the construction of electrified barriers designed to keep the fish out. This Asian carp was caught beyond the barriers in 2010. (AP/Wide World Photos)

the lamprey. Among the approaches used have been administration of a selective larval toxin and long-term community adaptation. The coho salmon (*Oncorhynchus kisutch*), another exotic species that is nonnative to the Great Lakes, has replaced the lake trout in commercial and sport fisheries.

Aquaculture has matured as an industry in part because marine fisheries resources have significantly declined. Many different exotic species have been brought from other countries or transplanted from one basin to another within North America in the ongoing effort to provide variety and find more efficient subjects on which to practice aquaculture, to provide broader culinary and sporting opportunities, or to satisfy aquarium enthusiasts. The rainbow trout (*Oncorhynchus mykiss*) has been moved from its native range in western North America to many points east, particularly to dam tailwaters, for sporting purposes. Brook trout (*Salvelinus fontinalis*) have been transplanted from their native streams in the northeastern United States to numerous Rocky Mountain lakes. Brown trout (*Salmo trutta*) were carried from Europe to North America and successfully added to the native salmonid fauna.

Goldfish (*Carassius auratus*), guppies (*Poecilia reticulata*), variable platy (*Xiphophorus variatus*), oscars (*Astronotus ocellatus*), and many other species have been imported to the United States for the aquarium

trade. The primary problem that has resulted from such importation has arisen from the ignorant and random release of exotic species simply because their owners have become tired of keeping them. The bighead carp (*Hypophthalmichthys nobilis*), ruffe (*Gymnocephalus cernuus*), and several cichlids (*Cichlasoma* spp. and *Tilapia* spp.) were introduced for possible food fish.

The importation of exotic species is sometimes done illegally, outside the supervision of any regulating agency. The potential for the disruption of biological communities by exotic species is serious enough that the American Fisheries Society was motivated to establish its Exotic Fishes Section, and many state game and fish agencies (or natural resource departments) have set up special committees or enacted regulations in attempts to control these importations.

OTHER EXOTIC SPECIES

Most ecologists are aware of the detrimental effects of the chestnut blight and Dutch elm disease in the United States. It is still unknown whether the targeted species will eventually adapt to resist these diseases or finally succumb to them. Other problems include the overpopulation of rabbits (*Sylvilagus* sp.), in the absence of adequate predators, following their introduction into Australia. After mongooses (*Herpestes* sp.) were introduced into some of the Caribbean islands to control snakes, they apparently ran out of snakes and started eating nonintended prey items.

An often-overlooked introduced species that has had a profound impact on the history and development of the Americas is the domestic horse (*Equus caballus*). Although originally native to North America, horses migrated to Asia (presumably across the Bering land bridge during the final Pleistocene glaciation), became extinct in America, and were reintroduced by the Spanish during early explorations. During the horse's absence from America, biotic communities would have certainly evolved horse-absent features, so that a reintroduction could well have caused ecological perturbations. On the other hand, the horse's actual role in a natural community as a primarily domesticated species is debatable.

An important part of biogeographic studies includes the determination of the native range of a species, which is then used to help explain dispersal routes and mechanisms and possibly predict future range extensions. Biogeographic studies are also correlated with continental drift patterns and distributions of the fossil record. When transplant records are never made, or are lost or inaccessible, it may be difficult for biogeographers to determine the status of certain species.

John Rickett

FURTHER READING

Enger, Eldon D., and Bradley F. Smith. "Biodiversity Issues." In *Environmental Science: A Study of Interrelationships.* 12th ed. Boston: McGraw-Hill Higher Education, 2010.

Mooney, Harold A., and Richard J. Hobbs, eds. *Invasive Species in a Changing World.* Washington, D.C.: Island Press, 2000.

Pimentel, David, ed. *Biological Invasions: Economic and Environmental Costs of Alien Plant, Animal, and Microbe Species.* Boca Raton, Fla.: CRC Press, 2002.

Rilov, Gil, and Jeffrey A. Crooks, eds. *Biological Invasions in Marine Ecosystems: Ecological, Management, and Geographic Perspectives.* New York: Springer, 2009.

Van Driesche, Jason, and Roy Van Driesche. *Nature Out of Place: Biological Invasions in the Global Age.* Washington, D.C.: Island Press, 2000.

SEE ALSO: Balance of nature; Ecosystems; Food chains; Kudzu; Predator management; Wild horses and burros.

Invasive species. *See* Introduced and exotic species

Iron fertilization

CATEGORY: Weather and climate

DEFINITION: Purposeful introduction of iron into the oceans on a large scale to encourage plankton blooms as a measure to decrease global warming by removing carbon dioxide from the atmosphere

SIGNIFICANCE: Small-scale experiments have tested the technique of iron fertilization, but the results were not conclusive, and conducting experiments on a larger scale is problematic for several reasons. How much of a role iron fertilization may eventually play in efforts to remove large amounts of carbon from the atmosphere remains unknown.

Phytoplankton are microscopic plants (algae) that can be found throughout the upper layers of the world's oceans. These tiny plants require light as well as inorganic material such as phosphate, nitrate, ammonium, carbon dioxide, or carbonate, along with some trace metals such as iron or zinc. Phytoplankton reproduce asexually, with a typical ocean doubling time of a population from a few hours to a few days. Under the right conditions and with access to abundant nutrients, plankton form blooms that can cover hundreds of square kilometers of ocean and are readily visible to satellites orbiting the earth. Such blooms can remove from the atmosphere large amounts of carbon dioxide, a greenhouse gas that has been linked to global warming. Among the natural causes of such large algal blooms are hurricanes, which bring up nutrients from lower ocean depths into the upper reaches of the oceans.

Since the late 1980's scientists have known that depositing particles of iron in warm ocean waters can trigger plankton blooms, and by the early 1990's twelve small-scale experiments involving the introduction of a ton or more of iron dust into a few hundred square kilometers of ocean had shown that massive blooms could be stimulated artificially. These experiments also raised a number of problems and complicated earlier views, however. First, the rate at which carbon was taken up by the phytoplankton was many times lower than had been predicted on the basis of laboratory experiments. Some of this difference was the result of the chaotic mixing that occurs in the oceans as well as the difficulty of placing iron dust in the oceans under just the right conditions to prevent the dust from sinking too quickly into lower depths. Second, the degree to which the carbon remained locked up in the plankton (which upon death would sink to the seafloor, thus relatively permanently sequestering the carbon) was shown to vary considerably. Many plankton would be consumed, and the carbon would begin a journey through the food chains of the ocean; ultimately much of it would reenter the atmosphere as carbon dioxide.

A few large-scale iron fertilization experiments have been undertaken, but no evidence has yet been produced that this method could substantially reduce the amount of carbon dioxide present in the atmosphere. Critics have argued that no large-scale experiment will be able to determine fully the effectiveness of this technique because of the very long time frame within which such a scheme would have to operate, as well as the potent technical difficulties involved in effectively monitoring the impacts of such experiments even in the short term over the vast reaches and depths of the world's oceans. Others have pointed out that the introduction of large amounts of iron dust into the oceans could trigger unintended and unanticipated consequences. Further study of these and other issues may lead to iron fertilization's playing some role in carbon sequestering, but it appears unlikely that the method offers a panacea to the problem of human introduction of greenhouse gases into the atmosphere.

Dennis W. Cheek

FURTHER READING

Goodell, Jeff. *How to Cool the Planet: Geoengineering and the Audacious Quest to Fix Earth's Climate.* Boston: Houghton Mifflin, 2010.

Thompson, Kalee. "Carbon Discredit." *Popular Science,* July, 2008, 54-59, 91.

SEE ALSO: Carbon dioxide; Environmental engineering; Environmental ethics; Geoengineering; Global warming.

Irrigation

CATEGORIES: Water and water pollution; agriculture and food

DEFINITION: Watering of land through human-created means

SIGNIFICANCE: Like many other human modifications to natural ecosystems, the use of water for irrigation achieves some remarkable but temporary advantages that are complicated by long-term environmental problems.

The demands of feeding and clothing the rapidly expanding world population require the production of increasing amounts of food and fiber. One important strategy for achieving the necessary levels of production has been the use of irrigation techniques to supply additional water to arid and semiarid regions where few, if any, crops could otherwise be grown.

Approximately 141.6 million hectares (350 million acres) of land worldwide are irrigated. In the United States more than 10 percent of all crops, encompassing approximately 20.2 million hectares (50 million acres), receive water through irrigation techniques; 80 percent of these are west of the Mississippi River. In certain other countries, including India, Israel, North Korea, and South Korea, more than one-half of food production requires irrigation. From 1950 to 1980, the amount of irrigated cropland doubled worldwide; increases since the 1980's have been more modest.

An often-cited example of irrigation success is that of the Imperial Valley of Southern California. The valley, more than 12,900 square kilometers (5,000 square miles) in size, was originally considered to be a desert wasteland. The low annual rainfall resulted in a typical desert, with cacti, lizards, and other arid-adapted plants and animals. In 1940, however, engineers completed the construction of the All-American Canal, which carries water 130 kilometers (80 miles) from the Colorado River to the valley. The project converted the Imperial Valley into a fertile, highly productive area where farmers grow fruits and vegetables all year.

Successful agriculture in Israel also requires irrigation. As a result of continuing settlement of the area throughout the twentieth century, large amounts of food had to be produced. To fulfill this need, a system of canals and pipelines was built to carry water from the northern portion of the Jordan Valley, where the rainfall is heaviest, to the arid south.

METHODS

All types of irrigation are expensive, requiring advanced technologies and large investments of capital. In many cases, irrigation systems convey water from sources hundreds of miles distant. In the United States, such vast engineering feats are largely financed by taxpayers. Typically, water from a river is diverted into a main canal and from there into lateral canals that supply each farm. From the lateral canals, various systems are used to supply water to the crop plants in the field.

Flood irrigation supplies water to fields at the surface level. Using the sheet method, land is prepared so that water flows in a shallow sheet from the higher part of the field to the lower part. This method is especially suitable for hay and pasture crops. Row crops are better supplied by furrow irrigation, in which water is diverted into furrows that run between the rows. Both types of flood irrigation cause erosion and loss of nutrients. However, erosion can be reduced in the latter type through the contouring of the furrows.

Sprinkler irrigation systems, though costly to install and operate, are often used in areas where fields are steeply sloped. Sprinklers may be supplied by stationary underground pipes, or a center pivot system may be used, in which water is sprinkled by a raised horizontal pipe that moves slowly around a pivot point. Aside from its expense, another disadvantage of sprinkler irrigation is loss of water by evaporation. In drip irrigation, in contrast, water is delivered by perforated pipes at or near the soil surface. Because water is delivered directly to the plants, much less water is lost to evaporation than is the case with sprinkler irrigation.

Much of the water utilized in irrigation never reaches the plants. It is estimated that most practices deliver only about 25 percent of the water to the root systems of crop plants. The remaining water is lost to evaporation, supplies weeds, seeps into the ground, or runs off into nearby waterways.

NEGATIVE IMPACTS

As fresh water evaporates from irrigated fields over time, a residue of salt is left behind. The process, called salinization, results in a gradual decline in pro-

A sprinkler irrigation system in use on a hay field. (©Steve Baxter/iStockphoto.com)

ductivity and can eventually render fields unsuitable for further agricultural use. Correcting saline soils is not a simple process. In principle, large amounts of water can be used to leach salt away from the soil, but in practice the amount of water required is seldom available, and if it is used, it may waterlog the soil. Also, the leached salt usually pollutes groundwater or streams. One way in which farmers address the problem of salinization is by using genetically selected crops adapted to salinized soils.

As the number of hectares of farmland requiring irrigation increases, so does the demand for water. When water is taken from surface streams and rivers, the normal flow is often severely reduced, changing the ecology downstream and reducing its biodiversity. Also, less water becomes available for other farmers downstream, a situation that often leads to disputes over water rights. In other cases water is pumped from deep wells or aquifers. Drilling wells and pumping water from such sources can be expensive and may lead to additional problems, such as the sinking of land over aquifers. Such land subsidence is a major problem in several parts of the southern and western United States. Subsidence in urban areas can cause huge amounts of damage as water and sewer pipes, highways, and buildings are affected. In coastal areas, depletion of aquifers can cause the intrusion of salt water into wells, rendering them unusable. In the United States, the federal government spends millions of dollars each year to repair damage to irrigation facilities.

Like many other human modifications to natural ecosystems, the use of water for irrigation achieves some remarkable but temporary advantages that are complicated by long-term environmental problems. Assessments of the total financial costs and environmental impacts of irrigation are continuously weighed against gains in production.

Thomas E. Hemmerly

FURTHER READING

Albiac, José, and Ariel Dinar, eds. *The Management of Water Quality and Irrigation Technologies*. Sterling, Va.: Earthscan, 2008.

Graves, William, ed. "Water: The Power, Promise, and Turmoil of North America's Fresh Water" (special issue). *National Geographic*, November, 1993.

Meiners, Roger E., and Bruce Yandle, eds. *Agricultural Policy and the Environment*. Lanham, Md.: Rowman & Littlefield, 2003.

Molden, David, ed. *Water for Food, Water for Life: A Comprehensive Assessment of Water Management in Agriculture*. Sterling, Va.: Earthscan, 2007.

Wescoat, James L., Jr., and Gilbert F. White. *Water for Life: Water Management and Environmental Policy*. New York: Cambridge University Press, 2003.

SEE ALSO: Agricultural revolution; Erosion and erosion control; Runoff, agricultural; Soil salinization; Sustainable agriculture; Water use.

Italian dioxin release

CATEGORIES: Disasters; human health and the environment; pollutants and toxins

THE EVENT: Explosion at a chemical factory in Seveso, Italy, that resulted in the release of dioxin

DATE: July 10, 1976

SIGNIFICANCE: Exposure to the dioxin released in the chemical plant explosion killed thousands of small animals and adversely affected the health of some five thousand people, many of whom continued to experience the negative effects for years following the incident.

On July 10, 1976, a chemical-reaction chamber exploded at a factory in Seveso, Italy, about 21 kilometers (13 miles) north of Milan. The factory, owned by Industrie Chimiche Meda Societa Anonima (ICMESA), produced the phenoxy-type herbicide trichlorophenol 2,4,5-T. Developed in 1945, this herbicide was used to control broad-leaved weeds and brush along highways and railroads, in rangeland and forests, and in wheat, rice, corn, and sugarcane fields. Agent Orange, a defoliant that received widespread use during the Vietnam War, is made from a combination of trichlorophenol 2,4,5-T and 4,4-D. The U.S. Environmental Protection Agency (EPA) restricted use of 2,4,5-T to rice fields and rangeland in 1979 and banned the herbicide entirely in 1985. One of the chief complaints was that 2,4,5-T contains dioxin, minute quantities of which cause cancer in animals. The toxic cloud that settled over Seveso contained an estimated 1.8 kilograms (4 pounds) of dioxin.

The disaster occurred when the chemical mixture in a reactor became seriously overheated and exploded, releasing a white plume that shot into the atmosphere to a height of about 45 meters (150 feet). The plume turned into a cloud that drifted in a southerly direction before cooling and settling over an area of approximately 280 hectares (700 acres), exposing some five thousand people. Those living closest to the ICMESA factory experienced nausea and skin irritation on exposed parts of their bodies. People who had been outdoors during the explosion developed burnlike sores on their faces, arms, and legs. Widespread cases of chloracne, an acnelike symptom that results from industrial exposure to dioxin, also occurred. All residents of the area developed headaches, dizziness, and diarrhea. Birds were especially vulnerable to the chemicals and died quickly, often before they could fly out of the area. Approximately 3,300 small animals such as rabbits, mice, chickens, and cats sickened and died soon after the accident.

Five days after the explosion, authorities instructed the population not to eat fruits or vegetables from the most affected areas. Nineteen children were hospitalized and later evacuated. Two weeks after the disaster

occurred, Milan's health administration determined that dioxin contamination from the cloud was serious enough to warrant evacuation of the population. This began on July 26 with the relocation of more than seven hundred people living in Zone A, a 108-hectare (267-acre) area directly south of the ICMESA plant. People in the surrounding 269-hectare (665-acre) Zone B were allowed to remain in their homes, but the area was sealed off from nonresidents. Italian army troops strung barbed wire around Zone A and patrolled the perimeter. To prevent toxins from entering the food chain, authorities ordered the emergency slaughtering of farm animals, and by 1978 some 77,000 animals had been killed. Scientific evidence that dioxin causes cancer in humans has been mixed, but an ongoing study of Seveso residents found elevated risk of several cancers following the dioxin exposure.

Peter Neushul

A state policeman wears a gas mask and protective clothing as he posts warning signs around the town of Seveso, Italy, in August, 1976, following the area's contamination by a toxic cloud containing dioxin. (©Hulton Deutsch Collection/CORBIS)

FURTHER READING

Ballarin-Denti, A., et al., eds. *Chemistry, Man, and Environment: The Seveso Accident Twenty Years On—Monitoring, Epidemiology, and Remediation.* New York: Elsevier, 1999.

Consonni, Dario, et al. "Mortality in a Population Exposed to Dioxin After the Seveso, Italy, Accident in 1976: Twenty-five Years of Follow-Up." *American Journal of Epidemiology* 167, no. 7 (2008): 847-858.

Schecter, Arnold, and Thomas Gasiewicz, eds. *Dioxins and Health.* 2d ed. New York: Taylor & Francis, 2002.

SEE ALSO: Agent Orange; Agricultural chemicals; Bhopal disaster; Chloracne; Dioxin; Pesticides and herbicides; Times Beach, Missouri, evacuation.

Ivory trade

CATEGORY: Animals and endangered species

DEFINITION: Buying and selling of the tusks of certain mammals, primarily elephants

SIGNIFICANCE: The international trade in ivory decimated the world's once-abundant elephant populations, but a 1990 international ban on the trade has done much to restore the species.

Ivory is made up of enamel, a resilient material found in mammalian teeth. The principal source of ivory is elephant tusks, although the teeth of the hippopotamus and walrus and of some types of whales and boars are also considered to be ivory. Tusks, also called "raw" ivory, grow on the male Indian, or Asian, elephant *Elaphus maximus* and on both sexes in the African elephant *Loxodonta africana*. Diverse human societies have used ivory for centuries to make crafts and medicines and in rituals. The impact on elephant populations was originally minimal. Trade in ivory in the eighteenth and nineteenth centuries was limited by the logistics of obtaining ivory and by low demand. The twentieth century witnessed an explosion in the ivory trade driven by the enormous profits that could be made in exporting ivory to Western and Eastern countries from India and, especially, Africa. Such profits soon precipitated a large illegal trade in ivory, and poaching became rampant.

In both Asia and Africa, elephant populations were decimated by ivory poaching. The African elephant used to roam most of the continent but has now been reduced to isolated populations in protected national parks and wildlife reserves. The African population was estimated at 1.3 million in 1979; it had dropped to 750,000 in 1988 and 600,000 by 1992; by 2010, experts estimated that the number had dwindled further, to 400,000.

THE IVORY MARKET

Massive hunting of African elephants started in 1970 in order to satisfy the great demand for raw ivory in consumer countries, where it was converted into expensive crafted items and medicines. Hong Kong, Japan, Taiwan, and China were the major markets. The price of raw ivory rose from $3 per pound in 1975 to $125 per pound by 1987. The slaughter of elephants in producer countries was accelerated by poverty; a single elephant could fetch up to $3,600, more than a typical person in many impoverished countries could make in five years. Moreover, most of the countries involved in the ivory trade did not have the personnel and equipment needed to protect elephants, and corrupt government officials exacerbated the problem by participating in the trade themselves.

The effects of the widespread hunting on the species were significant. Since poachers went for the elephants with the largest tusks, the average tusk size among surviving elephants decreased from 9.8 kilograms (21.6 pounds) to 4.7 kilograms (10.4 pounds). Elephants are slow to reproduce and take years to mature; the relentless killing reduced the average elephant's life span from sixty years to only thirty years. The herds' tight matriarchal social systems were also disrupted. Selective removal of large bulls created a dearth of breeding males and lowered genetic diversity.

BANNING THE IVORY TRADE

Scientists and nonscientists alike predicted the demise of the largest terrestrial animal unless immediate measures were implemented to save it, and in the 1970's a global effort was initiated to halt the decline in elephant populations. In 1978, the Convention on International Trade in Endangered Species (CITES) called for protection of elephants and their habitats, and elephants were placed on the agreement's list of threatened species. In June, 1989, President George H. W. Bush authorized a moratorium on all ivory imports into the United States, and in October, elephants were moved to the CITES list of endangered

National Parks Board officials carry a large elephant tusk in the ivory stockpile storeroom of Kruger National Park in South Africa. (AP/ Wide World Photos)

species. The U.S. Congress passed the African Elephant Conservation Act the same year. Other countries joined the moratorium, and a total ban on the ivory trade by 103 CITES signatory nations was declared in 1990. Botswana, Malawi, South Africa, Zambia, Zimbabwe, and Hong Kong voted against the ban, but Kenya, one of the countries where poaching was rampant, demonstrated its support of the measure by burning 12 tons of confiscated tusks, representing the deaths of two thousand elephants; the ivory destroyed was valued at three million dollars.

In the wake of the ban's adoption, the price of ivory plummeted to one dollar per pound. As profits declined, so did poaching, and elephant populations in southern African countries increased by as much as 25 percent, sometimes exceeding the local carrying capacities. Habitat destruction, especially in woodland areas, occurred because elephants strip and knock down trees in search of succulent bark and leaves. Increased elephant-human interaction also led to stresses, as rogue elephants sometimes tore

down fences, raided farms, and occasionally killed people. As a result, measures to cull herds were adopted in Zimbabwe and South Africa. The meat from culled elephants is distributed to local communities to supplement their food supplies.

The ban was not embraced by all countries. Japan and several ivory-producing southern African countries proposed reversing the ban and returning the elephant to the less restrictive classification of "threatened" during the 1992 CITES meeting in Kyoto, Japan. For example, representatives from Zimbabwe argued that their country's elephant population was healthy and that the ban was depriving the nation of vital resources required to protect wildlife. Proponents of the ban argued that any trade in ivory would spur illicit traffic in ivory, however, and the ban on the ivory trade was maintained.

OTHER MEASURES

In 1997, several southern African countries were allowed to sell their legal stockpiles of ivory to Japan.

The ivory was required to be coded and marked by country of origin, and the codes were entered into a database that would enable tracking of the ivory by the Ivory Monitoring Unit. However, genetic and isotope markers used on "worked ivory" are not foolproof, raising the possibility of a renewed illegal trade in ivory. In the former ivory-producing nations of Africa, efforts toward generating revenue shifted to tourism, and limited hunting of elephants, in which tourists pay up to ten thousand dollars for the opportunity, is allowed in Zimbabwe, Namibia, and South Africa.

Joseph M. Wahome

FURTHER READING

Bonner, Raymond. *At the Hand of Man: Peril and Hope for Africa's Wildlife.* New York: Alfred A. Knopf, 1993.

Duffy, Rosaleen. "The Ethics of Global Enforcement: Zimbabwe and the Politics of the Ivory Trade." In *Elephants and Ethics: Toward a Morality of Coexistence,* edited by Christen Wemmer and Catherine A. Christen. Baltimore: The Johns Hopkins University Press, 2008.

Meredith, Martin. "Ivory Wars." In *Elephant Destiny: Biography of an Endangered Species in Africa.* New York: PublicAffairs, 2001.

Oldfield, Sara, ed. *The Trade in Wildlife: Regulation for Conservation.* Sterling, Va.: Earthscan, 2003.

Walker, John Frederick. *Ivory's Ghosts: The White Gold of History and the Fate of Elephants.* New York: Grove/ Atlantic, 2009.

SEE ALSO: Africa; Convention on International Trade in Endangered Species; Ecotourism; Elephants, African; Hunting; Poaching.

J

Johannesburg Declaration on Sustainable Development

CATEGORIES: Philosophy and ethics; resources and resource management

IDENTIFICATION: Statement of principles for international action and debate regarding sustainable development

DATE: Adopted on September 4, 2002

SIGNIFICANCE: The goal of the 2002 World Summit on Sustainable Development was to find ways to bring together developed and developing nations in the pursuit of sustainable and equitable development. Most observers agree that the resulting Johannesburg Declaration on Sustainable Development made only small steps toward reaching that goal.

In Stockholm, Sweden, in 1972, the United Nations Conference on the Human Environment, which became known as the first Earth Summit, provided a forum for the first international discussions of development and the environment. In June, 1992, the United Nations Conference on Environment and Development (UNCED), commonly known as the Rio Summit, was convened in Rio de Janeiro with representatives from 172 nations. The purpose was to draw up binding agreements for reducing pollution and fossil-fuel emissions and for ensuring adequate access to clean water for all people. Ten years later, the World Summit on Sustainable Development, or Earth Summit 2002, was held in Johannesburg, South Africa, to further international discussion of sustainable development. The summit, which ran from August 26

James Connaughton, chair of the U.S. Council on Environmental Quality, speaks to the press at the 2002 World Summit on Sustainable Development in Johannesburg, South Africa. (AP/Wide World Photos)

through September 4, was attended by government representatives as well as by nongovernmental groups and individuals, including women and children, scientists and environmentalists, business leaders, and agriculturalists. The United States, however, under President George W. Bush, did not participate. The gathering resulted in several international agreements, the most significant of which was the Johannesburg Declaration on Sustainable Development.

Under the terms of the declaration, signers agreed to work toward "a humane, equitable and caring global society cognizant of the need for human dignity for all." The declaration did not create any specific programs or include provisions for the funding for such programs; rather, it served as a statement of principles for further action and debate. It articulated commitments to decreasing the gap between rich and poor people, protecting biodiversity and fragile ecosystems, addressing global climate change and the added pressures of increased population and globalization, promoting dialogue between people of different ethnicities and cultures, supporting women's empowerment and equality, combating war and crime, and using the tools of sustainable development to achieve these goals. The declaration also confirmed a commitment to support existing international documents, including Agenda 21, a plan created at the 1992 Rio Summit. Alongside the declaration, delegates to the World Summit on Sustainable Development agreed to carry out the Johannesburg Plan of Implementation, a lengthy and specific action plan produced in conjunction with the declaration.

Although the declaration was high-minded and beautifully worded, its adoption was met with controversy. Because the United States did not participate in drafting the declaration and did not sign it, many delegates to the summit saw it as toothless. Before full agreement on the declaration could be reached, the original draft's sixty-nine articles had been reduced to thirty-seven, with several of the more specific and restrictive articles removed. Protesters outside the Johannesburg convention center in which the summit was held created a circuslike atmosphere that drew media attention away from the serious issues being discussed inside. Additionally, delegates to the summit were criticized when it was learned that they had been treated to expensive catered food and luxurious housing while they discussed the world's need for better distribution of food, water, and other basics of life to the poor.

Cynthia A. Bily

FURTHER READING

Hens, Luc, and Bhaskar Nath. *The World Summit on Sustainable Development: The Johannesburg Conference.* New York: Springer Science & Business, 2005.

Speth, James Gustave, and Peter M. Haas. "From Stockholm to Johannesburg: First Attempt at Global Environmental Governance." In *Global Environmental Governance.* Washington, D.C.: Island Press, 2006.

Tully, Edward. *International Documents on Corporate Responsibility.* Northampton, Mass.: Edward Elgar, 2007.

SEE ALSO: Agenda 21; Earth Summit; Environmental law, international; Globalization; International Institute for Sustainable Development; Population growth; Rio Declaration on Environment and Development; Sustainable agriculture; Sustainable development; United Nations Commission on Sustainable Development; United Nations Conference on the Human Environment; World Summit on Sustainable Development.

John Muir Trail

CATEGORY: Preservation and wilderness issues

IDENTIFICATION: Wilderness path through the high country of California's Sierra Nevada range

SIGNIFICANCE: A unique hiking footpath, the John Muir Trail passes through what many backpackers regard as the finest mountain scenery in the United States and among the finest in the world. Its scenery includes spectacular rugged granite cliffs, lakes, streams, waterfalls, diverse wildlife, and abundant wildflowers.

The John Muir Trail runs south 340 kilometers (211 miles) from Yosemite Valley to the summit of Mount Whitney, the highest mountain in the continental United States. Without crossing a single road, the trail goes through Yosemite, Kings Canyon, and Sequoia national parks, as well as national forest land and Devils Postpile National Monument. After rising from the floor of Yosemite Valley it rarely drops below an altitude of 2,400 meters (8,000 feet) and crosses six passes at altitudes above 3,400 meters (11,000 feet), the highest of which is Forester Pass at 4,009 meters (13,153 feet).

The John Muir Trail

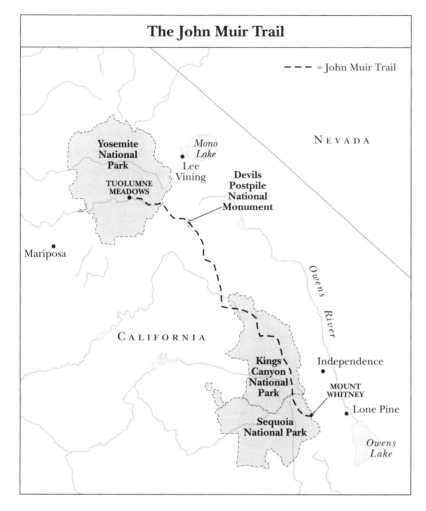

- - - = John Muir Trail

NEVADA

Yosemite
National
Park

TUOLUMNE
MEADOWS

*Mono
Lake*

Lee
Vining

Devils
Postpile
National
Monument

Mariposa

CALIFORNIA

Owens River

Kings
Canyon
National
Park

Independence

MOUNT
WHITNEY

Lone Pine

Sequoia
National Park

*Owens
Lake*

The mountain region through which the John Muir Trail passes was not explored until the late 1850's and the 1860's. Among the first to get deep into the backcountry was the Scottish American naturalist and environmentalist John Muir. A geologist, Josiah

Whitney, was tasked with making a geological survey of the state of California in 1864; that summer he led a team of scientists in exploring and surveying the area through which the trail would eventually wind. In 1884 a fourteen-year-old boy, Theodore Solomons, was the first to come up with the idea of a crest-parallel trail from Yosemite to Kings Canyon. Along with Solomons, during the 1880's and 1890's Joseph N. "Little Joe" LeConte and Bolton Brown explored routes and passes in the area.

In 1914 a member of the Sierra Club, after hiking the area, proposed that the club ask the California State Legislature to fund a high-mountain trail. The club took up the cause and gained the legislature's support for funding, and construction started in 1915, one year after John Muir died. The building of the trail was a very difficult task, as most of the country it would go through was extremely rugged, much of it consisting of very steep granite slopes. In addition, all of the materials and tools used had to be carried or packed in. The trail was completed in 1938.

C. Mervyn Rasmussen

SEE ALSO: Kings Canyon and Sequoia national parks; Muir, John; National Trails System Act; Sierra Club; Yosemite Valley.

K

Kaibab Plateau deer disaster

CATEGORIES: Disasters; animals and endangered species

THE EVENT: Eruption and crash of the population of mule deer on the Kaibab Plateau in Arizona

DATES: 1906-1939

SIGNIFICANCE: The Kaibab Plateau deer disaster is frequently cited as an illustration of the adverse impacts humans can have, particularly through predator control, on the natural balance between animal populations and their environments.

The Kaibab Plateau, part of the Kaibab National Forest, lies on the North Rim of the Grand Canyon in Arizona. In 1906 President Theodore Roosevelt set aside the Grand Canyon National Game Preserve, including the Kaibab Plateau, for the protection of game animals. As part of this protection, government trappers removed thousands of predators from the region between 1906 and 1931, including mountain lions, coyotes, bobcats, and the last remaining gray wolves. Livestock grazing by sheep, cattle, and horses was also greatly reduced between 1889 and 1931.

Following this reduction of predators and competing livestock, the mule deer (*Odocoileus hemionus*) population of the plateau was described as having increased from 4,000 in 1906 to 100,000 in 1924. After depleting its food supply and experiencing two consecutive harsh winters (1924-1925 and 1925-1926), approximately 60 percent of the herd purportedly died, and the population reportedly declined to fewer than 10,000 in 1939.

Researcher Graeme Caughley later evaluated the available information and challenged both the descriptions of the events that took place and their interpretation. Caughley noted that most of the population estimates were based on poor data. For example, estimates of the peak population in 1924 by different biologists and visitors to the area varied from 30,000 to 100,000 deer, and none of these estimates was grounded in sound scientific studies. The purported 60 percent decline (from 100,000 to 40,000 deer) was actually an extrapolation from one visitor's estimate of a 60 percent decline from 50,000 to 20,000 deer between 1924 and 1926. The only deer population estimates with any continuity over this period were made by the forest supervisor, who reported a general increase from 4,000 deer in 1906 to approximately 30,000 deer in 1923; notably, he also reported a stable population of approximately 30,000 deer over the time span (1924 to 1929) of the purported population crash. Thus many of the figures available on the Kaibab deer herd between 1906 and 1939 have been judged to be unreliable and inconsistent.

Overall, there is reasonable field evidence for a significant population increase, some starvation, and a population decline among the deer of the Kaibab Plateau between 1906 and 1932. In a general fashion, the increase in deer numbers coincided with a predator-control program and a reduction of livestock grazing in the region. Caughley argued that the Kaibab deer eruption, like similar eruptions observed in other ungulates (hoofed animals), was most likely linked to changes in food or habitat and was terminated by depletion of food supplies. In summary, however, the magnitude, timing, and causes of the Kaibab deer population eruption and crash are unclear. Few scientific conclusions about human impacts on ungulate populations can confidently be inferred from these events.

Richard G. Botzler

FURTHER READING

Rue, Leonard Lee, III. "A Historical View of Management Problems." In *The Deer of North America*. 1997. Reprint. Guilford, Conn.: Lyons Press, 2004.

Young, Christian C. *In the Absence of Predators: Conservation and Controversy on the Kaibab Plateau*. Lincoln: University of Nebraska Press, 2002.

SEE ALSO: Balance of nature; Grand Canyon; Hunting; Predator management.

Kalahari Desert

CATEGORIES: Places; ecology and ecosystems

IDENTIFICATION: Extensive sand-covered semiarid region in southern Africa

SIGNIFICANCE: The Kalahari is under increasing stress from growing human pressure on a drought-prone environment. Increasing commercial cattle raising has brought conflicts with wildlife and resulted in degradation of savanna vegetation. Surface and groundwater resources are under pressure from increasing human and livestock populations, and mining and tourism have had negative environmental impacts on local areas.

Bounded to the east by the highlands of eastern Botswana and western Zimbabwe and on the west by the uplands of central Namibia, the Kalahari Desert is an extensive sand-covered plain extending from the Orange River to the Caprivi Strip in northern Namibia. Most of the drainage in the region is internal, and pans (dry lakes or playas) are common. The Okavango, Kwando, and Cuíto rivers flow southeast from the humid highlands of Angola and form large inland deltas and swamps in northwestern Botswana and adjacent areas of Namibia and Zambia, where 96 percent of their discharge is lost by evaporation and transpiration. Some water spills over from the Okavango delta to the Makgadikgadi Pans.

Although often called a desert, the Kalahari is really a semiarid thirstland or edaphic desert. Rainfall occurs mainly as convectional storms in the austral summer and is highly variable in space and time, with quasi-periodic severe drought. Annual rainfall decreases from 500-800 millimeters (20-31 inches) in northern areas to 150-200 millimeters (6-8 inches) in the southwest. Vegetation cover in the Kalahari is tree and bush savanna, largely as a result of the high porosity of the surface sands.

The rural economy of the Kalahari is dominated by subsistence and commercial cattle raising, sustained by boreholes that tap deep groundwater sources. Traditional land use involved the coexistence of small numbers of cattle and extensive herds of wild ungulates (such as hartebeest, wildebeest, springbok, and eland), which provided for subsistence hunting by local peoples. The drilling of boreholes started during the 1930's and increased over time, and the provision of water enabled steady growth in numbers of livestock, mostly cattle. Many areas, however, especially those close to boreholes, have experienced degradation of natural rangeland vegetation resources, with a decrease in grass cover and encroachment by shrubs, leading to a decline in the quality and quantity of grazing, depletion of soil nutrients, and competition for resources between cattle and wildlife populations.

Traditionally, the Kalahari was home to large populations of wildlife, most of which were able to survive for long periods without surface water. Initially decimated by European hunters during the late nine-

A group of meerkats in the Kalahari. (©Peter Malsbury/iStockphoto.com)

teenth and early twentieth centuries, wildlife populations were severely affected from the 1950's onward by the construction of an extensive network of fences designed to control disease vectors in cattle populations, which blocked natural wildlife migration routes. The expansion of commercial ranching and continued growth of human populations has further reduced wildlife numbers and fragmented habitats. Protection to Kalahari wildlife populations is provided by national parks and wildlife reserves in Botswana, South Africa, and Namibia, but many of these are circumscribed by commercial and traditional farming and ranching activities. In the twenty-first century, however, the value of wildlife has been increasingly recognized as a sustainable land use and as a means to diversify national economies through ecotourism.

The environment and economy of the Kalahari are additionally threatened by climate change. Most climate models project little change in annual rainfall totals but increased temperatures, which will lead to higher evapotranspiration and therefore less effective rainfall. Some studies suggest that widespread decreases in vegetation cover as a result of climate change may result in remobilization of surface sands by wind erosion.

Nicholas Lancaster

FURTHER READING

Scholes, R. J., and D. A. B. Parsons, eds. *The Kalahari Transect: Research on Global Change and Sustainable Development in Southern Africa.* Stockholm: International Geosphere-Biosphere Programme, 1997.

Sporton, Deborah, and David S. G. Thomas, eds. *Sustainable Livelihoods in Kalahari Environments: A Contribution to Global Debates.* New York: Oxford University Press, 2002.

Thomas, David S. G., and Paul A. Shaw. *The Kalahari Environment.* 1991. Reprint. New York: Cambridge University Press, 2009.

SEE ALSO: Africa; Cattle; Desertification; Ecosystems; Ecotourism; Overgrazing of livestock; Savannas; Sustainable development; Wildlife management; Wildlife refuges.

Kesterson Reservoir

CATEGORIES: Water and water pollution; animals and endangered species

IDENTIFICATION: Series of holding ponds for drainage from irrigated agricultural lands in California's San Joaquin Valley

SIGNIFICANCE: After the transfer of subsurface drainage water from the San Joaquin Valley into the Kesterson Reservoir led to widespread death and deformity of migratory birds at the Kesterson National Wildlife Refuge, owing to toxic levels of dissolved selenium in the water, the U.S. Geological Survey set up programs to study the ecosystem effects of irrigation drainage water.

Selenium is a naturally occurring trace element that is essential for health in concentrations ranging from 0.05 to 0.3 parts per million (ppm) but becomes increasingly toxic at concentrations exceeding these low dietary levels. Selenium is found in igneous rocks, sedimentary rocks, and fossil fuels. Some soils contain naturally high selenium concentrations, and selenium occurs in the drainage from these soils because its ionic forms are readily soluble.

The Kesterson Reservoir in Merced County, California, was a series of holding ponds, spread over more than 486 hectares (1,200 acres), that collected the surface and subsurface drainage from irrigated agricultural land in the western San Joaquin Valley. Because development had eliminated more than 90 percent of California's wetlands and because of the scarcity and high price of water in California, the U.S. Bureau of Reclamation and the U.S. Fish and Wildlife Service mutually agreed to incorporate the reservoir and its waters as part of the National Wildlife Refuge System in the mid-1970's.

By 1982 virtually all the water entering the Kesterson Reservoir was from subsurface drainage, and in 1983 scientists found that aquatic and migratory birds nesting in the refuge had grossly deformed embryos and high embryo mortality rates. By 1984 dead adult birds were being discovered in the area in unusually high numbers. Agricultural chemicals were initially suspected to be the cause of the deaths and deformities, but tissue analysis of the birds revealed that they had selenium concentrations one hundred times in excess of normal concentrations because of selenium bioaccumulation in the food chain.

In 1986 discharge of drainage water into the Kesterson Reservoir was halted, and the reservoir was allowed to drain naturally. In 1988, to protect migratory waterfowl from nesting in periodically flooded zones of the reservoir, about 60 percent of the lowest-lying parts of the reservoir were filled with enough selenium-free backfill to ensure that the average seasonal groundwater level remained 15 centimeters (6 inches) below the soil surface. These steps were taken because wildlife biologists determined that the potential for selenium contamination of wildlife was much more limited in dryland environments than in wetland environments.

The Kesterson Reservoir has continued to be monitored because studies indicate that selenium concentrations in plants and wildlife in the area will remain elevated for decades. Bioremediation schemes were also investigated, and it was observed that microorganisms, particularly fungi, were able to volatilize up to 50 percent of the selenium in their vicinity within one year.

Mark Coyne

Further Reading

Byron, Earl R., et al. "Ecological Risk Assessment Example: Waterfowl and Shorebirds Feeding in Ephemeral Pools at Kesterson Reservoir, California." In *Handbook of Ecotoxicology*, edited by David J. Hoffman et al. 2d ed. Boca Raton, Fla.: CRC Press, 2003.

Fordyce, Fiona. "Selenium Deficiency and Toxicity in the Environment." In *Essentials of Medical Geology: Impacts of the Natural Environment on Public Health*, edited by Olle Selinus et al. Burlington, Mass.: Elsevier Academic Press, 2005.

See also: Agricultural chemicals; Biomagnification; Bioremediation; Food chains; Groundwater pollution; Runoff, agricultural; Wetlands; Wildlife refuges.

Khian Sea incident

Category: Waste and waste management
The Event: The around-the-world voyage of a ship carrying hazardous waste from Philadelphia that was originally intended to be disposed of in the Bahamas
Dates: 1986-1988
Significance: The *Khian Sea* incident is a clear example of an industrial country's efforts to transfer the hazardous by-products of its society to less industrialized, poorer nations. International concerns over actions such as this led to the signing of the 1989 Basel Convention on the Control of Transboundary Movements of Hazardous Wastes.

Unable to dispose of the toxic ash from a municipal incinerator, the city of Philadelphia contracted with Amalgamated Shipping to transport the waste to the Bahamas in its ship the *Khian Sea*. The ship left Philadelphia on September 5, 1986, for the Bahamas, but when it reached its destination, the Bahamian government turned the ship away. The *Khian Sea* subsequently tried and failed to get entry into ports in the Dominican Republic, Honduras, and the west African nation of Guinea-Bissau, as well as Puerto Rico, Bermuda, and the Netherlands Antilles.

In December, 1987, the crew of the *Khian Sea* unloaded 3,700 tons of ash on the beach at Gonaïves, Haiti, claiming it was fertilizer. The Haitian government prevented the crew from unloading more waste, but the *Khian Sea* left before the crew could be forced to reload the ash they had deposited on the beach. At least 10,000 tons of waste remained to be disposed of, and the ship visited Senegal, Cape Verde, Sri Lanka, Indonesia, and the Philippines in search of a dump site, only to be refused by each country. The cargo was never fully accounted for later, but it seems that 10,000 tons disappeared in the Indian Ocean between the Suez Canal and Singapore sometime in 1988.

Each of the countries where the *Khian Sea* tried to dispose of the waste is a poor, relatively unindustrialized country. The city of Philadelphia denied any responsibility for the waste once it was on board the *Khian Sea* as well as any responsibility for the waste dumped in Haiti. Haiti removed the ash from the beach and placed it in a sealed concrete bunker in a nearby hill. Although the owners of the *Khian Sea* were tried and convicted of federal perjury charges in connection with the incident, they faced no other punishment for their role.

Starting in 1992 the Haitian government began to pressure the United States to remove the waste from Haiti, as did various environmental groups, including Greenpeace. Philadelphia continued to deny any responsibility for the waste in Haiti, maintaining that the situation was the fault of the waste contractor. The successor company of the original firm agreed to pay $100,000 toward cleanup of the waste, largely because it wanted to gain another waste-handling contract. It was estimated that the cleanup would cost another $250,000, which Philadelphia mayor Ed Rendell refused to pay, as he did a lesser amount suggested by some activists. The issue remained mired in the courts for the next few years, and activists continued to try to embarrass both the city of Philadelphia and the U.S. government into helping to pay for the cleanup.

Finally, in April, 2000, the waste was removed from Haiti; the removal was paid for in part by the U.S. Department of Agriculture, the New York City Toxic Waste Commission, and the Haitian government. Several communities in the United States were approached about taking the waste, among them the Cherokee Nation of Oklahoma. In July, 2002, the waste was at last interred in a landfill in south-central Pennsylvania, 193 kilometers (120 miles) from Philadelphia.

The *Khian Sea* incident is a flagrant example of an industrial country's efforts to transfer the hazardous by-products of its society to less industrialized, poorer nations. Such actions were later prohibited by the 1989 Basel Convention on the Control of Transboundary Movements of Hazardous Wastes and the 1995 amendment to that convention known as the Basel Ban. Some commentators have asserted that this incident is also an example of environmental racism. The population of Haiti is largely nonwhite, as are the populations of many of the other places where the *Khian Sea* attempted to dump its cargo. Even the area of Pennsylvania where the waste was finally deposited is inhabited by a large poor, minority population.

John M. Theilmann

FURTHER READING

Clapp, Jennifer. *Toxic Exports*. Ithaca, N.Y.: Cornell University Press, 2001.
Pellow, David Naguib. *Resisting Global Toxics*. Cambridge, Mass.: MIT Press, 2007.

SEE ALSO: Basel Convention on the Control of Transboundary Movements of Hazardous Wastes; Environmental justice and environmental racism; Environmental law, international; Greenpeace; Hazardous waste; Incineration of waste products; Landfills.

Killer bees

CATEGORY: Animals and endangered species
DEFINITION: Aggressive genetic strain of honeybees
SIGNIFICANCE: The problem of killer bees—the unintended consequence of a failed experiment to produce a hybrid honeybee—is an example of the potential for negative environmental impacts when nonnative organisms are released into the wild.

Killer bees—or, more properly, Africanized honeybees—are a genetic strain of honeybees introduced to Brazil in 1957 in an attempt to hybridize the familiar European honeybee with an African subspecies. Although killer bees are no more venomous than ordinary honeybees, they exhibit lower average honey production, greater dispersal, and more aggressive behavior than their European counterparts. A person who disturbs a hive of Africanized honeybees may be pursued for distances of more than 90 meters (300 feet) by thousands of bees. Each bee stings only once, but an aggravated swarm can inflict a lethal number of stings. While the popular media have exaggerated their danger, killer bees can be as deadly as the name implies. They have caused hundreds of human casualties since 1957. Most of these victims were trapped or otherwise unable to run away from the bees. In South and Central America, wild bee colonies under bridges or near farm machinery are common sources of attacks. In the United States, where most honeybees are managed in hives, the Africanized bee problem has been less severe.

An individual Africanized honeybee appears so similar to the European type that the two are practically indistinguishable. Most of the differences are behavioral: In addition to their more aggressive nature, Africanized bees reproduce more often, depleting the hive's food reserves as they send out swarms to establish new colonies. Through frequent swarming, Africanized bees have expanded their range in the Americas by 400 kilometers (250 miles) per year. Be-

An Africanized honeybee, left, and a European honeybee on honeycomb. Although there are color differences between these two bees, for the most part the two types of bees cannot be distinguished by their appearance. (USDA)

cause they store less honey and do not cluster for warmth in winter, Africanized bees do not survive in cold climates. Since their introduction to Brazil, they have spread across much of South America, Central America, Mexico, and the southern United States. They have slowed their advance at the southern third of the United States, with the limit of their range fluctuating from year to year, depending on weather patterns.

The Africanized bee problem is a result of human introduction of a subspecies of honeybees from Africa to South America. Although the original goal was to improve honey harvests in the Tropics, the unintended consequences of this failed experiment stand as a warning against releasing nonnative organisms into the wild. Because Africanized traits are dominant, hybridization and subsequent back-crosses produce bees with essentially African characteristics. As a

result, the killer bee has not become more docile as its range has expanded.

The difficulty in managing Africanized bees has threatened the multimillion-dollar beekeeping industry in Mexico and the United States. Even more serious is the potential loss of pollination. Because managed bee colonies have replaced native pollinating insects in many agricultural and natural communities, threats to beekeeping represent a global ecological concern.

Eradication of the Africanized bee is unlikely, but steps can be taken to limit its impact. Beekeepers have responded by changing management practices and by maintaining tame varieties for breeding stock. Biologists continue to investigate honeybee genetics, and the public is learning to exercise caution around wild bee colonies.

Robert W. Kingsolver

FURTHER READING

Buchmann, Stephen, with Banning Repplier. *Letters from the Hive: An Intimate History of Bees, Honey, and Humankind.* New York: Bantam Books, 2005.

Caron, Dewey M. *Africanized Honey Bees in the Americas.* Medina, Ohio: A. I. Root, 2001.

Winston, Mark L. *Killer Bees: The Africanized Honey Bee in the Americas.* Cambridge, Mass.: Harvard University Press, 1992.

SEE ALSO: Balance of nature; Bees and other pollinators; Ecosystems; Food chains; Introduced and exotic species; Predator management.

Kings Canyon and Sequoia national parks

CATEGORIES: Places; preservation and wilderness issues

IDENTIFICATION: U.S. national parks in California's southern Sierra Nevada range

DATES: Established in 1890 (Sequoia) and 1940 (Kings Canyon)

SIGNIFICANCE: An ongoing environmental challenge can be seen in the struggle to maintain a balance between preserving the environment in Kings Canyon and Sequoia national parks and allowing the public to have access to the parks' lands for recreational purposes.

Sequoia National Park and Grant Grove were established in 1890 in California's southern Sierra Mountains. In 1940 Grant Grove was incorporated into the new Kings Canyon National Park. The impetus for the creation of both Kings Canyon and Sequoia came from the work of such preservationists as John Muir and the Sierra Club, as well as from the desires of local inhabitants of the San Joaquin Valley. When the National Park Service was created in 1916, it established the obligation of all national parks in the United States to provide public resources for recreation as well as to preserve the natural environment. These two aims have proved to be difficult to reconcile, however, particularly in Sequoia and Kings Canyon.

A hiker is dwarfed by giant redwood trees in Sequoia National Park. (©Jim Lopes/iStockphoto.com)

By the end of the 1920's it was apparent to some that environmental damage was already occurring in the Giant Forest area of Sequoia. Even before 1890 people had enjoyed camping among the redwoods, and it was inevitable that concessionaires would construct cabins and other buildings in the shadows of the great trees. Their fragile root systems were negatively affected by the buildings, the sewer system, and the human traffic that grew over time. It was not until the 1990's, however, that the buildings were actually torn down, to be relocated in areas where they would not have detrimental impacts on the redwoods.

Fire prevention is another concern of national park officials. Time and knowledge have radically altered the approaches taken toward fire prevention in Sequoia, as well as in other national parks. In the 1960's park officials realized that periodic fires are necessary to maintain the health of forests because they remove dead material and allow new seedlings to sprout and flourish. A living forest requires fire, and controlled burns became common in the park.

Although most visitors to Sequoia and Kings Canyon get no farther than the Giant Forest and General Grant areas, both similar in their redwood ambience, both parks are mainly wilderness areas of streams, meadows, and some of the nation's most rugged mountains. Before the parks were established, the areas had been overgrazed by sheep and cattle. Although this problem ended when the parks were founded, overuse by campers and backpackers continued to be a challenge. Eventually, access to the backcountry was limited through a permit system, the overused meadows were temporarily closed, and all visitors were required to pack out everything that they carried in to minimize damage to the fragile wilderness environment.

Kings Canyon has faced environmental challenges. Its dramatic central valley along the Kings River has been subject to numerous requests for dam construction, even within the park, and requests that the valley floor be developed for additional visitors. Here, too, preservation has taken precedence over recreation and other development, and human-built facilities have been limited intentionally.

Preservation of the environment and of ecosystems has continued to take precedence over mere recreation in both parks, with the public generally understanding and supporting the Park Service's priorities. However, the parks have been subject to threats by forces outside the Park Service's control, particularly air pollution wafting up from the floor of the valley below and, further afield, the possible long-term dangers of climate change.

Eugene Larson

FURTHER READING

Beesley, David. *Crow's Range: An Environmental History of the Sierra Nevada.* Reno: University of Nevada Press, 2004.

Eldredge, Ward. *Kings Canyon National Park.* Charleston, S.C.: Arcadia, 2008.

Noss, Reed F., ed. *The Redwood Forest: History, Ecology, and Conservation of the Coast Redwoods.* Washington, D.C.: Island Press, 2000.

SEE ALSO: Muir, John; National Park Service, U.S.; National parks; Wilderness Act; Yellowstone National Park.

Klamath River

CATEGORIES: Places; water and water pollution; animals and endangered species; resources and resource management

IDENTIFICATION: River that flows from south-central Oregon through northern California to the Pacific Ocean

SIGNIFICANCE: The Klamath River and its tributaries epitomize the conflicting economic, social, and environmental considerations involved in water management in a region of low rainfall and periodic drought. The central question is how best to reconcile the preservation of endangered fish species with local agricultural needs and regional demand for hydroelectric power.

The Klamath River is sometimes dubbed an "upside-down river" because it originates in a shallow, highly eutrophic (overly nutrient-enriched) lake in an area of intensive agricultural development and ends up as a swift and apparently pristine stream. In prehistoric times the marshes surrounding Upper and Lower Klamath lakes supported a great wealth of wildlife and large populations of Native American hunters. Despite the dispersal of tribal lands in the twentieth century, many of their descendants still occupy the area.

Development of the Klamath River basin has been steady since the beginning of white settlement but increased dramatically during the 1950's and 1960's with construction of hydroelectric dams on the river and its tributaries, diversion of increasing amounts of water for irrigation, increases in agricultural runoff, and conversion of wetland wildlife habitat to farmland. During this same period, logging and the manufacture of wood products, once mainstays of the local economy, declined in importance because of unsustainable logging practices on federal lands.

Environmental problems and conflicts surfaced during a drought in the 1990's after the U.S. Bureau of Reclamation, which allocates irrigation water, proposed a water management plan that threatened two endangered species of suckers, fish native to Klamath Lake, with extinction and also threatened a distinct population of coho salmon. Provisions of the Endangered Species Act effectively blocked irrigation, and the region experienced considerable economic damage.

Lake water levels and stream flow are not the only factors that have threatened fish populations in the Klamath River basin. Phosphates from agricultural runoff have caused massive blooms of blue-green algae, depleting oxygen in the water and raising pH levels. Increased alkalinity and higher water temperatures were implicated in a 2002 fungal epidemic that killed off large numbers of salmon. In addition, logging and recreational development have compromised spawning streams, and hydroelectric dams have created serious barriers to salmon migration.

The results of an integrated study showed that any plan that could effectively restore endangered fish populations in the Klamath River would require the removal of several hydroelectric dams. In November, 2008, officials of PacifiCorp, the power company that operates the dams, reached a tentative nonbinding agreement on water rights with representatives of the region's farmers, fishers, Native American tribes, and recreational users; the agreement, which called for the removal of four dams beginning in 2020, assumed more concrete form in February, 2010. In the meantime, dry years spell disaster for farmers in the upper Klamath basin, and it is unclear whether efforts that have been implemented will prevent extinction of threatened fish before 2020.

Legitimate questions have also been raised regarding whether it makes sense, from an overall point of view, to expend scarce environmental dollars to demolish sources of clean, carbon-neutral energy for which any possible replacement also has its acknowledged downsides. Replacing hydroelectric dams with wind farms, for example, would entail huge energy costs for start-up; in addition, critics argue that wind farms are eyesores and that they pose threats to migratory birds.

Martha A. Sherwood

FURTHER READING

Blake, Tupper Ansel. *Balancing Water: Restoring the Klamath Basin.* Berkeley: University of California Press, 2000.

Committee on Endangered and Threatened Fishes in the Klamath River Basin. *Hydrology, Ecology, and Fishes of the Klamath River Basin.* Washington, D.C.: National Academies Press, 2008.

Most, Stephen. *River of Renewal: Myth and History in the Klamath Basin.* Portland: Oregon Historical Society Press, 2006.

SEE ALSO: Dams and reservoirs; Endangered Species Act; Eutrophication; Fisheries; Hydroelectricity; Runoff, agricultural; Wetlands.

Kleppe v. New Mexico

CATEGORIES: Treaties, laws, and court cases; animals and endangered species; land and land use

THE CASE: U.S. Supreme Court ruling concerning wildlife on public lands

DATE: Decided on June 17, 1976

SIGNIFICANCE: The Supreme Court's decision in the case of *Kleppe v. New Mexico* remains its most definitive statement on the use of the U.S. Constitution's property clause to protect wildlife.

The U.S. Congress enacted the Wild Free-Roaming Horses and Burros Act in 1971 to protect wild horses and burros on federal public lands from capture, harassment, and death. After passage of the law, the New Mexico Livestock Board received a complaint from a rancher who held a permit to graze live-

stock on federal land; the rancher asserted that wild burros were interfering with his livestock operation. The board responded by using its authority under the state's Estray Law to remove the burros and sell them at public auction. The U.S. Bureau of Land Management (BLM) then demanded that the board recover the animals and return them to public lands, but the Livestock Board challenged the constitutionality of the federal statute. A three-judge federal district court held that the statute was an unconstitutional use by Congress of its power under the property clause.

The case was appealed to the U.S. Supreme Court, and in a unanimous decision the Court rejected New Mexico's narrow view of the U.S. Constitution's property clause as giving Congress only the power to dispose of and make incidental rules for the use and protection of federal property. The Court took a broad view of the property clause that granted Congress unlimited power over public lands. The Court held that Congress had both the power of a proprietor to protect its lands from trespass and injury and the power of a legislature to make needful rules respecting federal lands, including the power to protect wildlife living on those lands. In passing the Wild Free-Roaming Horses and Burros Act, the Court found, Congress had determined that the animals were an integral part of the natural system of public lands and that their management was necessary to the achievement of ecological balance on public lands.

The Court also rejected the state's argument that the federal statute encroached on state sovereignty and that Congress could not obtain exclusive jurisdiction over public lands within the state unless New Mexico consented. The Court acknowledged that the federal government did not have exclusive jurisdiction over its public lands within New Mexico; the state still had the authority to enforce its civil and criminal laws on federal lands. In this case, however, the New Mexico Estray Law violated the U.S. Constitution's supremacy clause, Article VI, section 2, because it conflicted with the Wild Free-Roaming Horses and Burros Act by authorizing the state Livestock Board to remove and sell federally protected animals.

Kleppe v. New Mexico is the Supreme Court's most definitive decision on the use of the property clause to protect wildlife on public lands. The Court did not decide whether the clause extended the reach of the Wild Free-Roaming Horses and Burros Act to protect those animals that stray onto private land, but its decision in *Camfield v. United States* (1897) suggests that

Congress can use its property clause power to make laws that regulate conduct on private land that affects public land. Together with the Supreme Court's treaty clause decision in *Missouri v. Holland* (1920) and its commerce clause decision in *Douglas v. Seacoast Products* (1977), the *Kleppe* decision provides broad constitutional authority for federal wildlife law.

The *Kleppe* decision and the Federal Land Policy and Management Act of 1976 granted the BLM increased power to protect wildlife on public lands, reflecting the growing importance of environmental values in land management practices in the American West during the 1970's. At the same time, these changes led to quarrels between federal agencies and ranchers who leased federal lands to graze their livestock. This discontent grew into the Sagebrush Rebellion, advocates of which used legal and political actions to assert, albeit unsuccessfully, state control over lands managed by federal agencies.

William Crawford Green

FURTHER READING

Alexander, Kristina, and Ross Gorte. *Federal Land Ownership: Constitutional Authority and the History of Acquisition, Disposal, and Retention.* Washington, D.C.: Congressional Research Service, 2007.

Merrill, Karen. *Public Lands and Political Meaning: Ranchers, the Government, and the Property Between Them.* Berkeley: University of California Press, 2002.

SEE ALSO: Bureau of Land Management, U.S.; Department of the Interior, U.S.; Federal Land Policy and Management Act; Multiple-use management; Privatization movements; Property rights movement; Sagebrush Rebellion; Wild horses and burros; Wildlife management.

Krakatoa eruption

CATEGORIES: Disasters; atmosphere and air pollution
THE EVENT: Massive volcanic eruption that took place in the Malay Archipelago
DATES: August 26-27, 1883
SIGNIFICANCE: The eruption of the volcano Krakatoa ejected tons of rock, ash, and gases into the atmosphere, cooling the planet and causing sea levels to fall.

The volcano Krakatoa (or Krakatau) erupted in a series of four spectacular explosions on August 26 and 27, 1883. The eruption was the fourth largest in recorded history and destroyed two-thirds of the island—also known as Krakatoa—on which the volcano stood. The cataclysm resulted in more than 36,000 deaths, most of them on the nearby islands of Java and Sumatra, and had profound worldwide consequences as well.

Krakatoa's eruption produced a shock wave that passed around the earth seven times, and its sound was clearly heard 4,777 kilometers (2,968 miles) away. The tsunamis that resulted from the eruption, which caused many deaths in the immediate vicinity, were

Krakatoa in Modern Indonesia

detected as ripples as far away as the English Channel. An estimated 21 to 25 cubic kilometers (5 to 6 cubic miles) of material was ejected into the atmosphere by the volcano, some of it as high as 48 kilometers (30 miles). Ash drifted down on ships thousands of miles away, but much of the finer material remained floating in the stratosphere and was carried around the earth. Vivid atmospheric effects of the eruption, including red and yellow sunsets and halos around the sun and moon, were reported from widely scattered locations for three years. Such observations provided scientists with their first proof of a worldwide system of high-altitude winds.

Of greater importance to the larger environment, sulfur dioxide (SO_2) gas emitted during the eruption combined with water vapor to create droplets of sulfuric acid (H_2SO_4) in the upper atmosphere. Together with floating ash, these droplets reduced the sunlight striking the earth by about 1 percent for two years, cooling the planet as much as 0.5 degree Celsius (0.9 degree Fahrenheit) for approximately five years. In 2006 P. J. Gleckler and five colleagues determined that the reduction in sunlight also caused the oceans to cool and, as a result, sea levels to fall. Sea levels regained their previous height only during the middle of the twentieth century.

The fragments of Krakatoa remaining after the eruption were covered by as much as 40 meters (131 feet) of ash, and it was assumed that every living organism on the island had died. For scientists concerned with the recolonization of such a landscape by plants and animals, the site became a living laboratory. A few months after the eruption a small spider was observed, and six years later a monitor lizard and a variety of insects were noted. Within a few decades, trees had sprouted and grown to more than 12 meters (39 feet) in height.

In late June, 1927, fishermen noticed steam rising from the sea above the collapsed portions of Krakatoa, indicating a renewal of volcanic activity. On January 26, 1928, a new island appeared above the surface of the sea and was soon named Anak (child of) Krakatoa. It has since grown steadily and has erupted several times. The Krakatoa area was declared a nature reserve in 1921, and it was later made a part of Indonesia's Ujung Kulon National Park. In 1991 the park was designated a World Heritage Site by the United Nations Educational, Scientific, and Cultural Organization (UNESCO).

Grove Koger

FURTHER READING

Gleckler, P. J., et al. "Krakatoa's Signature Persists in the Ocean." *Nature* 439 (February 9, 2006): 675.

Simkin, Tom, and Richard S. Fiske. *Krakatau, 1883: The Volcanic Eruption and Its Effects.* Washington, D.C.: Smithsonian Institution Press, 1983.

Winchester, Simon. *Krakatoa: The Day the World Exploded, August 27, 1883.* New York: HarperCollins, 2003.

SEE ALSO: Aerosols; Airborne particulates; Asia; Climate change and oceans; Global warming; Mount Tambora eruption; Sea-level changes; Sulfur oxides; Volcanoes and weather.

Kudzu

CATEGORY: Forests and plants

DEFINITION: Fast-growing perennial vine indigenous to Asia that has been imported to other areas worldwide and is often considered a weed

SIGNIFICANCE: As an introduced species in North America, kudzu has been both beneficial and detrimental to the environment. Initially planted widely to combat soil erosion, kudzu proved difficult to control, and its spread over vast areas of the United States had negative impacts on native plant and animal species as it added harmful chemicals to soil, water, and atmosphere. Efforts to find environmentally benign methods of eradicating kudzu are ongoing.

Kudzu (*Pueraria montana*) vines can reach more than 30.5 meters (100 feet) in length, growing by 30 centimeters (12 inches) daily in summers and attaining widths of 25 centimeters (10 inches). The mature roots of kudzu are often several inches thick and can reach depth of 3 meters (10 feet); they may weigh as much as 136 kilograms (300 pounds). Although kudzu flowers and produces seeds, the plant's resilience depends on root crowns that are the sources of most new vines.

North Americans began importing kudzu from Japan and China during the late nineteenth century. During the early twentieth century, the U.S. Department of Agriculture (USDA) and land-grant colleges studied kudzu and issued reports on the cultivation of kudzu for livestock fodder. By the 1920's, conservationists recognized kudzu's potential to prevent agricultural fields from losing topsoil. During the 1930's, the Soil Conservation Service and the Civilian Conservation Corps planted thousands of hectares of kudzu as an erosion-control measure. Within decades, however, it became clear that kudzu was having detrimental impacts on the environment. The USDA declared kudzu an invasive weed by 1953, and the transport of kudzu across state lines was not allowed without permission of the USDA secretary.

Kudzu overwhelms other plants, including crops, as it spreads into their territory, blocking sunlight and interfering with chlorophyll production. Trees are vulnerable to kudzu vines, which can damage tree branches and trunks. Kudzu also alters habitats, often causing indigenous plant species to die and displacing animals, reducing biodiversity. Further, the atmosphere is harmed when the isoprene released by kudzu becomes ozone. Kudzu has also been found to admit nitrogen pollutants, including ammonia.

By 2009, kudzu covered 30,000 square kilometers (11,583 square miles) of land in the United States. Annually, new kudzu growth expands by more than 500 square kilometers (193 square miles). Satellite images reveal kudzu growth around the world, and the International Union for Conservation of Nature includes kudzu in its Global Invasive Species Database. Yearly economic losses associated with kudzu total approximately $500 million.

Initially, the methods used to remove kudzu involved herbicides, burning, or mowing; most proved ineffective in the long run, as they killed leaves and damaged vines but left behind underground roots that produced new growth. Researchers and the U.S. Forest Service have worked to find methods of eradicating kudzu that do not harm the environment. With the realization that kudzu growth is controlled in Asia by indigenous predators absent in the United States, scientists began to investigate various biocontrols, such as exposing kudzu to viruses, bacteria, or fungi with diseases.

Entomologists have focused on attacking kudzu

Kudzu grows over abandoned vehicles in Tennessee. The invasive plant is extremely hardy and will grow over almost anything that is not moving. (©Roel Smart/iStockphoto .com)

with insects that will not damage other plants and disrupt ecosystems. In the late 1990's, after consulting with scientists at the Chinese Academy of Sciences to identify Asian insects that eat kudzu, one group of researchers at North Carolina State University began a study that involved inserting wasp eggs into soybean looper caterpillars, which feed on kudzu. The larvae increased the caterpillars' appetites and thus their consumption of kudzu leaves; because the plants had to replenish their foliage repeatedly, they used up the starch reserves in their roots and eventually died.

By the early twenty-first century, scientists began to study the potential for using kudzu in the production of biofuels. Research conducted by the USDA and the University of Toronto showed that more gallons of ethanol could be produced from the harvest of one hectare of kudzu than from the harvest of one hectare of corn, the crop most often used in ethanol production in the United States. Concerns have been voiced about the difficulty of controlling kudzu as a crop, but advocates of using kudzu for ethanol production note that harvesting the plant on a large scale would reduce harmful emissions while at the same time creating an opportunity for income generation. In 2010 the Tennessee company Agro-Gas Industries announced plans to explore using kudzu to make ethanol.

Elizabeth D. Schafer

FURTHER READING

Forseth, Irwin N., Jr., and Anne F. Innis. "Kudzu (*Pueraria montana*): History, Physiology, and Ecology Combine to Make a Major Ecosystem Threat." *Critical Reviews in Plant Sciences* 23, no. 5 (2004): 401-413.

Sage, Rowan F., et al. "Kudzu [*Pueraria montana* (Lour.) Merr. Variety *lobata*]: A New Source of Carbohydrate for Bioethanol Production." *Biomass and Bioenergy* 33, no. 1 (January, 2009): 57-61.

Sun, Jiang-Hua, et al. "Survey of Phytophagous Insects and Foliar Pathogens in China for a Biocontrol Perspective on Kudzu, *Pueraria montana* var. *lobata* (Willd.) Maesen and S. Almeida (Fabaceae)." *Biological Control* 36, no. 1 (January, 2006): 22-31.

SEE ALSO: Alternative fuels; Biodiversity; Ecosystems; Environmental law, U.S.; Erosion and erosion control; Forest Service, U.S.; Habitat destruction; Introduced and exotic species; Pesticides and herbicides; Soil conservation.

Kyoto Protocol

CATEGORIES: Treaties, laws, and court cases; weather and climate

THE TREATY: International agreement that commits nations to place legally binding limits on their emissions of six greenhouse gases

DATE: Adopted on December 11, 1997

SIGNIFICANCE: The Kyoto Protocol represents the first international, legally binding attempt to prevent human activity from causing significant adverse changes to the earth's climate.

In 1988 the World Meteorological Organization and the United Nations established the Intergovernmental Panel on Climate Change (IPCC), a team of more than two thousand leading scientists from around the world whose mission would be to assess scientific information on climate and the environmental impacts of climate change. In 1990 the IPCC released its first report, which concluded that human-made greenhouse gases would exacerbate the greenhouse effect, resulting in additional warming of the earth's surface by the twenty-first century unless measures were enacted to limit the emissions of these gases.

At the 1992 Earth Summit in Brazil, the United Nations Framework Convention on Climate Change (UNFCCC) was adopted. This treaty, signed by more than 150 nations, required each nation to limit its greenhouse gas emissions, with the industrialized nations taking the first step by voluntarily reducing their emissions to 1990 levels by the year 2000. U.S. participation in the treaty was ratified by the U.S. Senate in October, 1992, and eventually more than 160 nations joined in the agreement.

In 1995 the IPCC released a report stating for the first time that the balance of evidence suggested a "discernible human influence on global climate." The report noted that even if emission levels were to remain constant, the atmospheric concentrations of carbon dioxide would approach twice the preindustrial concentration by the end of the twenty-first century, changing the earth's climate in significant ways. Between 1992 and 1995 global greenhouse emissions continued to rise, and it was agreed that the voluntary approach had not been successful. At a meeting in Berlin, Germany, in 1995, negotiators agreed that the industrialized nations, which emit the majority of greenhouse gases, would have to take the lead in adopting stronger measures. In 1996 the United States announced that future emission targets should be legally binding and challenged other industrialized nations to agree. More than one hundred nations agreed to develop legally binding targets.

In a March, 1997, meeting in Bonn, Germany, the European Union took the lead by proposing that industrialized nations reduce emissions by 15 percent from 1990 levels by the year 2010. The U.S. government proposed a system of international trading of emissions rights, in which nations could buy and sell the rights to emit greenhouse gases. The United States also proposed a "joint implementation" program that would allow nations to earn emissions credits by implementing non-carbon-based "clean energy" projects. At the same meeting, the IPCC chairperson reported that reductions undertaken solely by industrialized nations would not be sufficient to limit global warming to environmentally sustainable levels.

As the December, 1997, Conference to the Parties to the UNFCCC in Kyoto, Japan, drew near, it appeared that reaching a consensus would be difficult, as proposed policies ranged from a 20 percent reduction (compared to 1990 levels) by 2005 to the U.S. proposal of merely stabilizing emissions at 1990 levels. Most developing nations stated that they would not commit to emissions controls until after the developed nations had acted.

As talks opened, the United States made a significant change in its position by announcing that it would support a system of flexible targets for different nations that would take into consideration the situation of each country. The United States maintained its position of zero reduction compared to 1990 levels until late in the conference, when Vice President Al Gore instructed the U.S. delegation to be more flexible in negotiations. Gore's involvement helped to break the logjam of deliberations, and the final days of the conference were marked by around-the-clock negotiations until an agreement was finalized during the last hours.

The Kyoto Protocol called for a 5.2 percent reduction in emissions of carbon dioxide, methane, and nitrous oxide from 1990 levels by the period from 2008 to 2012, with the United States reducing emissions by 7 percent, Japan by 6 percent, and the European Union by 8 percent. Three other greenhouse gases, all chlorofluorocarbon (CFC) substitutes, were to be cut by comparable levels, using 1995 rather than 1990 as the baseline. While the accords reached in Kyoto marked the beginning of a legal framework for reducing carbon emissions over the long run, the emission cuts required by the treaty fell far short of the levels that most climate scientists predicted would be needed to prevent significant climate change.

The Kyoto Protocol was criticized by many environmental leaders as having major loopholes, such as "flexibility measures" that would allow nations to circumvent their requirements for reductions in emissions. Rules for how industrialized nations could trade or sell emission rights were not formalized in the agreement. Russia was left with emission credits because its emissions plummeted during the collapse of its economy in the 1990's—these credits could be sold to countries such as the United States, with the result that there might be little or no reduction in emis-

Emissions Trading

Article 6 of the Kyoto Protocol, reproduced below, establishes the basic framework for parties to the treaty to trade pollution credits with one another, thereby employing market principles to drive international emission reductions.

1. For the purpose of meeting its commitments under Article 3, any Party included in Annex I may transfer to, or acquire from, any other such Party emission reduction units resulting from projects aimed at reducing anthropogenic emissions by sources or enhancing anthropogenic removals by sinks of greenhouse gases in any sector of the economy, provided that:

   (a) Any such project has the approval of the Parties involved;

   (b) Any such project provides a reduction in emissions by sources, or an enhancement of removals by sinks, that is additional to any that would otherwise occur;

   (c) It does not acquire any emission reduction units if it is not in compliance with its obligations under Articles 5 and 7; and

   (d) The acquisition of emission reduction units shall be supplemental to domestic actions for the purposes of meeting commitments under Article 3.

2. The Conference of the Parties serving as the meeting of the Parties to this Protocol may, at its first session or as soon as practicable thereafter, further elaborate guidelines for the implementation of this Article, including for verification and reporting.

3. A Party included in Annex I may authorize legal entities to participate, under its responsibility, in actions leading to the generation, transfer or acquisition under this Article of emission reduction units.

4. If a question of implementation by a Party included in Annex I of the requirements referred to in this Article is identified in accordance with the relevant provisions of Article 8, transfers and acquisitions of emission reduction units may continue to be made after the question has been identified, provided that any such units may not be used by a Party to meet its commitments under Article 3 until any issue of compliance is resolved.

sions by major polluters. Also not addressed in the agreement was how joint implementation would affect emissions reduction targets. Finally, no compliance mechanisms were included.

The most contentious issue resulting from the Kyoto conference concerned the exclusion of developing countries from the accords. At the conference, representatives from a united group of 130 developing nations, led by China and India, voiced strong objections to the process of emissions trading, claiming that such trades would allow rich polluters to buy their way out of reductions. At the conference, the United States demanded voluntary commitments from developing nations to reduce emissions, but the proposal was rejected in the final negotiations. On a per capita basis, emissions in developing nations were well below those of the industrialized nations. However, emissions were growing so rapidly in developing nations that they were projected to exceed those of the industrialized nations by 2025, with China projected to be the largest emitter of carbon dioxide by 2015. Many industrialized nations, including the United States, refused to ratify the treaty unless the developing nations were included.

The Kyoto Protocol represented the first international legally binding attempt to prevent human activity from causing significant adverse changes to the earth's climate. For the treaty to enter into force, however, it had to be ratified by governments representing at least 55 nations, including industrialized nations representing 55 percent of all 1990 carbon dioxide emissions. Successful implementation of the protocol thus awaited the resolution of issues of nationalism, ideology, and economics. Continued discussions among nations resulted in steady progress on these issues, and the requirement of 55 signatory nations was met in 2002. When Russia signed in 2004, the requirement that signatories include industrialized nations representing 55 percent of 1990 carbon dioxide emissions was satisfied, allowing the treaty to enter into force soon after, on February 16, 2005. By 2009 the treaty had been ratified by 186 nations and the European Union.

Craig S. Gilman

A Madagascar nursery grows seedlings that will be used to plant new forests. The forests' carbon-fixing ability will be converted to credits and sold on the global market created by the Kyoto Protocol. (AP/Wide World Photos)

FURTHER READING

Conkin, Paul K. "Greenhouse Gases and Climate Change." In *The State of the Earth: Environmental Challenges on the Road to 2100.* Lexington: University Press of Kentucky, 2007.

Dessler, Andrew E., and Edward A. Parson. *The Science and Politics of Global Climate Change: A Guide to the Debate.* New York: Cambridge University Press, 2006.

Press, Frank, et al. "Earth's Environment, Global Change, and Human Impacts." In *Understanding Earth.* 4th ed. New York: W. H. Freeman, 2004.

Zedillo, Ernesto, ed. *Global Warming: Looking Beyond Kyoto.* Washington, D.C.: Brookings Institution Press, 2008.

SEE ALSO: Air-pollution policy; Chicago Climate Exchange; Climate accommodation; Climate change and human health; Climate change and oceans; Climate change skeptics; Climate models; Intergovernmental Panel on Climate Change; United Nations Framework Convention on Climate Change; U.S. Climate Action Partnership.

L

Lake Baikal

CATEGORIES: Places; water and water pollution; ecology and ecosystems

IDENTIFICATION: Freshwater lake located in Siberian Russia

SIGNIFICANCE: When industrial water pollution began to threaten the unique nature of Lake Baikal, with its pure waters and distinctive plant and animal life, the Soviet Union experienced its first environmental protests.

Lake Baikal is the oldest and deepest lake on earth. Almost 650 kilometers (400 miles) long, its surface area is approximately the same as that of Lake Superior, but its water volume makes up one-fifth of all fresh water on the earth's surface. The lake itself is more than 1.6 kilometers (1 mile) deep. Below the surface lies a floor of sediment approximately 6.4 kilometers (4 miles) deep; the sediment has drifted down through the lake's waters over some twenty to thirty million years. Lake Baikal lies atop a rift where several tectonic plates meet along a little-studied geological fault. The activity of these plates has apparently widened the rift and deepened the lake over many millennia.

Lake Baikal essentially has a closed ecosystem. Several hundred rivers feed into it, but the watershed consists entirely of the mountain area surrounding the lake, unconnected to other river systems. Baikal has only one outlet, the Angara River, which flows out from its southeast corner, past the old frontier city of Irkutsk, and ultimately to the Arctic Ocean. Because of the lake's isolation, it is the habitat of many species that are found nowhere else on earth. It is a fascinating site for biological study.

Among the lake's intriguing fauna is the silver-furred nerpa, the smallest known seal. Its closest relative, the Arctic seal, lives some 3,200 kilometers (2,000 miles) away. How the nerpa reached Baikal and adapted to fresh water is one of the lake's many mysteries. Nerpas eat an oily fish also found only in Baikal, the golomyanka. This almost transparent fish gives birth to live young, then promptly sinks and dies.

Algae, plankton, and similar microscopic creatures form the bottom of the lake's food chain. These serve as prey for a tiny crustacean called epishura, which strains Baikal's waters to a pristine clarity.

The land around Lake Baikal—consisting of taiga, or northern woodlands—shelters a variety of Siberian wildlife. Mountains ring the lake, creating spectacular scenery. Olkhon Island, located near the western shore in the midsection of the crescent-shaped lake, has a dry, almost snowless climate and contains grasslands and sand dunes. The smaller islands are seal nesting grounds. The lake surface freezes to a depth of more than 1 meter (3.28 feet) during the long Siberian winter.

Despite its remote location, the region has not been immune to technological forces. The Trans-Siberian Railroad's builders left clear-cut areas and debris in the southern reaches, a problem repeated with the building of the Baikal-Amur Mainline paralleling the northwest shore decades later. By the early twentieth century, sable in the surrounding forests had been hunted almost to extinction. Sturgeon and sturgeon eggs, valued as luxury food items, were dangerously depleted by the 1950's.

Buryat Mongols, indigenous inhabitants of the region, lived unobtrusively on the land, much as Native Americans lived for thousands of years in the Americas. Russian settlers, arriving either by choice or by involuntary exile, had to make a living in the isolated region. Although fisheries could sustain operations without depleting populations of marine life, they did not always do so. Meanwhile, timbering and farming methods took a major toll in erosion.

ENVIRONMENTAL DAMAGE

The greatest environmental damage to the Lake Baikal region came with the establishment of large-scale industry. During Joseph Stalin's regime as the Soviet leader (1922-1953), the Soviet Union emphasized industrial production over every other goal. Stalin's immediate successors retained this policy. A huge cellulose production plant opened at Baikalsk on the southern shore of the lake in 1966 and began spewing toxic chemicals into the lake water and

Lake Baikal, Soviet Union

murky smoke into the air. A pulp plant at Selingiske, located on one of Lake Baikal's major tributaries, and factory wastes and sewage from Ulan-Ude farther upriver created another major pollution area around the Selenga River delta. Hydroelectric dams built near Irkutsk in the 1950's brought more heavy industry and noxious by-products.

The harm wrought by these factories was unquantifiable but quite visible. In many places along the southern shore of the lake, the formerly pure water became unfit to drink. Many square miles of lake area simply died, becoming unable to support aquatic life. Elsewhere, the populations of many native species shrank drastically. In some places newly introduced, hardier species began to replace them, with unpredictable consequences.

Efforts to Protect the Lake

Distrust of the Baikalsk manufacturing project spawned the first environmental protests in the Soviet Union. Although these did not prevent the plant's opening, a precedent and framework was created for the future. By the time of Mikhail Gorbachev's term as Soviet leader (1985-1991), the lake's pollution was obvious, and the political scene had become less repressive. Gorbachev pledged to convert the Baikalsk plant to nonpolluting activities, but political turmoil and the Soviet Union's breakup intervened, after which Lake Baikal became Russia's responsibility.

The biggest problems for Lake Baikal in the 1990's were those of inaction caused by Russia's political and economic problems. Antipollution laws and plans were adopted, but the money to implement them was scarce, and coordination among local political entities was difficult. Declining industrial activity caused by economic slowdown may have slowed pollution of the lake more effectively than any active measures. The Baikalsk plant was eventually closed down in 2008—not because of environmental concerns, but because it had become unprofitable. Although the owner had sworn it would not reopen, it began operating again in 2010.

In 1996, the United Nations Educational, Scientific, and Cultural Organization (UNESCO) named Lake Baikal a World Heritage Site, and many Russians have embraced the lake as a national treasure deserving of protection. In the early years of the twenty-first century, plans were drawn up to lessen or reverse damage to the lake and region, and Russia began to draw on outside resources, both scientific and business-oriented, for help in this effort. Investors also began to promote the relatively clean industry of tourism at Lake Baikal.

Emily Alward

FURTHER READING

Belt, Don. "Russia's Lake Baikal: The World's Great Lake." *National Geographic*, June, 1992.

Matthiessen, Peter. *Baikal: Sacred Sea of Siberia.* San Francisco: Sierra Club Books, 1992.

Minoura, Koji, ed. *Lake Baikal: A Mirror in Time and Space for Understanding Global Change Processes.* New York: Elsevier, 2000.

Thomson, Peter. *Sacred Sea: A Journey to Lake Baikal.* New York: Oxford University Press, 2007.

Venable, Sondra. *Protecting Lake Baikal: Environmental Policy Making in Russia's Transition.* Saarbrücken, Germany: VDM, 2008.

SEE ALSO: Ecosystems; Food chains; Pulp and paper mills; Water pollution; Watershed management; *Zapovednik* system.

Lake Erie

CATEGORIES: Places; water and water pollution

IDENTIFICATION: Large freshwater lake bordering Ontario, Canada, and the U.S. states of Michigan, Ohio, Pennsylvania, and New York

SIGNIFICANCE: In ecological trouble even before European settlement of its region, Lake Erie was badly damaged by human tampering with its ecosystem and the careless introduction of chemical fertilizers and industrial wastes. By the mid-twentieth century, the lake was severely polluted, but sustained efforts to rehabilitate it that began in 1965 gradually brought about a general recovery.

Lake Erie, the shallowest and southernmost of the Great Lakes, was far along in the process of eutrophication, or natural aging, prior to settlement by Europeans. Early settlers accelerated this process by draining coastal wetlands and stripping away vegetation, which increased the amount of sediment carried to the lake. With the advent of widespread agriculture, artificial fertilizers also began to wash into Lake Erie, which contributed to overenrichment of the lake's waters. Sewage and fertilizers, such as phosphorus and nitrogen, caused the rapid growth of surface algae scums, which affected the taste and odor of drinking-water supplies, clogged water intakes, and forced beach closures. More important, decaying algae consumed the water's oxygen, leading to the suffocation of bottom-dwelling organisms. Eventually, desirable fish stocking the lake were stressed by lack of adequate food, and populations declined.

Industries along the lake's main tributaries contributed to the problem by injecting industrial wastes, oil, floating solids, and heavy metals into the water supply. Heavy metals such as lead and mercury, as well as organic chemicals such as polychlorinated biphenyls (PCBs) and dioxin, magnify in concentrations as they pass up the food chain. These bioaccumulation and biomagnification effects may increase levels of toxic materials by one million times in fish such as salmon and trout. Fish consumption advisories were required for some species because of contamination levels. By 1965 Lake Erie had become so polluted that public indignation over its condition led to action by government officials.

Four U.S. states (Michigan, New York, Pennsylvania, and Ohio) and one Canadian province (Ontario) share in managing Lake Erie. The International Joint Commission was established by Canada and the United States in 1909 to arbitrate disputes over shared boundary waters. In 1972 the commission established the Great Lakes Water Quality Agreement, which provided for reduction in nitrogen and phosphorus discharge in Lake Erie by improving the municipal sewage treatment systems. The retention of sanitary and storm-sewer overflow for later treatment greatly reduced health problems in the region, and beach closures at Lake Erie became less frequent with improved treatment of sewage.

The International Reference Group on Great Lakes Pollution from Land Use, under the International Joint Commission Authority, prepared reports that provided the groundwork for the ecosystem approach to reducing pollution in Lake Erie promulgated in the 1978 revisions to the Water Quality Agreement. The agreement states that the programs must center on the physical, chemical, and biological relationships among air, land, and water resources. The ecosystem approach mandated by the 1978 agreement means that standards and monitoring methods must take into account the air, land, and water movement of pollutants and their risks to humans and other organisms.

The Great Lakes Fishery Commission, created in 1955 by Canada and the United States, is concerned with restoring and stocking lake fish. The population of whitefish in the lake has shown signs of recovery. Lake trout and coho salmon are stocked in the lake, and walleye and yellow perch are managed for recreational and commercial fishing. Tests have also revealed that the levels of PCBs in some Lake Erie fish have diminished over time.

In 1985 the Great Lakes Charter was inaugurated to resist the transfer of Great Lakes water to other areas. The charter authorized the development of an information database for surface-water and groundwater resources. In 1986 the Great Lakes Toxic Substances Control Agreement (GLTSCA) was formed to coordinate the actions of the Great Lakes states to reduce toxic substances in the basin. The 1987 protocol to the Great Lakes Water Quality Agreement called for forming specific ecosystem objectives and indicators. This approach has enhanced the evolution of full ecosystem management strategies, which incorporate mathematical modeling.

Ronald J. Raven

Lake Erie

FURTHER READING

Caldwell, Lynton Keith, ed. *Perspectives on Ecosystem Management for the Great Lakes: A Reader.* Albany: State University of New York Press, 1988.

Grady, Wayne. *The Great Lakes: The Natural History of a Changing Region.* Vancouver: Greystone Books, 2007.

McGucken, William. *Lake Erie Rehabilitated: Controlling Cultural Eutrophication, 1960's-1990's.* Akron, Ohio: University of Akron Press, 2000.

SEE ALSO: Aral Sea destruction; Biomagnification; Cultural eutrophication; Cuyahoga River fires; Dioxin; Great Lakes International Joint Commission; Lake Baikal; Polychlorinated biphenyls; St. Lawrence Seaway; Water pollution.